WORLD ENERGY OUTLOOK

2002

INTERNATIONAL ENERGY AGENCY

9, rue de la Fédération,
75739 Paris Cedex 15, France

The International Energy Agency (IEA) is an autonomous body which was established in November 1974 within the framework of the Organisation for Economic Co-operation and Development (OECD) to implement an international energy programme.

It carries out a comprehensive programme of energy co-operation among twenty-six* of the OECD's thirty Member countries. The basic aims of the IEA are:

- to maintain and improve systems for coping with oil supply disruptions;
- to promote rational energy policies in a global context through co-operative relations with non-member countries, industry and international organisations;
- to operate a permanent information system on the international oil market;
- to improve the world's energy supply and demand structure by developing alternative energy sources and increasing the efficiency of energy use;
- to assist in the integration of environmental and energy policies.

**IEA Member countries: Australia, Austria, Belgium, Canada, the Czech Republic, Denmark, Finland, France, Germany, Greece, Hungary, Ireland, Italy, Japan, the Republic of Korea, Luxembourg, the Netherlands, New Zealand, Norway, Portugal, Spain, Sweden, Switzerland, Turkey, the United Kingdom, the United States. The European Commission also takes part in the work of the IEA.*

ORGANISATION FOR ECONOMIC CO-OPERATION AND DEVELOPMENT

Pursuant to Article 1 of the Convention signed in Paris on 14th December 1960, and which came into force on 30th September 1961, the Organisation for Economic Co-operation and Development (OECD) shall promote policies designed:

- to achieve the highest sustainable economic growth and employment and a rising standard of living in Member countries, while maintaining financial stability, and thus to contribute to the development of the world economy;
- to contribute to sound economic expansion in Member as well as non-member countries in the process of economic development; and
- to contribute to the expansion of world trade on a multilateral, non-discriminatory basis in accordance with international obligations.

The original Member countries of the OECD are Austria, Belgium, Canada, Denmark, France, Germany, Greece, Iceland, Ireland, Italy, Luxembourg, the Netherlands, Norway, Portugal, Spain, Sweden, Switzerland, Turkey, the United Kingdom and the United States. The following countries became Members subsequently through accession at the dates indicated hereafter: Japan (28th April 1964), Finland (28th January 1969), Australia (7th June 1971), New Zealand (29th May 1973), Mexico (18th May 1994), the Czech Republic (21st December 1995), Hungary (7th May 1996), Poland (22nd November 1996), the Republic of Korea (12th December 1996) and Slovakia (28th September 2000). The Commission of the European Communities takes part in the work of the OECD (Article 13 of the OECD Convention).

SECOND EDITION
November 2002

FOREWORD

It is always a genuine pleasure to present the *World Energy Outlook*, the IEA's most ambitious and widely read publication. It is particularly gratifying to introduce *WEO 2002*: first, because of the circumstances of its launch and, second, because of the importance of the messages it has to convey.

We decided to release the book, almost two months ahead of the originally planned schedule, at the 2002 Ministerial Meeting of the International Energy Forum, Consumer-Producer Dialogue, in Osaka, Japan. This decision responds to a request from the government of Japan, one of our largest and most dedicated members. It also recognises the value we place on the increasingly confident and fruitful dialogue taking place between oil producers and consumers.

The *World Energy Outlook* is a compendium of thousands of numbers and hundreds of pages of detailed analysis. It is a rich quarry. According to his special interests, the reader may seize upon any one or more of its many facets:

- that world energy demand will grow by two-thirds in the next 30 years;
- that fossil fuels will continue to dominate the energy mix;
- that nearly two-thirds of the growth in energy demand will arise in developing countries;
- that financing the required new energy infrastructure is a huge challenge, depending largely on the framework conditions created by governments;
- that international energy trade will expand dramatically;
- that natural gas demand growth will outpace that of any other fossil fuel, but will itself be outpaced by demand growth for renewables;
- that transport will dominate the growth in oil use;
- that electricity use will grow faster than any other energy end-use;
- that the proportion of the world's population without access to electricity will fall by a third; or, conversely, that 1.4 billion people will still lack access to electricity in 2030;

- that, on the basis of present policies, carbon dioxide emissions from energy use will continue to grow steeply;
- that new technologies will emerge on the energy scene within 30 years; but that it will be much longer before they become dominant.

An Alternative Policy Scenario in this book serves two purposes: it reminds us how the basic picture painted depends on key assumptions, including continuity of present policies; and it indicates how, and to what extent, that picture might be changed by deliberate policy actions. Many changes are possible, for example in policies related to poverty alleviation, energy security, environmental priorities, the nuclear component of supply and many other issues.

The policy mix adopted by governments has to conform to today's standards of sustainable economic development. Economic development cannot be achieved without energy; and it cannot be sustained unless the energy supply is reliable, i.e. secure. But energy production and use also have to be environmentally sustainable – and meet social needs and expectations. Policy-makers have to find the right way to reconcile these requirements. No single element can override the others.

The last chapter of this book deals with energy and poverty. Energy policy-makers alone cannot solve this problem, still less energy analysts. Energy analysts can, however, define the problem properly – the first step towards its solution. That is what we have sought to do. The definition has shocked us. It is totally unacceptable – both morally and economically – that 1.4 billion people should still be without electricity 30 years into this millennium.

This work is published under my authority as Executive Director of the IEA and does not necessarily reflect the views or policies of the IEA Member countries.

Robert Priddle
Executive Director

Comments and questions are welcome and should be addressed as follows:

Fatih Birol
Chief Economist
Head, Economic Analysis Division
International Energy Agency
9, rue de la Fédération
75739 Paris Cedex 15
France

Telephone: (33-1) 4057 6670
Fax: (33-1) 4057 6659
Email: Fatih.Birol@iea.org

World Energy Outlook Series

World Energy Outlook – 1993
World Energy Outlook – 1994
World Energy Outlook – 1995
Oil, Gas & Coal Supply Outlook – 1995
World Energy Outlook – 1996
World Energy Outlook – 1998
World Energy Outlook – 1999 Insights
  Looking at Energy Subsidies: Getting the Prices Right
World Energy Outlook – 2000
World Energy Outlook – 2001 Insights
  Assessing Today's Supplies to Fuel Tomorrow's Growth
World Energy Outlook – 2002
World Energy Outlook – 2003 Insights (forthcoming)
  Global Energy Investment Outlook

ACKNOWLEDGEMENTS

This study was prepared by the Economic Analysis Division (EAD) of the International Energy Agency (IEA) in co-operation with other divisions of the IEA. The Director of the Long-Term Office, Olivier Appert, provided support and encouragement during the project. The study was designed and managed by Fatih Birol, Head of the Economic Analysis Division. Other members of the EAD who were responsible for bringing this study to completion include: Armando Acosta, Maria Argiri, Amos Bromhead, François Cattier, Laura Cozzi, Lisa Guarrera, Claudia Jones, Hiroyuki Kato, Teresa Malyshev, Trevor Morgan, Scott Sullivan and Michael Taylor.

The following colleagues were also part of the *Outlook* team: Carmen Difiglio, Fridtjof Unander, Sohbet Karbuz, Mike Ting (Energy Technology Policy Division), Miharu Kanai (Oil Market Division), Peter Fraser (Energy Diversification Division) and Kyung-Hwan Toh (Non-Member Countries Division).

Input was provided by other IEA staff, namely: Richard Baron, Xavier Chen, Sylvie Cornot, Sylvie Lambert D'Apote, Ralf Dickel, Mark Hammonds, Lew Fulton, Rebecca Gaghen, Mitsuhide Hoshino, Benoit Lebot, Shin Morita, Isabel Murray, Carlos Ocana, Jonathan Pershing, Riccardo Quercioli, Loretta Ravera, Klaus Rehaag, Rick Sellers and Mike Wittner.

The study also benefited from direct contributions from government bodies, international organisations and energy companies worldwide. These include: the Energy Research Institute of the State Development Planning Commission of China, the European Commission Directorate General for Transport and Energy, Electricité de France, Eni, the Ministry of Energy of Mexico, the Tata Energy Research Institute of India, the United Nations Food and Agriculture Organization and the World Bank.

This study required a complex analysis of different issues. We asked several experts to comment on the underlying analytical work of this book and to review early drafts of each chapter. Their comments and suggestions were extremely useful. We attempt to acknowledge all those contributions on the following page and apologize if we have inadvertently left anyone off the list.

All errors and omissions are solely the responsibility of the IEA.

Acknowledgement of Experts outside the IEA

African Energy Policy Research Network, Kenya, *Stephen Karakezi.*

Asian Development Bank, Philippines, *Carol Litwin.*

Asia-Pacific Energy Research Centre, Japan, *Yonghun Jung* and *Naoko Doi.*

Australian Greenhouse Gas Office, Australia, *Stephen Berry.*

Cambridge Energy Research Associates, United States of America, *Kevin Lindemer.*

Centre d'Etudes Prospectives et d'Informations Internationales, France, *Nina Kousnetzof.*

CONCAWE, Belgium, *Jean-François Larive.*

Constellation Energy, United States of America, *John Paffenbarger.*

Department of Energy, United States of America, *Andy Kydes, Bruce Bawks, John Cymbalski, Lynda Doman* and *Crawford Honeycutt.*

Department for International Development, United Kingdom, *Gill Wilkins.*

Directorate General Development, European Commission, Belgium, *Philip Mann.*

Directorate-General for Transport and Energy, European Commission, Belgium, *Manfred Decker, Pietro Menna, Paolo Bertoldi* and *Randall Bowie.*

Electricité de France, France, *François Verneyre* and *Prabodh Porouchottamin.*

Energy Conservation Centre, Japan, *Naohito Okumura.*

Energy Efficient Strategies, Australia, *Lloyd Harrington.*

Energy Research Institute, State Development Planning Commission, China, *Dadi Zhou, Yuan Guo* and *Yufeng Yang.*

Eni, Italy, *Alessandro Lanza.*

German Advisory Council on Global Change to the Federal Government, Germany, *Marc Ringel.*

Institute of Energy Economics, Japan, *Koukichi Ito* and *Yukari Yamashita.*

International Atomic Energy Agency, Austria, *Kee-Yung Nam.*

Jyukankyo Research Institute Inc, Japan, *Naoto Sagawa.*

Kumasi Institute of Technology and Environment, Ghana, *Abeeku Brew-Hammond.*

Lawrence Berkeley National Laboratory, United States of America, *Jonathan Sinton.*

Ministry of Commerce, Industry and Energy, Korea, *Il-Hwan Oh* and *Jae-Joon Kim.*

Ministry of Economy, Trade and Industry, Japan, *Akihiko Inomata.*

Ministry of Energy and Mineral Resources, Indonesia, *Sumiarso Luluk.*

Natural Resources Canada, *Bob Lyman* and *Valentin Konza.*
National Technical University of Athens, Greece, *Leonidas Mantzos.*
Organisation for Economic Co-operation and Development, France, *Ulrich Heimenz, Barrie Stevens, David O'Connor, Jean-Claude Berthelemy, Patrick Love, Riel Miller* and *Celine Kauffmann.*
Organization of Petroleum Exporting Countries, Austria, *Nadir Guerer.*
PW Consulting, United Kingdom, *Paul Wade.*
Simmons & Company, United States of America, *Matt Simmons.*
Swiss Federal Office of Energy, Switzerland, *Jean-Christophe Füeg.*
Tata Energy Research Institute, India, *Rajendra Pachauri.*
Technical University of Vienna, Austria, *Reinhard Haas* and *Gustav Resch.*
United Nations Environment Program, France, *Lawrence Agbemabiese.*
United Nations Food and Agriculture Organization, Italy, *Gustavo Best, Miguel Trossero and Adrian Whiteman.*
University of Sao Paulo, Brazil, *Edmilson Moutinho dos Santos.*
WEFA Inc., United States of America, *Michael Lynch.*
World Bank, United States of America, *Jamal Saghir, Mohamed Farhandi, Dominique Lallement, Shane Streifel* and *Nouredine Berrah.*
World Coal Institute, *Malcom Keay.*

TABLE OF CONTENTS

List of Figures in Text

List of Tables in Text

EXECUTIVE SUMMARY

This edition of the *World Energy Outlook*, which sets out the IEA's latest energy projections to 2030, depicts a future in which energy use continues to grow inexorably, fossil fuels continue to dominate the energy mix and developing countries fast approach OECD countries as the largest consumers of commercial energy. The Earth's energy resources are undoubtedly adequate to meet rising demand for at least the next three decades. But the projections in this *Outlook* raise serious concerns about the security of energy supplies, investment in energy infrastructure, the threat of environmental damage caused by energy production and use and the unequal access of the world's population to modern energy.

Governments will have to take strenuous action in many areas of energy use and supply if these concerns are to be met. The core projections presented here are derived from a Reference Scenario that takes into account only those government polices and measures that had been adopted by mid-2002. A separate Alternative Policy Scenario assesses the impact of a range of new energy and environmental policies that OECD countries are considering adopting as well as of faster deployment of new energy technologies. Both scenarios confirm the extent of the policy challenges facing governments around the world.

A key result of the *Outlook* is that energy trade will expand rapidly. In particular, the major oil- and gas-consuming regions will see their imports grow substantially. This trade will increase mutual dependence among nations. But it will also intensify concerns about the world's vulnerability to energy supply disruptions, as production is increasingly concentrated in a small number of producing countries. Supply security has moved to the top of the energy policy agenda. The governments of oil- and gas-importing countries will need to take a more proactive role in dealing with the energy security risks inherent in fossil-fuel trade. They will need to pay more attention to maintaining the security of international sea-lanes and pipelines. And they will look anew at ways of diversifying their fuels, as well as the geographic sources of those fuels. The OECD Alternative Policy Scenario demonstrates the strong impact that new policies to curb energy demand growth and encourage switching away from fossil fuels could have on import dependence. Governments and consumers are, nonetheless, likely to continue accepting a degree of risk in return for competitively priced energy supplies.

Necessary expansion of production and supply capacity will call for massive investment at every link in the energy supply chain. Investment of almost $4.2 trillion will be needed for new power generation capacity alone between now and 2030. Mobilising this investment in a timely fashion will require the lowering of regulatory and market barriers and the creation of an attractive investment climate – a daunting task in many countries in the developing world and the former Soviet Union. Most investment will be needed in developing countries, and it is unlikely to materialise without a huge increase in capital inflows from industrialised countries.

Energy-related emissions of carbon dioxide are set to grow slightly faster than energy consumption in the Reference Scenario, despite the policies and measures taken so far. In the Alternative Policy Scenario, however, new policies that many OECD countries are currently considering, together with faster deployment of more efficient and cleaner technologies, would achieve energy savings and promote switching to less carbon-intensive fuels. These developments would eventually stabilise CO_2 emissions in OECD countries, but only towards the end of the *Outlook* period.

More than a quarter of the world's population has no access to electricity, and two-fifths still rely mainly on traditional biomass for their basic energy needs. Although the number of people without power supplies will fall in the coming decades, a projected 1.4 billion people will still be without electricity in 2030. And the number of people using wood, crop residues and animal waste as their main cooking and heating fuels will actually *grow*. To extend electricity supplies to the energy poor and give them better access to other forms of modern energy, stronger government policies and co-ordinated international action will be essential.

Fossil Fuels Will Continue to Dominate Global Energy Use

World energy use will increase steadily through 2030 in the Reference Scenario. Global primary energy demand is projected to increase by 1.7% per year from 2000 to 2030, reaching an annual level of 15.3 billion tonnes of oil equivalent. The increase will be equal to two-thirds of current demand. The projected growth is, nevertheless, slower than growth over the past three decades, which ran at 2.1% per year.

Fossil fuels will remain the primary sources of energy, meeting more than 90% of the increase in demand. Global *oil* demand will rise by about 1.6% per year, from 75 mb/d in 2000 to 120 mb/d in 2030. Almost three-quarters of the increase in demand will come from the transport sector. Oil will remain the fuel of choice in road, sea and air transportation. As a result, there will be a shift in all regions towards light and middle distillate products, such as gasoline and diesel, and away from heavier oil products, used mainly in industry. This shift will be more pronounced in developing countries, which currently have a lower proportion of transportation fuels in their product mix.

Demand for *natural gas* will rise more strongly than for any other fossil fuel. Primary gas consumption will double between now and 2030, and the share of gas in world energy demand will increase from 23% to 28%. New power stations will take over 60% of the increase in gas supplies over the next three decades. Most of these stations will use combined-cycle gas turbine technology, a form of generation favoured for its high energy-conversion efficiency and low capital costs. Gas is also often preferred to coal and oil for its relatively benign environmental effects, especially its lower carbon content.

Consumption of *coal* will also grow, but more slowly than that of oil and gas. China and India together will account for two-thirds of the increase in world coal demand over the projection period. In all regions, coal use will become increasingly concentrated in power generation, where it will remain the dominant fuel. Power-sector coal demand will grow with the expected increase in gas prices. The deployment of advanced technologies will also increase coal's attractiveness as a generating fuel in the long term.

The role of *nuclear power* will decline markedly, because few new reactors will be built and some will be retired. Nuclear production will peak at the end of this decade, then decline gradually. Its share of world primary demand will hold steady at about 7% through 2010, then fall to 5% by 2030. Its share of total electricity generation will fall even faster, from 17% in 2000 to 9% in 2030. Nuclear output will increase in only a few countries, mostly in Asia. The biggest declines in nuclear production are expected to occur in North America and Europe. The prospects for nuclear power are particularly uncertain. Some governments have expressed renewed interest in the nuclear option as a means to reduce emissions and to improve security of supply.

***Renewable energy* will play a growing role in the world's primary energy mix.** Hydropower has long been a major source of electricity production. Its share in global primary energy will hold steady, but its share of electricity generation will fall. Non-hydro renewables, taken as a group, will grow faster than any other primary energy source, at an average rate of 3.3% per year over the projection period. Wind power and biomass will grow most rapidly, especially in OECD countries. But non-hydro renewables will still make only a small dent in global energy demand in 2030, because they start from a very low base. OECD countries, many of which have adopted strong measures to promote renewables-based power projects, will account for most of the growth in renewables.

Demand Will Rise Fastest in Developing Countries...

More than 60% of the increase in world primary energy demand between 2000 and 2030 will come from developing countries, especially in Asia. These countries' share of world demand will increase from 30% to 43%. The OECD's share will fall from 58% to 47%. The share of the former Soviet Union and Eastern and Central Europe (the transition economies) will fall slightly, to 10%. The surge in demand in the developing regions results from their rapid economic and population growth. Industrialisation and urbanisation will also boost demand. The replacement of traditional biomass by commercially traded energy will increase recorded demand. Higher consumer prices as energy subsidies are phased out and international prices rise, are not expected to curb energy demand growth.

China, already the world's second-largest energy consumer, will continue to grow in importance on world energy markets as strong economic growth drives up demand and imports. The Chinese economy will remain exceptionally dependent on coal, but the shares of oil, natural gas and nuclear will grow in China's energy mix. Increasing oil- and gas-import needs will make China a strategic buyer on world markets.

...and Transport Uses Will Outstrip All Others

Transport demand, almost entirely for oil, will grow the most rapidly of all end-use sectors, at 2.1% per annum. It will overtake industry in the 2020s as the largest final-use sector. Transport demand will increase everywhere, but most rapidly in the developing countries. OECD transport demand will grow at a slower pace, as markets become more saturated. Consumption in the residential and services sectors will grow at an average annual rate of 1.7%, slightly faster than in industry, where it will rise by 1.5% per year.

Electricity will grow faster than any other end-use source of energy, by 2.4% per year over the *Outlook* period. World electricity demand will double through 2030, while its share of total final energy consumption will rise from 18% in 2000 to 22% in 2030. The biggest increase in demand will come from developing countries. Electricity use increases most rapidly in the residential sector, especially in developing countries. But the huge difference in per capita electricity consumption between the OECD and developing countries will hardly change over the projection period. The shares of oil and gas in world final consumption will also remain broadly unchanged. Oil products will account for roughly half of final energy use in 2030. The share of coal will drop from 9% to 7%. Coal use will expand in industry, but only in non-OECD countries. It will stagnate in the residential and services sectors.

Fossil Energy Resources Are Ample, but Technologies and Supply Patterns Will Change

The world's energy resources are adequate to meet the projected growth in energy demand. Oil resources are ample, but more reserves will need to be identified in order to meet rising oil demand to 2030. Reserves of natural gas and coal are particularly abundant, while there is no lack of uranium for nuclear power production. The physical potential for renewable energy production is also very large. But the geographical sources of incremental energy supplies will shift over the next three decades, in response to cost, geological and technical factors. In aggregate, almost all the increase in energy production will occur in non-OECD countries, compared to just 60% from 1971 to 2000.

Increased production in the Middle East and the former Soviet Union, which have massive hydrocarbon resources, will meet much of the growth in world oil and gas demand. Most of the projected 60% increase in global oil demand in the next three decades will be met by OPEC producers, particularly those in the Middle East. Output from mature regions such as North America and the North Sea will gradually decline. More oil will become available from Russia and the Caspian region, and this will have major implications for the diversity of supply sources for oil-importing countries.

Global crude oil refining capacity is projected to increase by an average 1.3% a year, reaching 121 mb/d in 2030. The growth of capacity will be slightly less than that of demand for refined products, because of increased utilisation rates and the elimination of some refinery bottlenecks. Over 80% of new refining capacity will be built outside the OECD, much of it in Asia. Refineries will have to boost their yields of transportation fuels relative to heavier oil products, as well as improve product quality.

Production of natural gas, resources of which are more widely dispersed than oil, will increase in every region other than Europe. The cost of gas production and transportation is likely to rise in many places as low-cost resources close to markets are depleted and supply chains lengthen.

There are abundant coal reserves in most regions. Increases in coal production, however, are likely to be concentrated where extraction, processing and transportation costs are lowest — in South Africa, Australia, China, India, Indonesia, North America and Latin America.

New sources of energy and advanced technologies will emerge during the *Outlook* period. Non-conventional sources of oil, such as oil sands and gas-to-liquids, are set to expand, as their production costs decline. Fuel cells are also projected to make a modest contribution to global energy supply after 2020, mostly in small decentralised power plants. The fuel cells that are expected to achieve commercial viability first will involve the steam reforming of natural gas. Fuel cells in vehicles are expected to become economically attractive only towards the end of the projection period. As a result, they will power only a small fraction of the vehicle fleet in 2030.

International energy trade, almost entirely in fossil fuels, will expand dramatically. Energy trade will more than double between now and 2030. All oil-importing regions – including the three OECD regions – will import more oil, mostly from the Middle East. The increase will be most striking in Asia. The biggest growth markets for natural gas are going to become much more dependent on imports. In absolute terms, Europe will see the biggest increase in gas imports. Cross-border gas pipeline projects will multiply, and trade in liquefied natural gas will surge.

Rising Demand Will Drive Up Carbon Dioxide Emissions

Global energy-related emissions of carbon dioxide will grow slightly more quickly than primary energy demand. They are projected to increase by 1.8% per year from 2000 to 2030 in the Reference Scenario, reaching 38 billion tonnes in 2030. This is 16 billion tonnes, or 70% more than today. Two-thirds of the increase will come in developing countries. Power generation and transport will account for about three-quarters of new emissions.

The geographical sources of new emissions will shift drastically, from the industrialised countries to the developing world. The developing countries' share of global emissions will jump from 34% now to 47% in 2030, while the OECD's share will drop from 55% to 43%. China alone will contribute a quarter of the increase in CO_2 emissions, or 3.6 billion tonnes, bringing its total emissions to 6.7 billion tonnes per year in 2030. Even then, however, Chinese emissions remain well below those of the United States.

The steep rise in projected emissions in the Reference Scenario illustrates the challenge that most OECD countries face in meeting their commitments under the Kyoto Protocol. Emissions in those OECD countries that signed the Protocol will reach 12.5 billion tonnes in 2010, the middle of the Protocol's target period of 2008-2012. That is 2.8 billion tonnes, or 29%, above the target. Russia, like Central and Eastern Europe, is in a very different situation, with projected emissions considerably *lower* than their commitments. Under the Protocol, lower emissions in Russia, Ukraine and Eastern Europe, known as "hot air", can be sold to countries with emissions over their target. But even "hot air" will not suffice to compensate for over-target emissions in other countries. The overall gap will be about 15% above target in 2010. If the United States, which does not intend to ratify the Kyoto Protocol, is excluded, the gap falls to 2%.

Carbon sequestration and storage technologies hold out the long-term prospect of enabling fossil fuels to be burned without emitting carbon into the atmosphere. These technologies, however, are unlikely to be deployed on a large scale before 2030. They are at an early stage of development and are very costly. If their costs could be lowered more quickly than assumed here, this would have a major impact on the long-term prospects for energy supply.

Policies under Consideration in the OECD Would Curb Energy Demand and Emissions

In the Alternative Policy Scenario, implementation of policies that are already under consideration in OECD countries would reduce CO_2 emissions by some 2,150 Mt in 2030, or 16% below the Reference Scenario projections described above. This is roughly equal to the total emissions of Germany, the United Kingdom, France and Italy today. Energy savings achieved by the new policies and measures and by faster deployment of more efficient technologies would be 9% of projected demand in the Reference Scenario in 2030. CO_2 savings would be even bigger, because of the additional impact of fuel switching to less carbon-intensive fuels. Because of the slow pace at which energy capital stock is replaced, CO_2 savings in the early years would be relatively small – only 3% by 2010 and 9% by 2020.

The biggest reduction in CO_2 emissions in the Alternative Policy Scenario would come from power-generation, because of the rapid growth of renewables and savings in electricity demand. OECD governments are currently emphasising renewables and electricity in their long-term plans to curb CO_2 emissions and enhance energy security. Although the three OECD regions would still not individually reach the targets under the Kyoto Protocol, "hot air" could allow the targets to be met.

The Alternative Scenario projections show a marked reduction in import dependence in the major energy-importing regions. In 2030, OECD gas demand would be 260 bcm, or 13%, below the Reference Scenario. The percentage fall in imports would be even greater. The reduction in EU gas imports by 2030 would be slightly less than total current imports from Russia and Norway. The savings in oil demand would reach 10%, or 4.6 mb/d.

Providing Modern Energy to the World's Poor Will be an Unfinished Task

Some 1.6 billion people have no access to electricity, according to data compiled specially for this study. More than 80% of the people who currently lack electricity access live in South Asia and sub-Saharan Africa. The majority of them live on less than $2 per day, but income is not the only determinant of electricity access. China, with 56% of its people still "poor" by international definition, has managed to supply electricity to the vast majority of its population.

In the absence of major new government initiatives, 1.4 billion people, or 18% of the world's population, will still lack electricity in 2030, despite more widespread prosperity and more advanced technology. The number without electricity in 2030 will be 200 million less than today, even though world population is assumed to rise from 6.1 billion in 2000 to 8.3 billion. Four out of five people without electricity live in rural areas. But the pattern of electricity-deprivation is set to change, because 95% of the increase in population in the next three decades will occur in urban areas.

Poor people in developing countries rely heavily on traditional biomass – wood, agricultural residues and dung – for their basic energy needs. According to information specifically collected for this study, 2.4 billion people in developing countries use only such fuels for cooking and heating. Many of them suffer from ill-health effects associated with the inefficient use of traditional biomass fuels. Over half of all people relying heavily on biomass live in India and China, but the *proportion* of the population depending on biomass is heaviest in sub-Saharan Africa.

The share of the world's population relying on biomass for cooking and heating is projected to decline in most developing regions, but the *total number of people* will rise. Most of the increase will occur in South Asia and sub-Saharan Africa. Over 2.6 billion people in developing countries will continue to rely on biomass for cooking and heating in 2030. That is an increase of more than 240 million, or 9%. In developing countries, biomass use will still represent over half of residential energy consumption at the end of the *Outlook* period.

Lack of electricity exacerbates poverty and contributes to its perpetuation, as it precludes most industrial activities and the jobs they create. Experience in China and elsewhere demonstrates how governments can help expand access to modern sources of energy. But electrification and access to modern energy services do not *per se* guarantee poverty alleviation. A variety of energy sources for thermal and mechanical applications are needed to bring productive, income-generating activities to developing countries. Nonetheless, because biomass will continue to dominate energy demand in these countries in the foreseeable future, the development of more efficient biomass technologies is vital for alleviating poverty in rural areas. Renewable energy technologies such as solar, wind and biomass may be cost-effective options for specific off-grid applications, but conventional fuels and established technologies are more likely to be preferred for on-grid capacity expansion.

PART A

GLOBAL TRENDS TO 2030

CHAPTER 1: THE ANALYTICAL FRAMEWORK

HIGHLIGHTS

- Economic growth is the main driver of energy demand. The world's gross domestic product is assumed to grow worldwide by an average 3% per year over the period 2000 to 2030 – a modest slowdown compared to the past three decades. Growth is expected to pick up in 2003 and to remain steady through to 2010, but will then slow progressively over the next two decades as developing countries' economies mature and their population growth slows.
- The world's population is assumed to expand by one-third, from 6 billion in 2000 to 8.2 billion in 2030. The rate of growth will slow gradually from 1.4% in the 1990s to 1% over 2000-2030. Most of the increase in world population will occur in the urban areas of developing countries.
- Crude oil prices are assumed to remain flat until 2010 at around $21 per barrel (in year 2000 dollars) – their average level for the past 15 years. They will then rise steadily to $29 in 2030. Natural gas prices will move more or less in line with oil prices, with regional prices in Europe, the Asia-Pacific region and North America converging to some degree. Coal prices will be flat to 2010 and rise very slowly thereafter.
- Changes in government policies and technological developments, together with macroeconomic conditions and energy prices, are the main sources of uncertainty in the global energy outlook. These factors will affect both the demand for energy services and the rate of investment in supply infrastructure. Uncertainty is inevitably much greater in the last decade of the projection period.

The core projections in this World Energy Outlook *are derived from a Reference Scenario based on a set of assumptions about macroeconomic conditions, population growth, energy prices, government policies and technology. It takes into account only those government polices and measures that have been enacted, though not necessarily implemented, as of mid-2002. An OECD Alternative Policy Scenario considers the impact of a range of new energy and environmental policies that OECD countries* might *adopt and of a faster rate of deployment of new energy technologies. The time horizon for this edition has been extended from 2020 to 2030 so as to consider the possible impact of new technologies on energy supply and demand, since much of the energy-related equipment in use today will have been replaced by 2030. The first year of the projections is 2001, as 2000 is the last year for which historical data are available.*

The Reference and OECD Alternative Policy Scenarios

This *Outlook* uses a scenario approach to analyse the possible evolution of energy markets to 2030. The goal is not to present what the IEA believes will happen to energy markets. Rather, it is to identify and quantify the key factors that are likely to affect energy supply and demand. The IEA's own World Energy Model is the principal tool used to generate our detailed projections. The model has been revised substantially since the last *Outlook*, especially with regard to the treatment of technological developments, renewable energy sources and supply-side factors. A global refinery model has been added. Regional disaggregation has been increased, with the development of separate models for the European Union, Korea, Indonesia and Mexico.[1]

The *Reference Scenario* incorporates a set of explicit assumptions about underlying macroeconomic and demographic conditions, energy prices and supply costs, technological developments and government policies. It takes into account many new policies and measures in OECD countries, most of them designed to combat climate change. Many of these policies have not yet been fully implemented; as a result, their impact on energy demand and supply does not show up in the historical data, which are available in most cases up to 2000. These initiatives cover a wide array of sectors and a variety of policy instruments.[2]

The Reference Scenario does *not* include possible, potential or even likely future policy initiatives. Major new energy policy initiatives will inevitably be implemented during the projection period, but it is impossible to predict precisely which measures among those that have been proposed will eventually be adopted and in what form. For that reason, the

1. A brief description of the structure of the model can be found in Appendix 1.
2. A detailed inventory of new climate-related policies and measures implemented in 2000 and 2001 by IEA countries can be found in IEA (2001a) and IEA (forthcoming).

Reference Scenario projections should not be seen as forecasts, but rather as a baseline vision of how energy markets might evolve if governments individually or collectively do nothing more than they have already committed themselves to do.

The rate of technological innovation and deployment affects supply costs and the efficiency of energy use. The sensitivity of our projections to these assumptions varies by fuel and sector. Since much of the energy-using capital stock in use today will have been replaced by 2030, technological developments that improve energy efficiency will have their greatest impact on market trends towards the end of the projection period. Most cars and trucks, heating and cooling systems and industrial boilers will be replaced in the next 30 years. But most existing buildings, many power stations and refineries and most of the current transport infrastructure will still be in use. The high cost of building these facilities makes early retirement extremely costly. They will not be replaced unless governments provide strong financial incentives. The very long life of energy capital stock will limit the extent to which technological progress can alter the amount of energy needed to provide a particular energy service. In general, it is assumed that current technologies become more efficient, but that no new breakthrough technologies beyond those known today will be used.

Although the Reference Scenario assumes that current energy and environmental policies remain unchanged at both national and regional levels throughout the projection period, the pace of implementation of those policies and the approaches adopted are nonetheless assumed to vary by fuel and by region. For example, electricity and gas market reforms aimed at promoting competition in supply will move ahead, but at varying speeds among countries and regions. Similarly, progress will be made in liberalising energy investment and reforming energy subsidies, but faster in OECD countries than in others. In all cases, energy taxes are assumed to remain unchanged. Likewise, it is assumed that there will be no changes in national policies on nuclear power. As a result, nuclear energy will remain an option for power generation solely in those countries that have not yet officially abandoned it.

The key underlying assumptions about macroeconomic trends, population growth and energy prices are summarised below. Assumptions about technology and about how existing government policies are implemented are detailed in the chapters on energy markets (Chapters 2 and 3) and in the regional chapters (Chapters 4 to 11).

We have developed an *OECD Alternative Policy Scenario* to analyse the impact of different assumptions about government policies and technological developments in OECD countries on energy demand and supply (Chapter 12). Basic assumptions on macroeconomic conditions and population are the same as for the Reference Scenario. However, energy prices change as they respond to the new energy supply and demand balance. The Alternative Policy Scenario differs from the Reference Scenario by assuming that OECD countries will adopt a range of new policies on environmental problems, notably climate change, and on energy security, and that there will be faster deployment of new energy technologies. The purpose of the Alternative Policy Scenario is to assist in the formulation of future policies by providing insights into how effective they might be.

Chapter 13 analyses the link between energy and poverty. It provides the Reference Scenario projections for electrification rates and for biomass energy demand in developing countries.[3]

Key Assumptions

Macroeconomic Prospects

Economic growth is the single most important determinant of energy demand. In the past, energy demand has risen in a roughly linear fashion along with gross domestic product.[4] Since 1971, each 1% increase in GDP has yielded a 0.64% increase in primary energy consumption. Only the oil price shocks of 1973-1974 and 1979-1980 and the very warm weather of 1990 have altered this relationship to any significant degree (Figure 1.1). Demand for transport fuels and electricity follow economic activity particularly closely. Consequently, the energy projections in this *Outlook* are highly sensitive to the underlying assumptions about macroeconomic prospects. All GDP figures are expressed in purchasing power parities (PPPs) rather than market exchange rates.

The last year has seen a major slowdown of the global economy. In contrast to previous cyclical downturns, this slowdown has been highly

3. China, East Asia, South Asia, Latin America, Africa and the Middle East.

4. All GDP data cited in this report are expressed in 1995 dollars using purchasing power parities (PPPs) rather than market exchange rates. PPPs compare costs in different currencies of a fixed basket of traded and non-traded goods and services and yield a widely based measure of standard of living. This is important in analysing the main drivers of energy demand or comparing energy intensities among countries.

Figure 1.1: **World Primary Energy Demand and GDP, 1971-2000**

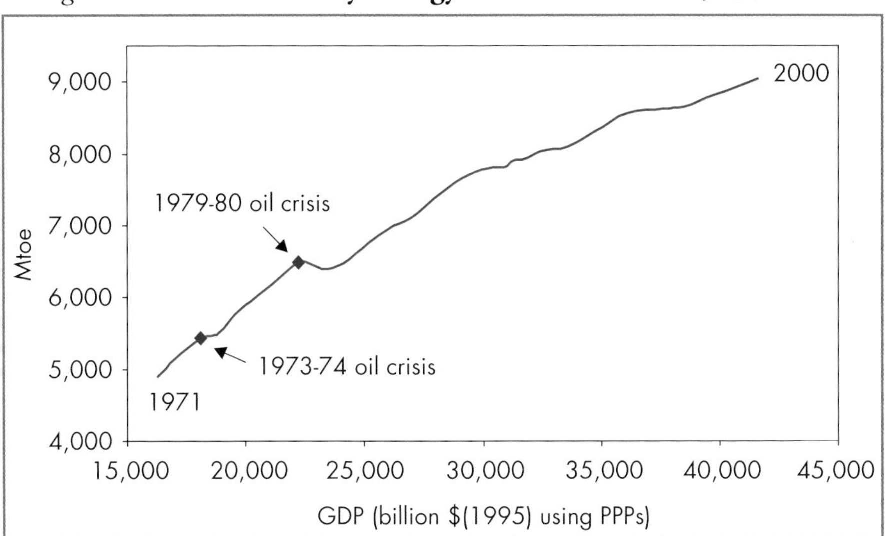

synchronised across regions. For the first time in over two decades, the major engines of the world economy – the United States, Europe and Japan – have slowed at the same time. The United States has led the global weakening. Its GDP was still growing at a year-on-year rate of almost 5% in the second quarter of 2000, before decelerating rapidly into 2001. The terrorist attacks of 11 September 2001 exacerbated this trend. The US economy grew by only 0.3% in 2001 after rising 3.8% in 2000. Global GDP growth, which was running at 4,6% in 2000, dipped to just over 2% in 2001, with a slump in international trade and investment. World trade was virtually flat in 2001 after surging by almost 13% in the previous year – one of the most severe decelerations in trade ever.

Signs that the bottom of the downturn had been reached and that the economic recovery had begun in the United States and several other OECD countries emerged in early 2002 as the causes of the slowdown started to dissipate. The recovery appears to be driven in most cases by private and public consumption. The replenishment of stocks of goods, which had been depleted in 2001, has boosted demand and stimulated output. A tentative revival of business and consumer confidence is expected to underpin a revival in business investment and rising consumer demand in several OECD countries. But the strength of the rebound remains very uncertain, given high levels of consumer and corporate debt, doubts about the sustainability of property values and the implications of lower share values, and the risk of higher energy prices.

This *Outlook* assumes that the OECD countries and most other regions will see higher growth from 2003.[5] The recovery is expected to be particularly strong in North America and in the OECD Pacific region. World GDP will grow by an average of 3.2% over the period 2000-2010 compared to 3% during the 1990s and 3.1% in the 1980s. Growth is assumed to slow slightly after 2010, averaging 2.8% over the two decades to 2030.

Figure 1.2: **Average Annual Real GDP Growth Rates by Region (%)**

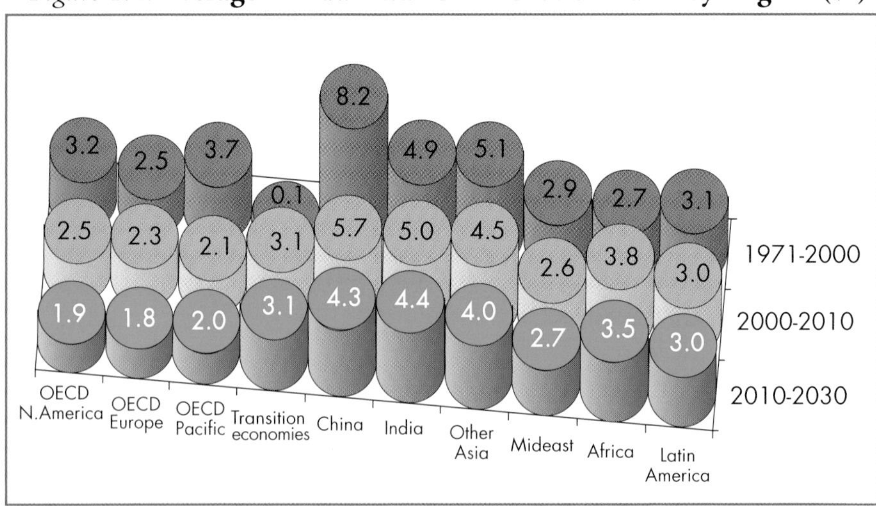

China, India and other Asian countries are expected to achieve the highest rates of economic growth. China tops the table with assumed growth of 4.8% per year. This is quite lower than the 8.2% of the past three decades, reflecting the fact that the Chinese economy becomes larger and more mature. The prospect of slower economic expansion also reflects doubts about the reliability of Chinese GDP data for recent years. China is nonetheless due to become the largest economy in the world in the last decade of the *Outlook* period. The industrialised countries are assumed to grow less rapidly in the next three decades than in the past. GDP in both

5. The economic growth assumptions in this *Outlook* are based on forecasts and studies by a number of organisations, including the OECD (2002), the World Bank (2001), the International Monetary Fund (2002) and the Asian Development Bank (2001). These assumptions can be found in Part D.

OECD Europe and OECD Pacific is assumed to grow by 2% per year over the projection period. OECD North America will see slightly higher growth, of 2.1%. Economic development in all major regions is expected to show a continuing shift away from energy-intensive, heavy manufacturing towards lighter industries and services.

International trade is expected to rebound rapidly once the economic recovery is in full swing. Trade in high technology products, which was severely hit by the recent downturn, will probably account for a large part of the increase – at least in the near term. Trade in energy itself is expected to rise in nominal terms and possibly as a share of overall trade too. Though still one of the largest categories of traded goods, at about 8% of the total, energy's share in world trade has fallen substantially since the early 1980s. An assumed reduction in trade barriers would stimulate trade in the long term. The decision taken at the ministerial meeting of the World Trade Organization in November 2001 to maintain the process of liberalisation and to launch a broader negotiating agenda was an important step in this process. An expansion of international trade is especially vital to sustained economic growth in the developing world. This will undoubtedly require a further major expansion of foreign direct investment (FDI), both in export industries and in infrastructure, including energy. At present, FDI is highly concentrated in industrialised countries and a small number of developing countries, notably Brazil and China. Globally, FDI surged in the 1990s, but it is believed to have fallen back in 2001. Energy makes up a major share of FDI, especially in developing countries (see Chapter 2).

Population Growth

Population growth affects the size and pattern of energy demand. The rates of population growth assumed for each region in this *Outlook* are based on the most recent United Nations projections.[6] Details of historical trends and growth rate assumptions by region are contained in Part D.

On average, the world's population is assumed to grow by 1.0% per year from 2000 to 2030. This is much slower than the 1.7% rate of the past three decades. Growth will decelerate progressively over the projection period, from 1.2% per year in 2000-2010 to 1.0% in 2010-2020 and 0.9% in 2020-2030. Global population will expand by more than a third, from 6 billion in 2000 to 8.2 billion in 2030.

6. United Nations Population Division (2001).

Population will continue to grow much faster in the developing countries, though more slowly than in the past, as birth rates drop. Africa's population is set to grow by around 2.1% per year over the period 2000-2030 – a sharp fall from the 2.7% rate from 1971 to 2000, in part due to the devastating effects of HIV/AIDS. Population will stagnate and the average age rise gradually in OECD Europe, OECD Pacific and the transition economies[7], despite high rates of immigration in some countries. The developing countries' share of world population will grow from 76% in 2000 to 81% in 2030. Most of the increase in global population will occur in urban areas.

Figure 1.3: **World Population by Region**

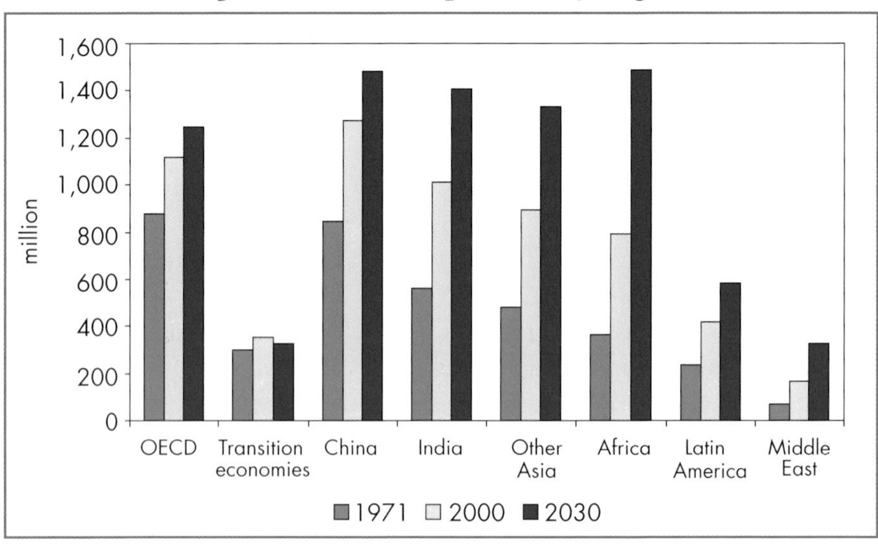

These assumptions, together with those for economic growth, imply that the developing countries will enjoy the fastest increase in per capita income, reaching $7,600 in 2030. This is still well below the level of OECD countries, whose income rises to $36,200, and for the transition economies, with income of $14,200 in 2030 (Figure 1.4).

Using similar GDP and population growth projections to those assumed here, the World Bank predicts that the number of people living on less than $2 per day (in year 2000 dollars) will fall from just under

7. Russia, other former Soviet Union republics and non-OECD Eastern and Central European countries.

Figure 1.4: **Per Capita Income by Region**

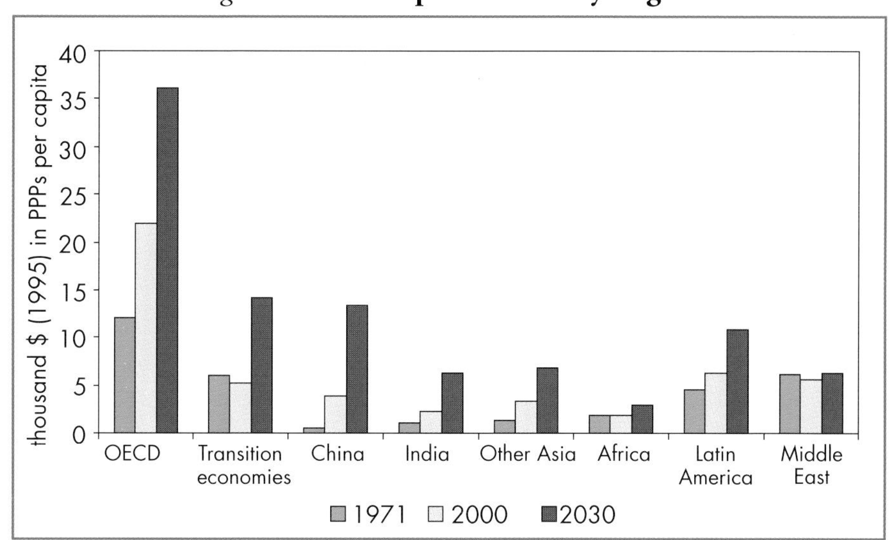

Figure 1.5: **Poverty in Developing Countries**

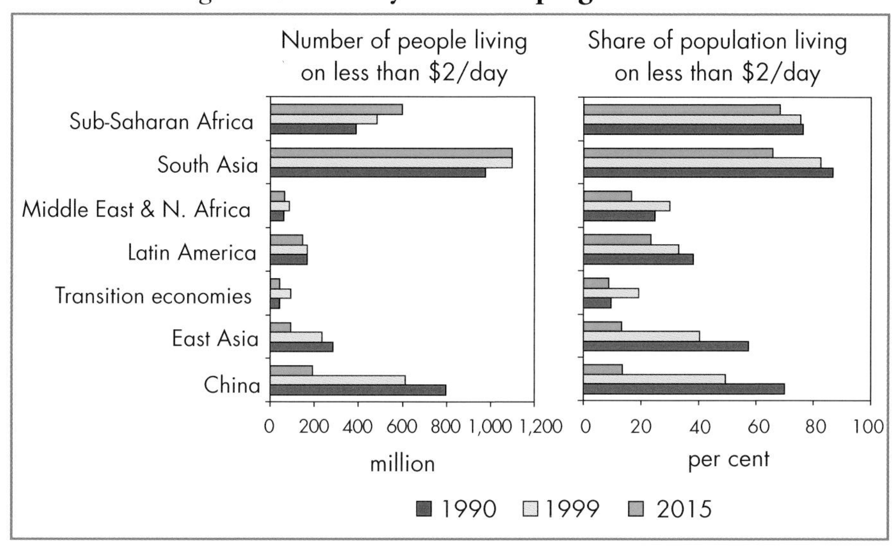

Source: World Bank (2001).

2.8 billion in 1999 to 2.2 billion in 2015.[8] The *share* of poor people in the total population of developing countries will fall even more rapidly, from

8. World Bank (2001).

55% in 1999 to 36% in 2015. China and other East-Asian countries are expected to see the biggest declines (Figure 1.5). Expanding access to modern energy services will play an important role in alleviating poverty.[9]

Energy Prices

Prices are important drivers of energy demand and supply, although the World Energy Model does not project or forecast their evolution. Average end-user prices are derived from assumed fossil fuel prices on wholesale or bulk markets. They take into account current tax rates, which are assumed to remain unchanged. Final electricity prices are derived from marginal electricity generation costs.

The assumed primary price paths reflect judgements about the prices needed to ensure sufficient supply to meet projected demand for each fuel in each region. The smooth price trends assumed should not be interpreted as a prediction of "stable" prices, but rather as long-term paths around which prices could fluctuate. Indeed, oil prices will probably remain highly volatile (Box 1.1). The underlying assumptions for wholesale fossil fuel energy prices in real terms are summarised in Table 1.1 (in fuel-specific units) and in Figure 1.7 (in comparable heat-equivalent units).

Box 1.1: **Oil Price Volatility**

Crude oil prices have become much more volatile on a monthly and annual basis since the mid-1990s (Figure 1.6). From 1987 to 1996, with the exception of a brief period during the run-up to the Gulf War in 1991, crude oil prices remained in a narrow range of $13 to $22 per barrel in nominal terms. Since 1996, prices have fluctuated over a much wider range – from a low of under $10 in February 1999 to a high of $33 in the autumn of 2000. Sharp day-to-day price movements have also become more common since the Gulf War.

Several factors have contributed to the recent increase in the volatility of oil prices, including:

- shifts in the production policies of OPEC and other producing countries and speculation over future production policies, especially those of Iraq;

9. The link between energy and poverty is discussed in more detail in Chapter 13.

- sudden shifts in demand, which have led to short-term supply imbalances;
- a decline in spare production capacity and inventories of crude oil and refined products, which have made prices more sensitive to shifts in demand;
- the fragmentation of product markets, caused by more stringent environmental regulations and product-quality requirements in some OECD countries, which has increased the frequency of localised supply shortfalls.

Sharp, unpredictable swings in oil prices complicate economic management and investment decisions. They also undermine investment in the oil industry, because the uncertainty they generate pushes up risk premiums. IEA analysis points to a robust inverse relationship between upstream oil investment and price volatility; an increase in volatility results in a decline in investments (and *vice versa*).[10] So, falling investment is both a cause and a consequence of price volatility. The level of prices is, nonetheless, the main determinant of investment.

Volatility may continue to increase in the future, given the growing share of oil from Middle East OPEC producers in the world total. Unless surplus capacity in crude oil production and refining increases, markets will remain sensitive to actual or feared swings or disruptions in supply, whether political or technical. Prices will need to be high enough to elicit the investment in oil production capacity needed to meet demand and maintain the energy security of oil-importing countries, but not so high as to curb economic growth and oil demand.

International Oil Prices

The average IEA crude oil import price, a proxy for international oil prices, is assumed in the Reference Scenario to average $21 per barrel over the period 2002 to 2010 (in year 2000 dollars). That price is roughly equal to the average for 1986-2001, when prices ranged on an average annual basis from $13 in 1998 to $28 in 2000. Prices are assumed to rise in a linear fashion after 2010, reaching $25 in 2020 and $29 in 2030.

10. IEA (2001b).

Figure 1.6: **Monthly Price of Dated Brent, 1987-2002**
(in nominal terms)

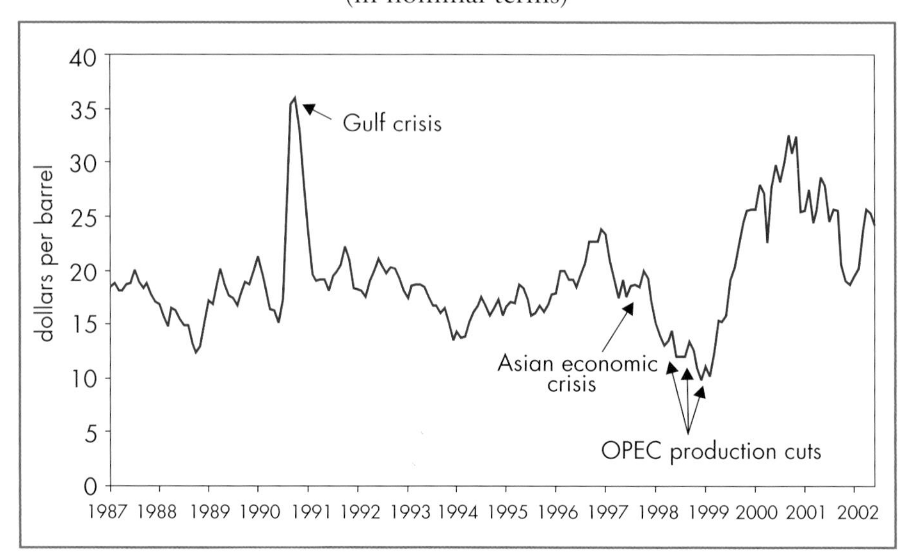

Note: Price is the monthly average spot price of Brent crude oil.

Table 1.1: **Fossil Fuel Price Assumptions** (in year 2000 dollars)

	2000	2010	2020	2030
IEA crude oil imports ($/barrel)	28	21	25	29
Natural gas ($/MBtu):				
US imports	3.9	2.7	3.4	4.0
European imports	3.0	2.8	3.3	3.8
Japan LNG imports	4.7	3.9	4.4	4.8
OECD steam coal imports ($/tonne)	35	39	41	44

Note: Prices in the first column are data. Gas prices are expressed on a gross calorific value basis (MBtu: million British thermal units).

The rising trend after 2010 reflects gradual changes in marginal production costs and supply patterns. Production from giant oil fields[11], which currently supply almost half of the world's crude oil output, is expected to decline and it will be increasingly necessary to obtain oil from smaller fields with higher unit costs of production. The share of giant fields in incremental production has fallen sharply since the 1950s. Rising marginal costs in high-cost producing regions with relatively modest

11. Fields producing 100,000 barrels a day or more.

resources, such as North America and the North Sea, are expected to lead to a decline in their production. Non-conventional oil, such as synthetic crude from oil sands and gas-to-liquids conversion is also expected to play a growing role in total oil supply over the period to 2030.

Figure 1.7: **Assumptions for International Fossil Fuel Prices**

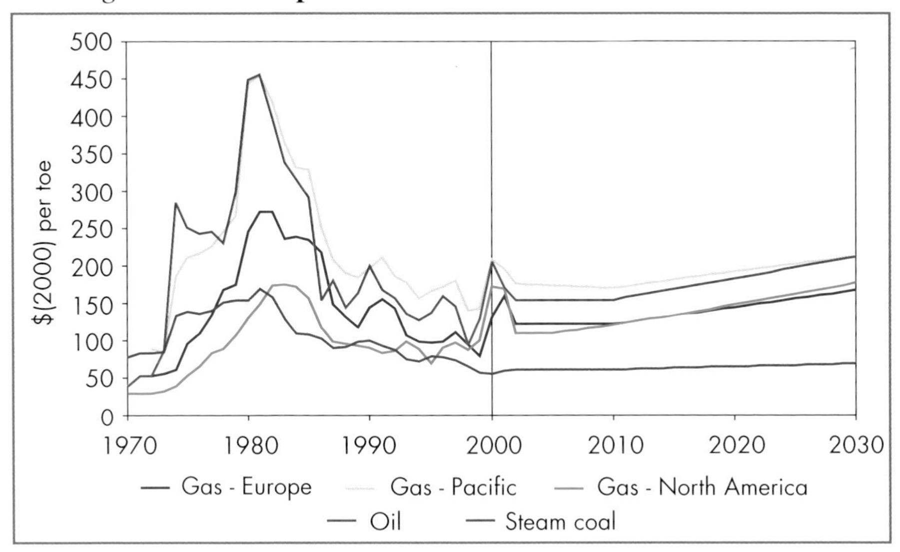

Because of these developments, oil-importing countries, including many IEA Members, are expected to rely increasingly on supplies from a few countries, mainly in the Middle East and the former Soviet Union. The bulk of the world's remaining reserves is found in these regions. Production costs in the Middle East are still the lowest anywhere, and that region is particularly well placed to meet much of the increase in oil demand. There are nonetheless considerable uncertainties about the ability of these countries to finance investment in upstream projects. The increasing market dominance of the biggest Middle East producers could lead to a shift in their production and investment policies in pursuit of higher crude oil prices in the longer term. But it is assumed that these producers will wish to avoid prices rising so much that they depress demand and encourage production of conventional and non-conventional oil in other countries.[12] Higher prices in 2001 curbed world demand growth and

12. See IEA (2001c) for an analysis of the sensitivity of oil production to price changes.

stimulated higher production in some non-OPEC countries, such as Russia.

Natural Gas Prices

Gas markets are highly regionalised, because it is expensive to transport natural gas over long distances. Prices can differ markedly across regions and can diverge within regions according to local market conditions. Nevertheless, prices usually move broadly in parallel with each other because of their link to the international price of oil, reflecting the close competition between gas and oil products.

Historically, Asian prices for liquefied natural gas have been the highest and North American prices for piped gas the lowest. Average prices in North America outstripped European import prices for the first time in 1999 as higher oil prices and dwindling US production sent prices in the United States and Canada soaring. European prices rose too on the back of the1999/2000 surge in oil prices, but with a lag. North American prices in early 2002 fell once again below those in Europe as a result of oversupply.

Assumed trends in regional gas prices from now to 2030 reflect the underlying trend in oil prices together with cost and market factors specific to each region. But increased short-term trading in LNG, which permits arbitrage between regional markets, will cause regional prices to converge to some degree over the next three decades:

- In *North America*, natural gas prices are assumed to average around $2.50/MBtu (in year 2000 dollars) in 2002 and to remain at that level until 2005. Prices will then start to rise, both in nominal terms and in relation to oil prices, as rising demand outstrips indigenous conventional gas production capacity. The region will become increasingly reliant on more costly sources. These include LNG imports, unconventional sources, such as coal-bed methane and gas from tight rock formations, and new conventional supplies from Alaska (if a pipeline is built). Prices will reach $2.7/MBtu by 2010 and continue to rise through to 2030 in line with increasing oil prices.
- In *Europe*, gas prices are assumed to remain flat from 2002 to 2010 at around $2.80/MBtu (in 2000 dollars). Gas-to-gas competition is expected to put some downward pressure on border-gas prices as spot trade develops. Lower downstream margins and efforts by national regulators to reduce access charges could further depress end-user prices. But the cost of bringing new gas supplies to Europe

could increase as the distances over which the gas has to be transported lengthen and project costs rise. These trends are assumed to offset the impact of growing competition. Projects to bring gas from new sources in Russia, the Middle East, West Africa and Latin America are likely to be costlier than past gas import projects closer to European markets. Prices are assumed to rise after 2010 in line with higher oil prices. The price link between oil and gas will be weakened, however, by a shift away from oil price indexation in long-term contracts and by more short-term gas trading.

- In the *Asia-Pacific* region, Japanese LNG prices are assumed to fall in relation to both oil prices and gas prices in North America and Europe over the entire projection period. The expiration of several long-term import contracts over the next few years will provide Japanese buyers with opportunities to press for lower LNG prices in new contracts and to seek out cheaper spot supplies. Prices in Asia will still be a little higher than in North America in 2030, reflecting the region's continuing heavy reliance on distant sources of gas, much of it in the form of LNG.

Figure 1.8: **Ratio of Natural Gas to Oil Prices**

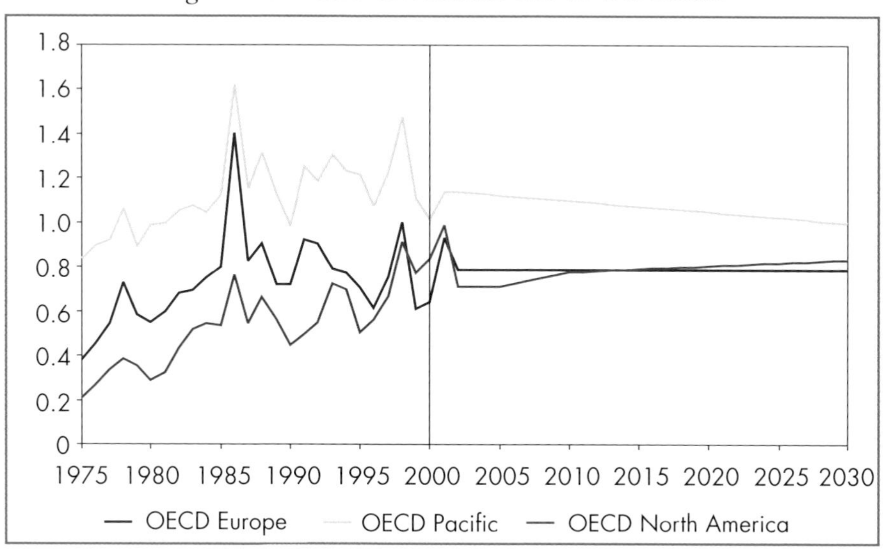

Figure 1.8 illustrates historical and assumed future trends in the ratio of regional gas prices to oil prices, on a heat-equivalent basis. Since the late

1980s, the ratio has fluctuated around its long-term average in Europe and Japan. Japanese gas prices have been slightly higher than oil prices, while European prices have been somewhat lower. The current ratio of gas prices to oil prices is lowest in the United States, but the gap with Europe and Japan narrowed in the 1990s. The ratio in the three regions is assumed to converge to some extent over the *Outlook* period, reflecting a tightening of gas supply in Europe and North America and a softening of supply in the Pacific. Gas is assumed to remain a little cheaper than oil in Europe and North America and slightly more expensive in the Pacific.

Steam-coal Prices

International steam-coal prices rebounded moderately from $34/tonne in 2000 to about $38/tonne in 2001 (in year 2000 dollars) after an almost uninterrupted decline since the early 1980s. A drop in US production was the primary reason for the price recovery. The surge in oil and gas prices in 1999 and 2000 prompted some industrial consumers and power generators to switch to coal, a move that helped to boost coal prices. This *Outlook* assumes that coal prices remain flat in real terms (in year 2000 dollars) over the period 2002 to 2010 at $39/tonne. This is equal to the average for 1997 to 2001. Thereafter, prices are assumed to increase very slowly and linearly, reaching $44/tonne by 2030. This assumption represents a slight change from *World Energy Outlook 2000,* which assumed flat prices at the higher level of $46.50/tonne throughout the period from 2000 to 2020.

The shallow rising trend after 2010 reflects a combination of competing factors. Higher oil prices from 2010 will tend to push up coal prices for two reasons:

- They increase the cost of transporting coal by land and sea, which on average represents more than half of the delivered cost of steam coal worldwide.[13]
- They will make coal relatively more competitive for industrial users and power generators, both in existing facilities and for new plants, thereby boosting the demand for and the value of coal.

On the other hand, the cost of *mining* coal is expected to fall as production is rationalised and concentrated in a small number of low-cost countries and regions. These include China, the United States, India, Indonesia, Australia and South Africa. Coal-mining productivity has

13. IEA (2001c).

improved markedly since the 1980s and there remains considerable scope for further gains (Figure 1.9). At the same time, environmental regulations restrict the use of coal or increase the cost of using it in many countries. These factors will tend to depress prices.

Figure 1.9: **World Coal-Mining Productivity**

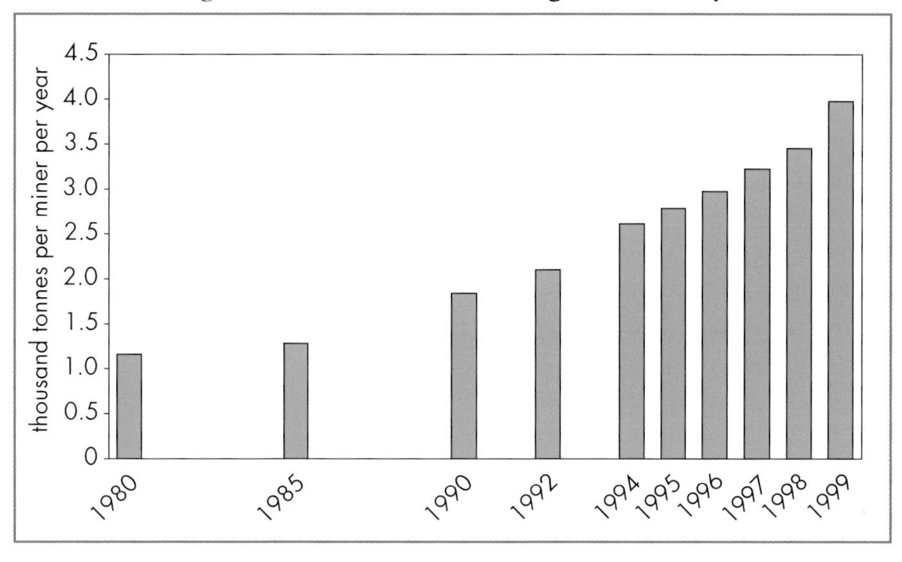

The assumed increase in coal prices after 2010 is much slower than that of oil and gas prices. The lowest-cost coal producers should see their earning rise. At the same time, the growing competitiveness of coal against gas could stimulate wider use of clean coal power-generation technologies.

Major Uncertainties

The *World Energy Outlook* projections are subject to a wide range of uncertainties. The reliability of our projections depends not just on how well the model represents reality, but on the validity of the assumptions that underpin them. How accurately causal relationships are represented in the World Energy Model depends partly on the quality and the availability of historical data. Shifts in these relationships in the future as a result, for example, of a technological breakthrough cannot be predicted today. Such imponderables add to the difficulties in projecting market trends.

Energy markets could evolve in ways that are much different from either the Reference Scenario or the OECD Alternative Policy Scenario. The main sources of uncertainty can be categorised as follows:

- *Macroeconomic conditions*: The pace of economic activity is the single most important driver of energy demand. If GDP growth is slower than assumed, demand will almost certainly grow less rapidly. Opinions among macroeconomists about the prospects for global economic growth in the next three decades typically range from 2% to 4% per year. Growth rates at the regional and country levels could be a lot different from those assumed, especially over short periods. Growth prospects for Russia, Eastern Europe and China are perhaps the least certain. The impact of structural economic changes, including the gradual shift from manufacturing to service activities and the growing role of high technology on demand for energy services (see Box 2.2 in Chapter 2) are also very uncertain, especially towards the end of the projection period.
- *Resource availability, supply costs and prices*: Resources of every type of energy are probably sufficient to meet demand for the next 30 years[14], but the future cost of extracting and transporting those resources is uncertain. This is partly because of a lack of information about geophysical factors. Oil and gas producers do not usually appraise reserves in detail until they are close to exploiting them. More difficult production conditions could, other things being equal, drive up costs and put upward pressure on prices.
- *Energy technology*: Improvements in the efficiency of current energy technologies and the adoption of new ones are major factors in the global energy outlook and key sources of uncertainty. Hydrogen-based fuel cells and carbon-sequestration technologies now under development could radically alter the energy-supply picture in the medium to long term. But commercialising them on a large scale will require sharp cost reductions and dramatic technical advances. The pace of these developments is very uncertain.
- *Energy and environmental policies*: Changes in government policies and new measures on energy security and environmental protection, especially climate change, could have profound

14. A detailed assessment of global energy resources and supply costs can be found in IEA (2001c).

consequences for energy markets. There are many sources of uncertainty, including the production and pricing policies of oil-producing countries, the impact of energy market reforms, taxation and subsidy policies. Regulations governing energy-related greenhouse gas emissions, the possible introduction of carbon-permit schemes and the role of nuclear power also affect the energy future.

- *Investment in energy supply infrastructure*: Massive investment in the production, transformation, transportation and distribution of energy will be needed to meet expected growth in demand in the coming decades. The bulk of this investment will be needed in developing countries, which will require major capital inflows in the form of loans and direct investment from the industrialised world. Mobilising this investment in a timely fashion will require the dismantling of regulatory and market barriers and deep reforms to corporate governance. Corrupt and inefficient practices must be stamped out and guarantees established to ensure foreign investors' rights over their assets and their right to repatriate earnings. Such moves would lower costs, increase productivity and boost the attractiveness of upstream projects to foreign investors.

CHAPTER 2: WORLD ENERGY TRENDS

HIGHLIGHTS

- World energy use will continue to increase steadily through 2030 in the Reference Scenario. Fossil fuels will remain the primary sources of energy and will meet more than 90% of the increase in demand to 2030. Among fossil fuels, natural gas will grow fastest, but oil will remain the most important energy source. Renewables will grow in importance, while the share of nuclear power in world energy supply will drop.
- Energy demand will increase most rapidly in developing countries, especially in Asia. The developing countries' share in world demand will increase from just 30% today to more than 40% by 2030. Per capita energy consumption, nonetheless, will remain much lower in developing countries.
- New sources of energy and advanced technologies, including oil sands, gas-to-liquids and fuel cells, will emerge during the *Outlook* period, especially after 2020.
- There will be a pronounced shift in the geographical sources of incremental energy supplies over the next three decades, in response to a combination of cost, geopolitical and technical factors. In aggregate, almost all the increase in energy production will occur in non-OECD countries.
- There will be a major expansion in international energy trade. The regions with most of the world's oil and gas resources, notably the Middle East and Russia, will greatly increase their exports. The OECD and the dynamic Asian economies, which already import large amounts of oil to meet their needs, will increase their oil imports still further. Trade in liquefied natural gas will surge. These developments will push supply security back to the top of the energy policy agenda.
- The expansion in production and supply capacity will call for a huge amount of investment at every stage of the energy supply chain — much of it in developing countries. Mobilising this investment in a timely fashion will require the lowering of regulatory and market barriers.

- Energy-related emissions of carbon dioxide will grow slightly more quickly than total primary energy supply. Power generation and transport will account for about three-quarters of the increase in emissions. More rigorous policies and measures than those so far adopted will be needed for the industrialised countries to meet their emissions reduction commitments under the Kyoto Protocol.

Energy Demand

Primary Energy

Global primary energy demand[1] in the Reference Scenario is projected to increase by 1.7% per year from 2000 to 2030, reaching 15.3 billion tonnes of oil equivalent (Table 2.1). The increase in demand will amount to almost 6.1 billion toe, or two-thirds of current demand. The projected growth is, nevertheless, slower than over the past three decades, when demand grew by 2.1% per year.

Table 2.1: **World Primary Energy Demand** (Mtoe)

	1971	2000	2010	2030	Average annual growth 2000-2030 (%)
Coal	1,449	2,355	2,702	3,606	1.4
Oil	2,450	3,604	4,272	5,769	1.6
Gas	895	2,085	2,794	4,203	2.4
Nuclear	29	674	753	703	0.1
Hydro	104	228	274	366	1.6
Other renewables	73	233	336	618	3.3
TPES	**4,999**	**9,179**	**11,132**	**15,267**	**1.7**

1. Total primary energy demand is equivalent to total primary energy supply (TPES). The two terms are used interchangeably throughout this *Outlook*. World primary demand includes international marine bunkers, which are excluded from the regional totals. Unless otherwise specified, world demand refers only to commercial energy and excludes biomass in non-OECD countries (see footnote 3). Primary energy refers to energy in its initial form, after production or importation. Some energy is transformed, mainly in refineries, power stations and heat plants. Final consumption refers to consumption in end-use sectors, net of losses in transformation and distribution. See Appendix 2 for detailed definitions.

Fossil fuels will account for just over 90% of the projected increase in world primary demand to 2030 (Figure 2.1). Their *share* in total demand actually increases slightly, from 87% in 2000 to 89% in 2030. **Oil** will remain the single largest fuel in the primary energy mix, even though its share will fall slightly, from 38% to 37%. Oil demand is projected to grow by 1.6% per year, from 75 mb/d in 2000 to 89 mb/d in 2010 and 120 mb/d in 2030. The bulk of the increase will come from the transport sector. No other fuel will seriously challenge oil in road, sea and air transportation during the projection period. In 2030, transportation will absorb 55% of total oil consumption, up from 47% now. Oil will remain a marginal fuel in power generation; a decline in the OECD area will offset a small increase in developing countries. Moderate increases are projected in industrial, residential and commercial oil consumption. Most of these increases will occur in developing countries, where competition from natural gas for space and water heating and for industrial processes will be limited.

Figure 2.1: **World Primary Energy Demand**

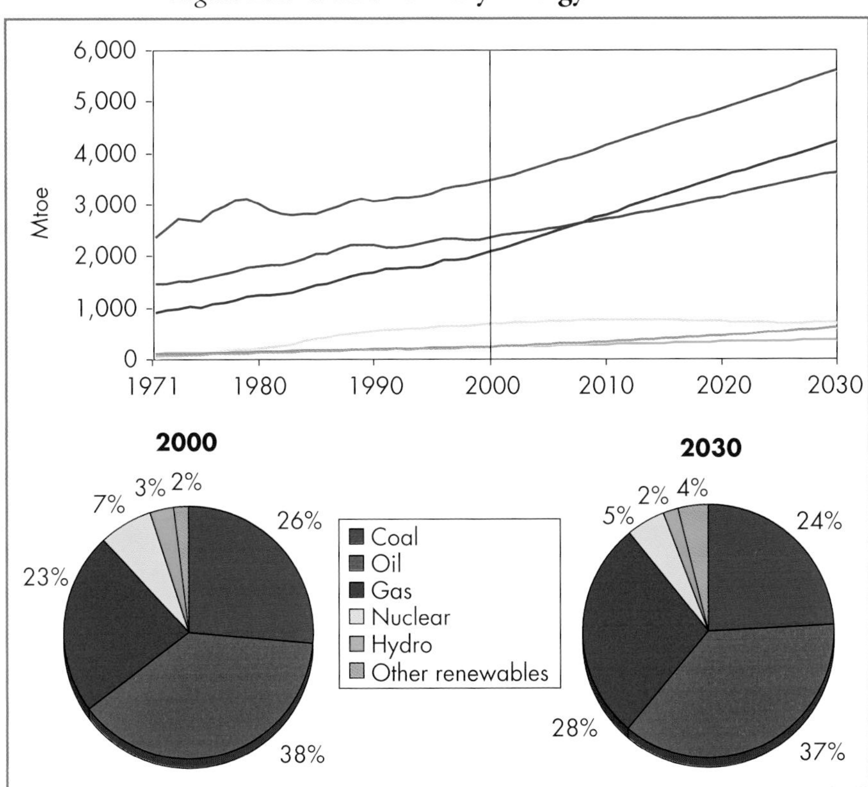

Demand for **natural gas** will grow faster than that for any other primary fuel, except for non-hydro renewable energy sources. With annual growth of 2.4% per year, gas will overtake coal just before 2010 as the world's second-largest energy source. Gas consumption will double between 2000 and 2030, and the share of gas in world demand will increase from 23% in 2000 to 28% in 2030, mostly at the expense of coal and nuclear energy. New power stations will account for over 60% of the increase in gas demand over the next three decades. Most of these stations will use combined-cycle gas turbine technology. This form of generation is often preferred to coal-based power technologies and nuclear power because of its high energy-conversion efficiency and low capital costs. Gas is also favoured over coal and oil for its relatively benign environmental effects, especially its lower carbon content. A small but growing share of natural gas demand will come from gas-to-liquids plants and from fuel cells for the production of hydrogen.

Demand for **coal** is projected to rise by 1.4% per year, but coal's share in world primary demand will still fall a little, from 26% in 2000 to 24% in 2030. China and India together account for almost three-quarters of the increase in coal demand in developing countries and two-thirds of the increase in world coal demand. Most of the increase in coal consumption will be in power generation. In OECD countries, an increase in power-sector demand for coal will offset a smaller decline in coal use in end-use sectors. The industrial, residential and commercial sectors in the transition economies and in developing countries will burn more coal, but power generation accounts for the bulk of the increase in overall coal demand in both groups.

The role of **nuclear power** will decline markedly over the *Outlook* period, because it is assumed that few new reactors will be built and several will be retired.[2] Nuclear production will peak in the next few years, then decline gradually. Its share of world primary demand will hold steady at about 7% through 2010, then fall to 5% by 2030. Nuclear output will increase in only a few countries, mostly in Asia. The biggest falls in nuclear production are expected to occur in North America and Europe.

2. Investment in new nuclear plants is projected to be limited on competitiveness grounds and because many countries have restrictions on new construction. Should governments enact strong policy measures to facilitate investment in nuclear plants, then the share of nuclear power in electricity generation could be significantly larger than projected here. Plant construction and extensions to the lifetimes of existing plants will depend critically on political decisions as well as economic factors.

Hydropower has long been a major source of electricity production, but its relative importance is set to diminish. Much of the OECD's low-cost hydro-electric resources have already been exploited and environmental concerns in developing countries will discourage further large-scale projects there. World hydropower production will grow slowly, by an average 1.6% a year through 2030, but its share of primary demand will remain almost constant at 2.5% over the *Outlook* period. The developing countries will account for most of the increase in hydropower production. Its share in global electricity generation will drop, from 17% to 14%.

Box 2.1: **Non-commercial Biomass Use in Developing Countries**

Non-commercial biomass represents one-quarter of total energy demand in developing countries. Biomass use in these countries is excepted to rise over the *Outlook* period, from 891 Mtoe in 2000 to 1,019 Mtoe in 2030, but its *share* in their total primary energy demand will fall. Demand will be stronger in the first decade, rising by 0.8% per year, but will slow to 0.1% per year in the last decade. The use of biomass, particularly wood products, in developing countries is likely to become more commercial over the projection period. Biomass will be increasingly traded in markets similar to those in many OECD countries today. The energy demand projections in this *Outlook* do not include traditional biomass use in developing countries. They are provided separately in the tables in Part D.

Other renewables[3], taken as a group, will grow faster than any other energy source, at an average rate of 3.3% per year over the projection period. But they will still make only a small dent in global energy demand in 2030, because they start from a very low base. Most of the increase in renewables use will be in the power sector. Their share in total generation will grow from 1.6% in 2000 to 4.4% in 2030. In absolute terms, the

3. This category includes geothermal, solar, wind, tidal and wave energy. It also includes biomass for OECD countries only. In this study, the term biomass includes traditional biomass energy, gas and liquid fuels from organic material, industrial waste and municipal waste. For developing countries, separate projections for traditional biomass use (wood, crop residues and animal waste), much of which is non-commercial, are included in the Annex to Chapter 13. Biomass is commercially traded in many developing countries, especially in South America. But IEA statistics do not distinguish between commercial and non-commercial biomass use. See Chapter 13 for a discussion of the link between biomass use and poverty in developing countries.

Figure 2.2: **Increase in World Primary Energy Demand by Fuel**

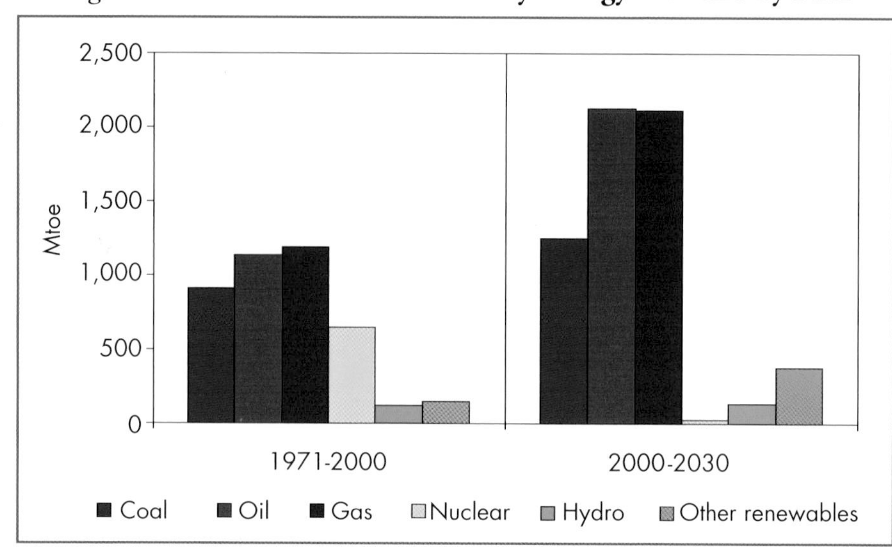

increase in the use of renewables will be much bigger in OECD countries, mainly because many of them have adopted strong promotional measures. Among non-hydro renewables, wind power and biomass will grow most rapidly, especially in OECD countries.

Regional Outlook

More than 60% of the increase in world primary energy demand between 2000 and 2030 will come from the developing countries (Figure 2.3). OECD countries will account for 30% and the transition economies for the remaining 8%. The OECD's share of world demand will decline, from 58% in 2000 to 47% in 2030, while that of the developing countries will increase, from 30% to 43%. The transition economies' share will fall slightly.

The increase in the share of the developing regions in world energy demand results from their rapid economic and population growth. Industrialisation, urbanisation and the replacement of non-commercial biomass by commercial fuels also boost demand. Increases in prices to final consumers, a result of the gradual reduction in subsidies and rising international prices, are not expected to curb energy demand growth in developing countries.

The developing regions will account for 29 mb/d of the 45-mb/d increase in global oil demand between 2000 and 2030. The developing

Figure 2.3: **Regional Shares in World Primary Energy Demand**

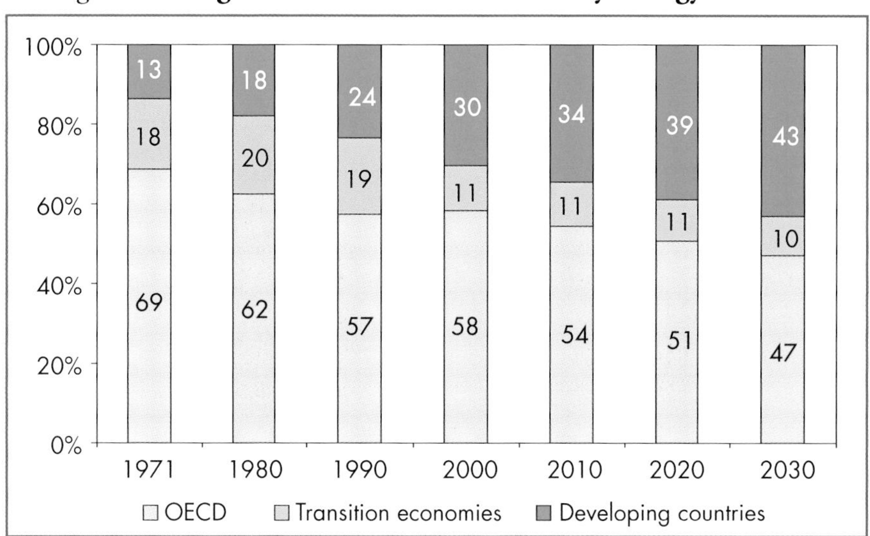

Asian countries will take the largest share. Oil demand in China will rise by 7 mb/d over the projection period to 12 mb/d in 2030. Other East Asian countries' oil demand will more than double, to 9.4 mb/d. Oil consumption in OECD North America will rise strongly too, from 22 mb/d in 2000 to almost 31 mb/d in 2030. Demand in other OECD regions will increase only modestly (Figure 2.4). North America remains by far the largest single market for oil.

Natural gas demand will grow strongly and the share of gas in the primary fuel mix will increase in every region. In volume terms, gas demand will increase the most in OECD North America and OECD Europe. But the fastest rates of growth will occur in China and South Asia, where gas consumption is currently very low. Coal demand will increase most in China and India, which have large, low-cost resources. Although coal's market share will decline a little, it will continue to dominate the fuel mix in those two countries. By 2030, China and India will account for 45% of total world coal demand, up from 35% in 2000. Nuclear power will fall in OECD North America, in OECD Europe and in the transition economies. It will increase in all other regions, but only marginally in most cases. The biggest increases in nuclear power production will occur in Japan, Korea and developing Asian countries.

Figure 2.4: **Increase in Primary Oil Demand by Region, 2000-2030**

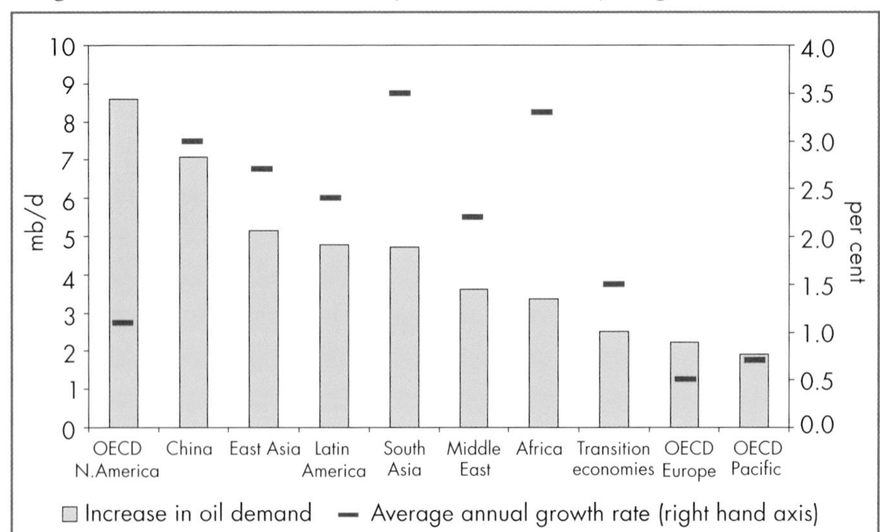

Energy Intensity

Energy intensity, measured as total primary energy use per unit of gross domestic product, is projected to decline in all regions. From 2000 to 2030, global energy intensity will fall by 1.2% per year. Intensity will fall most quickly in the non-OECD regions, largely because of improved energy efficiency[4] and structural economic changes towards lighter industry. The average rate of decline in energy intensity in these regions will accelerate from past trends. The transition economies, in particular, will become much less energy-intensive as more energy-efficient technologies are introduced, wasteful energy practices are tackled and energy prices are reformed. The shift to services is so far advanced in the OECD countries that their energy intensity is set to fall more slowly than in the past (Figure 2.5).

Energy-related Services

Energy is consumed in order to provide various services. Demand for energy is, in economists' parlance, a "derived demand". Identifying the drivers of demand for these services improves our understanding of long-term trends in energy consumption. These energy-related services are:

- mobility (non-electrical energy used in all forms of transport);

4. For a discussion of the link between energy intensity and energy efficiency and the role of pricing, see IEA (2000).

Figure 2.5: **Primary Energy Intensity by Region**

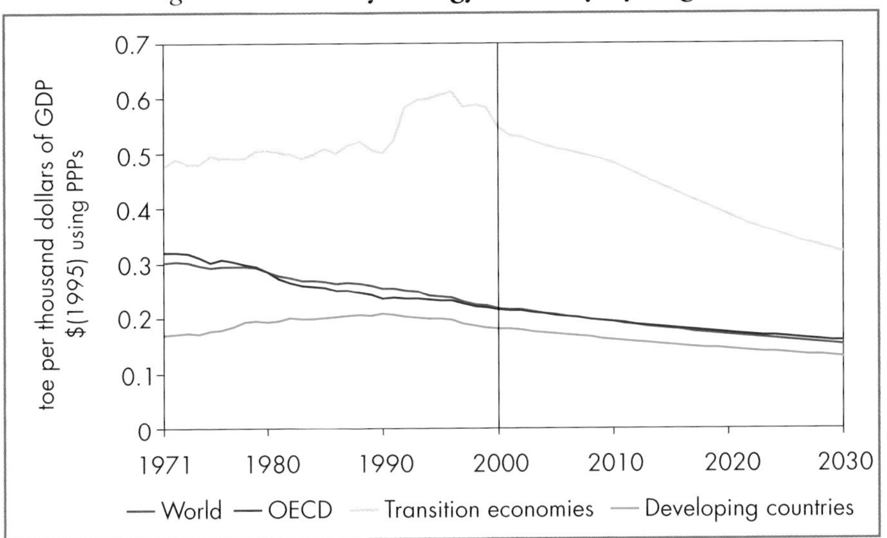

Figure 2.6: **World Energy-Related Services, 1971-2030**

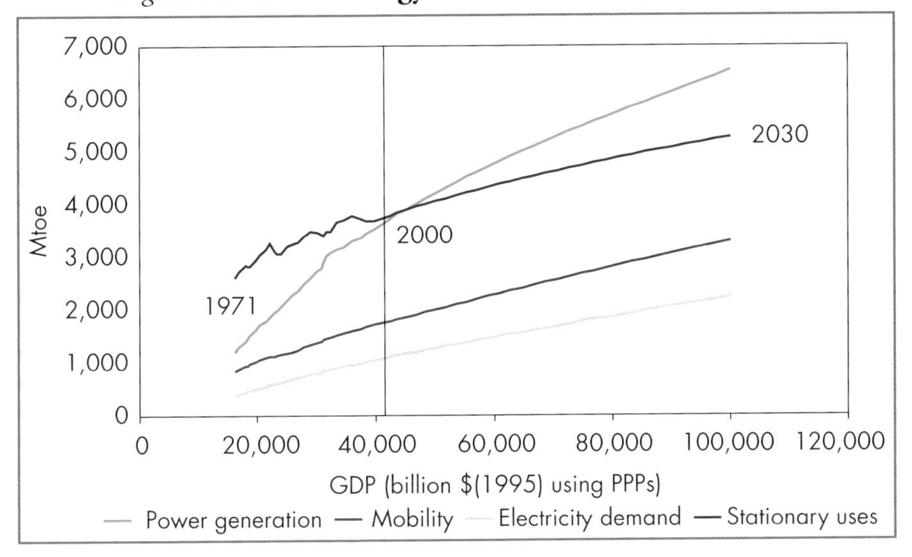

- stationary uses (fossil fuels used to provide heat in houses, commercial buildings and industrial processes);
- electrical uses (final electricity consumption in the residential, services, industrial and other end-use sectors); and
- fuel inputs to power generation (an intermediate energy-related service).

World energy use for both mobility and electrical services will grow along with GDP, but at a slightly slower rate (Figure 2.6). Over the last thirty years, electrical services expanded more rapidly than GDP, and mobility slightly less. Fuel inputs to power generation will also continue to rise with electrical services, but at a decelerating rate. Thermal losses in generation are expected to decline gradually as more efficient technologies, notably combined-cycle gas turbine plants, are deployed. Stationary uses of energy are less closely linked to economic growth than the other energy-related services. Having fluctuated around a moderately rising trend for the last two decades, demand for stationary uses is projected to grow slowly over the *Outlook* period, partly because of saturation effects and heavy industry's declining share of GDP in the OECD. Energy demand for stationary uses will grow more rapidly relative to GDP outside the OECD, partly because of strong growth in energy-intensive industries like iron and steel and chemicals.

Final Energy

Aggregate energy demand in final-use sectors (industry, transport, residential, services, agriculture and non-energy uses) is projected to grow by 1.7% per year from 2000 to 2030 – the same rate as primary energy demand. Transport demand will grow the most rapidly, at 2.1% per annum, overtaking industry in the 2020s as the largest final-use sector. Transport demand will increase everywhere, most rapidly in the developing countries, at 3.6% per year. OECD transport demand will grow at a more leisurely 1.4%, because of saturation effects. Residential and services consumption will grow at an average annual rate of 1.7%, slightly faster than industrial demand, which will rise by 1.5% per year. Information and communication technology introduces a major uncertainty into the prospects for final energy use (Box 2.2).

Among all end-use sources of energy, electricity is projected to grow most rapidly worldwide, by 2.4% per year from 2000 to 2030 (Table 2.2). Electricity consumption will double over that period, while its share in total final energy consumption will rise from 18% to 22%. Electricity use will expand most rapidly in developing countries, by 4.1% per year, as the number of people with access to electricity and per capita consumption increase. Demand will increase by 2% in the transition economies and by 1.5% in the OECD.

Box 2.2: The Implications of Information and Communication Technology on Energy Demand

The growth of information and communication technology is affecting energy demand in many ways, including:

- *Income effects*: To the extent that the production and use of ICT boost overall economic growth, they raise the overall demand for energy services.
- *Price effects*: ICT will increase productivity in various sectors by different amounts. These differences will alter the relative price of goods and services, thereby affecting patterns of energy use.
- *Structural effects*: Because these new technologies are used mostly in services, the service sector's share of GDP is growing faster than might otherwise have been the case. This phenomenon helps lower energy intensity, since services require less energy input per unit of output than does industry.
- *Efficiency gains*: Business-to-business and business-to-customer electronic commerce can reduce energy use by improving operational efficiency, partly through better supply chain and inventory management. Web-based retailing reduces the need for large inventories and shop space. If the overall demand for commercial premises falls, energy consumption in the construction sector would be lower. Part of the effect of improved efficiency on energy use would be offset by the impact of higher income (the "rebound effect").
- *Fuel mix effects*: ICT equipment is powered exclusively by electricity, and this tends to raise electricity's share in final energy use.

Recent studies give no clear indication of what the net effect of ICT on overall energy demand might be. But they suggest that the impact varies with structural factors within different countries, such as the relative size of the service sector.[5]

5. The IEA is continuing its quantitative analysis of this issue and is upgrading its models to capture the impact of ICT on energy-demand patterns and trends. The proceedings of the IEA workshop on the future impact of ICT on the energy system can be found at www.worldenergyoutlook.org.

Table 2.2: **World Total Final Consumption** (Mtoe)

	1971	2000	2010	2030	Average annual growth 2000-2030 (%)
Coal	630	554	592	664	0.6
Oil	1,890	2,943	3,545	4,956	1.8
Gas	604	1,112	1,333	1,790	1.6
Electricity	377	1,088	1,419	2,235	2.4
Heat	68	247	260	285	0.5
Renewables	66	86	106	150	1.8
Total final consumption	**3,634**	**6,032**	**7,254**	**10,080**	**1.7**

Figure 2.7: **Per Capita Electricity Consumption**

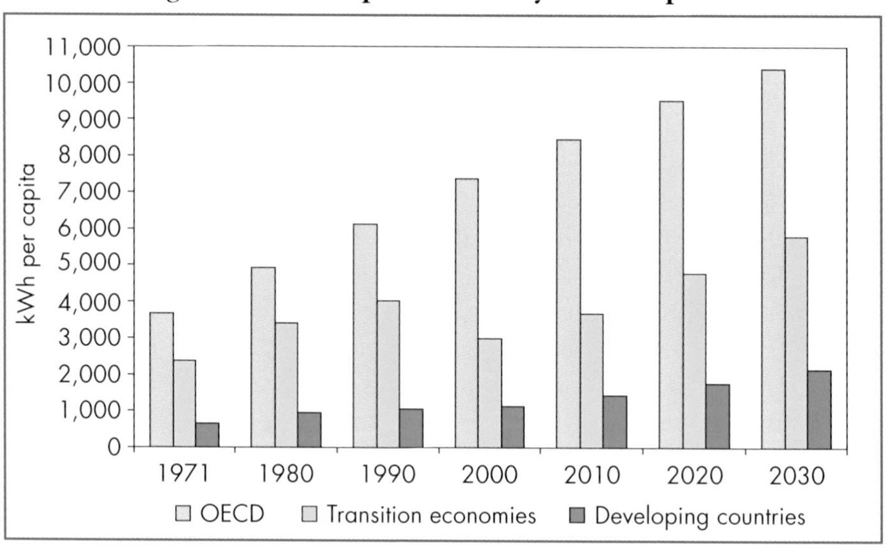

The share of electricity in total final consumption in developing countries catches up with that of the OECD by 2030. This is partly because continued electrification in the poorest developing countriesmore than keeps pace with population growth (see Chapter 13). In part, it is also because the share of natural gas in final uses is much lower in developing countries than in the OECD countries. Electricity's *apparent* share in these countries is also boosted by the fact that biomass is not included in our figures for total final consumption. However, per capita electricity consumption of people with access to electricity in developing countries remains well below that of OECD countries in 2030 (Figure 2.7).

The share of coal in world final consumption will drop from 9% to 7%. Coal use will expand in industry, but only in non-OECD countries, and will stagnate in the residential and services sectors. The shares of oil and gas in world final consumption will hardly change over the projection period. Oil products will account for roughly half of final energy use and gas for 18%. Oil demand will grow by 1.8% per year, with almost three-quarters of the increase coming from transport. The share of transport in incremental oil demand is over 90% in the OECD.

Energy Production and Trade

Resources and Production Outlook

The world's energy resources are adequate to meet the projected growth in energy demand. Global oil supplies will be ample at least until 2030, although additional probable and possible reserves will need to be "proved up" in order to meet rising demand. Unconventional oil will probably carve out a larger share of global oil supplies. Reserves of natural gas and coal are particularly abundant, while there is no lack of uranium for nuclear power production through 2030. Renewable energy sources are also plentiful.

There will be a pronounced shift in the geographical sources of incremental energy supplies over the next three decades, in response to a combination of cost, geological and technical factors. Their aggregate effect will be that almost all the increase in energy production will occur in non-OECD countries, compared to just 60% from 1971 to 2000 (Figure 2.8).

Increases in production in the Middle East and the former Soviet Union, which have massive hydrocarbon resources, are expected to meet much of the growth in world oil and gas demand. Latin America, especially Venezuela and Brazil, and Africa will also raise output of both oil and gas. Oil production will decline almost everywhere else. Production of natural gas, resources of which are more widely dispersed than oil, will increase in every region other than Europe. Although there are abundant coal reserves in most regions, increases in coal production are likely to be concentrated where extraction, processing and transportation costs are lowest. Coal production is likely to grow most rapidly in South Africa, Australia, China,

Figure 2.8: **Increase in World Primary Energy Production**

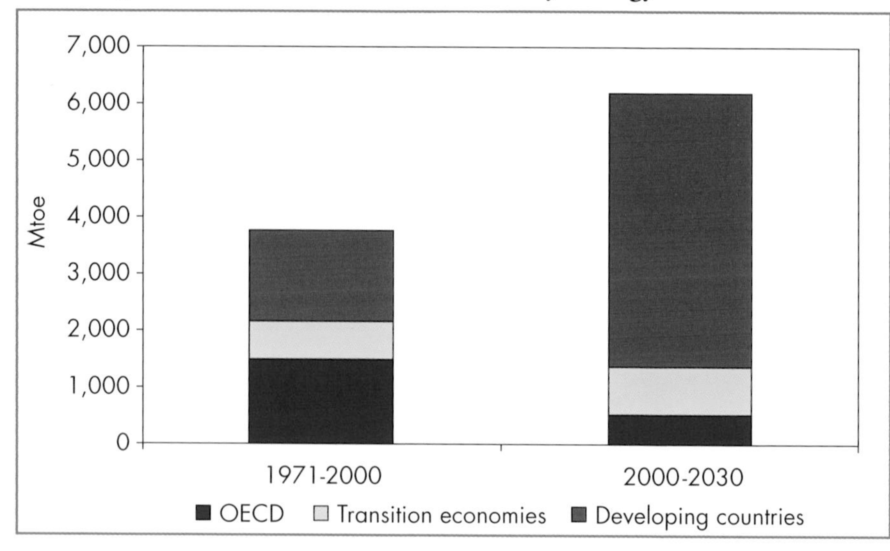

Indonesia, North America and Latin America. The production prospects for each fuel are discussed in more detail in Chapter 3.

Implications for International Energy Trade

International trade in energy will expand in both absolute terms and as a share of production to accommodate the mismatch between the location of demand and that of production (Figure 2.9). Growing trade, almost entirely in fossil fuels, will have major geopolitical implications. Dependence on Middle East oil will continue to grow in the net oil-importing regions, essentially the three OECD regions and some parts of Asia. This development will increase mutual dependence, but will also intensify concerns about the world's vulnerability to a price shock induced by a supply disruption. Maintaining the security of international sea-lanes and pipelines will become more important as oil supply chains lengthen.

Increasing dependence on imports of natural gas in Europe, North America and other regions will heighten those concerns. The disruption in liquefied natural gas supplies from Indonesia in 2001, caused by civil unrest, demonstrated the risks of relying on imports of gas from politically sensitive regions. On the other hand, the expected expansion of international LNG trade could alleviate some of the risks of long-distance supply chains if it leads to more diversified supplies. Increased short-term trading will also make LNG supplies more flexible.

Figure 2.9: **Share of Net Inter-Regional Trade in World Fossil-Fuel Supply**

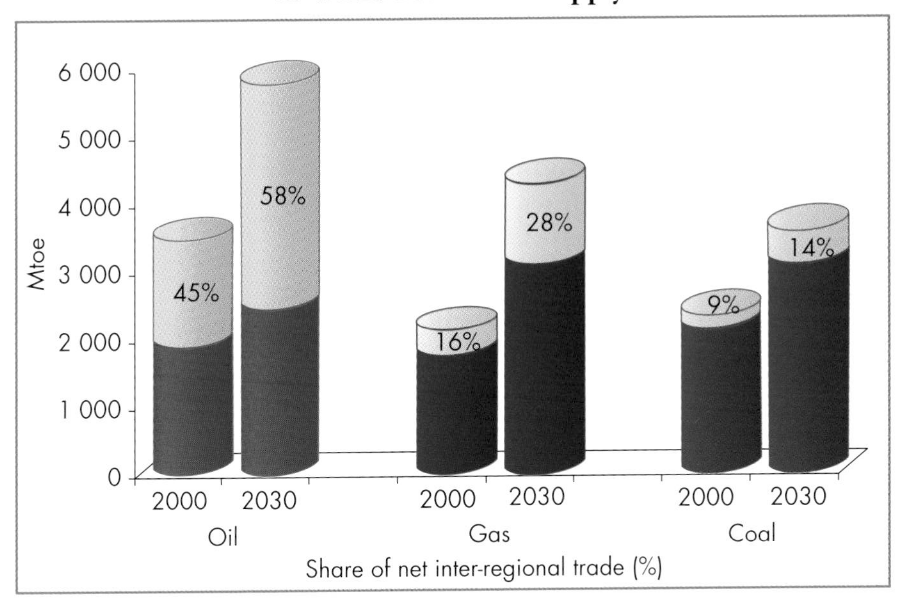

Note: Total international trade is even larger because of trade between countries within each *WEO* region and re-exports.

The governments of oil- and gas-importing countries are expected to take a more proactive role in dealing with the energy security risks in fossil fuel trade. They are likely to work on improving relations with energy suppliers. They will also step up measures to deal with short-term supply emergencies or price shocks. Governments and end-users are, nonetheless, likely to set a limit on the premium they are prepared to pay in order to enhance the security of energy supplies.

Implications for Investment

The projected increase in production and supply capacity will call for a huge amount of investment at every link in the energy supply chain.[6] For example, the Reference Scenario projections of power generating capacity call for cumulative investment of $2,100 billion in developing countries and $300 billion in the transition economies (see Chapter 3). Financing

6. The IEA is undertaking a major study of energy investment, to be published in 2003 *(WEO 2003 Insights: Global Energy Investment Outlook)*. It will quantify the amount of investment that will be needed globally to meet the increase in demand projected in this *Outlook* and will consider project-financing issues.

the building of new energy infrastructure will be a major challenge. Most of the required investment is needed in developing countries. This will call for a huge increase in capital inflows from industrialised countries. Private foreign investment in energy projects (excluding upstream oil and gas) in developing countries and transition economies boomed in the early to mid-1990s, peaking at nearly $51 billion in 1997. Investment slumped in the wake of the 1997-1998 economic crisis in emerging market economies to less than $18 billion in 1999, but recovered to some $30 billion in 2000 (Figure 2.10).[7]

Figure 2.10: **Private Foreign Energy Investment in Developing Countries and Transition Economies***

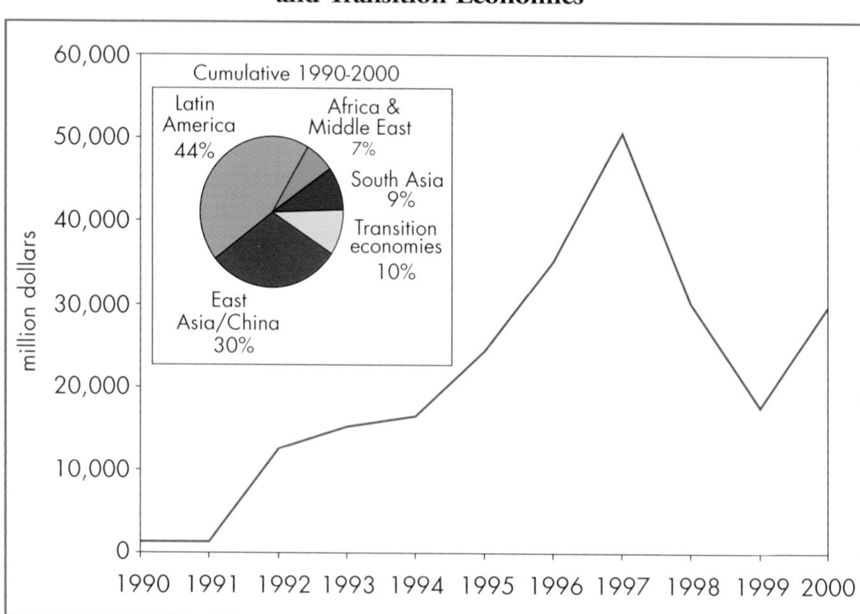

* Not including oil and gas exploration and production.
Source: World Bank's Private Participation in Infrastructure (PPI) database.

Mobilising this investment in a timely fashion will require the lowering of regulatory and market barriers. Most major oil and gas producers in Africa, the Middle East and Latin America, recognise the need for foreign involvement. Algeria, Egypt, Libya and Nigeria, for example,

7. See also Saghir (2002). Overseas development assistance from industrialised countries to developing countries declined in the1990s.

have changed their upstream policies and practices to attract joint-venture investment by international oil companies. Since 1992, Venezuela has sought private investment in the oil and gas sectors. Saudi Arabia has recently started to open its upstream gas sector to foreign companies. Key coal producers, including China and India, will need to attract huge amounts of capital to meet their medium-term production targets. Many developing countries are liberalising and restructuring their electricity industries in order to attract private domestic and foreign investment. Improvements in the way state-owned energy companies are run are also needed in many countries. A lack of transparency and consistency, together with weak judicial systems, have encouraged vested interests and led to corruption, fraud, theft and money laundering in some cases. These problems raise production costs and discourage private investment.

Implications for Global CO_2 Emissions

The Reference Scenario projections for energy demand imply that worldwide carbon-dioxide emissions will increase by 1.8% per year from 2000 to 2030.[8] They will reach 38 billion tonnes in 2030. This is 16 billion tonnes, or 70%, above current levels. Two-thirds of the increase will come from developing countries. By 2010, energy-related CO_2 emissions will be 36% higher than in 1990.

Figure 2.11: **Energy-Related CO_2 Emissions by Region**

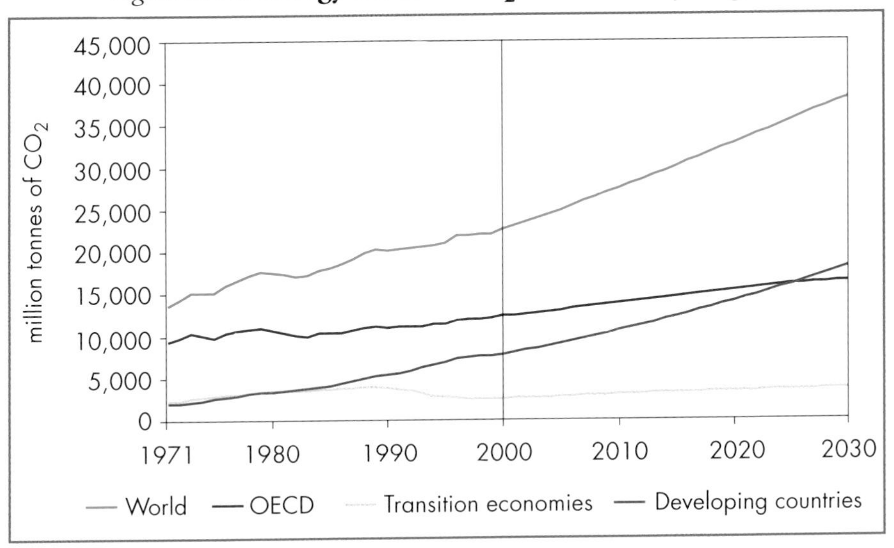

8. Detailed projections of energy-related CO_2 emissions can be found in Part D.

The geographical structure of new emissions will change drastically over the *Outlook* period (Figure 2.11). Historically, OECD countries have been the largest emitters of greenhouse gases. In 2000, they produced 55% of global carbon emissions; developing countries accounted for 34% and the transition economies for the remaining 11%. By 2030, developing countries will have become the most heavily emitting region, with 47% of global emissions. The OECD will account for 43% and the transition economies for 10%. OECD emissions will increase by 4 billion tonnes between 2000 and 2030. China's emissions alone will grow by 3.6 billion tonnes.

Over the past three decades, the mining and burning of coal accounted for 40% of the *increase* in world CO_2 emissions, while oil produced 31% and gas 29%. In the coming three decades, oil will produce 37% of new energy-related carbon emissions and coal 32%. As a result, coal's share in *total* emissions will fall by three percentage points, to 36% in 2030. The share of gas will rise to 25%.

The trend in the relationship between *total* CO_2 emissions and primary energy demand will reverse over the projection period. Over the past three decades, carbon emissions grew by 1.8% a year, but were outstripped by energy demand, which grew at 2.1%. Over the next 30 years, on present policies, emissions will continue to increase at 1.8%

Figure 2.12: **Average Annual Growth Rates in World Energy Demand and CO_2 Emissions**

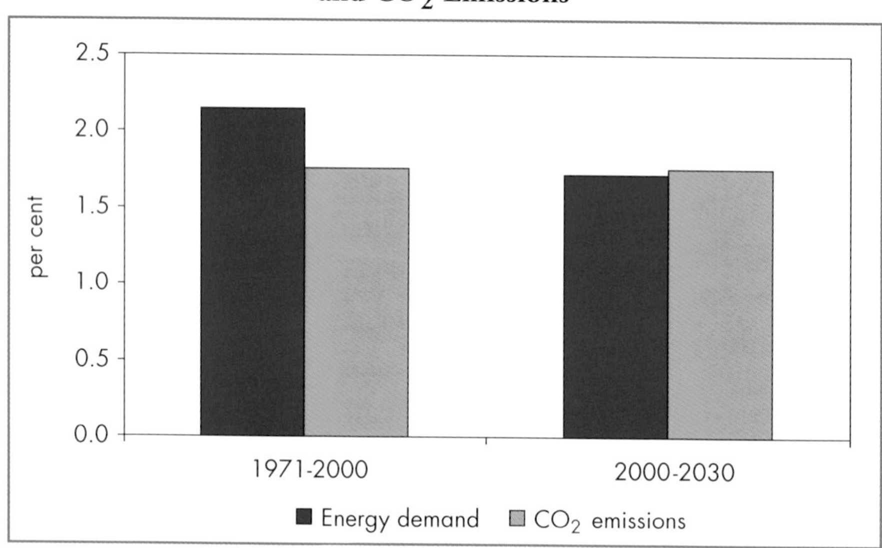

per year, while energy demand growth will be only 1.7% per year (Figure 2.12). As a result, the average carbon content of energy — CO_2 emissions per unit of aggregate primary energy consumption — will increase slightly, from 2.47 tonnes per toe to 2.50 tonnes. In 1971, the average carbon content was 2.7. The main cause of this reversal will be the declining share of nuclear power and hydroelectricity in the global energy mix. Non-hydro renewables will be increasingly used and technology will increase the efficiency of energy systems, but neither of these developments will make up for the increase in fossil fuel use required to replace nuclear energy and hydropower.

Box 2.3: **Trends in Carbon Intensity**

Carbon intensity is typically defined as the amount of CO_2 emitted per unit of GDP. From 1997 to 2000, global emissions grew by 1.1% a year, while the global economy grew, on average, by 3.5%. Consequently, carbon intensity *fell* by 2.2% annually over the period (Figure 2.13).

The bulk of this reduction can be attributed to China, which, according to official statistics, achieved a spectacular 7.4% decrease in carbon intensity. This was led by an annual decrease of 2.2% in CO_2 emissions[9] from coal. Improved energy efficiency in transition economies helped to lower global carbon intensity. So did the shift from industry to services in the OECD.

Carbon intensity will drop throughout the world over the next three decades, but not fast enough to avoid a net rise in CO_2 emissions. The decline in carbon intensity will be driven mainly by a global shift from manufacturing to the service sector, technological advances and by fuel switching.

Over the projection period, the greatest improvement will be in the transition economies, as old and inefficient capital stock is replaced and as natural gas substitutes for coal and oil in power generation. The economic recovery in Russia is expected to continue, and the country's carbon intensity, which is now six times as high as Japan's and three times as high as China's, will decline by more than 2% annually to 2030. China and India are also expected to see rapid improvements in their carbon intensity over the projection period.

9. In 1997, China's CO_2 emissions from coal accounted for 30% of the world emissions from coal. This figure is now down to 27%. See Chapter 7 for more details on Chinese coal consumption figures.

Regional Trends in CO_2 Emissions

The Reference Scenario sees CO_2 emissions in OECD countries reaching 16.4 billion tonnes in 2030. The largest increase will come in North America[10], where emissions will reach 9.1 billion tonnes, an increase of 1.1% per year, or 71% over 1990. In the OECD Pacific region, CO_2 emissions will grow by 0.9% annually, from 1.9 billion tonnes in 2000 to 2.5 billion tonnes in 2030, or 67% over 1990. Emissions in OECD Europe will rise by 23% over the projection period, a far more moderate rate than in North America and Pacific. This relatively successful performance can be attributed to the increased use of renewable energy and natural gas.

Among developing countries, China's emissions will increase by far the most. This will be due to China's strong economic growth, its rapid increase in electricity demand and its continuing heavy reliance on coal. Chinese emissions will more than double, from 3.1 billion tonnes in 2000 to 6.7 billion tonnes in 2030. China will contribute one-quarter of the increase in global CO_2 emissions.

Figure 2.13: **Carbon Intensity by Region**
(Index, 1971=1)

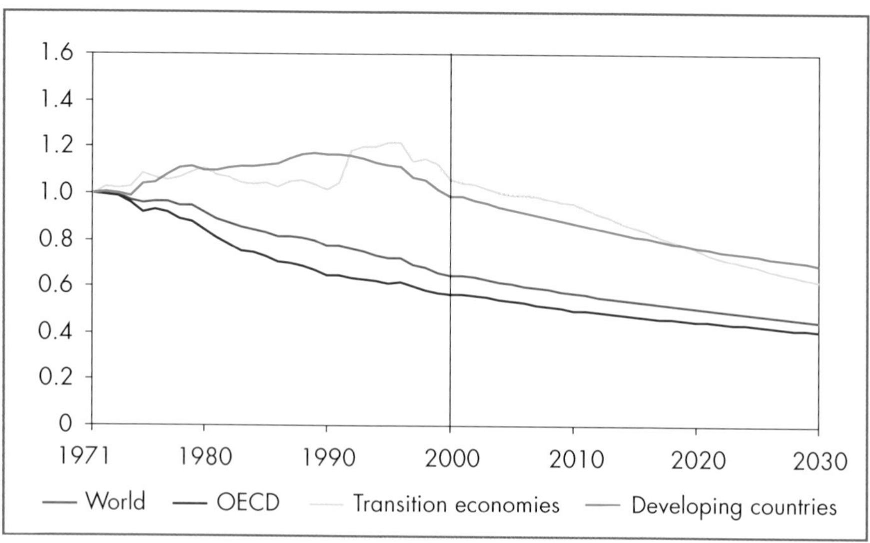

10. See Chapters 4 to 11 for a more comprehensive description of regional trends and Chapter 12 on the OECD Alternative Policy Scenario for an assessment of the potential for reducing CO_2 emissions in the OECD.

Coal will also be used to fuel new power generating capacity in other developing Asian countries and will account for a large proportion of the increase in their CO_2 emissions. In India, CO_2 emissions will grow by 3% a year, as most new power stations built in the period to 2030 will run on coal. Indian transport-related emissions will account for a quarter of the national increase in total emissions. In East Asia, emissions will rise from 1.1 billion tonnes in 2000 to 2.8 billion tonnes in 2030. Half of the increase will come from the power sector and more than a quarter from transportation.

CO_2 emissions in Latin America will rise by 3% per year over the projection period as a result of a rapid rise in fossil fuel demand. Emissions from power plants will rise faster than electricity generation, as the continent's hydropower potential is used up and it starts tapping its large natural gas reserves. Latin America's total emissions will rise from 0.9 billion tonnes in 2000 to 2.1 billion tonnes in 2030. Africa's contribution to global CO_2 emissions will remain small over the next three decades, as large segments of the population will continue to live without commercial energy.

Trends in Per-capita CO_2 Emissions

Per capita CO_2 emissions worldwide are expected to grow by 0.7% per year over the next three decades. They will reach 4.7 tonnes in 2030, up from 3.8 tonnes in 2000. Regional differences will remain very large. Per capita emissions will rise considerably in China, from 2.4 tonnes to 4.5 tonnes in 2030. In India, they will rise from 0.9 tonnes to 1.6 tonnes. They will more than double in Indonesia. In Africa, emissions per head are now very low at 0.9 tonnes per capita and will rise by half to 1.3 tonnes in 2030. Despite these increases, the OECD and the transition economies will still have much higher per capita emissions in 2030: 13 tonnes in the OECD and 11 tonnes in the transition economies (Figure 2.14).

Urbanisation will play a significant role in the growth in per capita emissions. Seven of the world's ten most populated cities are in developing countries. More than half the entire population of the developing world will live in urban areas in 2030, up from 40% today. Per capita emissions in cities are often two or three times those at the national level, because urban dwellers have better access to commercial energy than the rural population. They also have better access to transport services.

Figure 2.14: **Per Capita CO_2 Emissions**

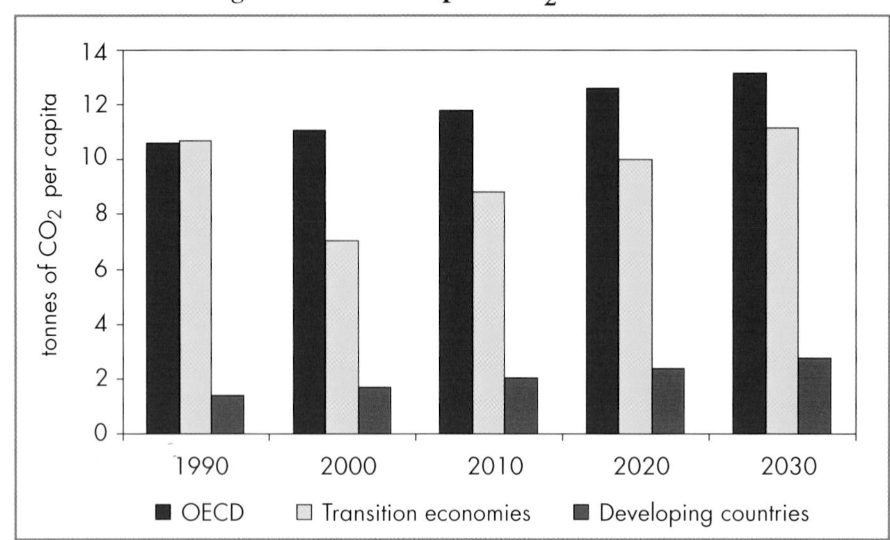

CO_2 Emissions by Sector

Power generation will contribute almost half the increase in global emissions between 2000 and 2030 (Table 2.3). Transport will account for more than a quarter. The residential, commercial and industrial sectors will account for the rest.

Table 2.3: **Increase in CO_2 Emissions by Sector** (million tonnes of CO_2)

	OECD		Transition economies		Developing countries		World	
	1990-2010	**2000-2030**	**1990-2010**	**2000-2030**	**1990-2010**	**2000-2030**	**1990-2010**	**2000-2030**
Power generation	1,373	1,800	44	341	2,870	5,360	4,287	7,500
Industry	11	211	-309	341	739	1,298	440	1,850
Transport	1,175	1,655	-52	242	1,040	2,313	2,163	4,210
Other*	244	363	-428	234	620	1,365	436	1,962
Total increase	2,803	4,028	-746	1,158	5,268	10,336	7,325	15,522

*Agriculture, commercial, public services, residential and other non-specified energy uses.

Power Generation Emissions

Electricity generation will be a growing source of total CO_2 emissions, rising from 40% of total CO_2 emissions in 2000 to 43% in 2030. The Reference Scenario sees the power sector becoming more dependent on fossil fuels over the projection period. It projects CO_2 emissions growing in closer synchrony with electricity generation than in the past. The expected increase of thermal efficiency in power generation, the greater use of natural gas and the growing use of non-hydro renewables will moderate the growth in emissions to some extent, but not decisively.

Developing countries will account for almost three-quarters of the incremental CO_2 emissions from power generation. Coal-fired power plants in these countries will still account for *more than half* the global increase in power generation CO_2 emissions in the next three decades. Power sector emissions in the OECD and in the transition economies will rise much more slowly, because renewables and natural gas will take market share from coal.

Emissions per unit of electricity are expected to decrease over time, but regional differences will remain high even on this point (Figure 2.15). The efficiency of power plants in the transition economies and in the developing countries could improve more quickly than projected here, but only if modern technology is deployed soon on a larger scale.

Figure 2.15: **CO_2 Emissions per kWh of Electricity Generated**

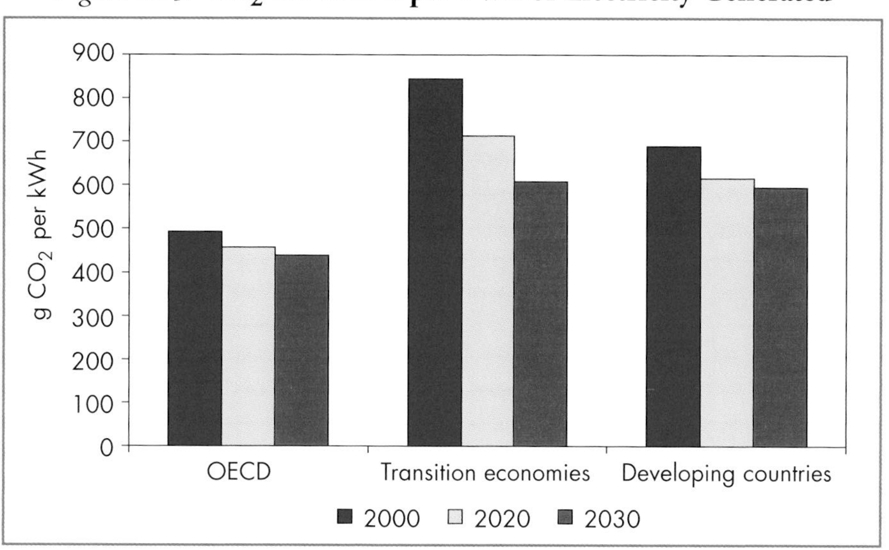

Note: Emissions in this chart include those from heat production. This overestimates somewhat the emissions per unit of electricity shown above.

Transport Emissions

Rising oil consumption in the transport sector, mainly by cars and trucks, is the second major source of increased CO_2 emissions in the Reference Scenario, after power generation. Global transport sector emissions are projected to rise by more than 85% from 2000 to 2030. In 2030, transport will account for roughly a quarter of global energy-related emissions, up from 21% in 2000. More than half of the increase is expected to take place in the developing countries. OECD countries will contribute about 40%. Most of the increase in the OECD will come from road transport.

The rapid increase of CO_2 emissions in developing countries can be largely attributed to a projected increase in both vehicle ownership and freight transport. The increase is particularly strong in Asia where per capita car and motorbike ownership is still low compared with the global average. In 2000, China averaged 12 vehicles per 1,000 persons; and India 8.4. In the United States and Canada, the figure is close to 700. Road freight is also expected to increase sharply. Because the developing Asian countries hold nearly half the world's population, an increase in road transport would have a huge impact on global emission levels. These projections make it clear that Asian countries will have to make enormous investments in their road infrastructure. They will also face a host of local environmental problems, especially traffic congestion and air pollution.

CO_2 Emissions Projections and the Kyoto Protocol

The CO_2 emissions projections in this *Outlook* have particular relevance to the commitments of developed countries ("Annex B") under the Kyoto Protocol.[11] Table 2.4 provides a regional breakdown of emissions projections for these countries and the gap remaining between them and the Kyoto commitments in both percentage and absolute terms. These figures reflect only energy-related CO_2 emissions, while the Protocol covers six gases and the contribution of forest sinks.

11. The 1997 Kyoto Protocol calls for industrialised countries listed in its Annex B to reduce their greenhouse gas emissions by an amount that would bring the total to at least 5% on average below 1990 levels over the 2008 to 2012 period. Annex B includes all OECD countries except Korea, Mexico and Turkey. To take effect, the Kyoto Protocol must be ratified by at least 55 nations, which together must represent at least 55% of developed countries' carbon dioxide emissions. By June 2002, 74 countries, including all European Union countries, had ratified the Protocol. These countries account for 36% of emissions.

Table 2.4: **CO_2 Emissions from Energy and Targets in Annex B Countries, 2010** (million tonnes of CO_2)

	Emission targets for 2010	WEO emissions 2010	Gap* (%)	Gap (Mt CO_2)
OECD Annex B countries**	9,662	12,457	28.9	2,795
Russia	2,212	1,829	-17.3	-383
Ukraine and Eastern Europe	1,188	711	-40.2	-477
Total	13,062	14,997	14.8	1,935

* The difference between target emissions and projected emissions as a percentage of the target emissions. In other words, it is the extent to which projected emissions exceed targets.
** This total covers all OECD countries with commitments under the Kyoto Protocol (Annex B countries). Turkey, Mexico and Korea are the only OECD countries not included in Annex B. However, Australia and the United States announced in 2001 that they would not ratify the agreement.
Note: The emission targets for 2010 differ from those in *WEO 2000* because emissions data for 1990 have been revised.

The steep rise in emissions in the Reference Scenario highlights the challenge that the Kyoto Protocol represents for most OECD countries, particularly in North America and the Pacific. For OECD European countries, energy-related CO_2 emissions are projected to be about 8% above target by 2010. Emissions in all OECD countries with commitments under the Protocol will be 12.5 billion tonnes, that is 2.8 billion tonnes, or 29%, above their target.

Russia, like Central and Eastern Europe, is in a very different situation, with projected emissions considerably *lower* than their commitments. Russia's emissions will be some 0.4 billion tonnes below its commitments. Emissions in Ukraine, with other Central and Eastern European countries, will also be below their commitments, by about 0.5 billion tonnes[12]. The Protocol allows for countries to offset mutually their emission commitments through a trading system. But lower emissions in Russia, Ukraine and Eastern Europe ("hot air") will not be enough to compensate for higher emissions in other Annex B countries. The overall gap will amount to about 15% above target in 2010. However, the gap is only 2% if the United States is excluded.

12. The Russian gap was 600 million tonnes of CO_2 in *WEO 2000*. The reduction in this edition comes mainly from revisions to data for the base year, 1990, as well as more favourable assumptions about GDP growth.

Technological Developments

Throughout the projection period, the supply and consumption of energy is expected to take place using broadly the same technologies that are already in use or are currently available. Technological advances are assumed to take place, but these will be incremental rather than revolutionary. Some technologies that exist today will become commercial during the next three decades. There will be a gradual shift towards less polluting technologies, particularly those based on renewable energy in power generation. Technological breakthroughs may well take place in some areas, but predicting their timing and magnitude is impossible. Government support of energy research and development will continue to play a key role in the pursuit of technological progress.

Demand-side Technologies

The Reference Scenario assumes that the efficiency of energy use – the amount of energy needed to provide a given amount of energy service – will continue to improve at a pace similar to that of the past three decades. Because most of the energy-using capital stock has a long life, technological advances can affect the average energy efficiency of equipment and appliances in use only very gradually (Figure 2.16).

Figure 2.16: **Lifetimes of Energy Capital Stock**

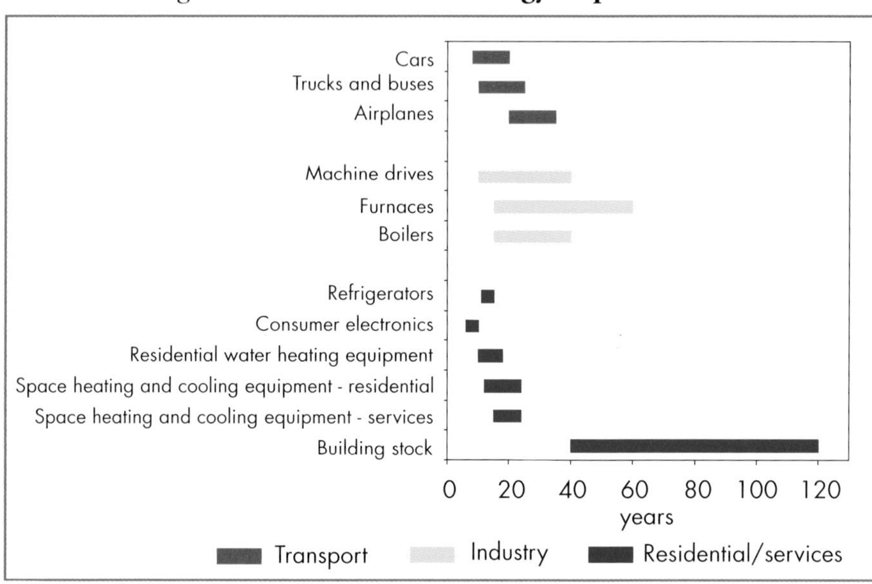

In the transport sector, vehicle fuel efficiency will continue to improve in most regions. Voluntary agreements with car manufacturers and standards are expected to lead to improvements in the fuel efficiency of new passenger vehicles of 30% between 2000 and 2030 in the European Union and 20% in Japan, Australia and New Zealand. However, the energy saved will be partly offset by an increase in the total number of kilometres driven. No improvement is expected in the United States and Canada, because technical advances in vehicle fuel efficiency will be offset by an increase in car size, weight and the number of appliances in each car. No changes in efficiency standards, known as CAFE in the United States, are assumed in the Reference Scenario. In all regions, hybrid vehicles that run on both conventional fuels and electric batteries will gain a foothold in the vehicle fleet. Fuel-cell vehicles are not expected to penetrate the fleet to a significant degree before 2030.

In stationary energy uses in the industrial, commercial and residential sectors, progressive improvements in energy efficiency are assumed to occur as a result of ongoing technological advances. For example, the growing deployment of integrated building designs, which incorporate efficient lighting, heating and cooling systems, will reduce energy consumption per square metre of office space in new office buildings. Energy efficiency standards and labelling programmes already in place will continue to encourage more efficient equipment and appliances in these sectors. However, these efficiency improvements will be very gradual, because of the slow rate of replacement of energy-capital stock, especially buildings.

Supply-side Technologies

Improvements will continue to be made in supply-side technologies, including cost reductions. Efforts will continue on reducing the cost of finding and producing oil and gas. Key new technologies in this area, such as advanced seismic techniques, will improve the identification of reservoir characteristics. Better drilling and production engineering can also be expected. Further advances will be made in deep-water technologies and enhanced oil recovery techniques. Major advances are also expected in high-pressure gas pipelines, LNG processing and gas-to-liquids production technology. The use of advanced coal-mining technology, together with an increase in the scale of individual mine projects, will continue to drive productivity gains and lower the cost of coal extraction and preparation.

Considerable progress is expected to be made in improving the fuel-conversion efficiency of existing power generation technologies. There will also be reductions in the capital costs of emerging fossil fuel- and renewables-based power technologies. The average efficiency of new combined-cycle gas turbine plants is assumed to rise from 55% in 2000 to 62% by 2030 in OECD countries. Coal-fired integrated gasification combined-cycle (IGCC) plants are expected to become competitive with gas-fired by the middle of the projection period. But this technology will come under renewed competitive pressure later from renewables. The average efficiency of IGCC technology is assumed to reach 52% in 2030 compared to 43% at present. The higher efficiencies of new gas- and coal-fired plant will push up the average efficiency of all plants in operation over the projection period (Figure 2.17). No breakthrough in nuclear power technology is assumed before 2030.

The capital costs of renewable energy technologies are expected to fall substantially, making electricity production from renewables increasingly competitive over the projection period. Capital and overall generating costs will continue to vary widely across regions according to local factors. Further reductions are expected in the generating costs of wind power from larger turbines, which improve performance, and from higher efficiencies in biomass conversion. The projected rate and extent of the decline in costs for each source are shown in Figure 2.18, but the figures are very uncertain.

Figure 2.17: **Average World Power Generation Efficiency by Fuel**

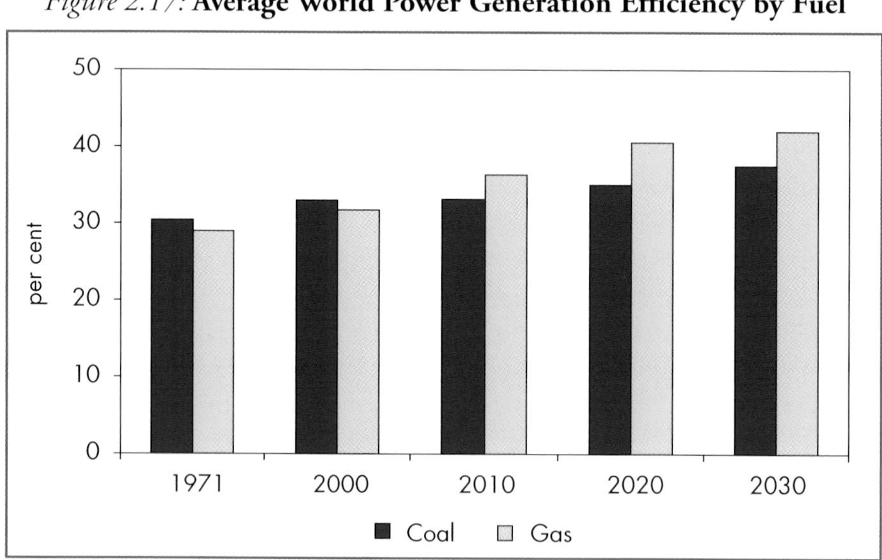

Figure 2.18: **Reductions in Capital Costs of Renewable Energy Technologies, 2000-2030**

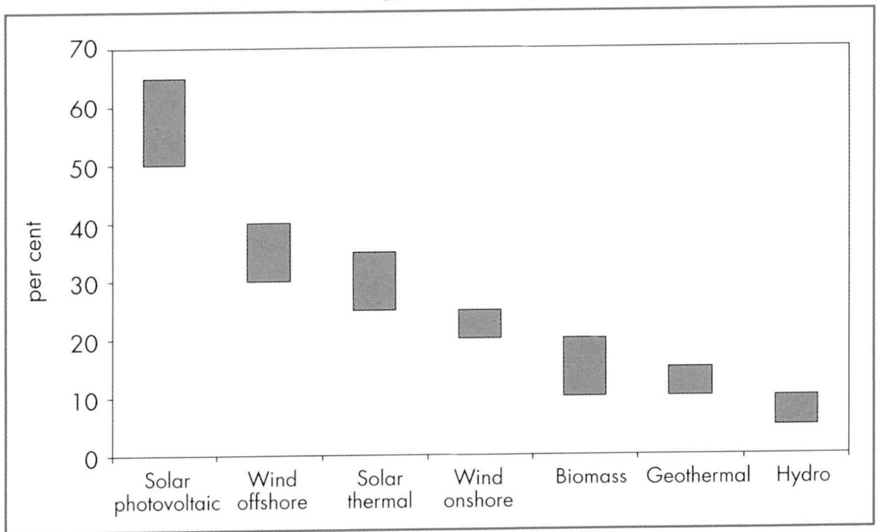

Fuel cells are projected to make a contribution to global energy supply after 2020, mostly in stationary applications. Fuel cells are battery-like devices that convert oxygen and hydrogen into electricity. Hydrogen can be extracted from hydrocarbon fuels using a process known as reforming, and from water by electrolysis. The fuel cells that are expected to achieve commercial viability first will likely involve the reforming of natural gas inside the fuel cell. Production of hydrogen from coal and biomass or using electrolysis is not likely to be economically feasible before 2030. Almost all the fuel cells in use by 2030 will be for distributed power generation. Fuel cells are expected to become competitive in distributed generation when capital costs fall below $1,000/kW, just over a quarter of current costs, and their efficiency approaches 60% (compared to less than 40% now). Fuel cells in vehicles are expected to become economically attractive only towards the end of the projection period. As a result, they will account for only a small fraction of the vehicle fleet in 2030.[13]

Carbon sequestration and storage technologies are not expected to be deployed on a large scale before 2030. It is by no means clear how soon these technologies could become economically and technically feasible. If their costs could be lowered sufficiently, they would increase the

13. See Barreto *et al.* (2002) for a long term scenario describing the role of hydrogen in the global energy system.

attractiveness of fossil fuels over renewable energy sources. This would revolutionise long-term prospects for energy supply (Box 2.4).

Box 2.4: **Capture and Storage of CO_2 from Fossil Fuels**

Technologies are being developed to capture the carbon dioxide emitted from fossil fuel-fired power plants and to store it underground in geological structures or in the ocean. The most common approach to capturing CO_2 exploits a reaction with amines to "scrub" the CO_2 from the gas stream. This process, already used in the chemical industry, could potentially be adapted to capture CO_2 from existing gas- and coal-fired power plants after the combustion process. The cost is projected to be approximately $30 to $50 per tonne of CO_2. Another approach under development aims to separate CO_2 pre-combustion.

Capturing the CO_2 is only part of the problem; the gas must then be transported and stored permanently. A number of options for storage have been identified:

- Reinjecting CO_2 into oil fields may lead to enhanced oil recovery, and this would offset part of the cost of dealing with the gas. Global storage potential in oil-producing reservoirs has been estimated at about 130 billion tonnes. Another 900 billion tonnes could be stored in depleted gas fields. But storing CO_2 in depleted oil or gas fields raises some new issues. Filling a reservoir with CO_2 would increase pressure. Injecting CO_2 into deep coal-beds could enhance methane production. Global coal-bed storage capacity is estimated at about 15 billion tonnes.
- Highly saline underground reservoirs could provide an enormous additional storage capacity, although they offer no offsetting revenue potential. Since 1996, a million tonnes of CO_2 separated from the gas produced from the Sleipner West field in the Norwegian sector of the North Sea has been injected annually into a saline undersea reservoir. Seismic monitoring suggests that the CO_2 is effectively trapped below the impermeable geological cap overlying the reservoir. However more experiments in injecting CO_2 into aquifers are needed to gain a better understanding of the process and

potential risks. Saline reservoirs throughout the world might store as much as 10 trillion tonnes of CO_2, equivalent to more than ten times the total energy-related emissions projected for the next 30 years.

- Disposal of CO_2 in the ocean might be the solution for regions with no depleted oil and gas fields or aquifers. The oceans potentially could store *all* the carbon in known fossil fuel reserves. Tests are underway on a small scale to assess the behaviour of CO_2 dissolved in the ocean and its impact on the ocean fauna.

It is not yet clear how geological and oceanic systems will react to large-scale injection of CO_2. Key technologies for capture and geological storage of CO_2 have all been tested on an experimental or pilot basis, but they will be deployed on a commercial scale only if the risks and costs can be sufficiently reduced and a market value is placed on reducing CO_2 emissions.

CHAPTER 3: THE ENERGY MARKET OUTLOOK

HIGHLIGHTS

- Most of the projected 60% increase in global oil demand in the next three decades will be met by OPEC producers, particularly those in the Middle East. Output from mature regions such as North America and the North Sea will decline. Resources of conventional crude oil are adequate to meet demand to 2030, but the role of non-conventional oil, such as oil sands and gas-to-liquids, is likely to expand, especially after 2020. All the oil-importing regions – including the three OECD regions – will import more oil. The increase in volume terms will be greatest in Asia.
- Over 80% of new crude oil refining capacity will be built outside the OECD, which will become more reliant on refined product imports. Crude oil refineries will have to boost yields of transportation fuels relative to heavier oil products, as well as improve product quality. The share of refined products in total oil trade will increase.
- Demand for natural gas is projected to rise more strongly than for any other fossil fuel, driven mainly by the power sector. Gas demand will reach 5 trillion cubic metres in 2030, double that of 2000. The biggest markets for gas will become much more dependent on imports. In absolute terms, Europe and North America will see the biggest increase in imports. Russia and the Middle East-Africa will be the biggest exporters in 2030.
- Demand for coal will also grow, but more slowly than for oil and gas. China and India together will account for two-thirds of the increase in world coal demand over the projection period. In all regions, coal will be increasingly concentrated in power generation.
- World electricity demand is projected to double between 2000 and 2030, with most growth in developing countries. The next three decades will see a pronounced shift to gas in the fuel mix for power generation. But coal will still be the main generating

fuel in 2030. Non-hydro renewables – notably wind power and biomass – will also grow rapidly, especially in OECD countries, where renewable energy receives active government support.

- To meet the projected increases in electricity demand, total investment of $4.2 trillion will be needed from 2000 to 2030 in power generating capacity alone. Just over half this amount will be needed in developing countries. In many countries, it is uncertain that enough financing will be forthcoming.

Oil Market

Oil Demand

Global oil demand will rise at an even pace of about 1.6% per year, from 75 mb/d in 2000 to 120 mb/d in 2030 (Table 3.1). This rate is slightly less than that projected in *WEO 2000* (Box 3.1). Economic growth will remain the principal driver of oil demand. Past trends in world oil demand have closely followed trends in global economic growth (Figure 3.1).

Figure 3.1: **Global Economic and Oil Demand Growth**

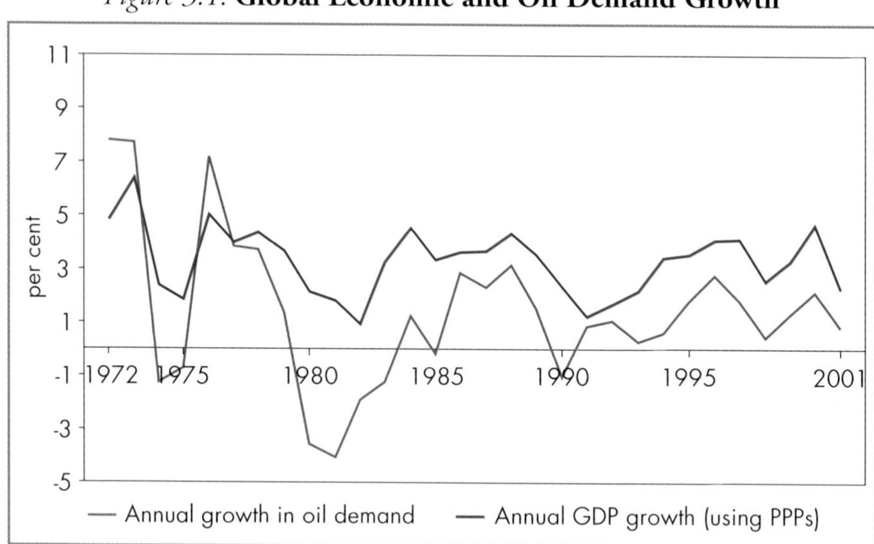

Box 3.1: **Comparison of Global Oil Projections**

This year's *WEO* projects a lower growth rate in world oil demand over the next twenty years than was anticipated in *WEO 2000.* This difference is mainly due to downward revisions to historical data and slower growth than expected in recent years. From 2002 to 2010, the average annual growth of 1.5 mb/d projected in this *Outlook* is close to that projected in *WEO 2000.*

The share of OPEC countries in world oil supply in this *Outlook* differs markedly from that in *WEO 2000.* Projected OPEC production is now 11.6 mb/d lower in 2020. This is partly because of lower expectations for growth in world oil demand, as discussed above. An even greater portion of the difference in projections can be explained by stronger expected growth in non-OPEC and non-conventional oil production. These increases are explained primarily by technological factors. Because OPEC production plays the role of the swing producer in the World Energy Model, the projected share of its production in world oil supply is lower than in *WEO 2000.*

Table 3.2 compares our projections of world oil demand from 2000 to 2020 with those of other organisations. *WEO 2002* projections fall in the middle of the range. Shell's forecast is much lower than the others, largely because its "Dynamics as Usual" scenario projects a high share of new technologies in the global fuel mix in 2020. The US Department of Energy and DRI/WEFA project stronger growth in world oil demand, partly because of their higher GDP assumptions and lower oil prices.

Consumption in developing countries and the transition economies will grow much faster than in the OECD. As a result, the OECD share of world demand will drop from 62% in 2000 to only 50% in 2030. The fastest growth will occur in the rapidly expanding Asian economies, particularly China and India. The largest increase in absolute terms will be in China and OECD North America, which will remain the biggest consuming region. Nonetheless, almost two-thirds of the total growth in oil demand will occur in non-OECD countries. Per capita oil consumption will remain much higher in the OECD.

Table 3.1: **World Oil Demand** (million barrels per day)

	2000	**2010**	**2020**	**2030**	**Average annual growth 2000-2030 (%)**
OECD North America	22.2	24.8	27.7	30.8	1.1
US and Canada	20.2	22.5	24.8	27.3	1.0
Mexico	1.9	2.3	2.9	3.5	2.0
OECD Europe	14.1	15.3	16.0	16.4	0.5
EU	12.3	13.2	13.7	13.9	0.4
Other OECD Europe	1.9	2.1	2.4	2.5	1.0
OECD Pacific	8.5	9.5	10.3	10.5	0.7
Japan/Australia/N.Zealand	6.4	6.9	7.2	7.0	0.3
Korea	2.1	2.6	3.1	3.4	1.6
OECD total	**44.8**	**49.6**	**54.0**	**57.6**	**0.8**
Transition economies	4.6	5.4	6.3	7.1	1.5
Russia	2.7	3.1	3.7	4.4	1.7
Other	1.9	2.3	2.5	2.7	1.1
China	4.9	7.0	9.4	12.0	3.0
Indonesia	1.1	1.5	2.0	2.4	2.7
Other East Asia	3.2	4.4	5.7	7.0	2.7
India	2.1	3.0	4.2	5.6	3.3
Other South Asia	0.5	0.8	1.2	1.8	4.0
Brazil	1.8	2.4	3.1	3.8	2.5
Other Latin America	2.7	3.4	4.3	5.5	2.4
Africa	2.0	2.9	3.9	5.4	3.3
Middle East	4.1	5.2	6.3	7.7	2.2
Non-OECD	**27.1**	**35.9**	**46.4**	**58.3**	**2.6**
Bunkers and stock changes	3.1	3.3	3.6	4.1	1.0
Total demand	**75.0**	**88.8**	**104.0**	**120.0**	**1.6**

Table 3.2: **Comparison of Oil Demand Projections, 2000 to 2020**

	GDP (average annual growth,%)	International oil price in 2020 (2000 $)	Oil demand (average annual growth,%)
Shell*	3.5	n.a.	1.1
PEL**	3.0	22	1.6
OPEC	3.3	20	1.7
WEO 2002	3.1	25	1.7
WEO 2000	3.1	27	1.9
US DOE 2002	3.2	25	2.2
DRI/WEFA***	3.2	23	2.2

* Dynamics as usual scenario, 2000-2025. ** 2000-2015. ***1999-2020
Note: The definition of the international oil price differs among studies.
Sources: Shell International (2001); Petroleum Economics Ltd. (2002); *OPEC Review* (2001); US Department of Energy/EIA (2002); DRI/Wharton Econometric Forecasting Associates (2001).

Table 3.3: **Refined Product Demand by Region and Product Category**

	Light	Middle	Heavy	Other	Total
Demand in 2000 (mb/d)					
OECD countries	20.7	15.6	3.9	3.2	43.4
Developing Asia	3.7	4.4	2.1	0.6	10.8
Rest of world	4.9	4.8	2.7	1.3	13.6
World*	**29.4**	**25.5**	**10.8**	**5.1**	**70.7**
Demand in 2030 (mb/d)					
OECD countries	28.8	20.1	3.6	3.8	56.4
Developing Asia	10.1	12.6	3.6	0.9	27.1
Rest of world	10.1	10.3	4.2	2.0	26.6
World*	**49.1**	**43.9**	**14.3**	**6.7**	**114.0**
Average annual growth 2000-2030 (%)					
OECD countries	1.1	0.8	-0.3	0.6	0.8
Developing Asia	3.4	3.5	1.8	1.6	3.1
Rest of world	2.4	2.5	1.3	1.4	2.2
World*	**1.7**	**1.8**	**0.8**	**0.9**	**1.5**

* Includes stock changes and international marine bunkers.
Note: See Appendix 2 for definitions of product categories.

Global demand for refined petroleum products is expected to increase by 1.5% per year, from 71 mb/d in 2000 to 114 mb/d in 2030. Almost three-quarters of this increase will come from the transport sector. As a result, there will be a shift in all regions towards light and middle distillate products and away from heavier oil products, used mainly in industry. By 2030 light and middle distillates will represent 82% of global refined product demand, up from 78% in 2000. This shift will be more pronounced in developing regions that currently have less transport fuels in their product mix (Figure 3.2 and Table 3.3).

Figure 3.2: **Change in Refined Product Demand by Product Category and Region, 2000-2030**

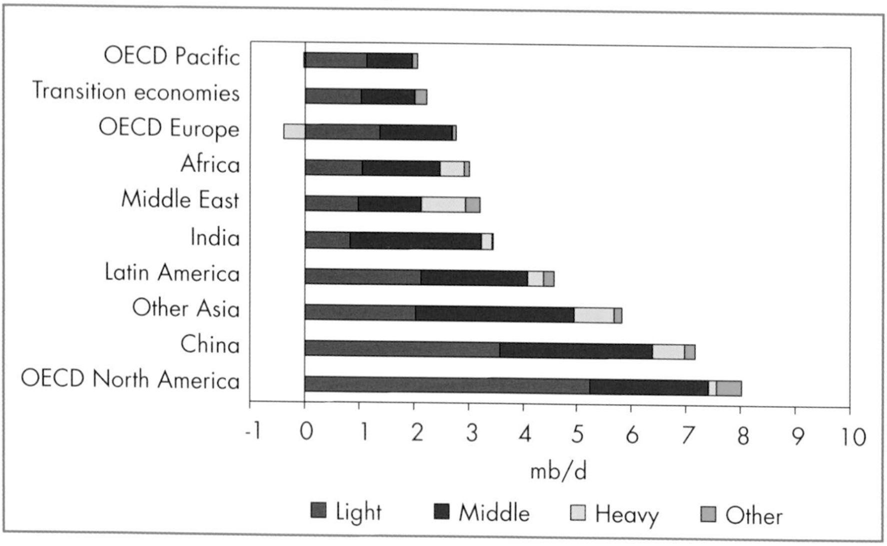

Crude Oil Production

Conventional Oil[1]

The production of oil will remain highly concentrated in a small number of major producers, although the role of some smaller producers will grow. Members of the Organization of Petroleum Exporting Countries,

1. "Conventional oil" is defined as crude oil and natural gas liquids produced from underground reservoirs by means of conventional wells. This category includes oil produced from deepwater fields and natural bitumen. "Non-conventional oil" includes oil shales, oil sands-derived oil and derivatives such as synthetic crude products, and liquids derived from coal (CTL), natural gas (GTL) and biomass (biofuels). It does not include natural bitumen.

particularly in the Middle East, will increase their share, as output from mature regions such as North America and the North Sea declines (Table 3.4). At present, only two of the seven largest oil producers are OPEC countries. Non-OPEC production is expected to peak at just under 48 mb/d around 2010 and decline very slowly thereafter. The Middle East holds well over half the world's remaining proven reserves of crude oil and NGLs, and nearly 40% of undiscovered resources (Table 3.5).[2]

Box 3.2: Methodology for Projecting Oil Production

The oil supply projections in this *Outlook* are derived from aggregated projections of regional oil demand, as well as projections of production of conventional oil in non-OPEC countries and non-conventional oil worldwide. OPEC conventional oil production is assumed to fill the gap.

Two methodologies are used to generate non-OPEC conventional output projections: a "bottom-up" or field-by-field approach for the period to 2010 (medium term) and a "top-down" or resource depletion model for the remaining twenty years (long term). The projected medium-term production profiles are derived from assumptions about the natural decline rates for fields already in production. Also factored in are expected investments aimed at increasing the recovery of oil in place as well as "proving up" reserves and adding new, small "satellite" fields as they are discovered. For the long term, the resource depletion model takes into account ultimate recoverable resources, which in turn depend on the recovery rate. The recovery rate, which generally increases slowly over time, reflects the assumed price of oil and advances in upstream technologies.

2. Oil that has been discovered and is expected to be economically producible is called a proven reserve. Oil that is thought to exist, and is expected to become economically recoverable, is called an undiscovered resource. Total resources include reserves and undiscovered resources. Comparison of reserves and resource assessments is complicated by differences in estimation techniques and assumptions among countries and companies. In particular, assumptions about prices and technology have a major impact on how much oil is deemed to be economically recoverable.

Table 3.4: **World Oil Supply** (million barrels per day)

	2000	2010	2020	2030	Average annual growth 2000-2030 (%)
Non-OPEC	**43.4**	**47.8**	**45.7**	**42.1**	**-0.1**
OECD total	21.2	19.8	16.3	12.8	-1.7
OECD North America	13.6	14.0	12.3	9.9	-1.1
US and Canada	10.1	9.9	8.3	7.1	-1.2
Mexico	3.5	4.1	4.0	2.7	-0.8
OECD Europe	6.7	5.2	3.5	2.5	-3.3
EU	3.3	2.3	1.6	1.1	-3.5
Other OECD Europe	3.4	3.0	1.9	1.4	-3.0
OECD Pacific	0.9	0.5	0.5	0.5	-1.8
Non-OECD	22.2	28.0	29.4	29.3	0.9
Russia	6.5	8.6	9.0	9.5	1.3
Other transition economies	1.6	4.1	4.9	5.4	4.1
China	3.2	2.8	2.5	2.1	-1.4
India	0.7	0.5	0.4	0.3	-2.5
Other Asia	1.6	1.4	1.1	0.7	-2.8
Brazil	1.3	2.3	3.2	3.9	3.7
Other Latin America	2.3	2.0	2.0	1.9	-0.5
Africa	2.8	4.5	4.9	4.4	1.5
Middle East	2.1	1.8	1.5	0.9	-2.7
OPEC	**28.7**	**35.9**	**50.2**	**64.9**	**2.8**
OPEC Middle East	21.0	26.5	37.8	51.4	3.0
Indonesia	1.4	1.5	1.7	1.7	0.6
Other OPEC	6.3	7.9	10.7	11.8	1.9
Non-conventional oil	1.1	3.0	5.6	9.9	7.7
Of which GTL	*0.0*	*0.3*	*1.1*	*2.3*	*14.2*
Processing gains	1.7	2.2	2.6	3.1	1.9
OPEC share (%)	38.4	40.4	48.3	54.1	1.2
OPEC Middle East share (%)	28.1	29.8	36.4	42.9	1.4
Total supply	**75.0**	**88.8**	**104.0**	**120.0**	**1.6**

Table 3.5: **Oil Reserves, Resources and Production by Country**

Rank	Country	Remaining reserves (billion barrels)	Undiscovered resources (billion barrels)	Total production to date (billion barrels)	2001 production (mb/d)
1	Saudi Arabia	221	136	73	8.5
2	Russia	137	115	97	7.0
3	Iraq	78	51	22	2.4
4	Iran	76	67	34	3.8
5	UAE	59	10	16	2.5
6	Kuwait	55	4	26	1.8
7	United States	32	83	171	7.7
8	Venezuela	30	24	46	3.0
9	Libya	25	9	14	1.4
10	China	25	17	24	3.3
11	Mexico	22	23	22	3.6
12	Nigeria	20	43	16	2.2
13	Kazakhstan	20	25	4	0.8
14	Norway	16	23	9	3.4
15	Algeria	15	10	10	1.5
16	Qatar	15	5	5	0.8
17	United Kingdom	13	7	14	2.5
18	Indonesia	10	10	15	1.4
19	Brazil	9	55	2	1.4
20	Neutral zone*	8	0	5	0.6
	Others	73	220	91	16.2
	Total	**959**	**939**	**718**	**75.8**

* Kuwait/Saudi Arabia.
Note: Estimates include crude oil and NGLs; reserves are effective 1/1/96; resources, effective 1/1/2000, are mean estimates. See footnote 2 for definitions of reserves and resources.
Sources: United States Geological Survey (2000); IEA databases.

Resources of conventional crude oil and NGLs are adequate to meet the projected increase in demand to 2030, although new discoveries will be needed to renew reserves. The importance of non-conventional sources of oil, such as oil sands and gas-to-liquids, is nonetheless expected to grow, especially after 2020. Conventional production is projected to increase from 72 mb/d in 2001 to 107 mb/d in 2030 (Figure 3.3). The approach used to generate these projections is described in Box 3.2.

Figure 3.3: **World Oil Production**

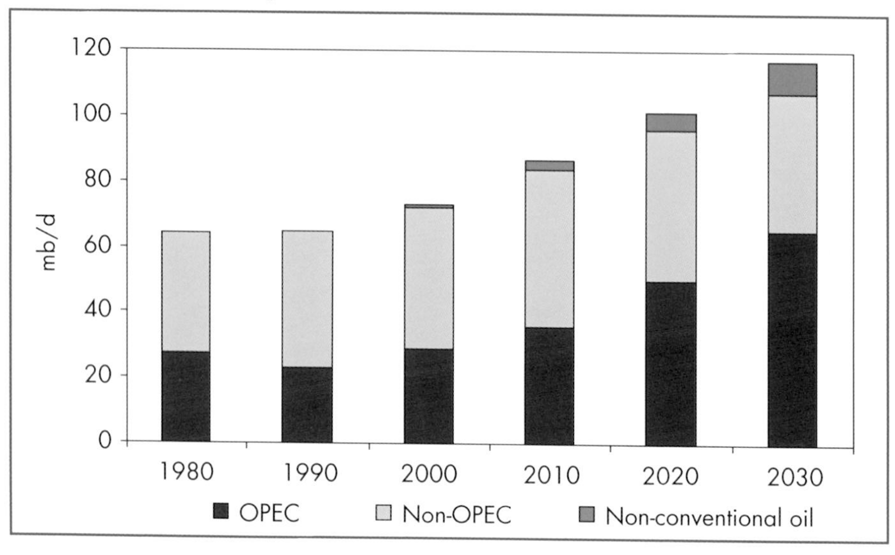

Output prospects outside OPEC diverge, depending primarily on the maturity of the basins already in production. Higher oil prices and drilling rates in recent years will boost near-term production in several mature regions, such as North America and the North Sea. But faster depletion will bring forward the time when production peaks. Higher recovery rates with better technology, however, will offset this factor to some extent. US production, which has been falling steadily since the early 1970s, is expected to rise in the near term with the start-up of some new large fields in the Gulf of Mexico. Canadian output will continue to rise until the second half of this decade, as new offshore East Coast fields come onstream. But it will begin to drop by 2010. Mexican production is expected to maintain its upward trajectory for the next decade or so and fall gradually after 2020. North Sea production is *already* in decline and this is unlikely to change. A brief pick-up in Norwegian production in the near term will not offset the fall in production in the United Kingdom and Denmark.

The only other non-OPEC producing countries that will see a significant increase in crude oil production in the medium term are Russia,

Kazakhstan, Azerbaijan, Brazil and Angola.[3] Russian prospects remain highly dependent on large amounts of investment in development drilling and pipeline construction. The country's output is projected to increase from 7 mb/d in 2001 to 8.6 mb/d by 2010 and continue to rise through to 2030, albeit at a more moderate pace. Higher output from the Caspian region would require new export pipelines, for which cross-border transit agreements and financing are still uncertain. On the assumption that those lines *are* built, the combined production of Kazakhstan and Azerbaijan is projected to jump from just over 1.1 mb/d in 2001 to over 3.5 mb/d in 2010. Large projected increases from major offshore finds in Brazil and Angola will depend on the successful deployment of advanced deep-water technologies and on the existence of stable regulatory and tax regimes in both countries.

A small number of OPEC countries with large reserves is expected to make up most of the shortfall between non-OPEC production and global demand (Figure 3.4). The list includes Saudi Arabia, Iran, Iraq, Kuwait and the United Arab Emirates, as well as Venezuela and Nigeria. The assumption used here is that the international oil price remains flat until 2010, at $21 per barrel in real terms. It then rises in a linear fashion to $29 in 2030. This is judged to be consistent with the ambition of OPEC producers to increase their market share in the long term and with an expected increase in marginal production costs in non-OPEC countries. It is further assumed that the OPEC countries will also find the capital needed to increase their installed production capacity. At around $4 per barrel, the total cost of developing new supplies in the Middle East is the lowest in the world, and well below assumed price levels.[4] OPEC's price and production policies and the financing of capacity additions are, nonetheless, extremely uncertain. The political risks of investing in the Middle East are high, and some producers have been reticent about accepting investment and operational participation by foreign companies.

Production prospects for OPEC and non-OPEC countries alike are subject to uncertainty about the impact of short-term price volatility on potential investors and spending on upstream research and development (see last section of Chapter 1). In many non-OPEC countries, the production outlook will depend critically on new technologies that improve exploration-drilling success rates, that lower engineering costs and

3. See IEA (2001) for a discussion of oil production prospects in these and other major producing countries.
4. See IEA (2001) for a detailed analysis of oil supply costs.

Figure 3.4: **Change in World Oil Production**

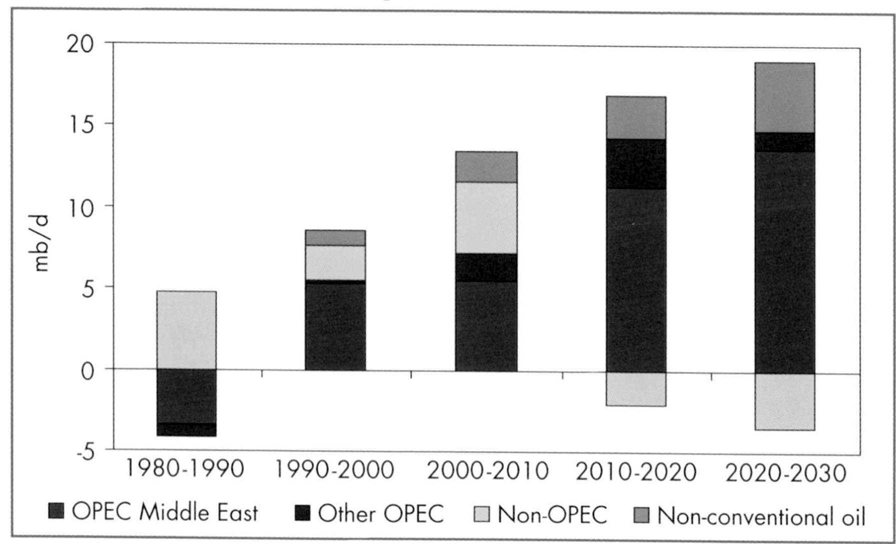

that permit production in such extreme environments as ultra-deep water and the Arctic.

Another factor related to how much is invested in new oil production capacity and where that investment is directed is the natural decline rate. This term refers to the rate of decline in production over time in the absence of new investment in drilling additional wells and in enhanced recovery techniques. Experience in North America and the North Sea shows that advanced technology can help arrest declines in production. But there is growing evidence that decline rates are becoming steeper in some regions. Older giant fields are no longer able to sustain plateau production and, paradoxically, some newer and smaller fields exhibit faster decline rates once they pass their peak due to more efficient extraction techniques. The average age of the world's 14 largest oilfields, which together account for more than a fifth of total oil output, is more than 43 years.[5] Of the other 102 giant oilfields that each produce more than 100,000 b/d and which contribute about half of total world supply, most have been in production for more than twenty years. Production from many of these fields is declining with increasing rapidity as their reserves are depleted. Some experts believe that the natural decline rate in many regions, including North America, now exceeds 10% per year. The number of discoveries of

5. Simmons (2002).

giant oilfields and their average production have fallen sharply since the 1960s (Figure 3.5). The percentage of oil supply coming from non-giant fields will continue to grow in the future.

Figure 3.5: **The World's Giant Oilfields**

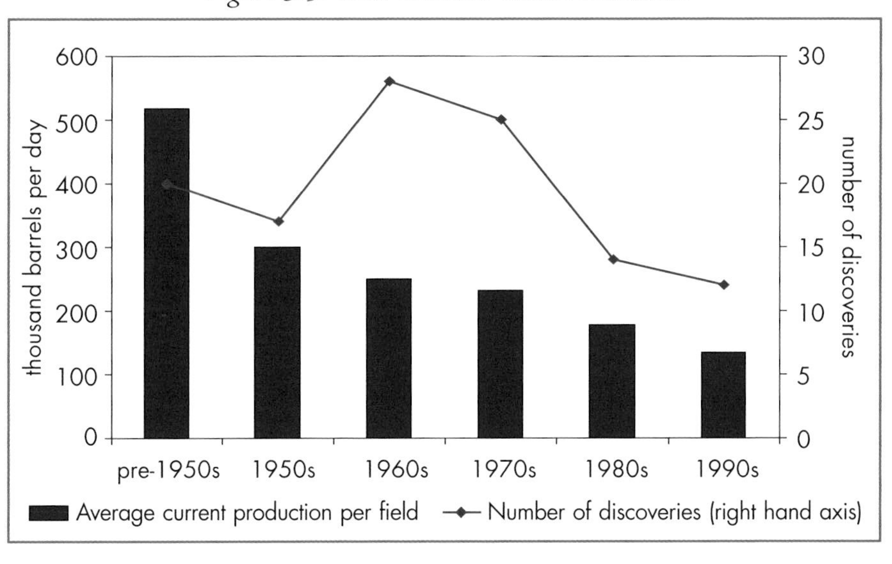

Non-conventional Oil

Non-conventional oil is expected to contribute to just over 8% to total world oil supplies by 2030. This represents production of 9.9 mb/d – a sharp increase over the 1.1 mb/d of 2000. This increase results mainly from technological improvements that reduce the cost of extracting and upgrading non-conventional resources. Gas-to-liquids plants will make a growing contribution to non-conventional oil supplies. GTL production is projected to rise from 43 kb/d today to 300 kb/d by 2010 and 2.3 mb/d by 2030.

The greater part of future non-conventional oil will come from Canadian oil sands and Venezuelan extra-heavy bituminous crude. It is estimated that these two regions contain 580 billion barrels of recoverable reserves – more than the entire reserves of conventional crude oil in the Middle East. The proximity of Canada and Venezuela to the US market will help these sources compete against lower-cost producers further afield. They will provide the United States and other countries with an opportunity to reduce their dependence on Middle-Eastern crude oil, as OECD conventional oil production declines.

Canadian non-conventional oil production is centred in the province of Alberta. This region produces diluted bitumen and upgraded crude. In both cases, the primary hydrocarbon content, known as natural bitumen, is extracted from mined oil-sand deposits. This bitumen is then diluted with lighter hydrocarbons and transported to a refinery. Alternatively, it is processed onsite into upgraded crude. The quality of this upgraded oil is almost as good as that of West Texas Intermediate, the most widely traded crude in North America. In 2001, Canadian production of non-conventional oil totalled 350 kb/d, not including 300 kb/d of the natural bitumen that is diluted. Volumes of upgraded crude are expected to increase to 1.2 mb/d by 2010 and 3.2 mb/d by 2030.

Venezuelan non-conventional oil is produced mainly in the Orinoco Belt region. This area contains massive reserves of extra-heavy oil that has to be treated to reduce its viscosity. The oil is emulsified with water to produce a liquid fuel known as Orimulsion, which is used in power stations, or it is upgraded into synthetic oil. Venezuelan production of non-conventional oil totalled 307 kb/d in 2001. Output is projected to increase to 1 mb/d in 2010 and 2.9 mb/d in 2030.

The main factors that will influence the volume of non-conventional oil production during the *Outlook* period are international oil prices, capital investment and environmental policies. Recent technological improvements have drastically reduced the cost and energy intensity of exploiting non-conventional oil resources. But their costs remain higher than those of conventional oil, and significant investment will be required to make them fully competitive. New pipelines, particularly from Canada into the US, will also be necessary to open up new markets.

Oil Refining

Crude Distillation Capacity

At the start of 2000, the world's crude oil distillation capacity[6] totalled 81.5 mb/d. About a quarter of current refinery capacity is located in OECD North America, a similar amount in the Asian region and just under 20% in OECD Europe. During 2000 the average utilisation rate of refinery capacity was about 96% in both OECD North America and Europe and was 90% in the Asian region.

To meet demand for refined products of 114 mb/d, global refining capacity is projected to increase by an average 1.3% a year, reaching 121 mb/d in 2030. Over 80% of this additional capacity will be built in non-OECD countries (Table 3.6). The projected rate of growth of refinery capacity is less than that of refined product demand as de-bottlenecking[7] and increased utilisation rates are expected at existing facilities. GTL plants also account for a growing share of refined product supply.

Table 3.6: **Global Crude Oil Distillation Capacity** (mb/d)

	1990	2000	2010	2020	2030	Average annual growth 2000-2030 (%)
OECD						
North America	19.4	20.7	21.4	22.9	24.5	0.6
Europe	15.3	15.3	15.3	16.2	17.0	0.4
Pacific	6.1	8.4	9.0	10.0	10.4	0.7
Non-OECD						
Transition economies	n.a.	11.3	11.3	11.3	11.8	0.1
China	2.2	4.3	5.9	8.0	10.3	2.9
India	1.1	1.9	2.3	3.2	4.2	2.8
Rest of Asia	3.2	5.0	6.3	8.3	10.3	2.5
Latin America	5.5	5.7	6.2	7.6	9.3	1.6
Africa	2.9	3.0	3.8	5.3	7.3	3.0
Middle East	5.0	5.9	10.0	12.6	15.6	3.3
Total capacity	**60.6**	**81.5**	**91.5**	**105.2**	**120.6**	**1.3**

The largest expansion will occur in Asia. Despite the Asian economic crisis of the late 1990s, refinery capacity has increased strongly with such recent additions as Formosa Petrochemical's 450 kb/d refinery in Chinese Taipei and Reliance Petroleum's 540 kb/d refinery in India. In recent years, additions have exceeded incremental demand growth; so the region now has an excess of refining capacity. Further expansion is likely to be limited until refinery utilisation rates have increased. The highest *rate* of refining capacity growth is projected to occur in the Middle East.

6. Projections on the size and location of additions to crude oil distillation unit (CDU) capacity are based on past trends in refinery construction, currently announced plans for additional CDU capacity, estimates of growth in demand for refined products and existing surplus CDU capacity. It is assumed that product specifications will not provide a barrier to inter-regional trade. For more information, see Appendix 1.
7. Upgrading one or more parts of a refinery that permits fuller use of other parts of the refinery without making any direct changes to them.

Additional capacity will outstrip new local demand and the region will become an increasingly important exporter of refined products.

Future growth in refining capacity will be modest in all OECD regions. Sluggish increases in demand for refined products will be met by capacity creep at existing refineries and by an increased reliance on imports. Since 1990, many OECD countries have restructured their refining industries and have closed their less efficient refineries. De-bottlenecking and increasing utilisation rates of remaining facilities, rather than construction of new refineries, has largely offset lost capacity. Little capacity expansion is expected in the transition economies as the region currently has a massive oversupply, with refinery utilisation rates averaging just 48%. The region's refineries will need to be upgraded as they are outdated and no longer efficiently produce the type of products required.

Upgrading and Treatment Capacity

Throughout the *Outlook* period, refinery complexity[8] will have to be increased in order to raise yields of light and middle distillate products so that refinery output continues to match the changing profile in market demand. This challenge will be exacerbated by the expected deterioration in the quality of refinery feedstocks during the *Outlook* period, due to increased reliance on oil from the Middle East and on non-conventional crude oils, such as extra heavy crude. Additional investment will be needed to accommodate these trends, as heavier oils such as these require more intensive and costly processing.

The fastest growth in demand for light and middle distillates is expected to occur in Asia. These products currently represent 75% of the region's total consumption, but this will increase to 84% by 2030. In general, Asian refineries have low complexity, reflecting the region's large market for heavier oil products. Asian refineries will need large investments in conversion capacity to keep up with the falling share of heavy fuel oil demand. Similarly, refineries in the transition economies were designed to supply fuel oil for heavy industry and so have limited upgrading and conversion capacity. As the demand for fuel oil has dried up, refinery output in the region no longer matches the needs of the market. In some cases, the best option may be to close entire refineries, or at least the older parts of existing refineries, and to rebuild completely.

8. A refinery's complexity or "sophistication" is defined by the number and type of upgrading processes it boasts. Such processes include catalytic cracking and hydrocracking, which are used to convert the "heavy end of the barrel" into transportation fuels.

The most complex refineries are found in North America and, to a lesser extent, in OECD Europe, which have high proportions of light products in their product mix. Only minor improvements in conversion capacity will be required in these markets. Any such investment in Europe is likely to be in the form of additional hydrocracking capacity in order to maximise yields of middle distillates, which are in short supply.

Since the early 1990s, most investment in the refining sector, particularly in OECD countries, has been aimed at complying with fuel-quality legislation, rather than at increasing capacity. This trend is expected to continue in the coming decade, as timetables for further improvements in fuel quality have already been announced in the major OECD markets. In 2000, global distillation capacity increased only marginally and yet investment by the global refining industry totalled $44 billion.[9] In addition, during the 1990s, no major new refinery was built in the United States, yet the industry spent $47 billion on environmental compliance and is expected to spend $12 billion more in the next several years just to meet reduced sulphur requirements.[10]

The biggest investments will go to reducing sulphur in diesel and gasoline and reducing the aromatic, benzene and olefin content of gasoline whilst maintaining existing octane levels. Profits have been low recently in the refining sector in many OECD countries. Some smaller and less sophisticated refineries may be unable to justify the necessary investment and will have to shut down.

The variation in fuel quality standards between countries and, in some cases, within countries could hamper inter-regional trade over the next few years. Export refining centres, such as those in the Middle East, Singapore and the transition economies, will need to improve both their upgrading and treatment capability dramatically if they are to produce large volumes of fuel suitable for markets in North America and Europe.

Possible Variations to CDU Projections

Future requirements for refinery capacity could vary from these projections if oil demand is either higher or lower than expected or if there is an increase in products from sources other than conventional oil refineries. An overall reduction in oil demand could follow a major advance in vehicle technology such as a breakthrough in fuel cells. Transport fuels

9. Institut Français du Pétrole (2002).
10. American Petroleum Institute (2001).

that could displace output from conventional refineries include compressed natural gas and renewable fuels (particularly if renewable fuel mandates are introduced) and the output from gas-to-liquids processes.

Oil Trade

International oil trade is set to grow considerably, as the gap between indigenous production and demand widens in all *WEO* regions. Net inter-regional trade rises from 32 mb/d in 2000 to 42 mb/d in 2010 and 66 mb/d in 2030 (Figure 3.6).[11] All the net oil-importing regions will import more oil at the end of the projection period, both in absolute terms and as a proportion of their total oil consumption. The increase is most dramatic for Asia, where imports jump from 4.9 mb/d (42% of demand) in 2000 to 24 mb/d (83%) in 2030. Net imports in China alone rise from 1.7 mb/d (35%) in 2000 to 10 mb/d (83%) in 2030.

Among the three OECD regions, Europe's dependence grows most rapidly, from 52% to 85%, while the Pacific remains the most import-dependent at almost 94% (Figure 3.7). Rising production in Canada, especially from oil sands, and in Mexico will help to temper the increase in North America's imports. On average, the OECD will import 69% of its oil needs in 2030 compared to 51% in 2000.

The Middle East will see the biggest increase in net exports, from 19 mb/d in 2000 to 46 mb/d in 2030. The bulk of these additional exports are expected to go to Asia, with China emerging as the single largest market, followed by India. Oil exports from Africa, Latin America and the transition economies will also grow, but much less rapidly.

Refined Product Trade

OECD countries will become increasingly reliant on imported refined products during the *Outlook* period. Imports supplied just over 2% of total product demand in OECD countries during 2000 but this is expected to increase to 11% by 2030. This change will be due largely to the increased imports of OECD North America, which are projected to reach a fifth of the region's product demand by 2030.

In addition to regional shortages in refinery capacity, the increase in product trade will be due to the widening gap between refinery output and

11. Total international trade is even larger because of trade between countries within each *WEO* region and re-exports.

Figure 3.6: **Net Inter-regional Oil Trade, 2030** (mb/d)

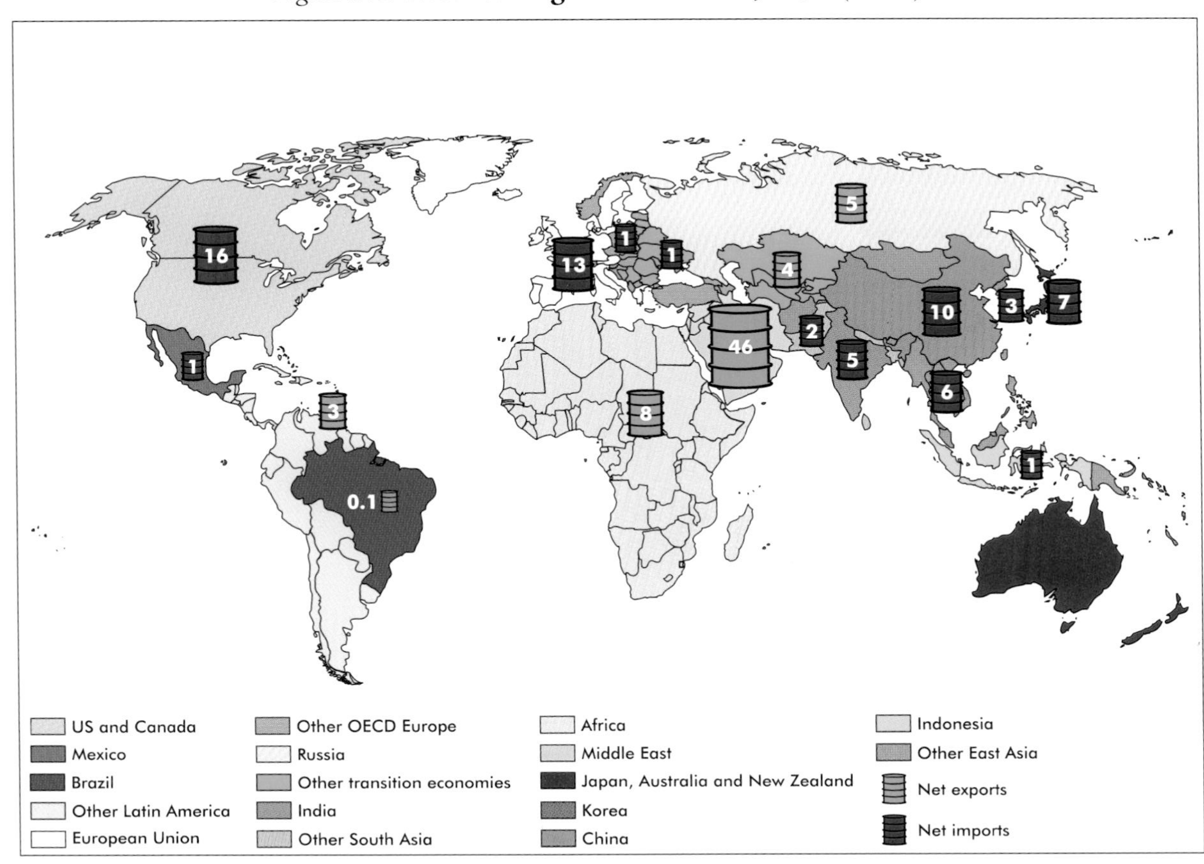

Figure 3.7: **Oil Import Dependence by Region**

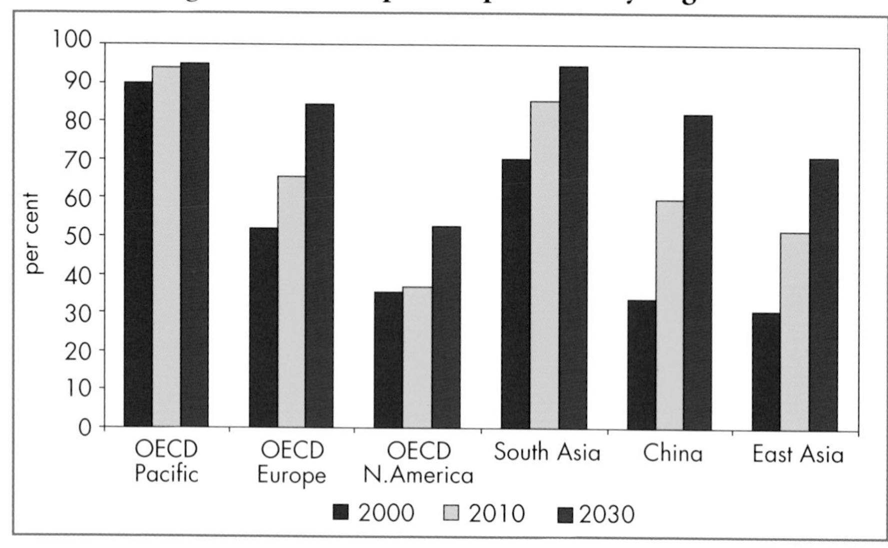

market demand within particular markets. For example, the European refining industry produces too much gasoline and too little diesel for its local market. This imbalance is expected to widen in the future due to the increased share of diesel cars in Europe's fleet, increased road haulage and increased demand for jet fuel (which will compound the shortage of middle distillates). Europe may prefer product trade rather than investment in costly new refining processes to solve this imbalance. The likely suppliers of Europe's diesel shortfall are refineries in the Middle East and in the transition economies. However, this will depend on the development of suitable refinery configurations. Higher than expected economic growth in the transition economies could also limit the volume of refined products available for export. Europe's excess gasoline output is likely to be exported to the United States.

By 2030, the largest net exporters are expected to be the Middle East and the transition economies. North America and the Asian region will be the largest net importers (Figure 3.8). The share of refined products in total oil trade will increase over the projection period.

Figure 3.8: **Net Inter-regional Trade in Refined Oil Products, 2030** (mb/d)

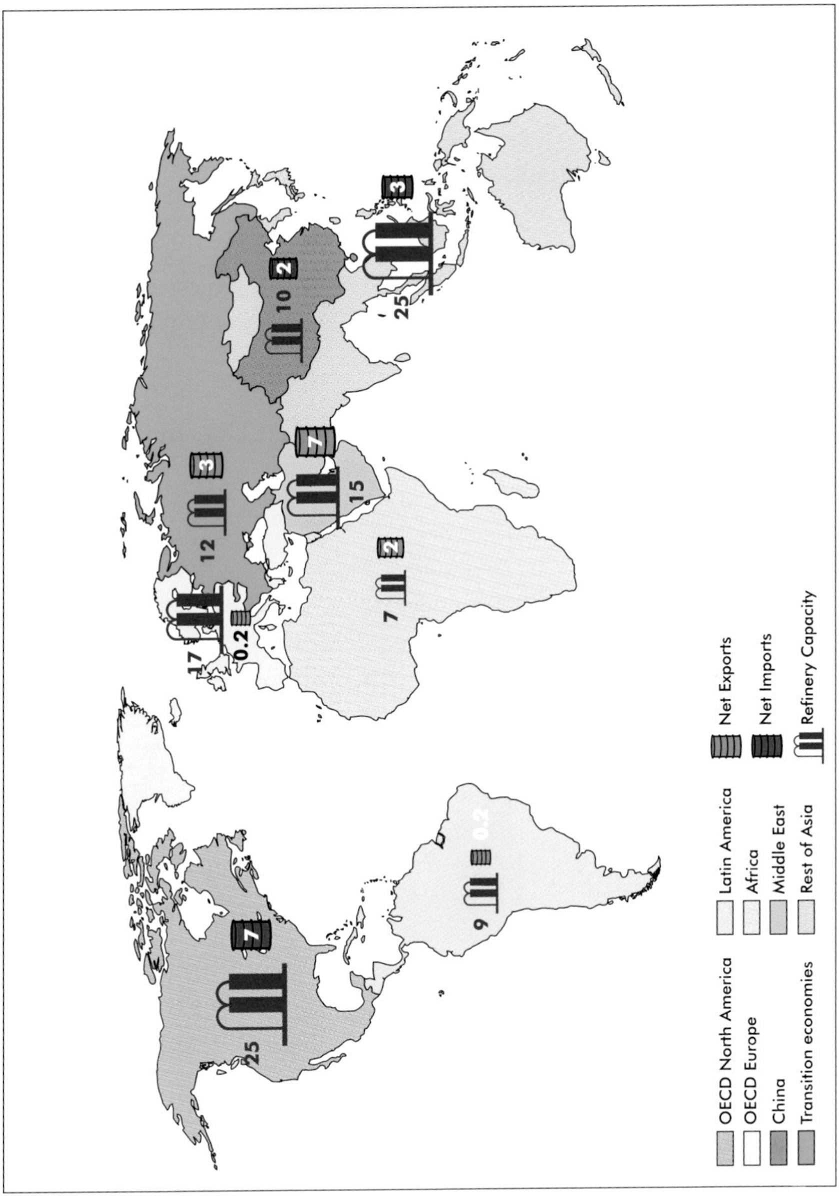
25
7
9
0.2
17
0.2
12
3
7
2
15
7
10
2
25
3
OECD North America
OECD Europe
China
Transition economies
Latin America
Africa
Middle East
Rest of Asia
Net Exports
Net Imports
Refinery Capacity

Natural Gas Market

Natural Gas Demand

Gas consumption is projected to rise strongly in most regions over the next three decades, driven chiefly by demand from power generators. Natural gas has relatively low carbon content and is, in many cases, priced competitively. Globally, gas consumption increases by an average 2.4% per year from 2000 to 2030, less rapidly than the 3% of the past three decades.[12] Demand grows most rapidly in the fledgling markets of developing Asia, notably China, and in Latin America. Nonetheless, North America, Russia and Europe remain by far the largest markets in 2030 (Table 3.7). The share of gas in the global primary energy mix will increase from 23% in 2000 to 28% in 2030.

Table 3.7: **World Primary Natural Gas Demand** (bcm)

	2000	2010	2020	2030	Average annual growth 2000-2030 (%)
OECD North America	788	992	1161	1,305	1.7
OECD Europe	482	640	799	901	2.1
OECD Pacific	122	168	201	243	2.3
Transition economies	609	748	876	945	1.5
China	32	61	109	162	5.5
East Asia	83	139	200	248	3.7
South Asia	51	96	153	205	4.7
Latin America	105	167	251	373	4.3
Middle East	201	272	349	427	2.5
Africa	53	95	155	239	5.2
World	**2,527**	**3,377**	**4,254**	**5,047**	**2.4**

The power sector will account for a growing share of total primary gas consumption worldwide (Figure 3.9). Its use of gas will increase by 3.5% per year from 2000 to 2030. In most regions, gas will account for the bulk of incremental generation, because gas is assumed to be competitively priced. It will also be chosen for its inherent environmental advantages over other fossil fuels. Distributed generation, which will continue to grow, will

12. Gas demand grows much less rapidly in the OECD Alternative Policy Scenario (see Chapter 12).

Figure 3.9: **World Natural Gas Demand by Sector**

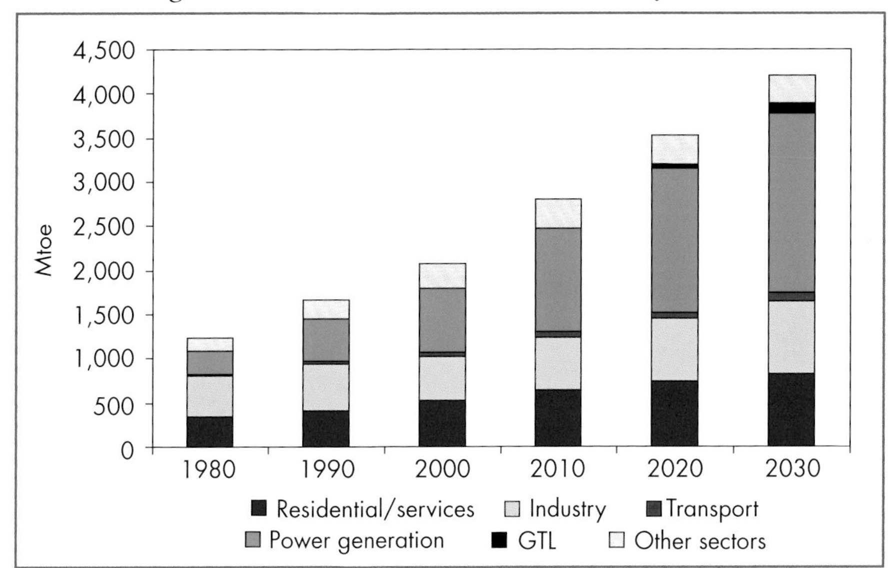

also favour the use of gas (see Box 3.5). In absolute terms, gas demand in the power sector increases most in North America. In percentage terms, the rate of increase in gas use for power production will be highest in China, Brazil and Africa.

A key feature of our primary gas consumption projections is the emergence of gas-to-liquids (GTL) plants as a new market (Box 3.3). Global GTL demand for gas is projected to increase from 4 bcm in 2000 to 29 bcm in 2010 and 233 bcm in 2030. Much of this gas will be consumed in the conversion process. The rate of increase in GTL production is nonetheless subject to enormous uncertainty, particularly after 2010.

Box 3.3: **Outlook for Gas-to-Liquids Conversion**

Gas-to-liquids (GTL) conversion is on the brink of commercial viability. GTL will provide a major new alternative to liquefied natural gas as a way of exploiting gas reserves that cannot be piped to markets economically. Interest in developing GTL projects is booming thanks to improved technology, growing reserves of "stranded gas" and higher oil prices. All the plants already in operation, under construction or planned are based on the Fischer-Tropsch technology originally developed in the 1920s. This technology converts natural gas feedstock into synthetic gas (syngas) and then, through catalytic reforming or synthesis, into conventional oil products. Recent technical advances, including improved catalysts, have enhanced liquid yields and reduced both capital and operating costs.

The economics of GTL processing are highly dependent on plant construction costs, product types and yields and the energy efficiency of the plant, as well as the market prices of the liquids produced and the cost of the gas feedstock. GTL plants are complex and capital-intensive. They require large sites and construction lead times of two-and-a-half to three years. They are also very energy-intensive, consuming up to 45% of the gas feedstock. This characteristic means that CO_2 emissions from GTL plants are much higher than from oil refineries. On the other hand, GTL plants usually produce a range of middle distillates with very *good* environmental qualities, demand for which is rising.

GTL production costs have fallen sharply in recent years, largely due to improved yields and thermal efficiency. The latest GTL technologies being developed by Shell and Sasol, a South African energy company, are thought to involve capital costs of around $20,000 per barrel per day of capacity.[13] A 75,000-b/d plant would, therefore, cost about $1.5 billion, nearly twice as much as a modern oil refinery. But GTL can yield a better return on investment than can oil refining if the cost of the natural gas feedstock is significantly lower than that of crude oil. Shell claims that its Middle Distillate Synthesis technology is profitable at a crude oil price of $14/barrel, assuming low gas-field development costs and no penalty for carbon emissions.

With the exception of the 30,000 b/d Mossgas plant in South Africa, which was built in 1990 in response to trade sanctions during the apartheid era, all GTL plants currently in operation are pilot projects. Several oil companies are now planning to build large-scale commercial plants. Shell plans four 75,000-b/d plants, possibly in Egypt, Indonesia, Iran and Trinidad and Tobago. Output of liquids from GTL plants is projected to jump from 43,000 b/d now to around 300,000 b/d by 2010 and 2.3 mb/d by 2030. Higher oil prices will contribute to the rapid increase in GTL production after 2010. These projections are highly dependent on oil price developments and the successful demonstration of emerging technologies.

13. See IEA (2001).

Final gas consumption will grow less rapidly than primary gas use – by 1.7% a year in industry and 1.4% in the residential, services and agricultural sectors. Final gas consumption will slow in the OECD because of saturation effects, sluggish output in the heavy manufacturing sector and a slowdown in population growth. Demand will grow more strongly in developing countries and transition economies along with rising industrial output, commercial activity and household incomes. Several oil-producing developing countries are encouraging switching to gas in order to free up more oil for export.

Natural Gas Supply

Gas resources are more than sufficient to meet projected increases in demand. Proven reserves were 165 tcm at the start of 2001, exceeding slightly the world's total proven reserves of oil in energy-equivalent terms. Half of global gas reserves are found in two countries, Russia and Iran (Figure 3.10). The number of countries known to have significant reserves has risen from around 50 in 1970 to nearly 90 today. The ratio of global reserves to production is around 60 years at present rates, compared to less than 44 years for oil. Remaining gas resources, including proven reserves, reserve growth[14] and undiscovered resources[15], are estimated by the US Geological Survey at 386 tcm (mean). Cedigaz, an international centre for gas information, estimates remaining ultimate resources[16] at 450-530 tcm. The latter estimate is equivalent to between 170 and 200 years of supply at current rates. Undiscovered gas resources total 147 tcm and reserve growth 104 tcm according to the USGS. Cumulative production to date amounts to less than 12% of total resources.

Proven gas reserves have doubled over the past twenty years, outpacing oil reserves, in large part because gas reserves are being depleted more slowly than those of oil. Strong growth in gas reserves has occurred in the former Soviet Union, the Middle East and the Asia/Pacific region. Further major discoveries will no doubt be made, but finding huge new fields in well-explored basins is unlikely. Exploration now leads increasingly to upward revisions of existing reserves, but also to smaller discoveries. Most of today's gas reserves were discovered in the course of

14. Increases in known reserves that commonly occur as gas fields are developed and produced.
15. Non-identified resources outside known fields that are thought to exist on the basis of geological information and theory.
16. The difference between the Cedigaz and USGS estimates is due to the adoption by Cedigaz of an unlimited forecast period instead of the 30-year forecast span used by USGS.

Figure 3.10: **Remaining Proven Natural Gas Reserves and Total Resources**

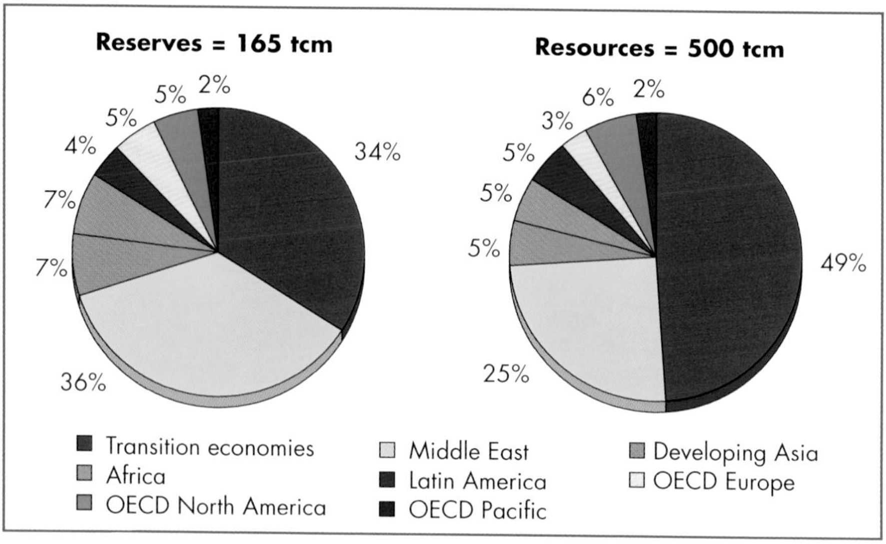

Reserves = 165 tcm

34% 36% 7% 7% 4% 5% 5% 2%

Resources = 500 tcm

49% 25% 5% 5% 5% 3% 6% 2%

Transition economies
Africa
OECD North America
Middle East
Latin America
OECD Pacific
Developing Asia
OECD Europe

Source: Cedigaz (2001).

exploration for oil. However, as much as a third of the world's gas reserves are currently "stranded". In other words, the costs of producing and transporting them to market are too high to make exploiting the reserves profitable.[17] Stranded gas is found in places a long way from markets, in deep-water reservoirs, in inaccessible places like the Arctic and in very small marginal fields.

The projected trends in regional gas production reflect to a large extent the proximity of reserves to the major markets. Production will grow most in absolute terms in the transition economies and the Middle East (Figure 3.11). Most of the incremental output will be exported to Europe and North America. Output will also increase quickly in Africa and Latin America. The projected 2,500-bcm increase in production between 2000 and 2030 will require massive investment in production facilities and transport infrastructure. In general, the share of transportation in total supply costs is likely to rise as reserves located closest to markets are depleted and supply chains lengthen. Technology-driven reductions in unit production and transport costs will, however, offset the effect of distance on total supply costs to some extent. Gas prices are expected to rise

17. Cedigaz (2001).

Figure 3.11: **Natural Gas Production by Region**

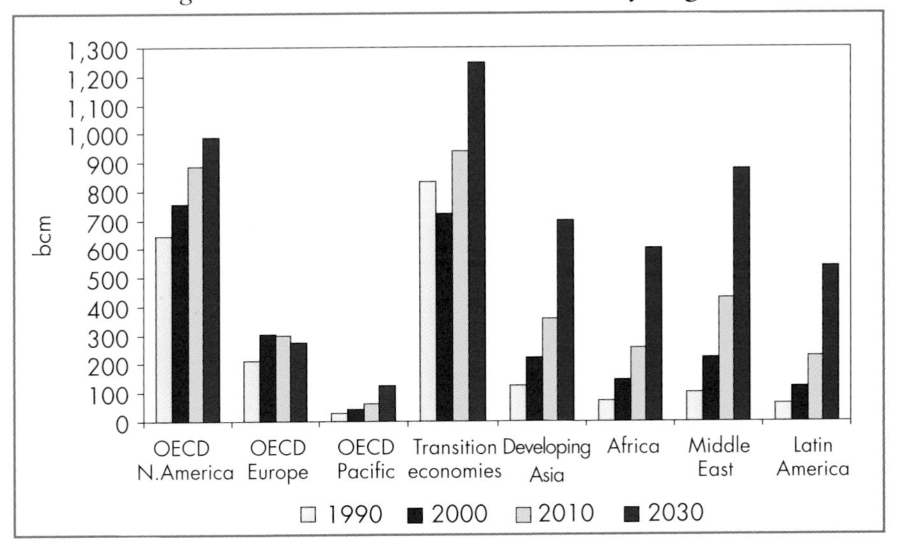

moderately after 2005 in North America and after 2010 elsewhere, as costs and oil prices increase. Pipelines will remain the principal means of transporting gas in North America, Europe and Latin America. Liquefied natural gas trade is also set to expand rapidly in the next three decades, mainly in the Asia/Pacific and Atlantic Basin regions.

Non-conventional gas could supplement conventional supplies, notably in North America. This type of gas includes coal-bed methane (CBM) and gas extracted from low permeability sandstone (tight sands) and shale formations (gas shales). The United States is currently the biggest producer of non-conventional gas, mainly tight sands and CBM from the Rocky Mountains. These sources already account for 25% of total US gas output. Improved extraction methods are likely to reduce the high cost of production. Ample reserves are expected to underpin a gradual increase in US production of both CBM and tight gas. CBM could also play an increasingly important part in gas supplies from Western Canada, where reserves are around 2 tcm. In other regions, little is known about the size of non-conventional gas resources. In some cases, there is no incentive to appraise these resources, as conventional gas resources remain large.

Box 3.4: **Uncertainties about Gas Supply Prospects**

Although there is little doubt that gas production could be increased to meet projected demand, our regional production projections are subject to uncertainty arising from several factors, including:

- the cost of developing reserves and transporting them to market;
- technological developments that may permit projects such as ultra-deep-water pipelines that are not technically feasible at present;
- gas prices, which are expected to remain closely linked to oil prices, and taxes;
- oil and gas depletion rates;
- geopolitical factors, which affect project risk and may prevent cross-border pipelines from being built;
- environmental regulations affecting the siting of gas production and processing facilities and pipelines;
- the pace of liberalisation of gas markets.

Natural Gas Trade

The geographical mismatch between resource endowment and demand means that the main growth markets for gas are going to become much more dependent on imports (Figures 3.12 and 3.13). In absolute terms, the biggest increase in imports is projected to occur in OECD Europe (Table 3.8).

Europe's import dependence will continue to rise, from 36% in 2000 to 69% in 2030. The Middle East will emerge as a major new supplier of gas to Europe, while Latin America (Trinidad and Tobago and Venezuela) will greatly increase its exports to Europe. Russia, together with other former Soviet Union republics, will remain the largest single supplier to Europe. OECD North America, which is more or less self-sufficient in gas at present, is expected to have to import 10% of its needs by 2010 and 26% by 2030. All of these imports will be in the form of LNG, from Latin America, Africa, the Middle East and Asia. India and China will become gas-importing countries in the near term, with most of the gas coming from the Middle East and other Asian countries. Russia will also export gas to China and Korea in the longer term. The Middle East and Africa – already big exporters – will see the biggest increases in exports.

Table 3.8: **Natural Gas Import Dependence**

	2000		**2030**	
	bcm*	**%****	**bcm***	**%****
OECD North America	5	1	345	26
OECD Europe	186	36	625	69
OECD Pacific	83	67	121	50
Transition economies	-112	-18	-277	-29
Africa	-69	-130	-299	-125
China	0	0	47	29
Other Asia	-60	-36	-94	-19
Latin America	-10	-9	-103	-28
Middle East	-23	-11	-365	-85

* Net imports in bcm. ** Per cent of primary gas supply.

There are few physical connections now between the main regional markets of North America, Europe, Asia/Pacific and Latin America. But these are expected to increase considerably, with a rapid expansion in LNG trade and the construction of new long-distance and undersea pipelines. LNG shipping capacity will increase by a at least 40% just between 2002 and 2005. At the beginning of 2002, 53 carriers had been ordered, with options for an additional 23. Only one carrier was added to the world's fleet in 2001, taking the total to 128.

Prices in connected regional markets are likely to converge, as suppliers exploit opportunities to switch volumes between supply routes and markets. There are already signs that the LNG market is becoming more flexible, as international trade grows and as downstream markets gradually open up to competition. Buyers are increasingly looking for short-term supply flexibility. Spot trade in LNG will continue to grow in the medium term, with trade flows changing in response to regional market factors. Spot trade grew by 50% in 2001, accounting for about 8% of total LNG trade.

Figure 3.12: **Net Inter-regional Natural Gas Trade Flows, 2000** (bcm)

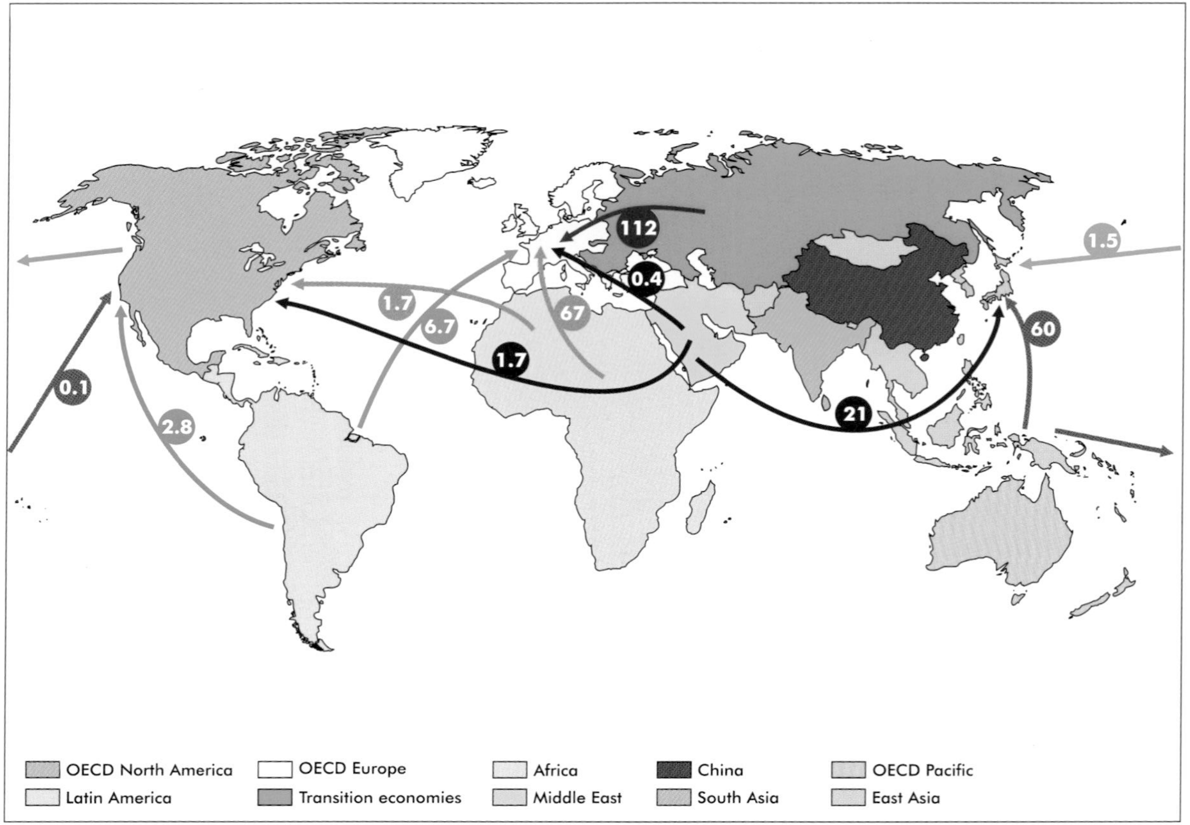

112
0.4
1.5
1.7
6.7
67
60
1.7
0.1
2.8
21

OECD North America
Latin America
OECD Europe
Transition economies
Africa
Middle East
China
South Asia
OECD Pacific
East Asia

Figure 3.13: **Net Inter-Regional Natural Gas Trade Flows, 2030** (bcm)

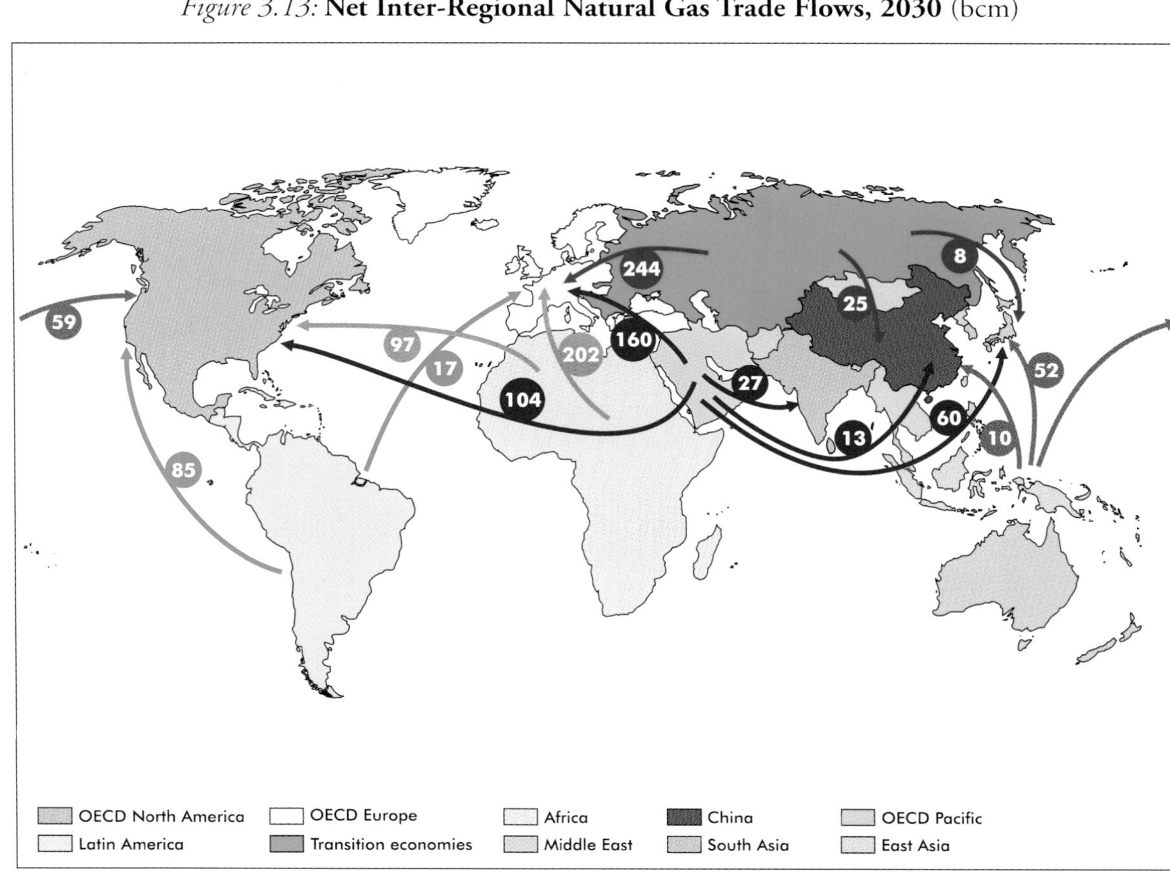

Coal Market

Coal Demand

Demand for coal is expected to grow more slowly than that for oil and gas. Global primary coal consumption will rise at an average annual rate of 1.4% over the *Outlook* period (Table 3.9). Its share in total energy consumption will drop from 26% in 2000 to 24% in 2010, and then remains almost stable through 2030.

Table 3.9: **World Primary Supply of Coal** (Mtoe)

	2000	2010	2020	2030	Average annual growth 2000-2030 (%)
OECD North America	579	586	651	685	0.6
OECD Europe	319	298	287	283	-0.4
OECD Pacific	184	205	221	215	0.5
Transition economies	213	252	248	260	0.7
Africa	91	105	131	174	2.2
China	659	854	1,059	1,278	2.2
Other Asia	281	366	487	655	2.9
Latin America	23	27	33	44	2.3
Middle East	7	9	12	14	2.6
World	**2,355**	**2,702**	**3,128**	**3,606**	**1.4**

In all regions, coal use becomes increasingly concentrated in power generation, which will account for almost 90% of the increase in demand between 2000 and 2030 (Figure 3.14). Coal demand in the power sector will be lifted by the assumed fall in the price of coal relative to that of gas and the gradual development and deployment of advanced coal technologies over the long term. But the anticipation of tougher environmental regulations and new measures to combat climate change may discourage investment in coal-fired capacity in industrialised countries. Industrial coal consumption will increase by 1.2% per year in developing countries and by 1.3% in transition economies from 2000 to 2030. These gains will be underpinned by heavy manufacturing, especially iron and steel. Industrial coal demand will decline in the OECD, by 0.4%

per year. Consumption in the residential and service sectors will fall, most sharply in OECD countries.

Figure 3.14: **World Primary Coal Demand by Sector**

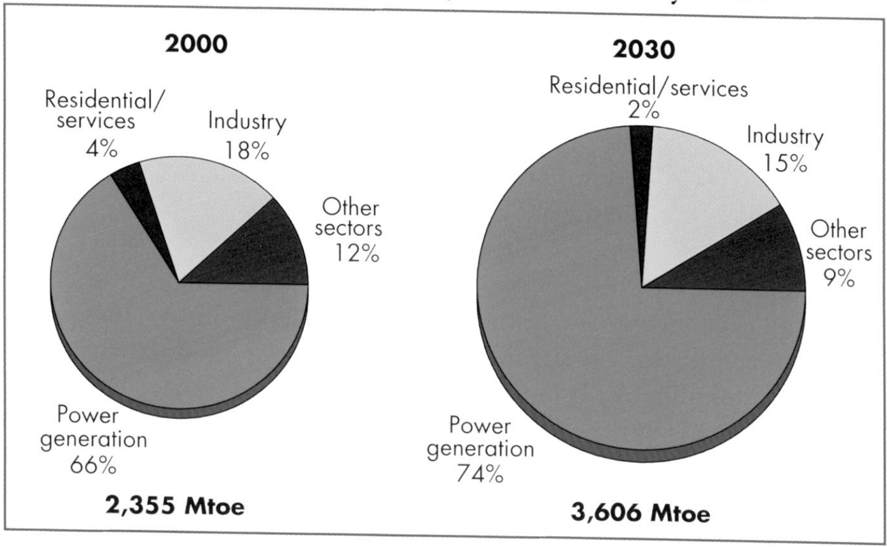

Coal demand is expected to be strongest in the developing world and the transition economies, where local supply is ample and production costs are low. The lack of indigenous gas resources will bolster coal use in several countries, particularly India and China. These two countries alone will account for close to two-thirds of the increase in world coal use over the period 2000 to 2030. Coal demand will increase slowly in OECD North America and the Pacific, but will fall in OECD Europe as gas elbows coal out of all end-use sectors and, to a slightly lesser extent, power generation.

Coal Supply

World reserves of coal are enormous. Compared with oil and natural gas, they are widely dispersed. Economically recoverable coal reserves are estimated at close to one trillion tonnes, or about 200 years of production at current rates. Almost half the world's reserves are located in OECD countries. In practice, the quality and geological characteristics of coal deposits are more important to the economics of production than the actual size of a country's reserves. Quality varies from one region to another. Australia, Canada and the United States all have high-quality coking coal. Australia, China, Colombia, India, Indonesia, Russia, South Africa and the United States have very large reserves of steam coal.

Coal production is likely to increase in China, the United States, India, Australia, South Africa, India, Indonesia, Canada, Colombia and Venezuela. Production will continue to decline in OECD Europe. The Asia/Pacific market will probably be supplied mainly by Australia, Indonesia and China. South Africa, the United States, Colombia and Venezuela will be the primary suppliers to the Europe-Atlantic market. South Africa's geographic location enables it to supply Europe, Asia and the Americas. Its role in transmitting price signals between regional markets will remain an important component of international coal trade.

These projections imply a need for sustained investment in both production and transportation infrastructure. This is especially true in China and India, where coal remains an important component of energy supply and the locomotive of future economic development. Investment will be accompanied by significant gains in labour productivity as the average size of mines will continue to increase, more advanced extraction, preparation and transport technology will be adopted and working practices will be improved. These factors are expected to offset the negative impact of depletion of reserves in well-worked mines on the delivered unit cost of internationally traded coal. They will also compensate for the growing cost of transportation, as more coal is shipped over longer distances to the main markets in Asia and Europe. In major coal-producing countries, growth in labour productivity averaged between 5% and 10% per year in the 1980s, and from 10% to 15% per year in the 1990s.[18] Productivity is expected to grow at rates equal to, or higher, than this in the future.

Several countries continue to subsidise their coal industries, but overall subsidies have fallen over the past decade. In 2000, only 7% of OECD coal production was still subsidised. Support fell by 55% in nominal terms between 1991 and 2000, to $5.8 billion. Major producers in developing countries such as China and India have also reined in coal subsidies in recent years.[19] The Reference Scenario assumes that remaining coal subsidies will be phased out over the course of the next three decades as part of industry-restructuring programmes.

18. See IEA (2001).

19. See IEA (1999) for estimates of the size of subsidies on coal and other fuels in the largest developing countries. China claims to have closed 47,000 inefficient coal mines between 1998 and 2000, while India removed coal-price controls in 2000.

Coal Trade

With coal reserves widespread geographically, coal demand is usually met on a regional basis. Internationally traded coal, of which two-thirds is steam coal or hard coal[20], accounts for only 12% of total world demand. Trade has grown since the 1970s, and is expected to go on expanding. Steam coal will continue to gain market share in world coal trade over the next two decades, stimulated mainly by strong demand in the power generation sector in major importing markets.

Power Generation

Electricity Demand

World electricity demand is projected to double between 2000 and 2030, growing at an annual rate of 2.4%. This is faster than any other final energy source. Electricity's share of total final energy consumption rises from 18% in 2000 to 22% in 2030. Electricity demand growth is strongest in developing countries, where demand will climb by over 4% per year over the projection period, tripling by 2030. Consequently, the developing countries' share of global electricity demand jumps from 27% in 2000 to 43% in 2030. Electricity demand increases most rapidly in the residential sector, especially in developing countries. Industrial electricity demand increases by 2.2% a year, but its share in total demand declines (Table 3.10).

The Fuel Mix

Electricity generation is projected to rise at 2.4% per year over the period 2000-2030. Today, coal is the most widely used fuel for generation, with a 40% share of electricity output. Natural gas, nuclear power and hydroelectricity – with almost equal shares – account for most of the remainder. The next three decades will see a pronounced shift in the generation fuel mix in favour of natural gas. The main changes in the fuel mix are as follows:[21]

20. Hard coal has a higher energy value than brown coal and so is more likely to be transported over longer distances. Hard coal is used both for heat and power generation and for coke production in integrated steel manufacturing. Lignite and brown coal have higher ash and moisture contents and lower energy values and are thus more likely to be burned near their point of mining, solely for heat and power generation.

21. The prospects for the electricity generation mix are considerably different in the OECD Alternative Policy Scenario (see Chapter 12).

Table 3.10: **World Electricity Balance**

	2000	2010	2020	2030	Average annual growth 2000-2030 (%)
Gross generation (TWh)	**15,391**	**20,037**	**25,578**	**31,524**	**2.4**
Coal	5,989	7,143	9,075	11,590	2.2
Oil	1,241	1,348	1,371	1,326	0.2
Gas	2,676	4,947	7,696	9,923	4.5
Hydrogen-fuel cells	0	0	15	349	–
Nuclear	2,586	2,889	2,758	2,697	0.1
Hydro	2,650	3,188	3,800	4,259	1.6
Other renewables	249	521	863	1,381	5.9
Own use and losses (Mtoe)	235	304	388	476	2.4
Total final consumption (Mtoe)	**1,088**	**1,419**	**1,812**	**2,235**	**2.4**
Industry	458	581	729	879	2.2
Residential	305	408	532	674	2.7
Services	256	341	440	548	2.6
Other*	68	89	111	133	2.3

*Includes transport, agriculture and non-specified uses of electricity.

- Coal's share in total generation declines in the period from 2000 to 2020, but recovers slightly thereafter. Coal remains the largest source of electricity generation throughout the projection period.
- Oil's share in total generation, already small, will continue to decline.
- The share of natural gas is projected to increase significantly, from 17% in 2000 to 31% in 2030, because the majority of new power plants will be gas-fired. The rate of growth in power sector demand for gas will slow in the second half of the *Outlook* period, because prices increase.
- Nuclear power production increases slightly, but its share in total generation is reduced by half because very few new plants are built and many existing reactors are retired.

- Hydroelectricity increases by 60% over the projection period but its share falls.
- Non-hydro renewables grow faster than any other source, at an annual rate of 6%, and the total output from renewables increases almost sixfold over the period 2000 to 2030. They will provide 4.4% of the world's electricity in 2030. Wind and biomass will account for 80% of the increase.
- Fuel cells using hydrogen from reformed natural gas are expected to emerge as a new source of power generation, especially after 2020. They will produce a little more than 1% of total electricity output in 2030.

Coal-fired generation is projected to increase from 5,989 TWh in 2000 to 11,590 TWh in 2030. But coal's share in the global electricity mix declines by more than 2 percentage points, to 37%, because its increase in developing countries is more than offset by a decline in the OECD. Coal-based generation in developing countries will more than triple by 2030, with most of the increase occurring in India and China. Coal will remain the dominant fuel in power generation in those countries because they have large, low-cost reserves. In the OECD, coal-fired generation increases at a much slower pace and its share in total generation drops from 39% now to 31% in 2030.

Natural gas-based electricity production is projected to increase by 4.5% per year over the *Outlook* period, reaching 9,923 TWh by 2030. The rate of increase slows after 2010, because gas prices rise more rapidly than coal prices, making coal a more competitive option in a growing number of countries.

In most countries, gas-fired combined-cycle gas turbine (CCGT) plants will remain the preferred option for new power generation plants for their economic and environmental advantages. There have been major advances in the efficiency and reliability of CCGT technology since it was first deployed commercially in the 1980s. CCGT plants achieve much higher efficiencies than traditional steam-boiler plants, they cost much less per kW of capacity and are quicker to build. The current plans of utilities in the OECD confirm that most new plants commissioned by 2010 will be gas-fired (Figure 3.15).

Gas price volatility, which results from tight supplies and from oil price volatility[22], is not expected to affect investment in gas-fired capacity so long as prices on average remain low. It may, however, have a transitory

22. See Box 1.1 in Chapter 1 on oil price volatility.

Figure 3.15: **Planned Power Generation Capacity Additions to 2010**

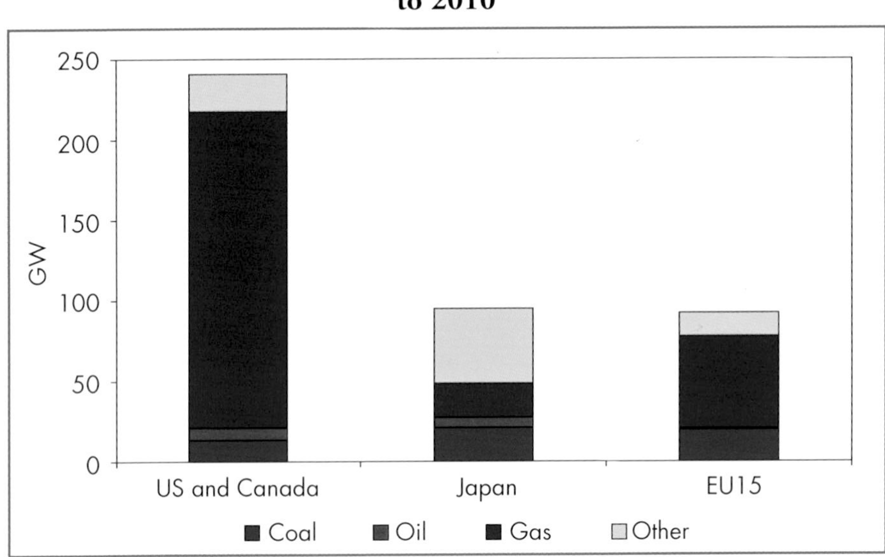

Source: Platts – UDI (2002), World Electric Power Plants database. Includes capacity under construction.

effect on the amount of electricity that a gas-fired power station produces at any given time. High gas prices in 2000 led to a temporary surge in output from coal-fired stations in many countries.

Fuel cells in distributed generation applications are expected to start contributing to power production around 2020, on the assumption that costs fall sufficiently to make this technology competitive (see Box 3.5 and Chapter 2). The Reference Scenario projects that 100 GW of fuel cells will be installed for power production by 2030, entirely in OECD countries. The most economic method for hydrogen production is expected to be steam reforming of natural gas.

The share of *oil* in electricity generation, which averaged 8% worldwide in 2000, is projected to fall to 4% in 2030. The decline will be steepest in the OECD and the transition economies, where both the amount of oil-fired generation and its share fall over time. By the end of the projection period, oil-fired generation will be concentrated in distributed generation applications in industry and remote areas (Box 3.5).

Box 3.5: **Distributed Generation**[23]

Distributed generation (DG) plants produce power on a customer's site or at the site of a local distribution utility and supply power directly to the distribution network. Although their capacity is usually small, they account for a significant proportion of total power supply in many parts of the world. In some developing countries, industrial firms have installed oil-fired engines or turbines to ensure a reliable supply of electricity. In the OECD, distributed generation includes combined heat and power plants in industry and commercial activities. It is used for backup power, for high reliability applications or for power supply inside a distribution network to meet peak loads. In Japan, DG plants are displacing more costly grid electricity in some cases.

Most DG systems in commercial operation today consist of diesel and natural gas reciprocating engines and gas turbines. Such systems have increased rapidly over the last decade. More than 10 GW of these types of capacity were installed in 2000.

Increasing demand for reliable power supply and government policies that encourage the use of combined heat and power and renewable energy will propel the growth of DG in the coming years. The Reference Scenario assumes that DG electricity output will grow by 4.2% annually between 2000 and 2030. New DG capacity will amount to 521 GW, excluding photovoltaics, in 2030. This is equal to 11% of total new generating capacity. The main sources of uncertainty for DG growth are fuel costs, the rate of cost reductions in DG technologies and how easily distribution networks can accommodate larger amounts of distributed power generation.

Demand for distributed generation will provide opportunities for new power generation technologies. Engines and small turbines will dominate DG orders in the short term. By 2020, fuel cells are likely to emerge as the primary DG technology, as their costs fall. Installed fuel cell capacity is expected to reach 100 GW by 2030. In total, annual capacity additions for fuel cells, photovoltaics and other DG technologies could reach 35 GW by 2030.[24]

23. See IEA (2002) for a detailed discussion of distributed generation.

24. The DG projections do not include biomass and wind, although some of the projected increases in capacity of these sources will probably fall into the category of DG.

Nuclear power plants produced 2,586 TWh of electricity in 2000, or 17% of total world electricity. Nuclear output is projected to increase over the next ten years, then to start falling slowly as several existing reactors are retired. Our projections assume that more than 40% of existing nuclear capacity will be retired by 2030. Nevertheless, nuclear electricity generation in 2030 is expected to be somewhat higher than now, reaching 2,697 Twh, or 9% of total generation. Nuclear plant capacity factors have been increasing and are projected to remain high over the next two decades. As plants age, however, their performance is expected to deteriorate somewhat.

Most of the projected growth in nuclear power occurs in Asia. Japan and Korea have the largest nuclear construction programmes in the region. Nuclear power is also projected to increase in China and India, although it remains a marginal source of generation in both countries in 2030. Nuclear capacity in the transition economies will decline, because new capacity built over the next three decades will not be sufficient to offset plant retirements. It is expected that these countries will retire three-quarters of their existing capacity by 2030. Lithuania, Slovakia and Bulgaria have agreed with the European Union to shut down some of their older reactors within the next ten years. Russia has the most ambitious nuclear programme in the region, but financing new plants will be difficult.

Hydroelectric power is expected to increase quickest in developing countries, where the remaining potential is still high. Growth in the OECD is limited to 0.6% per year, as much of the hydro potential has already been exploited and environmental considerations prevent the development of large-scale hydro plants. Globally, hydropower's share in electricity generation falls from 17% now to 14% by 2030.

Non-hydro renewable energy accounts for a small but rapidly growing share of global electricity (Figure 3.16). It reached 1.6% in 2000 and is projected to rise to 4.4% by 2030. Most of the growth will be in OECD countries, where renewable energy receives financial and regulatory support from governments in their efforts to reduce dependence on fossil fuels.[25] The share of non-hydro renewables in OECD electricity generation is projected to grow from 2.2% in 2000 to 7.1% in 2030. In developing regions, non-hydro renewables are expected to play a growing role in providing electricity to remote, off-grid locations as part of rural

25. Projections of OECD renewables-based electricity production are based on supply curves provided by the Technical University of Vienna. See also Huber *et al.* (2001).

electrification programmes. Several countries are showing interest in exploiting their substantial renewable energy potential. Non-hydro renewables are projected to meet 2% of total electricity generation in developing countries in 2030.

Wind and biomass will account for most of the projected growth in renewables-based power production. Wind power is projected to increase by 10% a year over the 30-year projection period, to reach 539 TWh in 2030, more than 80% of that amount in OECD countries. The cost of producing electricity from wind power is high compared with gas-fired plants, but declining capital costs and improved performance are expected to make wind more competitive. Biomass is projected to increase by 4.2% per year to reach 568 TWh in 2030.

The role of other types of renewables is also expected to grow, especially after 2020. Solar power will grow by nearly 16% a year over the projection period, reaching 92 TWh in 2030. Geothermal power will increase by 4.3% per year, and its contribution to total generation will double to 0.6%. Electricity production from tide and wave energy is projected to take off towards the end of the projection period, although its contribution remains very small.

Figure 3.16: **World Renewables-Based Electricity Generation**

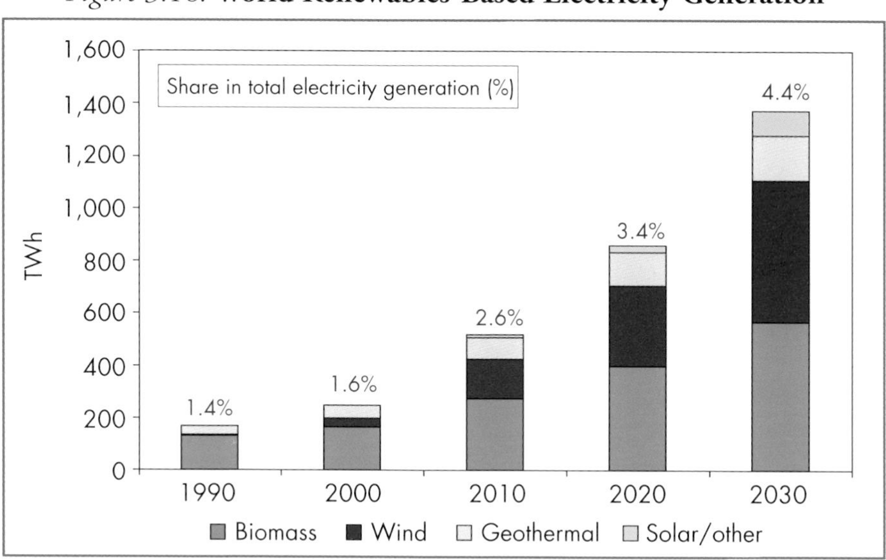

Capacity Requirements and Implications for Investment[26]

Over the period from 2000 to 2030, nearly 5,000 GW of generating capacity is expected to be built worldwide. Total installed capacity will rise from 3,397 GW in 1999 to 7,157 GW (Figure 3.17). By 2030, almost two-thirds of installed capacity will have been built after 2000. About a third of new capacity will be in developing Asia. OECD countries will require more than 2,000 GW to replace old plants and to meet rising demand.

More than 40% of new capacity is projected to be gas-fired (Figure 3.18). Gas-fired capacity jumps from 677 GW in 1999 to over 2,500 GW in 2030. Almost half of this increase occurs in the OECD, where gas is widely available and environmental restrictions limit the use of coal. Big increases are expected in the transition economies and developing countries too. But it remains very uncertain whether these countries will be able to find the capital required to build the infrastructure, including upstream facilities, pipelines and LNG terminals, to support new gas-fired projects.

Figure 3.17: **World Installed Electricity Generation Capacity**

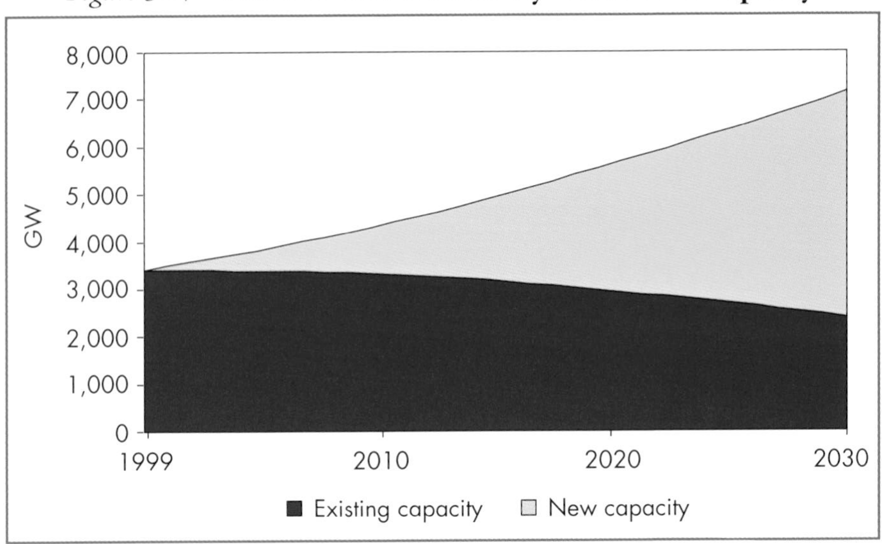

26. Investment requirements are calculated in the World Energy Model by multiplying the projected capacity requirements by the capital cost of each technology. Capital costs may vary between regions and over time.

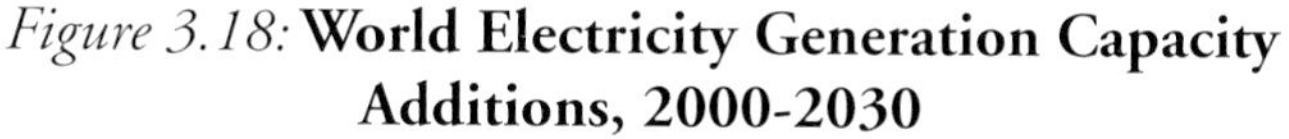
Figure 3.18: **World Electricity Generation Capacity Additions, 2000-2030**

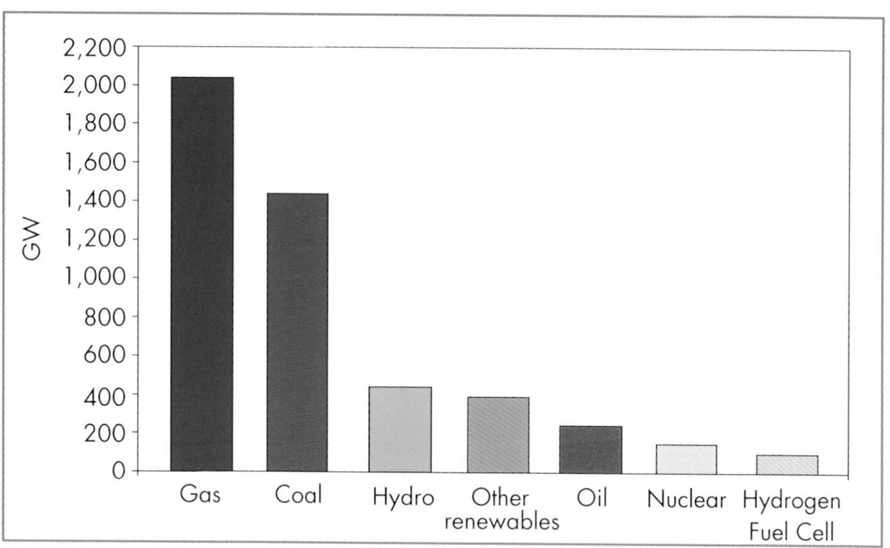

The projected capacity requirements would cost more than $4 trillion in current dollars.[27] Developing countries will need half of this amount. Table 3.11 summarises the projections of capacity additions and investment requirements by region.

27. Investment requirements and financing issues will be analysed in detail in *World Energy Outlook 2003: Insights*, which is due to be published in 2003.

Table 3.11: **New Electricity Generating Capacity and Investment by Region**

	Installed capacity (GW)		Additional capacity (GW)	Cumulative investment (billion $)
	1999	**2030**	**2000-2030**	
OECD Europe	698	1,109	786	656
European Union	*573*	*901*	*658*	*531*
OECD North America	961	1,595	942	715
US and Canada	*922*	*1,473*	*853*	*646*
OECD Pacific	356	591	340	369
Japan, Australia, New Zealand	*304*	*447*	*242*	*281*
OECD	**2,015**	**3,294**	**2,068**	**1,740**
Russia	*216*	*360*	*205*	*157*
Transition economies	**404**	**624**	**371**	**298**
China	300	1,087	800	827
East Asia	159	524	385	338
Indonesia	*36*	*124*	*90*	*73*
South Asia	132	459	345	315
India	*108*	*366*	*274*	*272*
Latin America	168	492	339	331
Brazil	*68*	*188*	*123*	*158*
Middle East	120	277	191	101
Africa	100	400	322	217
Developing countries	**979**	**3,238**	**2,382**	**2,130**
World	**3,397**	**7,157**	**4,821**	**4,168**

PART B

REGIONAL OUTLOOKS TO 2030

CHAPTER 4: OECD NORTH AMERICA

HIGHLIGHTS

- The Reference Scenario shows an average annual rate of growth of 1% in primary energy demand in the United States and Canada. Demand will rise more slowly after 2010, due to a gradual slowdown in economic growth, saturation effects and rising energy prices.
- The United States and Canada will remain heavily dependent on oil, which they use predominantly for road and air transport. But gas will grow in relative importance, because many new power plants will be gas-fired. The supply of renewables expands rapidly, though their share in primary supply will still be less than 10% in 2030.
- If the US and Canadian governments take no new action to rein in demand and boost production, net imports of oil will continue to rise, reaching 15.5 mb/d, or 57% of the region's consumption, in 2030. A large and growing share of these additional imports will come from OPEC countries. Gas imports, predominantly liquefied natural gas, will grow from very low levels now to around 30% of demand in 2030, as domestic supplies tighten and gas prices rise. New policies to promote switching to other fuels or curb gas demand, not taken into account here, would reduce gas-import dependence.
- New policy initiatives, including those recently proposed under the US National Energy Policy, could alter demand and supply trends substantially, as well as the outlook for energy-related carbon-dioxide emissions. In the absence of any new actions, emissions would rise by 1% per year from 2000 to 2030.
- Mexico's primary energy use will expand by 2.5% per year over the *Outlook* period, more than twice as fast as demand in the US and Canada. Oil will still dominate the fuel mix, but there will be a substantial increase in the use of gas. The development of Mexico's abundant energy resources and the expansion of its supply infrastructure hinge on the continuation of the government's reform programme.

OECD North America includes the United States, Canada and, for the first time in the World Energy Outlook, *Mexico. For this edition, Mexico has been modelled separately. Projections for Mexico, along with the discussion of current market trends and assumptions, are described at the end of this chapter.*

The United States and Canada

Energy Market Overview

The United States and Canada together account for just over a quarter of world energy demand. The region is homogeneous economically, with strong trading links between the two countries. The United States accounts for most of the region's wealth and energy consumption (Table 4.1). It has the largest economy in the world, accounting for a fifth of global GDP (in PPP terms) and around 37% of the OECD's GDP, although it has only a quarter of the OECD's population. Per capita GDP in both countries is well above the OECD average.

Fossil fuels dominate the region's energy supply. Oil and gas account for over 60% of primary demand, coal for a little less than a quarter and nuclear power for much of the rest. Renewables contribute 6%. The region is rich in fossil fuel and renewable resources. Indigenous production meets a little over 80% of total energy needs. Virtually all net imports are oil, supplemented by small volumes of liquefied natural gas. The United States is a net exporter of coal. Intra-regional trade is extensive: the United States imports large amounts of oil and natural gas and small quantities of electricity from Canada.

Table 4.1: **Key Economic and Energy Indicators of the United States and Canada, 2000**

	Canada	US	OECD
GDP (in billion 1995 US$, PPP)	818	8,987	24,559
GDP per capita (in 1995 $, PPP)	26,604	32,629	21,997
Population (million)	31	275	1,116
TPES (Mtoe)	251	2,300	5,291
TPES/GDP*	0.31	0.26	0.22
Energy production/TPES	1.5	0.7	0.7
Per capita TPES (toe)	8.2	8.3	4.7

*Toe per thousand dollars of GDP, measured in PPP terms at 1995 prices.

Energy intensity – measured either as energy use per unit of GDP or energy use per capita – is high compared to other OECD countries. This results from a combination of factors, including low energy prices, large distances between population centres, sprawling cities and pronounced seasonal swings in climate. Low population density also helps explain why transport accounts for as much as 39% of final energy use.

Energy use has been growing steadily in the last decade, driven by one of the longest periods of sustained economic growth in the region's history. Total primary energy supply rose at an average annual rate of 1.8% from 1990 to 2000. This was, however, much slower than GDP growth, and so the region experienced a marked reduction in energy intensity. This happened because of the rapid expansion of the services sector and the information and communication technology (ICT) equipment industry, which are less energy-intensive than heavy manufacturing.[1] Preliminary data suggest that energy demand was flat in 2001, as a result of the economic slowdown and higher energy prices.

Current Trends and Key Assumptions

Macroeconomic Context

Following four years of exceptionally rapid growth, the US economy began to slow in late 2000 and 2001. This was accompanied by a brutal correction in high-tech equity values, a slump in business investment and a drawdown of inventories. The terrorist attacks on 11 September 2001 further undermined business and consumer confidence, leading to additional lay-offs and delayed investment. Signs of recovery began to emerge in late 2001 and early 2002, including rebounds in profits and manufacturing output and strong retail sales.

The rapid easing of monetary policy, an expansionist fiscal policy and continuing gains in labour productivity are expected to underpin an upturn in the economy in 2002. The OECD currently projects US GDP to grow by almost 3.5% in 2003 – up from 2.5% in 2002. However, there are major downside risks to this projection. Prospects will depend on how strongly business investment and corporate profits recover and how well household spending holds up. All of these factors could be dampened by high business and consumer debt.

1. Box 2.2 in Chapter 2 discusses the impact of ICT on energy demand.

Unsurprisingly in view of the two countries' strong linkages, the Canadian economy has broadly followed that of the United States, albeit with a slight lag. Preliminary data show that the Canadian downturn has been a little less pronounced than in the United States. Canada benefited from tax cuts, which helped to keep consumer spending up, aggressive reductions in interest rates and higher world commodity prices. Rising export demand and healthy government finances are expected to contribute to a significant improvement in Canada's economic conditions in 2003.

The combined GDP of the United States and Canada is assumed to increase at an average of 2% annually over the projection period (Table 4.2). This compares to 3.2% between 1971 and 2000. GDP will grow more quickly in the current decade, at 2.5%, but will slow to only 1.6% between 2020 and 2030. This slowdown is partly due to falling population growth, from 1% over the past three decades to 0.8% over the projection period.

Table 4.2: **Reference Scenario Assumptions for the United States and Canada**

	1971	2000	2010	2030	Average annual growth 2000-2030 (%)
GDP (in billion 1995 $, PPP)	3,927	9,804	12,506	17,693	2.0
Population (million)	230	306	334	387	0.8
GDP per capita (in 1995 $, PPP)	17,100	31,993	37,487	45,787	1.2

Energy Liberalisation and Prices

Wholesale and retail energy prices have been extremely volatile over the last four years. Rising international oil prices in 1999 and 2000 brought big increases in gasoline and diesel prices at the pump. In most other OECD countries, pump prices rose much less in percentage terms because their fuel taxes are higher (Figure 4.1). The retail prices of refined oil products, and the taxes levied on them, are assumed to remain flat in real terms through to 2010. Pre-tax prices are assumed then to rise steadily to 2030 on higher crude oil prices, but taxes will remain unchanged.

Figure 4.1: **Retail Gasoline Prices in Selected OECD Countries, 2001**

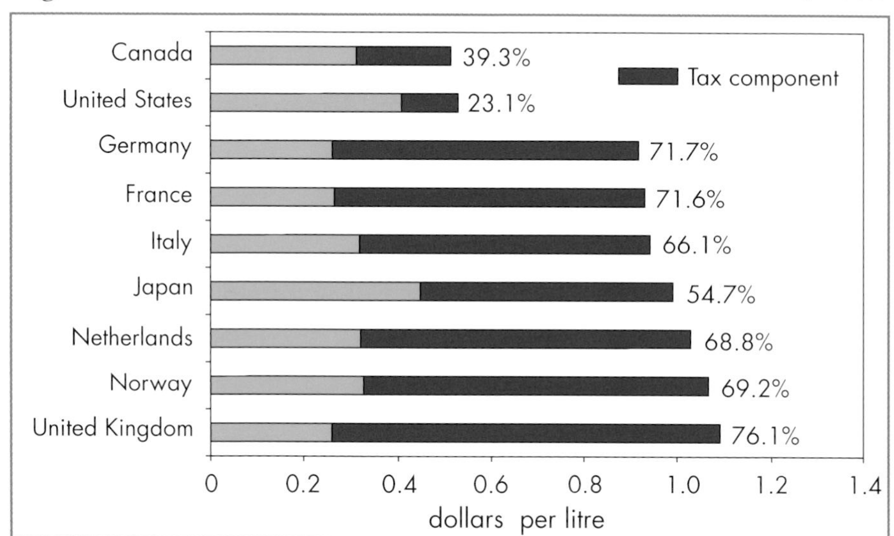

Source: IEA (2002).

Higher oil prices, together with tighter gas supply, drove natural gas prices across North America up to very high levels in the winter of 2000-2001 (Figure 4.2). A cold winter was a major factor, exacerbated by a fall in the drilling of exploration and development wells due to very low prices in the late 1990s. The size and speed of the price surge was unprecedented in the US market. Prices topped $10/MBtu[2] on some days in December 2000 and January 2001. Prices started to fall back in early 2001, as the slowing economy weakened demand and oil prices fell back slightly. By the end of the year they were generally below $2/MBtu.

High energy prices in 2000 and early 2001, together with the crisis in California's electricity market, have slowed moves by some US states to introduce full retail competition in natural gas. These states have chosen not to expand pilot programmes or have deferred adopting legislation opening up the retail market. The Reference Scenario assumes that full retail competition in gas will eventually be introduced across the United States and Canada, but more slowly than was anticipated in *WEO 2000.* Wholesale natural gas prices are assumed to remain flat in real terms to

2. All prices in this chapter are in US dollars.

2005 and then to rise steadily to 2030, as production falls behind demand and North America grows more dependent on imports of LNG.

Figure 4.2: **Wellhead Natural Gas Prices in the United States**

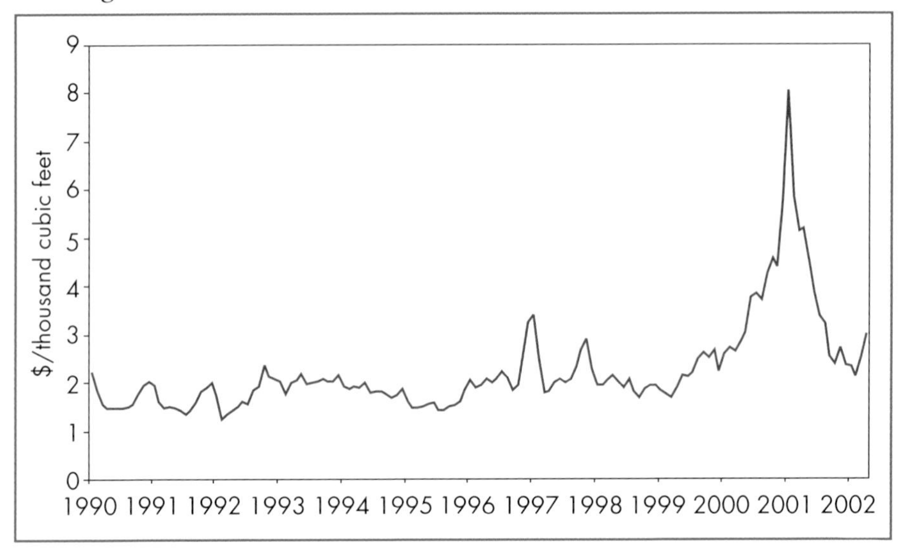

Source: US Energy Information Administration.

In early 2001, wholesale electricity prices also surged in many parts of the United States, reaching record heights in California. This resulted from the very tight supply of natural gas, low hydroelectric generation levels in the wake of a drought, other generation problems and surging electricity demand (Box 4.1). Power prices have since fallen, due to the slowing economy, capacity additions and lower natural gas prices. The problems in California and the state's abandonment of retail competition have slowed the trend to restructuring across the United States. At the beginning of August 2002, 21 states and the District of Columbia had programmes underway that allow residential consumers and other small-volume users to choose their electricity supplier. The *Outlook*'s projections assume a gradual transition to fully competitive pricing of electricity, a development which should put downward pressure on prices and could partly offset the effect of higher gas input prices on generation costs.

Box 4.1: **The Crisis in the Californian Electricity Market**

A combination of strong demand and the loss of a part of California's customary supply led to a massive increase in wholesale prices and frequent blackouts and brownouts in the second half of 2000 and early 2001. It bankrupted that state's largest utility and pushed the others to the verge of bankruptcy.

In 2000, lower rainfall than normal cut hydropower output in the north-west by 14%. Increased gas-fired generation made up for some of this loss, but pipeline bottlenecks caused gas prices to surge. This drove up wholesale electricity prices to unusually high levels. Transmission constraints between the north and the south of the state hindered supply. The design of the wholesale electricity market and the lack of competition within it are also thought to have contributed to higher electricity prices.

Under interim regulatory rules imposed in the late 1990s, retail electricity prices for the state's two largest utilities were capped. Prices in 2000 were actually *lower* than in 1999, and this led to a 6% surge in consumption. Hot weather also boosted demand. A physical shortfall in supply occurred, because power transmission capacity from other states and new in-state generation capacity were unable to make good the difference. In fact, net capacity decreased between 1990 and 1999 and no new capacity was brought online in the second half of 2000. A mere 1.9 GW was added in the first nine months of 2001, but a further 8.2 GW is to be completed by mid-2003. Difficulties in obtaining authorisation for new plants and new transmission lines, along with the price caps, deterred investment in the 1990s.

Market rules forbade the utilities from buying more than 20% of their wholesale requirements under bilateral long-term contracts with generators. They were obliged to buy the rest on the California Power Exchange. The gulf that opened up between wholesale and retail prices caused the utilities' debts to balloon. Southern California Edison subsequently negotiated a deal under which the state assumed the utility's debts from unrecovered charges and issued bonds to pay them off. Pacific Gas and Electricity declared bankruptcy, hoping for a better deal from the courts.

The state has taken several steps to deal with the supply shortages and to get the industry back onto a sound financial footing. In March 2001, retail prices were raised by almost 50% and new conservation measures were introduced to reduce demand. Wholesale price caps were introduced as a temporary measure, and the utilities were given the opportunity to make bilateral contracts outside the power exchange (which in any event went out of service in early 2001). Licensing procedures for new plants have also been speeded up, and California abandoned its retail choice programme entirely in September 2001.

In Canada, the electricity supply industry is being restructured, although the pace has slowed because of problems in Alberta and California. Alberta was the first Canadian province to introduce competitive electricity pricing, with the establishment of a wholesale market in 1995 and the opening up of the retail market to full competition at the beginning of 2001. Prices have risen sharply in recent months after the disappearance of what had been a surplus in generation capacity. The surplus was exhausted by strong demand growth and a lack of investment in new plant, partly caused by uncertainty about the impact of deregulation. Ontario began to restructure its electricity industry in 1998, but plans to extend competition to retail markets have been delayed several times.

Other Energy and Environmental Policies

Major changes in energy and environmental policies have been made or proposed in the United States and Canada in the past two years. They reflect heightened concerns about energy security in the wake of the Californian electricity crisis, higher oil prices and the 11 September terrorist attacks. The collapse of the energy-trading firm, Enron, contributed to worries about the reliability of supply, although no physical disruptions occurred. The Bush Administration has rejected the Kyoto Protocol on climate change and has proposed an alternative approach to limiting greenhouse gas emissions.

In the United States, a high-level group appointed by the president to draw up a national energy policy released its report in May 2001.[3] The report contains a number of recommendations for new Federal measures to promote energy conservation, to modernise and expand infrastructure, to

3. United States National Energy Policy Development Group (2001).

improve environmental protection and to enhance energy security. Most are aimed at boosting indigenous energy supplies, especially oil and gas, nuclear energy and renewable energy sources, and their reliability. One of the most contentious proposals would allow oil and gas exploration on Federal lands that are currently out of bounds, including the Arctic National Wildlife Reserve. This proposal is strongly opposed by environmental groups.

The recommendations that involve a continuation of current policies, or that have already been implemented, are taken into consideration in the Reference Scenario. Other recommendations that need Congressional action to become effective, including authorising drilling on Federal lands, are *not* taken into account. Two rival energy bills were presented to Congress in 2002. The House of Representatives passed a comprehensive bill that provided for opening up the Arctic Reserve to drilling, but the Senate rejected the proposal, as well as a plan to tighten fuel efficiency requirements for new vehicles (known as CAFE standards). Both bills included tax breaks to promote energy supply. The Senate bill concentrated on renewables and the House bill on fossil fuels and nuclear power. Legislators from both houses were working on a compromise bill as this report went to print.

The Bush Administration announced in early 2001 that it would not ratify the Kyoto Protocol that had been negotiated by the previous Administration. In February 2002, he announced a new climate change policy aimed at reducing greenhouse gas intensity by 18% between 2002 and 2012. Greenhouse gas intensity is defined as the amount of emissions per unit of GDP. It is likely to fall sharply anyway. Indeed, the US Department of Energy currently projects the intensity of CO_2, the main greenhouse gas, to fall by around 14% over the same period without any change in policies.[4] The new plan includes the following measures:

- *Tax incentives for renewables and co-generation*: The Administration proposes $4.6 billion in tax breaks over the next five years for renewables (solar, wind, and biomass), for hybrid and fuel-cell vehicles, for co-generation and for landfill gas.
- *Improvements in the voluntary emissions-reduction registry*: The plan would enhance the accuracy, reliability and verifiability of measurements of emissions reductions under voluntary agreements between the government and industry.

4. DOE/EIA (2001a).

- *Transportation programmes*: The Administration plans to step up its efforts to promote the development of fuel-efficient cars and trucks, to finance research on cleaner fuels and to implement energy efficiency programmes. Planned measures include research partnerships with industry, market-based incentives and standards. The proposed budget for Fiscal Year 2003 calls for more than $3 billion in tax credits over 11 years to help consumers buy fuel-cell and hybrid vehicles.
- *Carbon sequestration*: The plan sets out measures to enhance the capture and storage of carbon in natural sinks through conservation of farmland and wetlands. It also contains a commitment to develop accounting rules and guidelines for crediting sequestration projects.

President Bush also announced, in February 2002, a "Clear Skies Initiative" that will involve legislation setting new limits on power plants' emissions of sulphur dioxide, nitrogen oxides and mercury.[5] These reductions are to be achieved through a tradable permit programme that would give generators flexibility in meeting the requirements.

In October 2000, the provincial and federal governments (with the exception of Ontario) published their National Implementation Strategy on Climate Change, which sets out the country's broad policy framework and priorities. They also unveiled the First National Climate Change Business Plan, setting out the first round of unilateral and joint initiatives for achieving the national commitment to reduce greenhouse gas emissions to 6% below 1990 levels by 2008-2012. The federal government's Action Plan calls for switching to low or non-carbon emitting fuels, working with the United States to tighten fuel efficiency standards for vehicles, investing in refuelling infrastructure for fuel cells, researching the potential for storing carbon underground, increasing inter-provincial trade of hydropower, and the voluntary reporting of emissions reductions.

5. The aim is to cut sulphur dioxide emissions by 73% from their current level, nitrogen oxides emissions by 67% and mercury emissions by 69%, all by 2018. Interim emissions limits would be set for 2010. More details can be found at www.whitehouse.gov/news/releases/2002/02.

Results of the Projections

Overview

The Reference Scenario projects an average annual growth rate of 1% in primary energy supply in the United States and Canada from 2000 to 2030 (Table 4.3). Demand will grow most rapidly in the period to 2010, at 1.1% per year, then slow to 1% in 2010-2020 and to 0.8% in 2020-2030. Demand grew by 1.3% per year from 1971 to 2000. The projected deceleration in demand is due mainly to a gradual slowdown in economic and population growth, and to rising energy prices. Saturation effects also put a brake on demand in some sectors.

Table 4.3: **Primary Energy Demand in the United States and Canada**
(Mtoe)

	1971	**2000**	**2010**	**2030**	**Average annual growth 2000-2030 (%)**
Coal	295	572	578	675	0.6
Oil	801	976	1,085	1,316	1.0
Gas	548	620	771	979	1.5
Nuclear	12	227	230	169	-1.0
Hydro	37	52	55	58	0.3
Other renewables	43	104	134	224	2.6
TPES	**1,736**	**2,551**	**2,854**	**3,420**	**1.0**

Oil remains the main primary fuel. Its share in total demand, at 38% in 2000, will not change over the projection period (Figure 4.3). Natural gas use will increase the most, mainly due to surging demand for power generation. Gas's share in total primary energy supply will rise from 24% in 2000 to 29% in 2030. Renewables, including hydroelectricity, will also grow rapidly, but their share in primary energy consumption will reach only 9% in 2030. Nuclear energy supply will decline steadily after 2010, as plants are retired and average capacity factors decline with the ageing of reactors still in use. Although the government may take action to facilitate the building of new reactors, the Reference Scenario assumes that no new nuclear plants will be built on the grounds of competitiveness and public acceptability.

Figure 4.3: **Total Primary Energy Demand in the United States and Canada**

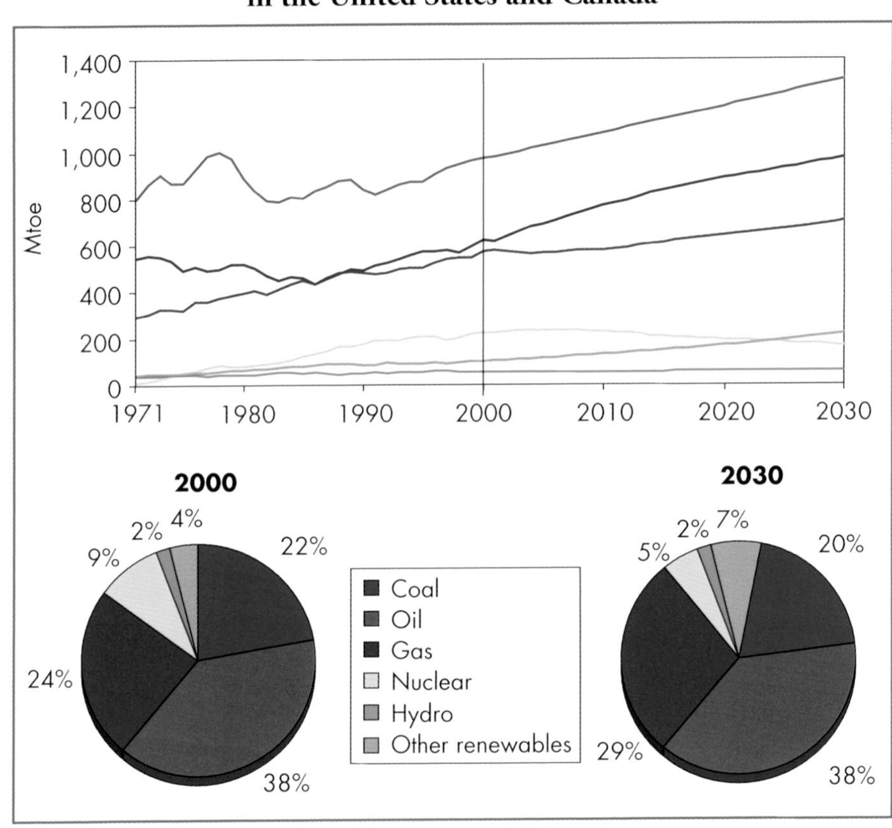

Primary energy intensity – the amount of energy needed to produce a unit of GDP – will continue its downward trend, but at a slower rate than that of the last three decades. The continuing structural shift towards less energy-intensive activities, together with further gains in energy efficiency, is the main reason for this projected decline. The rate of decline will pick up slightly after 2010, because of higher oil and gas prices. Energy use per person, however, is projected to increase slightly because the increase in demand for energy services more than offsets efficiency gains (Figure 4.4).

Demand by End-use Sector

Total final energy consumption will increase slightly faster than primary demand as the efficiency of power stations increases. The fuel mix in final consumption will change very little, with a slight increase in the shares of electricity and renewables and a decline in the shares of coal and

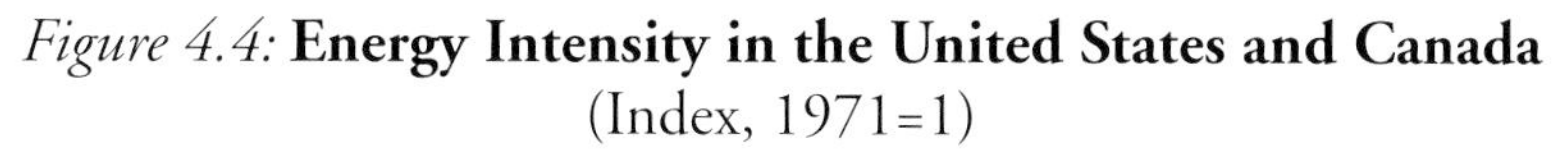

Figure 4.4: **Energy Intensity in the United States and Canada**
(Index, 1971=1)

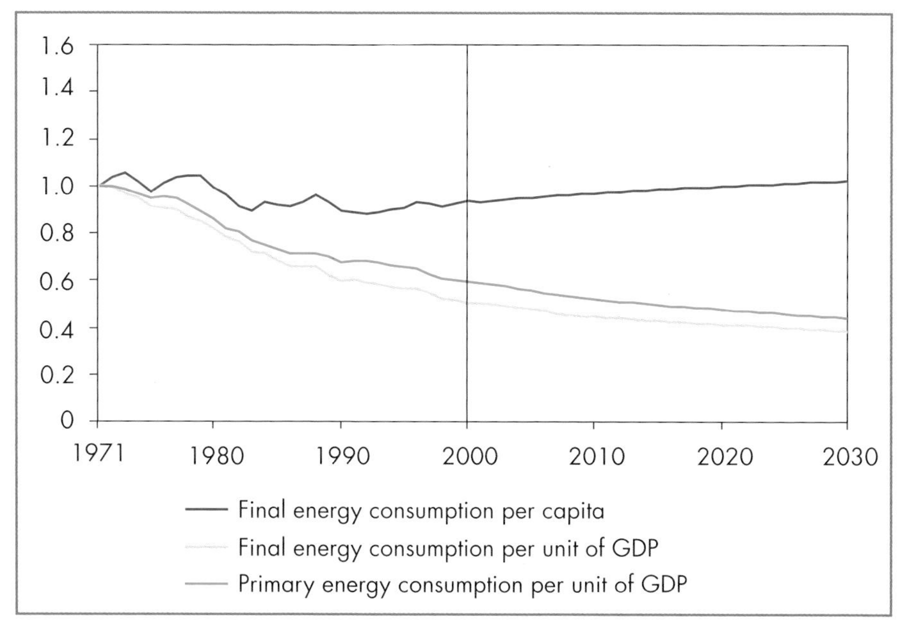

natural gas. The increase in final oil consumption occurs almost entirely in transport.

Transport demand increases most rapidly, by 1.3% per year on average from 2000 to 2030, although this is less than the 1.7% average over the past three decades (Figure 4.5). The share of transport, already the largest end-use sector, increases from 39% of final consumption in 2000 to 42% in 2030. Road freight is the fastest growing transport mode, followed by air travel and road passenger vehicles (Figure 4.6). The fuel efficiency of new heavy and light duty vehicles will improve with the deployment of advanced engine technologies such as direct fuel injection, variable valve timing and electric hybrids of gasoline and diesel engines. But the effect of this factor on fuel demand will be more than offset by rising traffic and the continued shift towards less efficient sports utility vehicles. Although a tightening of Corporate Average Fuel Efficiency (CAFE) standards is under consideration in the United States, this measure is not assumed in the Reference Scenario. Rising pump prices after 2010 will have only a marginal effect on demand growth. These projections assume that breakthrough alternative engine technologies such as hydrogen fuel cells are *not* adopted on a large scale before 2030. Their potential impact on energy use is assessed in Chapter 12.

Figure 4.5: **Total Final Consumption by Sector in the United States and Canada**

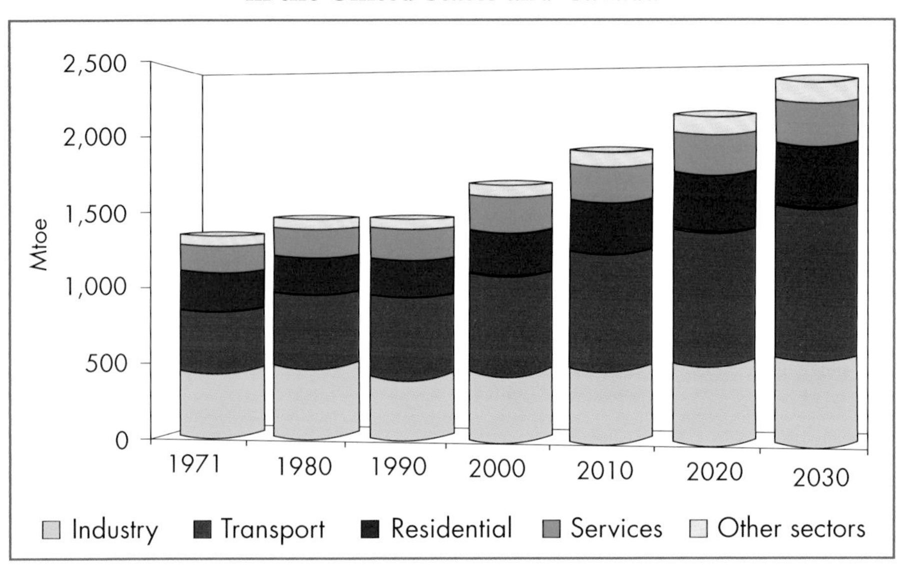

The residential, services and agricultural sectors combined currently account for just under a third of final energy consumption. This share changes little over the projection period. A number of factors are pushing energy use in this sector in different directions. More commercial activity, higher disposable incomes, more households and bigger homes are underpinning increased demand for energy services. But these factors are partly offset by an improvement in the energy efficiency of appliances and of heating and cooling equipment, by the effect of a recent tightening of building codes and by saturation in demand for some types of energy services. Electricity will be the fastest growing fuel in these sectors, at 1.5% per annum over the *Outlook* period, increasing its share to half of total final energy consumption, driven largely by growing demand for appliances and electronic equipment.

Industry's share of final consumption will continue to decline in the Reference Scenario, but less rapidly than in the past, from 25% in 2000 to around 23% in 2030. Industrial consumption of electricity increases most rapidly – by just over half from 2000 to 2030 – reflecting a shift to less energy-intensive specialised processing and manufacturing sectors, which use less process heat but more electricity. Coal use in industry declines, with the adoption of new steelmaking processes and a continuing preference for natural gas in boilers. Energy consumption is projected to grow most rapidly in the less energy-intensive industrial sectors.

Figure 4.6: **Transport Fuel Consumption in the United States and Canada by Mode**

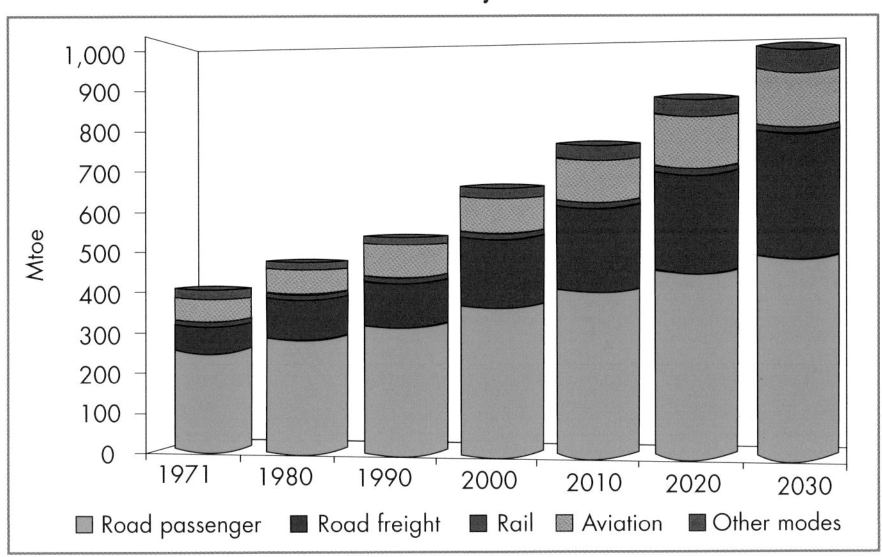

Oil

North American primary oil consumption grows by 1% per year on average in the Reference Scenario, from 20.2 mb/d in 2000 to 27.3 mb/d in 2030. As in most other regions, the bulk of the increase in demand is for transportation. Imports, which now meet around 46% of the region's needs, continue to grow as indigenous production stabilises (Figure 4.7).

Indigenous production of crude oil and natural gas liquids (NGLs) will be broadly flat over the projection period. Onshore production in the 48 continental states, the most mature oil-producing region in the world, is likely to fall rapidly, as drilling and new discoveries decline and production costs rise. The continued deployment of new technology will compensate for these factors to some extent. On the other hand, production in Canada and offshore Gulf of Mexico is expected to increase at least over the next two decades. Production of raw bitumen and synthetic crude oil from the oil sands of Alberta in western Canada reached 650 kb/d in 2001, and is set

Figure 4.7: **Oil Balance in the United States and Canada**

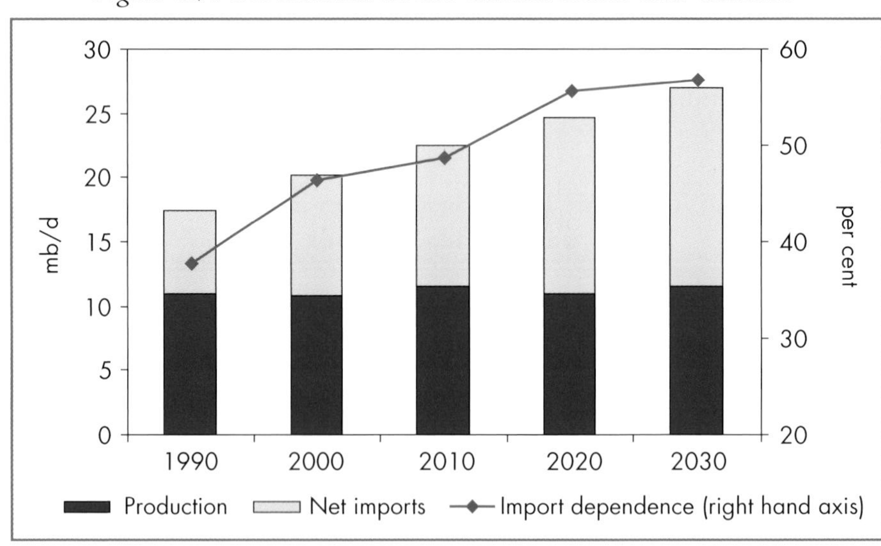

to grow very strongly. The National Energy Policy recommends a number of actions to boost US oil and gas production in the long term (Box 4.2).

Box 4.2: **Main Recommendations for Oil and Gas of the US National Energy Policy**

- Review land status and expedite the current study of barriers to oil and gas leasing for exploration and production on Federal lands.
- Consider new incentives for oil and gas development, such as royalty reductions for enhanced recovery, reduction of risk for frontier areas or deep gas formations and for the development of small fields that would otherwise be uneconomic.
- Review the regulation of energy-related activities and the siting of facilities in coastal areas and on the Outer Continental Shelf and make leasing and development approvals more predictable.
- Consider further lease sales in the Alaskan National Petroleum Reserve, including areas in the north-east corner not currently leased.
- Authorise exploration and possible development of the 1002 Area of the Arctic National Wildlife Reserve.

The United States is the world's largest oil-importing country. Gross imports of crude oil and refined products reached a record 11.5 mb/d in 2000 – or 53% of total US consumption. Import dependence for North America overall is a little lower, at 46%, because Canada is a net oil exporter. Net imports into the region are projected to rise to 10.9 mb/d by 2010 (when they will meet 49% of demand) and to 15.5 mb/d by 2030 (57% of demand). Much of this additional oil will undoubtedly come from OPEC countries, especially the Middle East. Saudi Arabia was the single biggest external supplier to the US and Canada in 2000, exporting almost 1.6 mb/d to the United States. Refined products are likely to account for a growing share of oil imports, most of which are currently in the form of crude oil, as little new refinery capacity will be built in the United States and Canada over the projection period.

Natural Gas

The region's primary supply of gas is projected to grow by an average 1.5% per year from 2000 to 2030. The biggest increase in gas use is expected to come from power generation, especially in the period from now to 2010. Most new power plants built during that period are expected to be gas-fired. Gas demand will also increase in the residential, services and industrial sectors, but at more pedestrian rates (Figure 4.8).

Figure 4.8: **Natural Gas Demand in the United States and Canada**

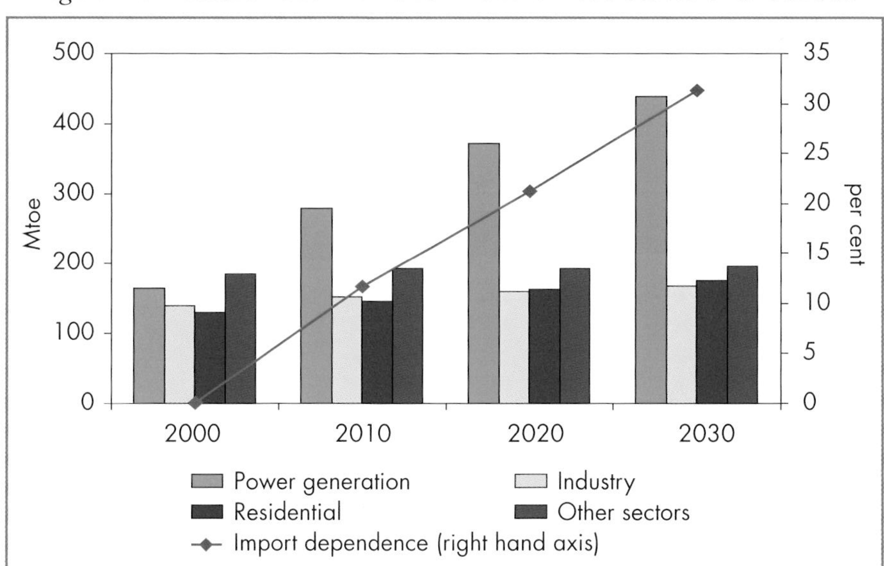

There is much uncertainty over gas production prospects in the US and Canada. Proven reserves are modest, at 6.55 trillion cubic metres (4.85 tcm in the United States) or 4% of the world total.[6] The cost of production at existing basins, concentrated in the southern and central United States and in western Canada, is likely to rise as mature fields are depleted. The surge in drilling induced by high gas prices in 2000 and 2001 brought only meagre increases in production and the natural rate of decline in output from new wells once they are brought onstream appears to have accelerated markedly.[7] Higher output in the future will require even more drilling in established producing basins in the continental US states and Canada, as well as new greenfield projects. If environmental concerns can be met, new indigenous supplies could come from the following sources:[8]

- *Current producing basins:* Attention is expected to shift to deeper water sites in the Gulf of Mexico.
- *New conventional gas basins:* These include offshore Atlantic basins in Labrador, Newfoundland and Nova Scotia (where small-scale production started in 2000) and in the Mackenzie River Delta and Beaufort Sea region in northern Canada. Development of frontier regions such as the Arctic Islands and the Northwest Territories will depend on further advances in drilling technology in extreme weather conditions.
- *Unconventional resources:* Large volumes of unconventional gas could be tapped from new basins in the United States and Canada, although generally at higher cost than conventional gas. Unconventional reserves include tight-formation and shale gas deposits, and coalbed methane, large reserves of which are found in the Rocky Mountains. Production from these sources is expected at least partly to offset declines in conventional output elsewhere.
- *Alaskan gas:* LNG and pipeline projects based on the large gas reserves on the Alaskan North Slope, largely unexploited so far, are now under consideration. A 35-bcm/year high-pressure pipeline has been proposed to link North Slope reserves to the US market

6. At 1 January 2001. Cedigaz (2001).
7. IEA (2001) discusses the relationship between drilling and production rates. See also Simmons (2002).
8. There is a detailed discussion of medium-term production prospects in IEA (2001). See also DOE/EIA (2001b).

via western Canada, but the recent fall in gas prices and the high cost of building the line have dampened interest in the project.

Aggregate production in the United States and Canada is projected to climb slowly from 732 billion cubic metres in 2001 to 823 bcm in 2010 before beginning to decline around 2020, to 812 bcm in 2030. With demand projected to grow beyond this, imports of LNG will play a growing role in North American gas supply in the long term. Capacity expansions at the four existing facilities on the Gulf and East Coasts and potential investment in *new* LNG terminals in the United States, or neighbouring countries, will depend on prices and project-development costs. The rising cost of gas from domestic sources, together with continuing reductions in the costs of LNG supply, is expected to boost US LNG imports, directly or via Mexico. There may also be potential for large-scale imports of Mexican gas by pipeline into the United States. But this will probably not happen before 2020, as Mexico is expected to meet its domestic needs first. Net imports of gas into the United States and Canada are projected to reach 109 bcm in 2010 and 371 bcm in 2030. This particular projection is subject to a very high degree of uncertainty. The increase in imports would be much less pronounced if production holds up better or demand rises less rapidly than expected. New policies to promote switching to other fuels and to curb gas demand, not taken into account in the Reference Scenario, could also reduce gas-import needs.

If producers are to drill more wells and marketers are to import more LNG, wellhead prices will have to rise to the point where they provide an incentive for doing so. The Reference Scenario assumes that import prices remain flat at around $2.50/MBtu (in 2000 dollars) from 2003 to 2005 and then begin to rise gradually, reaching $2.70 by 2010 and $4.00/MBtu by 2030.

Substantial investment will be needed in new transmission and distribution capacity in North America, including new lines to bring gas from Canada into the United States. How much new capacity will be needed will depend on several factors, including the load factor[9] of incremental demand (higher load factors require less capacity), the location of new supplies and trends in regional demand. Pipeline capacity is inadequate in some parts of the country, including California. Since much of the increase in demand will come from the power sector, the need for increased transmission capacity is likely to be proportionately less than the

9. Average daily system throughput (or consumption) divided by peak daily throughput.

overall growth in the market. This is because the load factor of power-sector demand is generally higher than average. Most expansion projects over the next three decades are likely to involve looping and added compression; so incremental capacity costs per km will probably be lower than in the past. But increasing reliance on gas located far from major markets – in Alaska, northern Canada and Nova Scotia/Newfoundland – could increase the need for new transmission capacity and the size of the required investment in the longer term.

Coal

Primary coal demand will increase by 0.6% per annum to 2030. Power generation, which already accounts for more than 90% of the region's coal use, will drive the increase (Figure 4.9). Demand will grow more rapidly after 2010, when coal becomes more competitive against natural gas in power generation. In most cases, though, gas will remain the preferred fuel in new power plants, because of the lower capital costs involved and the high cost of meeting emission limits on new coal-fired plant. Any tightening of current environmental restrictions on coal use, including higher penalties for sulphur and nitrogen oxide emissions, would lead to lower coal demand. But new clean coal technologies, if they can be made competitive, could boost coal demand: the National Energy Policy calls for $2 billion for research in this area over the next ten years.

The United States and Canada have enormous coal resources and are both expected to remain major producers and exporters of hard coal throughout the projection period. Proven reserves of hard coal and lignite in the United States alone amount to 250 billion tonnes – more than a quarter of the world total. Productivity in the US coal industry has improved by 6.7% per year on average since 1979, the result of advanced technology, economies of scale, improved mine design and management, and the closure of high-cost mines.[10] This trend is expected to continue in the future. In 2000, net exports amounted to around 5% of production in the United States – the world's second-largest producer after China – and just over a quarter of Canadian output. US exports have nonetheless fallen sharply in recent years because of competition from other countries.

10. DOE/EIA (2001a).

Figure 4.9: **Change in Primary Coal Demand in the United States and Canada**

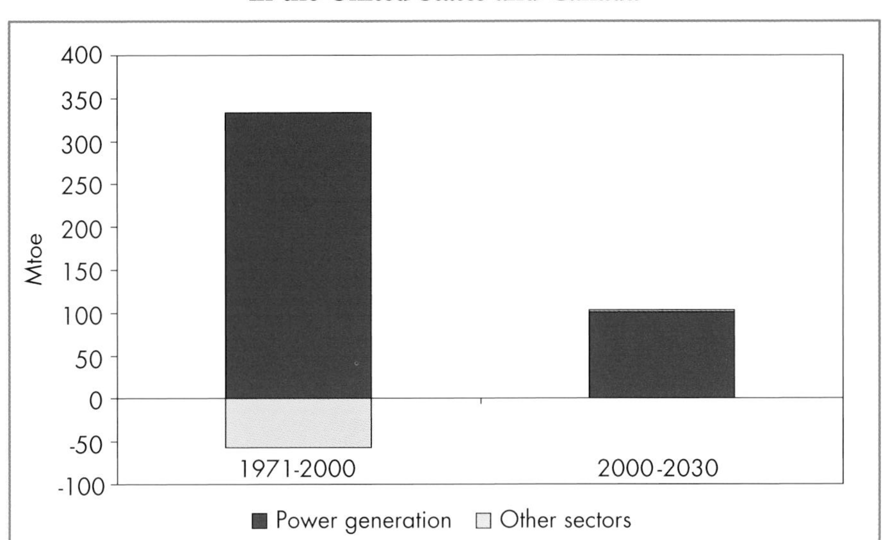

Electricity

Electricity demand in the United States and Canada is projected to grow at 1.5% per year from 2000 to 2030. Demand growth will slow progressively over the projection period as GDP decelerates.

Coal, which provided just under half the power produced in the United States and Canada in 2000, will remain the most important fuel in power generation in 2030. The absolute amount of coal-based output increases, but its *share* declines over the projection period, to 41% in 2030, because of faster growth in gas-fired generation (Table 4.4). Existing plants will provide most of the rise in coal-fired output in the period to 2010, but more new coal-fired plants will be built thereafter, as gas prices increase. In the absence of tighter environmental regulations, most new plants will use conventional coal technology. Some advanced technology plants, such as those using integrated gasification combined-cycle gas turbines, will be built towards the end of the projection period on the assumption that their capital costs are reduced.

Table 4.4: **Electricity Generation Mix in the United States and Canada** (TWh)

	1990	2000	2010	2030	Average annual growth 2000-2030 (%)
Coal	1,782	2,228	2,240	2,882	0.9
Oil	147	140	136	99	-1.1
Gas	391	664	1,398	2,123	4.0
Hydrogen-fuel cell	0	0	0	122	-
Nuclear	685	873	882	647	-1.0
Hydro	570	607	644	670	0.3
Other renewables	89	98	165	473	5.4
Total	**3,664**	**4,609**	**5,464**	**7,016**	**1.4**

More than half of all the new generating capacity brought on line over the projection period will be natural gas-fired (Figure 4.10). The share of gas in new capacity will be even higher in the period from now to 2010. Over 90% of capacity already under construction or planned to be built in the next ten years is gas-fired. Consequently, the share of gas in total

Figure 4.10: **Power Generation Capacity Additions in the United States and Canada, 2000-2030**

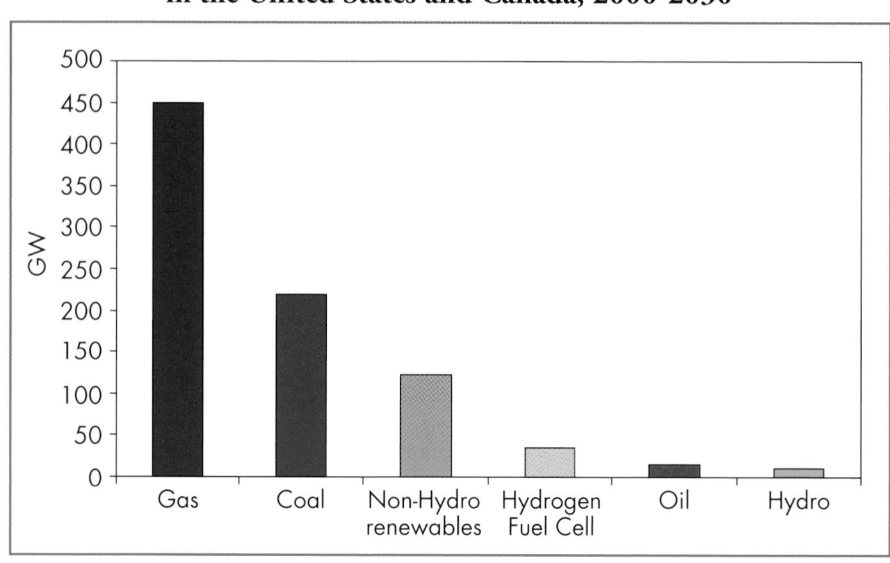

generation will rise from 14% in 2000 to 26% in 2010 and to 30% in 2030.

Nuclear generation will decline because it is assumed that no new plants are built and some existing ones are retired. Most existing nuclear plants will still be operating in 2030, because it is assumed that plant licences, virtually all of which expire before that date will be renewed. Regardless of the public acceptability of nuclear power, it is not expected to be competitive with other types of capacity over the projection period. The upgrading of some nuclear plants should, nonetheless, help to offset the impact of plant retirements. Six closed reactors in Ontario, four at Pickering A and two at Bruce A, are assumed to reopen soon.

A few new hydroelectric plants will be built in Canada, giving a modest boost to total hydro output in the region, although its share in total generation will decline. Non-hydro renewables share of generation will grow rapidly as their costs fall, but their overall contribution will still be small in 2030. Most of the increase will occur in the United States, where Federal and state incentives promote a large increase in investment in new wind and biomass capacity. Wind becomes increasingly competitive with fossil fuels in areas where wind speeds are high and regular, and where markets are close and easily accessible. New gasification technologies will probably boost the development of biomass projects. There is also significant potential for installing photovoltaic panels in buildings in the sunniest parts of the United States.

Distributed generation, based largely on natural gas, is projected to expand quickly. Fuel cells, producing electricity from steam-reformed natural gas, are expected to grow in importance after 2020.

Energy-related CO_2 Emissions

The projected increase in fossil fuel consumption will inevitably result in higher emissions of CO_2. The biggest increase in emissions will come from power generation, already the largest emitter, followed by transport (Figure 4.11). The declining shares of nuclear power and hydroelectricity in the generation fuel mix and the growing competitiveness of coal after 2010 will contribute to this trend.

The *carbon intensity* of both the US and Canadian economies, measured by the amount of carbon emitted per unit of GDP, is projected to continue to fall, by about 13% between 2000 and 2010. This compares to the 18% reduction that the Bush Administration has targeted for the *greenhouse gas intensity* of the United States over the period from 2002 to

Figure 4.11: **CO_2 Emissions and Intensity in the United States and Canada**

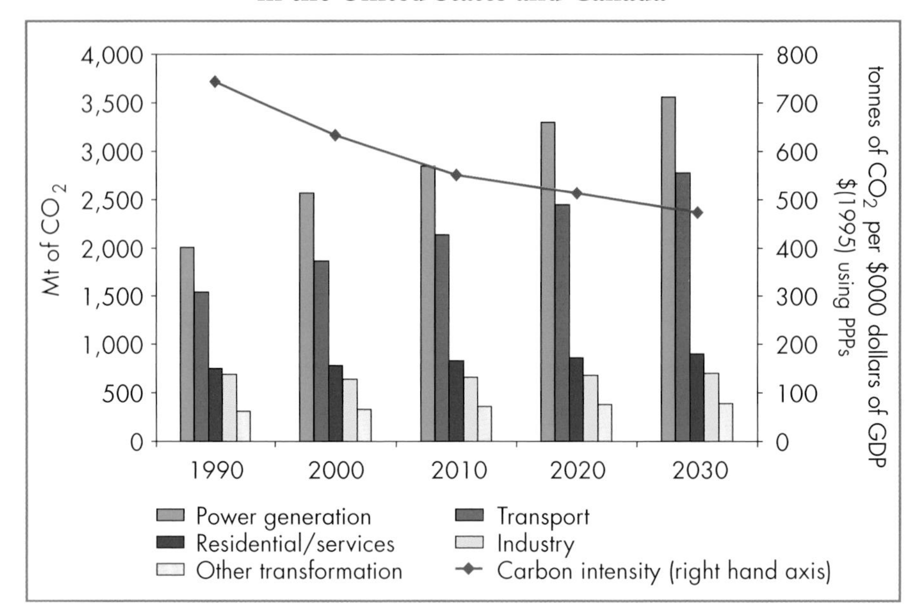

2012. The fall in *carbon* intensity is due almost entirely to the projected drop in *energy* intensity. Carbon intensity will continue to decline after 2010, but more slowly than before. This is mainly because nuclear power production will decline and demand for natural gas, which is less carbon-intensive than coal and oil, will slow.

Mexico

Energy Market Overview

Mexico is the world's twelfth-largest economy and eighth-largest exporter of goods and services. In 2000, the oil sector provided 3% of Mexican GDP, 8% of total exports by value and 37% of total government revenues.[11] Mexico's main trading partner is the United States, which took 90% of its energy exports and provided more than 70% of its energy imports in 2000 (Box 4.3). Trading links with the United States have increased since the North American Free Trade Agreement came into force in 1994. Mexico is now making efforts to diversify its export base by

11. Mexican Ministry of Energy (2001a).

negotiating other free trade agreements. By the end of 2001, it had signed such agreements with 31 countries.

Mexico is well endowed with oil and gas resources. It is one of the largest oil producers in the world. With almost 24 billion barrels, Mexico has the third-largest proven oil reserves in the Americas, after the United States and Venezuela. Its 940 bcm of natural gas reserves are the Americas' fourth-largest after the United States, Venezuela and Canada.

Mexico accounts for about 5% of total primary energy supply in OECD North America. In 2000, 65% of Mexico's primary supply came from oil and 23% from natural gas. Per capita energy demand is a little higher than the average for Latin America, but is much lower than in the United States and Canada. Oil production (including NGLs) averaged 3.6 mb/d in 2001, accounting for a quarter of OECD North American production.

Box 4.3: **Mexico's Energy Trade with the United States**

Mexico trades large amounts of energy, mainly with the United States:

- In 2001, Pemex, Mexico's state-owned oil and gas company, exported 1.71 mb/d of crude oil, worth $11.6 billion. Around three-quarters went to the United States and another 10% to Central and Latin America. Mexico's oil exports to the US have been growing in recent years, and Mexico is now the fourth-largest supplier of the US market, behind Saudi Arabia, Canada and Venezuela.
- Mexico has become a net importer of oil products because of a lack of domestic refining capacity. In 2001, net imports of gasoline averaged 60,000 b/d, mostly from the United States. Mexico also imported 97,000 b/d of LPG and 81,000 b/d of heavy fuel oil.
- Mexico's imports of natural gas from the US totalled 3.7 bcm in 2001 — almost entirely via pipeline. Exports in the other direction amounted to less than 250 mcm.
- Mexico is a small net importer of electricity from the United States. Net purchases to supply markets in the north of Mexico, where generating capacity is inadequate to meet demand, amounted to 524 GWh in 1999. There are nine interconnectors between the two countries.

Table 4.5: **Comparative Energy Indicators, 2000**

	Mexico	**OECD**	**World**
GDP (in billion 1995 $, PPP)	813	24,559	41,609
GDP per capita (in 1995 $, PPP)	8,219	21,997	6,908
Population (million)	99	1,116	6,023
TPES (Mtoe)	144	5,291	9,179
TPES/GDP*	0.2	0.2	0.2
Energy production/TPES	1.6	0.7	-
TPES per capita (toe)	1.5	4.7	1.5

*Toe per thousand dollars of GDP, measured in PPP terms at 1995 prices.

The energy sector in Mexico is highly regulated and politicised. It is dominated by two vertically integrated public companies: the oil and gas company, Pemex, and the Federal Electricity Commission (CFE). The Mexican Energy Regulatory Commission (CRE) regulates the electricity and natural gas industries, including pricing in the natural gas industry, licensing of transport and gas distribution. CRE also regulates the licensing of the private power generation and sales by private generators to the CFE.

Private investment is small and is limited to downstream gas and electricity generation. Mexico's government is planning structural reforms that will allow greater private participation in the energy sector, because the public sector will not be able to pay for the needed upgrading and expanding of Mexico's supply capacity (Box 4.4). Enticing private investment in oil and gas production will, however, require some legal changes. Under its constitution, only state companies can explore for and produce hydrocarbons in Mexico. Under the Regulatory Act, exploration, production and the initial sale of oil and gas are designated as strategic activities that can be handled only by public companies. The state has exclusive rights over the transmission, distribution and marketing of electricity, which are regarded as public service activities. The private sector *may* already take part in construction, operation, transportation, storage and final sales, including exports.

Box 4.4: Financing Mexico's Energy Industry

Mexico urgently needs investment to meet rising energy demand and to modernise its ageing energy infrastructure. The Ministry of Energy estimates that in the period 2000 to 2009, the country will require $139 billion — $40 billion for the exploration and production of crude oil, $19 billion for oil refining, $21 billion for natural gas production and distribution, and $59 billion for the electricity sector.[12] Over the next five years, Pemex needs $33 billion in new financing to revive its stalled oil and gas exploration programme. If these investments are not made, Mexico's crude oil production could plunge by a third. Because of other pressing social and economic needs, the public sector will not be able to provide all this funding.

Current Trends and Key Assumptions

Macroeconomic Context

After five years of strong growth, Mexican economic conditions began to deteriorate in late 2000, following the sharp slowdown in the United States. GDP fell slightly in 2001. Investment and employment growth dropped and consumer confidence wavered. The economic cycles in Mexico and the United States have become more synchronised with the integration of Mexico into NAFTA. In contrast to previous downturns, Mexican financial markets and capital inflows have remained stable.

In the past 20 years, Mexico experienced contrasting cycles of activity, with phases of dynamic growth interrupted by economic crises and recessions, such as the 1982-1983 debt crisis and the 1994-1995 peso crisis. Real GDP grew by about 2% a year between 1981 and 2000, just keeping pace with population growth. Economic growth has picked up in recent years, averaging around 3.5% a year from 1994 to 2000. Inflation has also been brought under control, down to 4.4% by the end of 2001. Fiscal and monetary discipline has contributed to macroeconomic stability and provided a platform for economic growth. In addition, the floating

12. Mexican Ministry of Energy (2001a).

exchange rate adopted after the 1994 crisis has helped the Mexican economy to absorb external shocks.[13]

The modernisation of the Mexican economy is reflected in its shift towards service activities. Services now account for 68% of Mexico's GDP, industry for 28% and agriculture for 4%.[14] Mexican labour productivity is, however, still less than a third of that in the United States. Despite strong growth in employment and considerable investment in manufacturing and services, job creation has not been rapid enough to reduce the size of the informal or "grey" economy. Geographical differences in output are large, resulting in large regional disparities in household incomes. In the northern states, where many export-oriented activities are located, per capita GDP has grown at an annual rate of about 1% since 1994. In the southern states, it has *fallen* by 1%.

Mexico's economy is expected to recover in 2002. The Ministry of Finance predicts growth of around 1.7%. Buoyant private consumption and investment, underpinned by lower interest rates and by increased government spending, are expected to drive the recovery.[15] The *Outlook* assumes that the economy will grow at an average rate of 3.4% per year from 2000 to 2030. GDP growth will average 3.4% from 2000 to 2010. It will rise to 3.6% in the second decade of the projection period and then fall back to 3.1% in the third decade. The population is assumed to increase by just over 1% per year, reaching 135 million by 2030.

Energy Sector Reforms and Prices

In 2000, the government announced plans to restructure the energy sector and reform its legal and institutional framework to boost efficiency and investment. The reforms will grant operational and financial autonomy to the public energy companies and lower their high tax burdens to enable them to reinvest.[16] The role of the private sector is to be enlarged. The government envisages an expanded role for private investors in power generation, non-associated gas production, gas processing and distribution, petrochemicals and renewable energy. This will require changes to the Mexican Constitution. The government does not, however, plan to privatise the existing state companies — an action that would meet strong political opposition and public resistance.

13. OECD (2002).
14. World Bank (2001).
15. Mexican Ministry of Finance and Public Credit (2002).
16. In 2000, Pemex had 274 billion pesos of taxable income, but paid *294 billion pesos* in federal taxes.

The main elements of the planned reforms are:

- Private oil and gas companies will be invited to tender bids to develop non-associated gas fields on behalf of Pemex under long-term service contracts.
- Private and foreign companies will be allowed to participate in building and operating LNG import terminals.
- The current restriction on majority private ownership of basic petrochemical assets, all of which are in the hands of Pemex, will be lifted, although a restriction may be placed on *foreign* ownership.
- Independent power producers will be allowed to sell directly to end-users any power not bought by the Federal Electricity Commission.
- Pemex's gasoline stations will be allowed to sell all brands of motor oil, not just the Pemex brand, Mexlub.
- International companies will be allowed to invest in LPG distribution and marketing.
- Energy subsidies will be phased out, and all fuels will be priced on the basis of full supply costs.

It is uncertain whether these reforms will be adopted. The government does not hold a majority in Congress, and its initial attempts to push through legislation have come up against stiff opposition. Major reforms are likely to be postponed until after mid-term legislative elections in 2003. The projections here assume that these reforms *will* eventually be implemented, but only very gradually. Once final energy prices rise to the world market level, they are assumed to follow international trends.[17]

Results of the Projections

Overview

Mexico's total primary energy demand will increase by 2.5% per annum over the *Outlook* period. This is well below the 4.6% rate of increase from 1971 to 2000. Per capita consumption will rise strongly from 1.5 toe in 2000 to 2.3 toe in 2030, but it will still be only 26% of the average for the rest of OECD North America. Energy intensity, which increased marginally over the last three decades, will fall by an average of almost 1% a year from 2000 to 2030.

17. See Chapter 2 for international price assumptions.

The primary fuel mix will change markedly over the *Outlook* period. Oil's share in primary demand will drop sharply, from 65% in 2000 to 55% in 2030. Natural gas grows fastest, by 3.9% per year. Driven by a 5.4% increase in power-sector demand, the share of gas in primary demand jumps from 23% now to 34% by 2030. Coal demand, which will grow by 1.3% per annum, will contribute proportionately less to TPES in 2030 (Figure 4.12).

Figure 4.12: **Total Primary Energy Demand in Mexico**

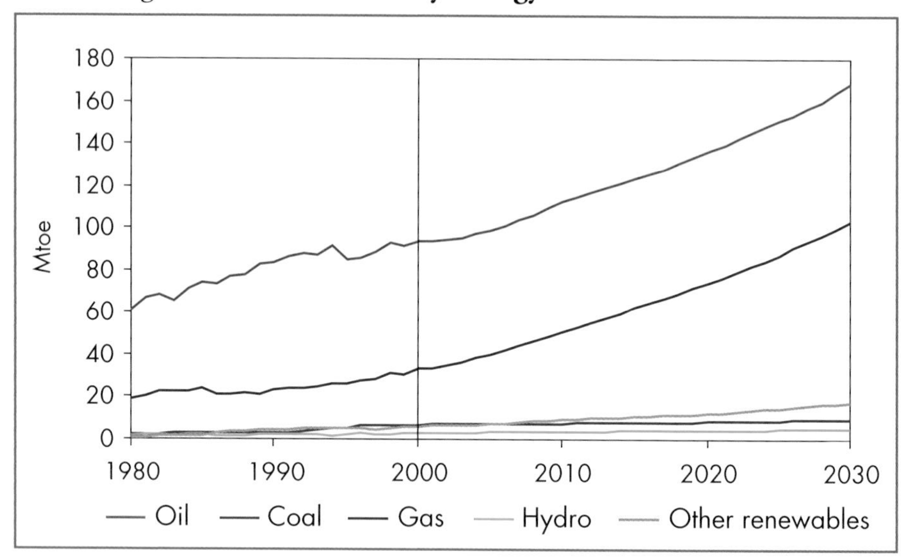

Trends in Sectoral Demand

Final energy consumption, excluding biomass, is projected to increase by 2.7% per year from 2000 to 2030, led by 3% growth in transport demand (Figure 4.13).[18] Gasoline used in private cars will account for most of this increase, but there will be some switching from gasoline to LPG (autogas). This assumes that LPG continues to be taxed less than gasoline for environmental reasons. Although industry's share in final consumption remains high in 2030 at 33%, it nonetheless falls slightly over the projection period. Natural gas and electricity account for virtually all the increase in industrial consumption. Energy demand in the residential, services and agricultural sectors combined rises by 2.6% per year. Natural

18. Biomass demand in Mexico was 7 Mtoe in 2000, some 7% of total final consumption. This share will be negligible in 2030.

Figure 4.13: **Total Final Consumption by Sector in Mexico**

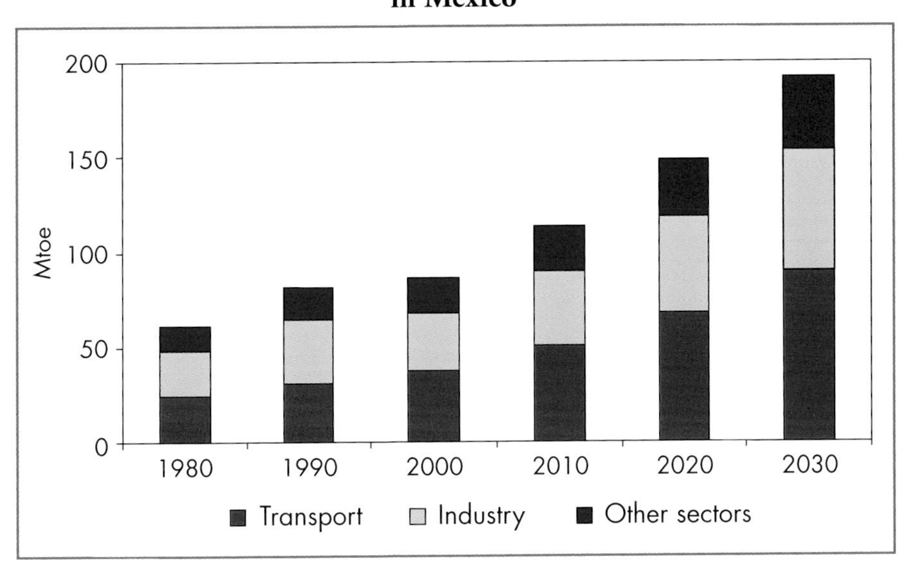

Note: "Other Sectors" include residential, commercial, agriculture and non-energy use.

gas consumption in these sectors rises most rapidly, as it gradually takes the place of LPG.

Oil

Primary oil consumption is projected to rise by 2% per year over the projection period, reaching 3.5 mb/d by 2030. As in every other region, the transport sector accounts for most of this growth. Although oil will remain Mexico's dominant fuel, its share in total demand will drop by ten percentage points.

Mexico has important oil resources and is a major producer. Although not a member of OPEC, it regularly attends that organisation's meetings as an observer. It has repeatedly participated in agreements to limit production to help support crude oil prices. In 2001, the country produced about 3.5 mb/d of crude, of which net exports were about 1.6 mb/d. The heavy Maya crude blend makes up more than two-thirds of exports by volume. This share has risen in recent years, and the oil produced by Mexican has grown heavier on average.

At the start of 2001, Mexico's proven crude oil reserves, excluding condensates and NGLs, stood at 23.7 billion barrels, according to official data.[19] But reserves have declined steadily over the past 17 years, reflecting a sharp contraction in exploratory activity due to financial and technological constraints. In 1998, Pemex introduced a new reserves-accounting methodology to bring its estimates into line with international practice. This change reduced official proven reserves by 40%. At the current rate of production, Mexico's reserves would last 21 years.

About three-quarters of Mexican oil production comes from Campeche Bay in the Gulf of Mexico, notably from the Cantarell fields . At the end of 2001, Cantarell was producing a little less than 1.9 mb/d. That figure is expected to rise to 2.2 mb/d by 2007.[20] Remaining reserves in the four fields that make up the Cantarell complex are estimated at 14 billion barrels.

Figure 4.14: **Oil Balance in Mexico**

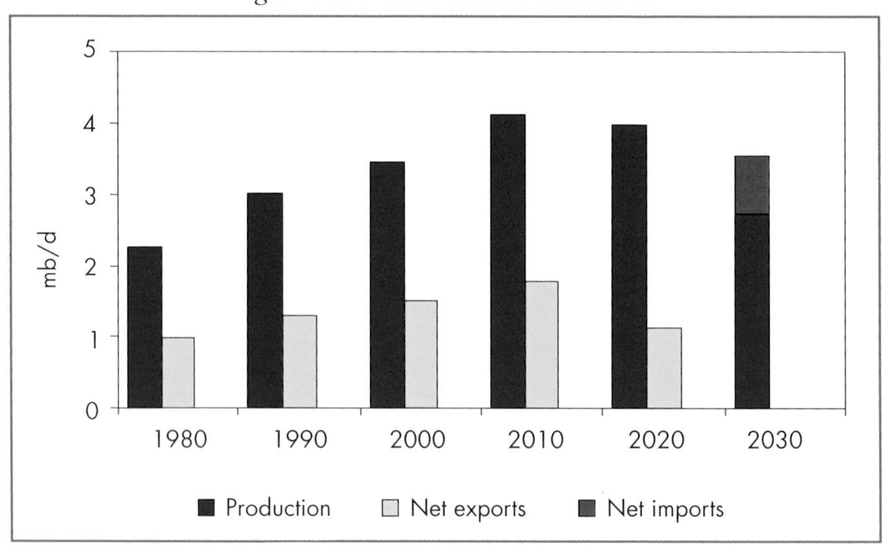

Pemex, the state-owned company, was created in 1938 when the Mexican oil industry was nationalised. It is by far the largest company in Mexico and is regarded as a symbol of national sovereignty. The company, the world's third-largest oil producer, has monopoly rights over oil and natural gas exploration and production in Mexico. Legislation introduced in the 1980s permits private investment in the petrochemical industry. But, despite the pressing need for new investment, the government has no

19. Mexican Ministry of Energy (2001a).
20. Baker (2002).

plans to allow private investors into the upstream oil sector. Private participation in refining is precluded by law.

Mexican crude oil production is projected to peak at 4.1 mb/d around 2010. Production will remain flat for about a decade, and then decline sharply, reaching 2.7 mb/d in 2030. New discoveries will not compensate for the decline in production from the large mature fields, such as Cantarell. Net exports of crude oil and products are expected to decline even more quickly than production, as domestic demand will continue to grow. By the third decade of the *Outlook* period, Mexico will become a net importer of crude oil (Figure 4.14).

Gas

Gas use in Mexico has expanded rapidly in the last decade, reaching 39 bcm in 2000 — or 23% of TPES. About a third of natural gas is reinjected into petroleum reservoirs to enhance oil production. Power generation is the next biggest use for gas. Most of the rest is consumed in industry.

Primary demand for gas is expected to grow by 3.9% a year over the *Outlook* period, to 122 bcm in 2030. The bulk of additional demand will come from power generation, followed by industry and the upstream oil and gas sector (Figure 4.15).[21] Switching to gas in power generation and industry will be impelled by environmental regulations that discourage fuel oil and coal use. Increasing volumes of gas will be needed for re-injection to help mitigate the projected long-term decline in oil production. Demand for gas in the residential and services sectors will increase rapidly as the gas-distribution network expands, but this market remains small compared to the other sectors.

In the long term, Mexico will probably be able to meet most of the increase in gas demand from its own resources. At the start of 2001, Mexico's proven natural gas reserves stood at 940 bcm, more than 80% of which were associated with oil reserves.[22] More than half the reserves are located in the north of Mexico. The US Geological Survey estimates undiscovered resources at 1.4 tcm.[23] In 2000, Mexico produced some 37 bcm of gas, nearly three-quarters associated with oil production. The

21. According to a government projection, the power sector will account for 57% of incremental gas demand over the period 2001 to 2010. Its share in total gas demand will rise from 16% in 2000 to 42% in 2010 [Mexican Ministry of Energy (2001b)].
22. Mexican Ministry of Energy (2001a).
23. USGS (2000). The estimate is for 1 January 1996.

Figure 4.15: **Change in Primary Natural Gas Demand by Sector in Mexico**

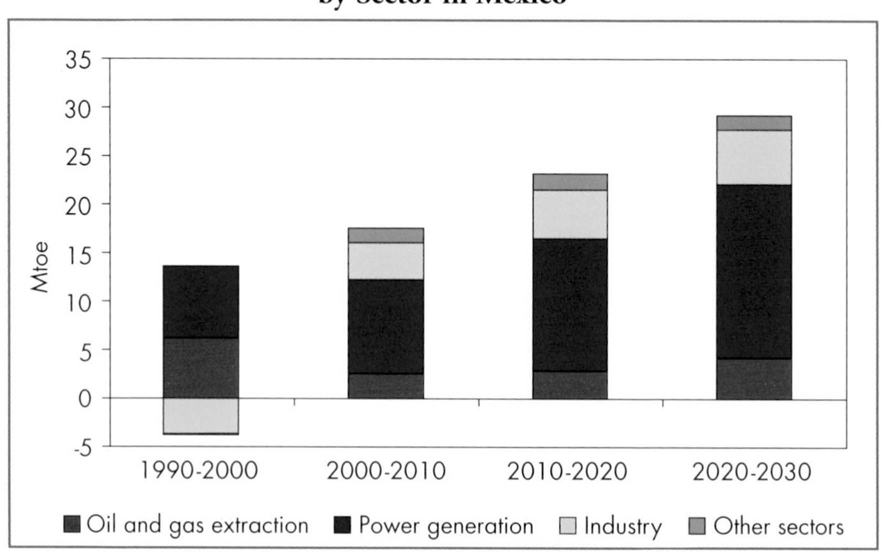

Burgos field in north-eastern Mexico contains large volumes of non-associated gas. Production at Burgos has increased at an annual rate of 27% since 1997, reaching 10 bcm in 2000. It is estimated that the field has the potential to double its current production. In addition, the Cantarell field holds a large volume of associated gas.

Despite Mexico's ample gas reserves, production has failed to keep pace with demand in recent years, because of inadequate upstream facilities and a lack of pipelines connecting associated gas fields to centres of demand. Consequently, Mexico has become a small net importer of US gas. The removal of import duties on gas from the United States in 1999 has strengthened this development.

The Mexican government is seeking to encourage investment in the gas industry. It plans to open up production of non-associated gas to private investment. The introduction of multiple service contracts (MSCs), announced in December 2001, would be a first step in that direction. MSCs would allow private firms to provide a range of services to Pemex, which would retain the rights over any gas produced and the profits from its sales. Pemex plans to offer MSCs first at the Burgos field. Their introduction will require a modification of the laws now governing the oil sector, but probably not a change in the constitution.

Natural decline rates are apparently high, especially at the non-associated gas fields where most of the projected increase in production is to occur. Major investment will be needed just to maintain current production. Mexico will, therefore, have to import some gas in the near term. The gas will either be piped over the border from the United States or be shipped in from West Africa or Latin America in the form of liquefied natural gas. At least four LNG receiving terminals, one on the Gulf of Mexico and three on the Pacific Coast, are now under consideration, although it is unlikely that all four will be built. Each terminal would have an annual capacity of up to 10 bcm. Some of the new capacity on the Pacific Coast could be earmarked for export to western US markets.

In the medium term, the introduction of MSCs will allow the more rapid development of Mexico's large gas resources. Domestic demand growth and the proximity of the expanding US market will provide an incentive to raise production. Mexico's gas production is projected to grow from 37 bcm in 2000 to 148 bcm in 2030. Indigenous production will gradually outstrip domestic consumption, and, in the second decade of the *Outlook* period, Mexico will become a net exporter. Mexican gas exports will rise to 11 bcm in 2020 and to 26 bcm in 2030.

The North American gas market is becoming increasingly integrated through new pipeline interconnections between Mexico, the United States and Canada (Box 4.5). This has led to a degree of convergence in prices across the region.

Box 4.5: **Development of Mexico's Natural Gas Infrastructure**

Mexico's natural gas network covers a large part of the country. It crosses 18 states, from Cactus in the south to Los Ramones in the north-east (Figure 4.16). Most of the network is owned and operated by Pemex, but several foreign gas companies have built transmission lines and distribution grids in recent years. There are eleven interconnections with the US gas network, including the recently completed 3-bcm per year Kinder Morgan line at Argüelles. The construction of an additional 5-bcm of new cross-border capacity is under consideration, as well as expansions to compression stations. Several other projects are planned, including a $230-million, 341-km line in North Baja connecting south-eastern California to Tijuana.

A major expansion of the national gas network will be needed to cope with rising traffic. According to recent studies by the Mexican

Energy Regulatory Commission (CRE), between $15 and $18 billion must be invested if supply is to meet the 66 bcm of gas demand it projects for 2010. Extra gas processing capacity, new pipelines from the United States and LNG terminals will be needed, as well as an entire sub-system in the Rio Bravo area to link Texas to a cluster of power plants in north-eastern Mexico. New pipelines connecting the Gulf of Mexico to power plants in the north-central region and to load centres in central Mexico will also have to be built. Growing markets in the south-east will probably be served from fields in the south, initially by boosting compression on the existing Mayakan pipeline.

Although Pemex still has a monopoly on gas exploration and production, the downstream business has been open to private investment since the Natural Gas Law was adopted in 1995. But the opening is limited; the law prohibits any company from owning more than one function within the industry. The law also liberalised international gas trade and established a regulatory framework for building and operating pipelines and storage facilities. The Energy Regulatory Commission is drawing up regulations concerning the approval procedure for LNG import terminals, which will probably be built by foreign companies.

Coal

Mexico's primary coal demand amounted to just over 6.8 Mtoe in 2000. Two-thirds was consumed for power generation and most of the rest in industry, mainly iron and steel. Demand is projected to continue to grow over the projection period, but more slowly than in the past, because coal will be less able to compete against gas in power generation. Coal demand will expand to 9.6 Mtoe by 2030, after average annual increases of 1.3%.

Mexico's coal reserves are small, some 1.2 billion tonnes, of which about two-thirds is hard coal.[24] Most of the reserves are located in the state of Coahuila in the north-east of the country. They are of low quality, with high ash content. To upgrade fuel quality, indigenous coal is mixed with lower-ash coal imported from the United States, Canada and Colombia.

24. World Energy Council (2001).

Figure 4.16: **Mexico's Natural Gas Interconnections** (mcm/d)

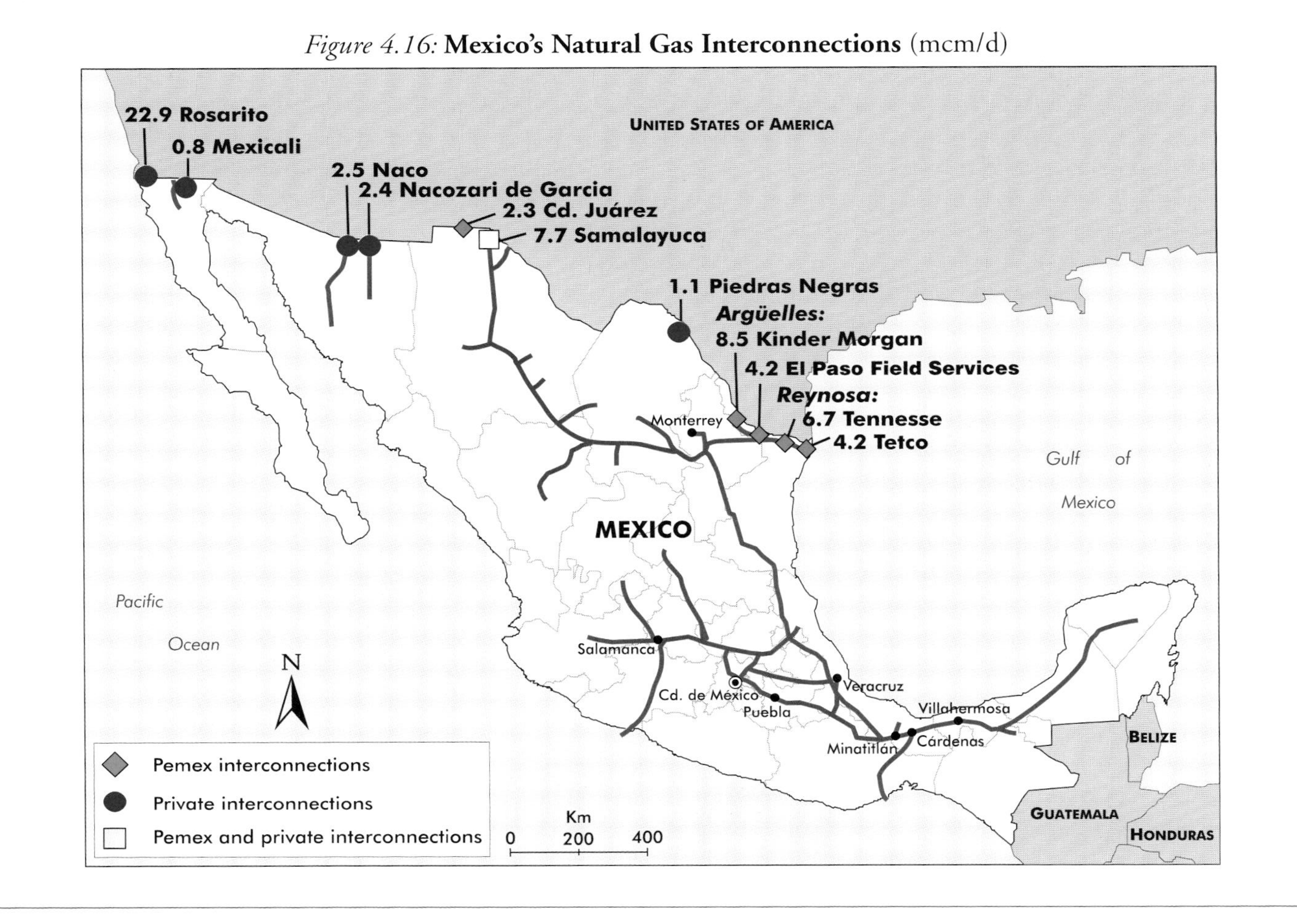

Electricity

Some 95% of Mexico's 100 million inhabitants now have access to electricity, and the residential sector is expected to be fully electrified by the end of the *Outlook* period. Electricity consumption is projected to increase by 3.5% per year from 2000 to 2030, fractionally faster than GDP. The share of electricity in final consumption will grow strongly, from 16% in 2000 to 21% in 2030.

In 2001, the government spent more than $3.5 billion subsidising household electricity bills. In February 2002, the Mexican government decided to reduce this subsidisation, by forcing those who consume the most to pay more for power. The government hopes to save more than $1 billion in this way in 2002. The money saved will be spent on investment projects in electricity generation, transmission and distribution.

Box 4.6: **CFE and Deregulation in Mexico's Electricity Supply Industry**

The Federal Electricity Commission (CFE) is the world's sixth-largest electricity company. It has a monopoly on transmission and distribution in most of Mexico, and is the dominant generator with exclusive rights over generation for public supply. A much smaller public utility, Luz y Fuerza Centro, has exclusive rights over public supply in Mexico City and surrounding areas. Limited private involvement in power generation has been allowed since 1992. Independent power producers (IPPs) accounted for almost all the generation capacity built in the last decade, although the total amount is small. CFE still generates more than 90% of the electricity produced in Mexico.

Further deregulation of the electricity sector is a contentious issue in Mexico. President Vincente Fox made it a priority when he entered office in 2000, but the process has stalled in the face of resistance from Congress. Although the government has not yet announced any firm proposals, a number of options are under discussion. These include introducing competition in generation and marketing based on third-party access to the grid. This would require vertical separation of CFE's generation, transmission and distribution activities. No progress in implementing any of these reforms is expected before elections in 2003.

Mexico now relies on oil to meet nearly half of its power generation demand. Over the *Outlook* period, oil-fired generation will continue to increase, but its share will fall to less than a quarter. Gas will become the dominant fuel. Gas-fired power generation will grow from 40 TWh in 2000 to 296 TWh in 2030. The share of gas in electricity generation is projected to jump from 20% to 51% of total generation over the *Outlook* period (Table 4.6). As in most other OECD countries, gas will be the preferred fuel for new electricity capacity, particularly in combined-cycle gas turbine (CCGT) plants. Coal will see its share in power generation fall over the projection period. The share of nuclear power will decline if, as expected, no new plants are built.

Table 4.6: **Electricity Generation in Mexico** (TWh)

	1990	**2000**	**2010**	**2030**	**Average annual growth 2000-2030 (%)**
Coal	8	19	22	32	1.7
Oil	70	97	109	141	1.2
Gas	13	40	97	296	6.9
Nuclear	3	8	10	10	0.8
Hydro	23	33	43	58	1.9
Other renewables	6	6	13	41	6.7
Total	**123**	**204**	**294**	**578**	**3.5**

Mexico has extensive renewable energy resources, and the government is committed to increasing the output of renewables-based power. Renewables can be economically viable in remote off-grid locations. They offer an opportunity to diversify Mexico's energy potential and reduce its dependency on oil. There is thought to be considerable technical potential for hydro, geothermal and wind power. Electricity output from hydro is projected to increase by 1.9% per year over the *Outlook* period. Power generation from non-hydro renewables is projected to grow much more quickly, although from a very low base. Demand for non-hydro renewables in generation will grow by 6.7% per year and will represent some 7% of electricity generation by 2030.

Table 4.7: **Potential and Installed Renewables Production Capacity by Source** (MW)

	Potential	**Installed capacity in 2000**	**Potential utilised in 2000 (%)**
Hydro	53,000	10,000	19
Wind	5,000	57	1
Geothermal	2,500	800	32

Nearly all of Mexico's power generation capacity is owned by the Federal Electricity Commission (CFE). Most plants are old and need rehabilitation: 17% are over 30 years old, 16% between 20 and 30, and 43% between 10 and 20. Some 11 GW are already under construction or firmly committed by CFE. Gas-fired plants will account for about 60% of the increase from now to 2030 (Figure 4.17). It is uncertain how this investment programme will be funded. It is likely to cost some $70 billion over the period 2000-2030 for power plant construction alone. Part of the additional capacity will be built by the CFE and part by independent power producers.[25]

Figure 4.17: **Power Generation Capacity Additions in Mexico**

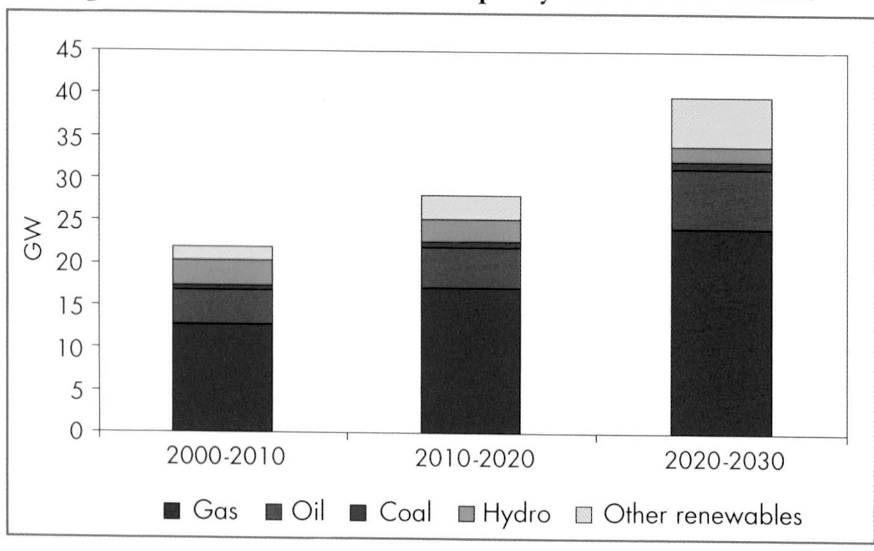

25. Mexican Ministry of Energy (2001c).

Environmental Issues

Energy-related CO_2 emissions are projected to reach 748 Mt by 2030 – a doubling over 2000 – boosting Mexico's share of global emissions from 1.6% to 2%. Per capita emissions will grow steadily but, at 5.5 tonnes in 2030, they will remain well below the OECD average of 13 tonnes.

Air pollution in large cities is a major concern, especially in Mexico City, Guadalajara and Ciudad Juarez. Mexico City's air quality is among the worst in the world, due to heavy traffic, to the city's high altitude and to the fact that it lies in a geographical basin. The Mexican government has adopted a number of measures to fight pollution, including incentives for using cleaner transport fuels. In major cities, for example, private cars must be fitted with catalytic converters. The government has also used subsidies to promote retrofitting vehicles to run on LPG, which produces less noxious emissions than gasoline and diesel. Planning regulations in the northern industrial regions have also been modified to take account of environmental concerns.

CHAPTER 5: OECD EUROPE

HIGHLIGHTS

- Primary energy demand in the European Union will rise by 0.7% a year to 2030, underpinned by GDP growth of 1.9%. Demand will rise slightly more rapidly in the rest of OECD Europe. In both sub-regions, oil and gas will still dominate the fuel mix, while the share of coal will continue to fall.
- The power sector will account for a growing share of EU primary energy use. Most of the projected increase in capacity will be gas-fired, but non-hydro renewables will grow quickly from a low base. The importance of nuclear energy will diminish as few new plants are built and some older ones are retired.
- The European Union will need to import progressively more fossil fuels, given coal, oil and gas production declines. The share of net imports in the Union's oil supply will climb from 73% in 2000 to 92% in 2030. Net imports of gas will also expand, from 44% now to 81% of total EU gas supply in 2030.
- Carbon dioxide emissions will rise at the same rate as primary energy use. Emissions will rise more quickly than in the past three decades. Without major new initiatives, the European Union will need to rely heavily on flexibility mechanisms in order to achieve its greenhouse gas emission target under the Kyoto Protocol.
- The possible introduction of new policies to curb rising energy imports and CO_2 emissions is a critical uncertainty in Europe's energy outlook.

For the first time, this year's Outlook *models the European Union separately from the rest of OECD Europe.*[1] *The European Union dominates the*

1. The 15 current EU members are Austria, Belgium, Denmark, Finland, France, Germany, Greece, Ireland, Italy, Luxembourg, the Netherlands, Portugal, Spain, Sweden and the United Kingdom. The rest of OECD Europe includes the Czech Republic, Hungary, Iceland, Norway, Poland, Switzerland and Turkey. The other countries of Central and Eastern Europe are included in the transition economies.

regions' energy consumption today, accounting for over 80% of primary energy use. The planned enlargement of the Union will take in four of the six countries in the rest of OECD Europe – the Czech Republic, Hungary, Poland and Turkey. The first three countries could join as early as 2004.

European Union

Energy Market Overview

The European Union represents around 16% of the world energy market and is the largest net energy-importing region in the world in absolute terms, importing close to half its needs. Despite continuing economic integration, national energy profiles and trends among the 15 EU member states remain very diverse. This diversity reflects varying stages of economic development and differences in policy priorities, economic structure, taxes, climate and local resources. Nuclear power, for example, is a source of primary energy in Belgium, Finland, France, Germany, the Netherlands, Spain, Sweden, Switzerland and the United Kingdom, but plays no role in the other six EU states. The share of renewable energy in each country's energy mix also varies markedly, depending on the local availability of resources and government efforts to promote their use.

Oil is the predominant energy source, although its share in primary energy use has declined since the 1970s. Coal use, in both relative and absolute terms, has fallen sharply and is now largely confined to power generation. The share of gas has risen steadily, from 8% in 1971 to 23% in 2000. Nuclear energy now accounts for about 15% of primary energy demand, having grown rapidly between the late 1970s and the early 1990s. In 1998, both capacity and output declined, albeit marginally, for the first time.

Indigenous production accounts for around 56% of the Union's natural gas consumption. The region produces half the coal it burns and 27% of its oil. Production of oil, most of it from the UK sector of the North Sea, has increased over the last decade or so. Gas production has also risen. Coal production has fallen by more than half since 1990. Subsidies to high-cost coal mines have been slashed in several countries, notably France, Germany, Spain and the United Kingdom.

Current Trends and Key Assumptions

Macroeconomic Context

Economic growth in the European Union fell from 3.4% in 2000 to 1.7% in 2001 in line with the slowing global economy. The slump has been most severe in Germany and Italy, while the United Kingdom and France have continued to enjoy moderate growth. The initial weakness of the euro after its launch in 1999, combined with higher oil prices, intensified inflationary pressures and reduced the European Central Bank's scope for cutting interest rates to stimulate economic activity. Nonetheless, euro notes and coins were introduced successfully at the beginning of 2002. GDP growth in EU countries is expected to average 1.5% in 2002 with a recovery in activity in the second half of the year.

In the Reference Scenario, the EU economy is assumed to grow by 1.9% a year on average from 2000 to 2030 (Table 5.1). Growth is fastest in the period to 2010 (2.3%), based on a prompt recovery from the current economic slowdown. Growth slows to 2% per year from 2010 to 2020 and to 1.6% from 2020 to 2030. The differences in growth rates between countries are expected to shrink with the macroeconomic convergence intended to result from economic and monetary integration. The European population is assumed to decline slightly over the projection period. It follows from these assumptions that GDP per capita will be more than 80% higher by 2030. If so, the European Union would remain the third-richest region in per capita terms, behind the US and Canada and OECD Pacific.

Table 5.1: **European Union Reference Scenario Assumptions**

	1971	2000	2010	2030	Average annual growth 2000-2030 (%)
GDP (in billion 1995 $, PPP)	4,096	8,241	10,326	14,689	1.9
Population (million)	343	377	378	367	-0.1
GDP per capita (in 1995 $, PPP)	11,958	21,889	27,320	39,994	2.0

Table 5.2: **Opening of the EU Electricity and Gas Markets (%)**

	Electricity			**Gas**		
		Actual switching (% of load)			**Actual switching (% of load)**	
	Declared market opening (%)	**Large customers**	**All customers**	**Declared market opening (%)**	**Large customers**	**All customers**
Austria	100	5-10	n.a.	49	<5	<5
Belgium	35	5-10	n.a.	59	<5	<2
Denmark	90	n.a.	n.a.	30	0	0
Finland	100	30	2-3	n.a.	n.a.	n.a.
France	30	5-10	n.a.	20	10-20	3
Germany	100	10-20	<3	100	<5	2
Greece	30	0	0	0	0	0
Ireland	30	30	10	75	20-30	25
Italy	45	10-20	n.a.	96	10-20	16
Luxembourg	56	n.a.	n.a.	51	0	0
Netherlands	33	10-20	n.a.	45	>30	17
Portugal	30	<5	n.a.	0	0	0
Spain	45	<5	<2	72	5-10	7
Sweden	100	100	5	47	<5	0
UK	100	80	<20	100	100	90

n.a.: not available.
Note: "Large customers" include large industrial energy users and, for gas, power stations.
Source: European Commission (2001a and 2001b).

Energy Liberalisation and Prices

Liberalisation of the European electricity and gas sectors is still underway. It is assumed here that the process will continue progressively well into the projection period. Success in opening up those markets to competition and privatising utilities has been mixed (Table 5.2). The United Kingdom and Sweden introduced competition in electricity generation and in supply, based on third-party access, several years ago. Other countries have only recently implemented reforms in response to a 1998 EU directive. The directive calls for at least 26% of each country's power market to be open to competition by 2000 and 33% by 2003. Some countries have gone beyond these requirements, but the overall pace of switching away from the former monopoly suppliers has, in most cases, been slow. Germany, for example, has already introduced full retail competition, but only a few customers, who account for less than 5% of electricity consumption, had changed their electricity supplier by the beginning of 2002.

Competition in gas is less advanced, mainly because the EU gas directive took effect two years after the electricity directive. At least 20% of each national market must now be open, the share to rise to 28% by 2005 and to 33% by 2010. The United Kingdom, which introduced partial gas competition in the late 1980s and full retail competition in 1998, still has by far the most competitive gas market in Europe.

In March 2001, the EU Commission proposed a directive aimed at speeding up the development of competition in the electricity and gas sectors. EU prime ministers and heads of state reached agreement on several of the key principles in March 2002 (Box 5.1). Negotiations on specific aspects of the directive are expected to be completed before the end of 2002. The EU leaders also called on the Council of Energy Ministers to agree as soon as possible a tariff-setting system for cross-border electricity transactions, including congestion management.

Box 5.1: **Agreed Elements of the Proposed EU Electricity and Gas Directive**

- The freedom to choose a supplier will be extended to all non-household consumers as of 2004 for both electricity and gas. Those eligible consumers must cover at least 60% of the total market for each fuel. A decision on further market-opening measures will be taken before the spring European Council in 2003.
- Transmission and distribution functions have to be separated (unbundled) from production and supply.
- Non-discriminatory access to networks will be made available to eligible consumers and producers on the basis of published tariffs.
- Every member state will establish a regulatory function to ensure, in particular, effective control over tariff-setting.

Competition is one of several factors that affect electricity and gas prices to final consumers. Taxes are another factor, as are oil prices, which affect the competitiveness of gas against substitute fuels and the price of imported gas under long-term contracts. Coal prices, which affect the variable cost of generation, are also important. The price assumptions for oil and coal track those for international markets. Oil prices will remain the main influence on gas prices in Europe through inter-fuel competition, but their influence will diminish with the coming shift away from oil price indexation in long-term contracts and more short-term gas trading. Gas prices are assumed to remain flat at an average of $2.80/MBtu (in year 2000 dollars) from 2003 to 2010. Gas-to-gas competition is expected to intensify across the Union. Lower downstream margins and pressures from national regulators to reduce access charges would be expected to depress end-user prices, but the rising cost of bringing new gas supplies to Europe is expected to counter the effects of competition on price. Gas prices pick up after 2010, in the wake of rising oil prices. Retail electricity prices are assumed to edge lower from now until 2010, due to efficiency gains and lower margins in generation, transmission and distribution. Electricity prices are assumed to rebound slowly after 2010, as gas prices increase.

The Reference Scenario assumes no major changes in existing taxes, nor any new taxes beyond those already agreed. Public anger over high transport fuel prices in 2000 and 2001 led several EU countries to trim gasoline and diesel tax rates. End-user energy prices in the European Union are among the highest in the world, and it is widely regarded as politically perilous to raise taxes further. Some tax increases and new taxes, such as carbon levies, are, nonetheless, possible.

Other Energy and Environmental Policies

The Reference Scenario takes into account EU and national policies that had already been announced and approved as of mid-2002, including those aimed at meeting emissions-reduction commitments under the Kyoto Protocol. Existing policies are assumed to be continued, but measures that *might* be adopted and implemented in the future have not been included in our assumptions[2].

Some new initiatives likely to be implemented by EU members will certainly have implications for EU energy markets[3]:

- A Green Paper on energy security sets out options for reducing the supply risks linked to rising import dependence and the need to meet the Kyoto commitments (Box 5.2).

2. Although the European Union recently ratified it, the Kyoto Protocol has not yet come into force. The Reference Scenario does not take into account the flexibility mechanisms, such as emissions trading, provided for under the Protocol.
3. The impact of the policies in the transport action plan, the White Paper and some of the policies in the Green Paper is assessed in the Alternative Policy Scenario (Chapter 12).

Box 5.2: **EU Green Paper on Energy Security**

In November 2000, the European Commission released a Green Paper called *Towards a European Strategy for the Security of Energy Supply*. The report concludes that the Union's dependence on external energy sources will rise from 50% now to 70% by 2030 and that the Union will not be able to meet its Kyoto commitments. It raises the following key questions:

1. What strategy should the Union adopt for dealing with dependence on external energy sources?
2. What policies to adopt for the liberalised energy market in Europe?
3. What is the role for tax and state aid?
4. What type of relations should the Union maintain with producer countries?
5. What should EU policy be on energy stocks?
6. How can energy supply networks be improved?
7. What support should be given to renewable energy sources? Should the traditional energy industries contribute financially to this support?
8. How can the Union find a solution to the problems of nuclear waste, reinforce nuclear safety and develop research into reactors of the future?
9. How should the Union combat climate change? What role will energy saving play?
10. Should there be an EU policy on biofuels? What form should it take?
11. Should incentives or regulatory measures be used to increase energy saving in buildings?
12. How can energy savings in transport be achieved? What measures are needed to encourage switching of freight from road to rail and to reduce the use of cars in towns and cities?
13. How can the Union, member states, regions, producers and consumers pull together to develop a sustainable system of energy supply? How best could that process be organised?

Source: European Commission (2000).

- An action plan on transport has been tabled, together with two proposals for directives to foster the use of alternative transport fuels.
- A White Paper on transport policy has been issued, aimed at improving the efficiency and quality of transportation services and reducing the environmental damage caused by growing mobility. It proposes revitalising European railways, promoting the use of inland waterways and better linking up different modes of transport.

Results of the Projections

Overview

Total primary energy demand in the European Union will rise by 0.7% per year from 2000 to 2030 in the Reference Scenario, slower than the 1.2% rate of 1971-2000. The fuel mix will change markedly (Figure 5.1). The share of coal in total primary energy use will continue to decline, from 15% in 2000 to 10% in 2030. The share of gas will increase, from 23% in 2000 to 28% in 2010 and 34% in 2030, by which time the use of gas will be almost as extensive as that of oil. The share of non-hydro renewables also rises steadily, overtaking nuclear just before 2030. Nuclear output is projected to fall away after 2010.

Aggregate primary energy production declines continuously over the projection period. Higher output from renewable sources is not sufficient

Figure 5.1: **Total Primary Energy Demand in the European Union**

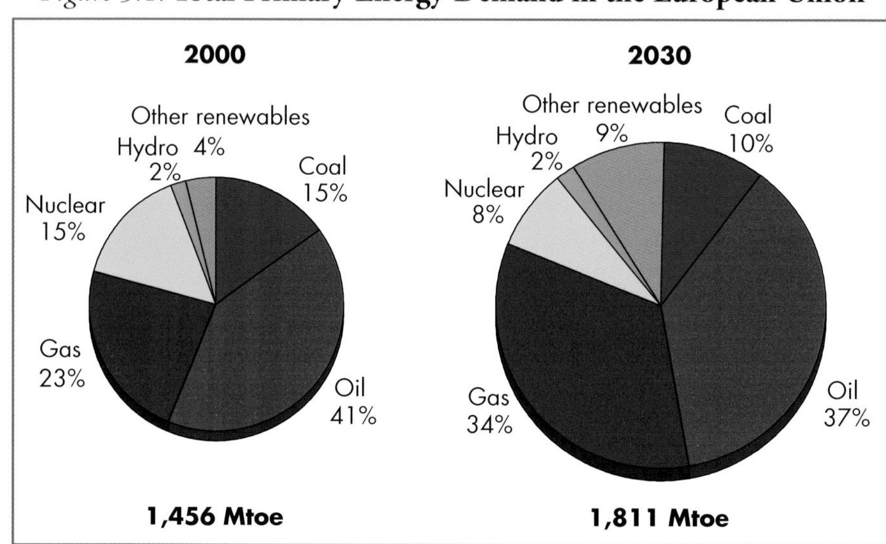

to compensate for the gradual decline in coal, oil, gas and — after 2010 — nuclear energy. The European Union's high import dependence will increase further as demand rises. EU net imports of oil and gas will jump from 566 Mtoe in 2000 (39% of primary consumption) to 1,116 Mtoe by 2030 (62%). This is in line with the projections in the European Commission's Green Paper on energy security. By the end of the projection period, the European Union will account for 20% of global net inter-regional trade in oil and 34% of that in gas.

Energy intensity will continue to fall in line with past trends, at an average annual rate of 1.2% from 2000 to 2030. Autonomous improvements in energy efficiency (improvements unrelated to new government policies) will drive energy intensity down. So will the continuing structural shift of the EU economy to less energy-intensive activities.

Demand by End-use Sector

Final energy consumption will grow by 1.2% per annum between 2000 and 2010 and by 0.7% between 2010 and 2030. The transport sector remains the fastest growing sector in final energy demand (Figure 5.2). From 30% in 2000, its share of total final consumption reaches 33% in 2030. The residential and service sectors will increase by about 0.9% a year.

Energy use in industry is projected to grow slowly in the coming three decades. But the mix of fuels used is expected to go on changing, with gas and electricity replacing coal and, to a lesser extent, oil products (mostly heavy fuel oil and heating oil). By 2030, gas and electricity will account for 63% of industry's total energy consumption.

Passenger traffic and freight will continue growing along with rising household incomes and business activity. Expected increases in both the number and size of passenger vehicles will partly offset improvements in vehicle fuel efficiency. But the pace of demand growth will decelerate after 2010, mainly due to saturation effects. Car ownership is already very high. Worsening traffic congestion in cities will also discourage extra driving. Air transport fuel use will continue to grow as passenger traffic increases, stimulated partly by liberalisation of the EU air travel market. Oil is projected to remain the primary fuel for transportation, although the role of biofuels[4] is set to grow in response to national initiatives.[5]

4. Fuels obtained by processing or fermenting non-fossil organic sources such as plant oils, sugar beet, cereals and other crops and wastes, that can be used on their own or in a mixture with conventional fuels.

5. Additional EU initiatives are assessed in the Alternative Policy Scenario (see Chapter 12).

Figure 5.2: **Total Final Consumption by Sector in the European Union**

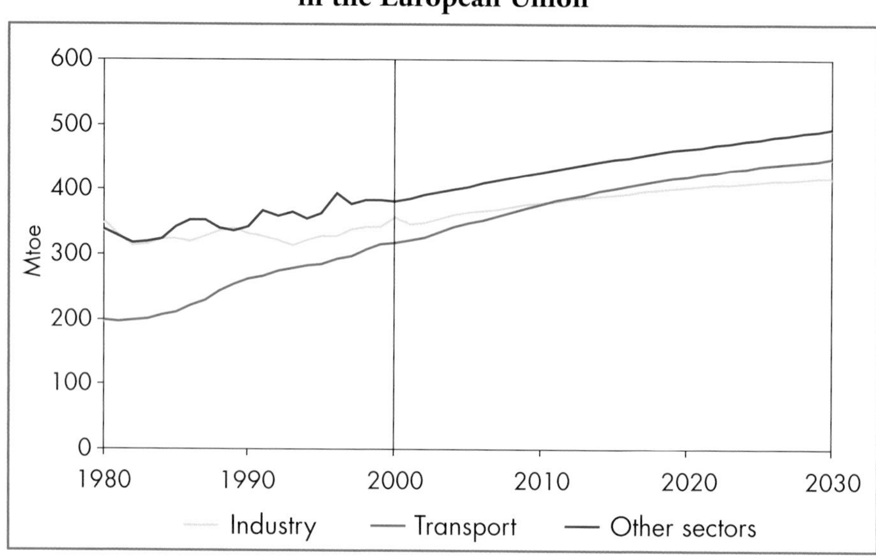

Despite ongoing improvements in the energy efficiency of household and commercial appliances and equipment and better building insulation, residential and services energy demand will rise steadily over the next three decades. Driven by growing business activity and rising living standards which will stimulate demand for larger homes, more office space and new appliances. Much of the growth in energy demand will be in the form of electricity.

Oil

Oil will remain Europe's largest energy source, with primary oil demand increasing by 0.4% per year from 2000 to 2030. Oil's weight in primary energy supply will fall slightly, from 41% in 2000 to 37% in 2030. Almost all the increase in demand will come from the transport sector. Aviation fuel demand will grow fastest of all. Demand for diesel will increase faster than for gasoline, because of continuing growth in road freight and a continuing trend towards using diesel in passenger cars. No tax changes are assumed and hence most countries will carry on taxing diesel more lightly than gasoline.

EU oil production (including NGLs) averaged about 3.2 mb/d in 2001, almost all of it coming from the UK (77%) and Danish (11%) sectors of the North Sea. Output grew steadily in the 1990s, peaking in 1999. Production is expected to decline gradually over the coming years, although new discoveries and the development of small marginal fields could arrest the decline, if only temporarily. EU oil production is projected to fall to 2.3 mb/d in 2010 and 1.1 mb/d in 2030 (Figure 5.3).

The European Union's net oil import requirements will rise sharply as indigenous production dwindles and demand rises. Oil from OPEC countries will meet a large part of these additional needs. OPEC currently accounts for 42% of the Union's oil imports.

Figure 5.3: **Oil Balance in the European Union**

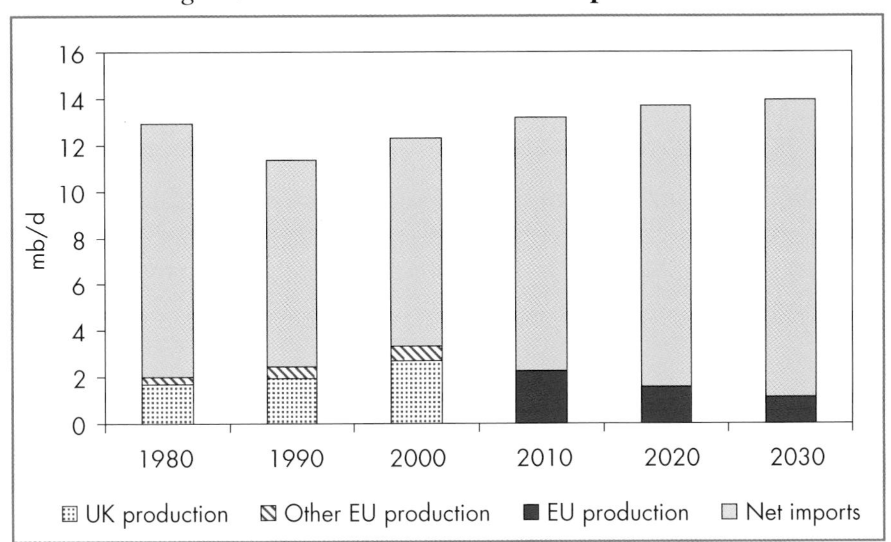

Natural Gas

Natural gas consumption has grown more in absolute terms than that of any other fuel over the past three decades, and this trend will continue to 2030. Primary gas use will grow by 2.9% per year from 2000 to 2010 and by 1.6% from 2010 to 2030. Demand will increase in all end-use sectors, but most dramatically so in power generation.

Natural gas resources in EU countries, a little more than 2% of the world's total, are expensive to exploit. Output from the North Sea – a mature producing region – will dwindle over the coming decades. The projected increase in demand will, therefore, have to be met by increased imports. The European Union already imports 44% of its gas needs.

Norway will undoubtedly meet some of these needs as well as some of those of EU applicant countries in Central and Eastern Europe. Currently contracted supplies from Norway to EU countries will probably plateau at about 75 bcm/year in 2005 or soon after. But there is scope for further increasing sales, even without adding capacity to the offshore pipeline network. With additional compression, current pipeline capacity of 86 bcm/year could probably be increased to 100 bcm/year.

Box 5.3: **Norwegian Energy Exports to the European Union**

Norway is the leading external supplier of energy to the European Union. Its net energy exports to EU countries rose from 97 Mtoe in 1990 to a record 131 Mtoe in 2000 – equivalent to 9% of the Union's total energy consumption. Just under 84 Mtoe (1.7 mb/d) of this was in the form of oil, 45 Mtoe (47 bcm) was gas and 1.7 Mtoe (19.3 TWh) was electricity. All electricity exports went to Sweden, Denmark and Finland.

Norway must abide by EU rules in exporting to EU countries, although it is not itself an EU member state. In July 2002, Norway agreed to incorporate the EU gas directive into the Agreement of the European Economic Area, to which Norway belongs, and to transpose it into national law. In an earlier move, the government abolished the GFU, a gas-marketing organisation led by the state-owned firm, Statoil, that negotiated all export contracts for gas not linked to single fields.

The rest of the Union's gas imports are likely to come from its two main current suppliers, Russia and Algeria, and a mixture of piped gas and liquefied natural gas from elsewhere. Sources will probably include Libya (via pipeline), Nigeria, Trinidad and Tobago, Egypt and possibly Qatar (LNG). Venezuela may also emerge in the longer term as a supplier of LNG, while spot shipments of LNG from other Middle East producers may also increase if a global short-term market in LNG develops (Figure 5.4).

There is undoubtedly enough gas in these countries to meet EU needs. But the unit costs of getting that gas to market will probably rise as more remote and costly sources are tapped. Piped gas from North Africa and the Nadym-Pur-Taz region in Russia are the lowest cost options, but supplies from these sources will not be sufficient to meet projected demand after 2010. Pipeline projects based on fields in the Yamal Peninsula and the Shtokmanovskoye field in the Barents Sea in Russia are among the most

Figure 5.4: **Net Imports of Gas by Origin in the European Union**

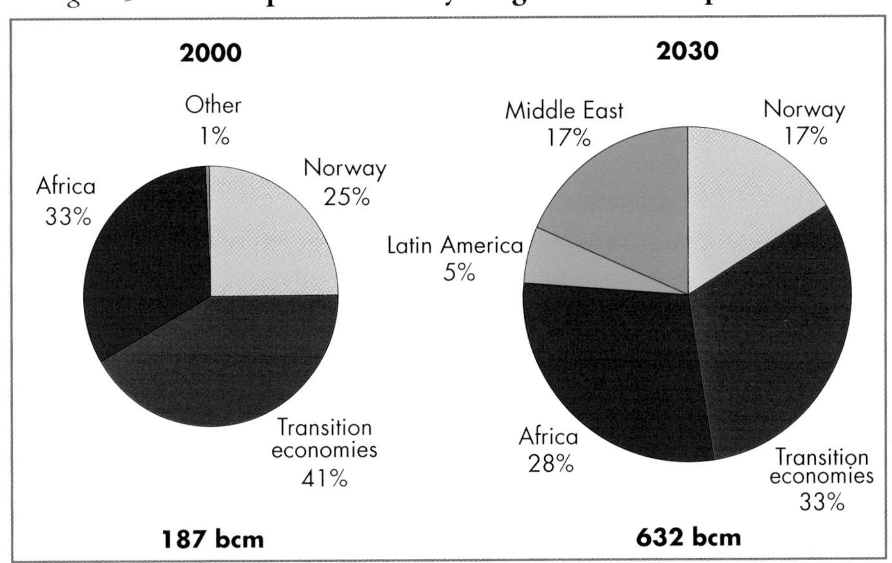

expensive longer-term options. So are pipelines from the Middle East and the Caspian region.

LNG, traded both under long-term contracts and on spot markets, could play a much more important role in supplying the European gas market if supply costs continue to fall. LNG would become especially important if there turns out to be less Russian gas than expected. This could occur if investment in new fields is insufficient to compensate for the decline in production from existing fields. In any event, the distances over which LNG imports from new sources need to be shipped may well drive costs and prices up (Figure 5.5).

Coal

Primary coal demand in the European Union is expected to go on falling, at an average of 0.6% per year over the projection period. Coal consumption will become increasingly concentrated in power generation and specialised industrial uses, such as steel-making. The power sector will account for 80% of primary coal use by 2030 compared to 76% in 2000, and the industrial sector will take almost all of the rest.

Despite the continuing contraction of the EU coal market, demand will be increasingly met with imports, as indigenous production declines even more rapidly. The amount of EU hard coal production receiving government support has fallen over the past decade, both in absolute and

Figure 5.5: **Indicative Costs For New Sources of Gas Supply Delivered to the European Union, 2010** ($/Mbtu)

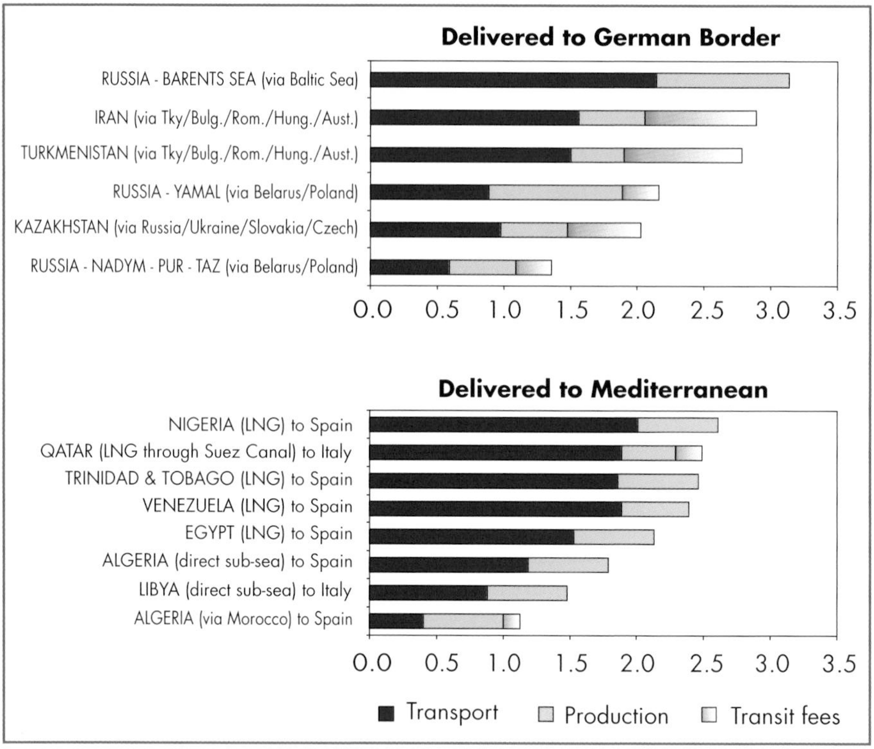

Source: IEA (2001).

percentage terms. Aid per tonne for the four EU countries that still subsidise coal production is shown in Table 5.3. Subsidised production is now concentrated in Germany and Spain. There are plans to reduce subsidies in all four countries. France plans to close its domestic industry by 2005. Germany is expected to reduce subsidies and subsidised output by a third by the same year. The United Kingdom reinstated subsidies to its coal industry in April 2000 but they only ran until July 2002. Spain expects to reduce production by a further 20% by 2005. By 2006, only Germany and Spain are expected to still offer subsidies. Recent EU rules allow coal subsidies to continue, in some cases, on energy security grounds. The European Union adopted a regulation in July 2001 that establishes rules for the continuation of state aid after the expiration of the European Coal and Steel Community Treaty in 2002. The regulation will remain in force until 2010.

Table 5.3: **Subsidies for Coal Production in EU Countries**

Country	Production (million tonnes of coal equivalent)		Aid per tonne of coal equivalent ($)	
	1999	2000	1999	2000
France	4.1	3.2	91.8	97.1
Germany	40.0	34.0	118.2	115.4
Spain	10.3	10.4	72.9	70.3
United Kingdom	32.1	27.5	-	3.2

Source: IEA (2001).

Electricity

The European Union's final electricity consumption will climb by 1.4% per year from 2000 to 2030, compared with 2.9% in 1971-2000. Electricity use will expand most rapidly in the residential and services sectors.

Installed generation capacity is projected to increase from 573 GW in 1999 to 679 GW in 2010 and 901 GW in 2030. Over half of existing plants are expected to be retired over the projection period. Most new capacity is expected to be gas-fired, particularly in combined-cycle gas turbine (CCGT) plants. In 2030, 41% of capacity will be gas-fired, compared to 16% in 1999. Over the first half of the projection period, very few new coal plants are expected to be built and several existing ones will probably be decommissioned. As a result, the share of coal in generation will drop sharply. In the second half of the projection period, higher gas prices and improvements in coal technologies will make new coal-fired generation more competitive. The share of nuclear power will more than halve on the assumption that few new plants are built and that many of the existing plants are retired. Installed nuclear capacity is projected to fall from 124 GW in 1999 to 76 GW in 2030. Fuel-cell capacity is projected to reach 30 GW in 2030.

Non-hydro renewables-based electricity, mainly wind and biomass, will increase rapidly (Figure 5.6). Wind's share of generation will grow from 0.9% in 2000 to almost 5% in 2030. The share of biomass in total generation will also increase rapidly, from 1.8% to 5%. Most biomass, essentially wood, agricultural residues and municipal waste, will be used in co-generation plants, initially in boilers or gas turbines. Gasification technology will become more widespread towards the end of the *Outlook* period.

Figure 5.6: **Electricity Generation Mix in the European Union**

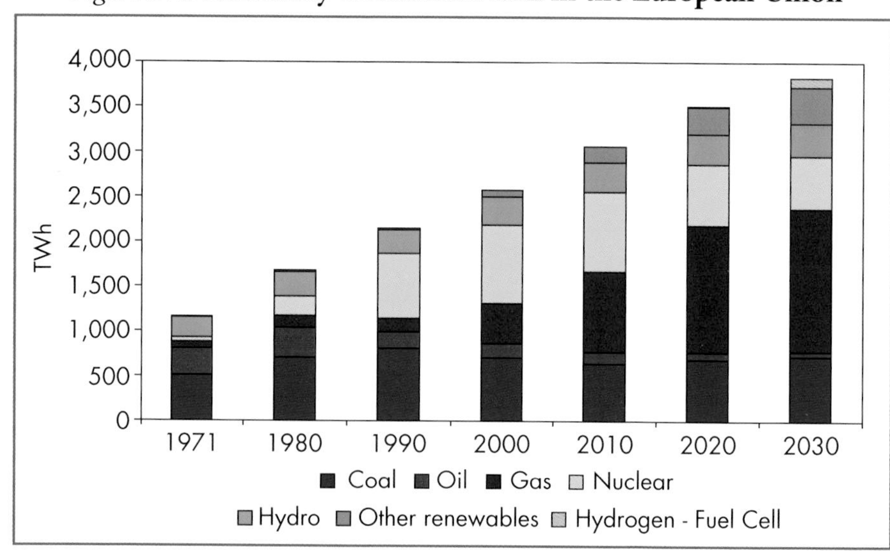

The Reference Scenario projections for EU electricity generation are subject to several uncertainties. The chief sources of uncertainty are future fuel price trends, national policies on nuclear power and renewables, technological developments and new environmental regulations. Such regulations could include limits on, or financial penalties for, particulate, SO_2, NO_x and CO_2 emissions. Whether new nuclear reactors are built and how long existing reactors operate will make a big difference to future developments. Nuclear policies differ among EU countries at present (Box 5.3). Renewables could take a much larger share of the fuel mix if the EU and national governments apply stronger policies and measures to promote them than those they have adopted so far (see Chapter 12).

Energy-related CO_2 Emissions

CO_2 emissions will rise more rapidly over the projection period than in the past three decades. From 3,146 Mt in 2000, emissions reach 3,422 Mt in 2010 and 3,829 Mt in 2030 — an average annual increase of 0.7%.

Power generation remains the single biggest CO_2-emitting sector in 2030 (Figure 5.7). Its share rises from 31% in 2000 to 34% in 2030, even though projected CO_2 emissions per unit of electricity output are due to fall slightly, thanks to the use of more gas and less coal. The share of transportation expands more rapidly, due to rapid growth in transport fuel consumption and the sector's continuing dependence on oil.

Box 5.3: **Recent Developments in Nuclear Policy in EU Countries**

- In Finland, the parliament approved in May 2002 an application by the power company TVO to build a fifth reactor.
- France is proceeding with research on the European Pressurised Reactor, a design that could be used for the next generation of nuclear plants.
- In Germany, the government and the electricity industry agreed in June 2000 to a phase-out of existing nuclear stations. For each plant, a maximum amount of generation is allowed on the basis of a standard operating lifetime of 32 years. These quantities can be transferred among plants to give operators flexibility in optimising their generation.
- In Sweden, the closure of a second reactor planned for 2001 has been delayed until at least 2003, because not enough non-nuclear capacity had been built to replace it. The government has proposed negotiating a nuclear phase-out agreement similar to Germany's.
- The Belgian government has announced plans to phase out nuclear power once existing reactors have come to the end of their 40-year lifetimes.
- In the United Kingdom, a government-sponsored energy review released in February 2002 recommends that the nuclear option be kept open.

Without major new initiatives to limit growth in energy demand and stimulate switching to less carbon-intensive fuels, the European Union will need to rely heavily on flexibility mechanisms such as emissions trading in order to achieve its emissions-reduction target under the Kyoto Protocol. Energy-related CO_2 emissions are expected to increase by 10% over the period from 1990 to 2010. The overall EU commitment is to maintain greenhouse gas emissions 8% *below* the 1990 level by the period 2008-2012.

Figure 5.7: **CO_2 Emissions and Intensity in the European Union**

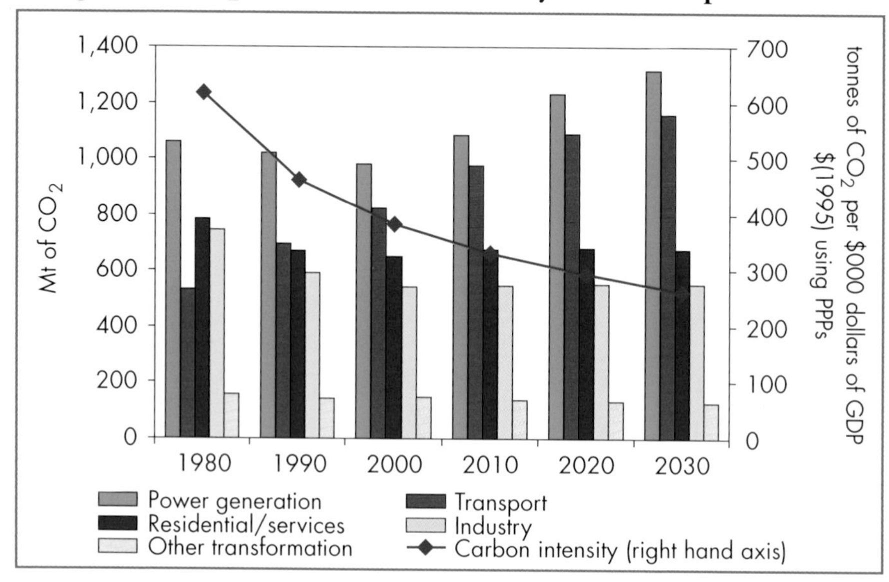

Rest of OECD Europe

Energy Market Overview

Primary energy demand in the non-EU countries of OECD Europe — the Czech Republic, Hungary, Iceland, Norway, Poland, Switzerland and Turkey — amounted to 292 Mtoe in 2000, or 17% of the total for the region. These countries have very diverse economies, climates and patterns of energy use. Per capita incomes in Switzerland and Iceland are almost three times those in Poland and Turkey (in PPP terms), but the gap has been narrowing and will probably continue to do so. Energy intensity also varies substantially among these countries: Turkey has the least energy-intensive economy and Iceland the most intensive, reflecting differences in both income levels and climate.

Norway, with large resources of oil, gas and hydroelectricity, is the only country in this group that produces more energy than it consumes. Switzerland, the Czech Republic and Hungary have nuclear power industries. The Czech Republic and Poland are major producers of coal, while Switzerland is a large hydropower producer.

Current Trends and Key Assumptions

GDP growth in this group of countries fell from 4.7% in 2000 to *minus* 1% in 2001. In the Reference Scenario, their economies are assumed

Figure 5.8: **Total Primary Energy Demand in Other OECD Europe**

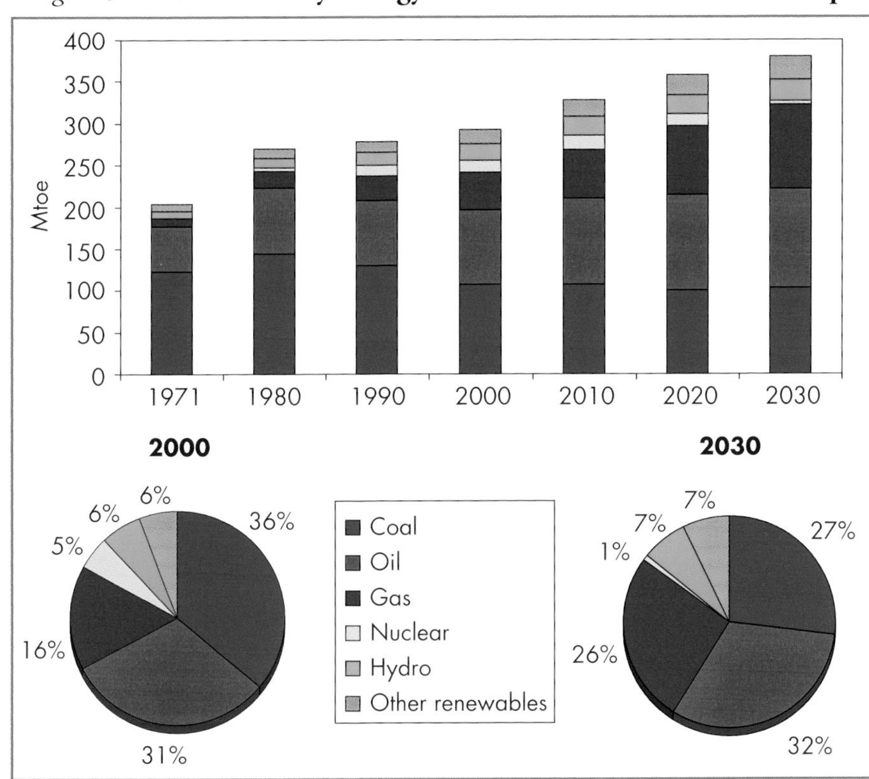

to grow by 2.7% a year on average from 2000 to 2010 and by 2.2% per year over the entire projection period. Turkey, the Czech Republic and Poland will achieve the fastest rates as they catch up, to some extent, with the income levels of the other countries. This process is likely to accelerate as some of these countries achieve EU membership. Population growth, at 0.4% per annum in 2000-2030, is a little faster than that assumed for the European Union, mainly thanks to Turkey.

Results of the Projections

The Reference Scenario projects primary energy use in the region to expand by 0.9% per year from 2000 to 2030, as against 1.2% from 1971 to 2000 (Figure 5.8). Demand growth will be most rapid in the first decade. Both the amount of coal consumed and its share in total energy use will continue to decline, to be replaced progressively by natural gas, especially in power generation. Coal's share falls from 36% of primary supply in 2000 to 33% in 2010 and to 27% in 2030. The share of gas increases from

16% in 2000 to 26% in 2030, by which time it is almost as large as that of coal. Transport and power generation, as in the EU countries, account for most of the growth in primary energy use.

Energy intensity will continue to fall in line with past trends, at a projected average annual rate of 1.3% from 2000 to 2030. Energy-related CO_2 emissions grow by 11% to 2010 and 27% to 2030 compared to 2000.

Electricity generation in these countries is projected to rise by 1.6% per year from 2000 to 2030. The rate is higher than in the European Union countries because of strong demand growth in Turkey, Poland, Hungary and the Cezch Republic. Several old, inefficient coal-fired power stations are expected to shut down during the *Outlook* period, especially in Poland and the Czech Republic. Most new capacity is projected to be gas-fired, pushing up the share of gas in total generation from 10% in 2000 to 31% in 2030. The share of nuclear power is projected to fall from 9% now to less than 2% in 2030, because most existing power plants are assumed to be retired. The only new nuclear capacity in the region is expected to be the second unit at the Temelin plant in the Czech Republic.

Hydropower output is projected to increase from 220 TWh (37% of total generation) in 2000 to 293 TWh (31%) in 2030. Nearly all the increase will be in Turkey, where many large hydropower stations are under construction. Any increases in hydro output in Norway and Switzerland, which have already exploited most of their potential, are likely to come from small stations. The share of other renewable sources in the generation mix, notably wind, biomass and geothermal, will increase, from 0.8% now to 2.2% in 2030.

Norway's large oil and gas resources will enable it to remain an important player in the European energy market. Oil production in the non-EU countries of OECD Europe is projected to decrease from 3.5 mb/d now to 1.4 mb/d in 2030, while gas production will more than double to 126 bcm in 2030.

CHAPTER 6: OECD PACIFIC

HIGHLIGHTS

- Primary energy demand in Japan, Australia and New Zealand will grow by 0.8% per annum from 2000 to 2030. But growth in demand decelerates over the period due to a gradual slowdown in economic growth, a continued shift to less energy-intensive activities, stagnating population and saturation effects in the transport, residential and services sectors.
- In this group of countries, the shares of natural gas, nuclear energy and renewable energy sources will grow at the expense of coal and oil. This trend results partly from government measures to promote less carbon-intensive fuels. Nonetheless, their oil import dependence will rise steeply, reaching 92% in 2030.
- Energy-related carbon dioxide emissions will increase broadly in line with primary energy use for the first decade of the *Outlook* period. As a result, these countries will not meet their Kyoto commitments unless they adopt vigorous new policies.
- Korea's primary energy demand will grow by 2.3% per annum over the projection period – much slower than in the past thirty years. Oil will continue to dominate Korea's fuel mix, but the shares of gas and nuclear energy will expand further. With virtually no indigenous fossil-fuel resources, Korea's share of international energy trade will continue to expand.

OECD Pacific includes Japan, Australia, New Zealand and, for the first time in the World Energy Outlook, *Korea, which became a member of the IEA in 2001. Korea has been modelled separately for this* Outlook. *Projections for Korea are presented at the end of this chapter, along with a discussion of current market trends and assumptions. Projections for the other three countries were prepared in aggregate.*

Regional Summary

The Reference Scenario projects a continuing slowdown in the growth of energy consumption in the OECD Pacific region (Japan, Korea, Australia and New Zealand) as a whole. This is mainly because of sluggish demand in Japan. Japan's population is expected to start declining from the middle of the projection period. Primary energy use in all four countries in aggregate will grow on average by 1.2% per year over the next three decades (Table 6.1). This compares to a brisk 3.1% per year from 1971 to 2000. Oil will still dominate the primary fuel mix in 2030, but its share will have declined to 42% from nearly 50% in 2000. Natural gas, as in most other regions, will be the most rapidly growing fossil fuel, even though most of it will have to be imported as liquefied natural gas (LNG) into Japan and Korea. These two countries are among only a few in the world where the role of nuclear power is expected to increase. The OECD Pacific region's share in global nuclear power supply will reach 30% in 2030.

Table 6.1: **Primary Energy Demand in OECD Pacific** (Mtoe)

	1971	**2000**	**2010**	**2030**	**Average annual growth 2000-2030 (%)**
Coal	84	184	205	215	0.5
Oil	240	412	458	505	0.7
Gas	5	106	145	210	2.3
Nuclear	2	112	150	210	2.1
Hydro	9	11	13	14	0.8
Other renewables	5	21	29	47	2.7
TPES	**346**	**847**	**1,001**	**1,200**	**1.2**

Japan, Australia and New Zealand

Energy Market Overview

Japan, Australia and New Zealand have diverse energy structures. *Japan*, with the world's third-largest economy (measured by purchasing power parity), is the fourth-largest energy-consuming country, accounting for 7% of world demand and 80% of demand in the region. Japan is highly dependent on imported energy, especially oil. The country has the least energy-intensive economy in the region and is one of the least energy-intensive in the OECD. This is due to historically high energy prices, highly developed service and light manufacturing sectors and low residential energy use. *Australia*, by contrast, is a major energy producer and exporter of natural gas and coal. It is the world's largest exporter of hard coal, most of which goes to Japan. *New Zealand* has ample gas, coal, hydropower and geothermal energy, but imports some oil. Both Australia and New Zealand are relatively energy-intensive.

After falling by 1% in 1998 in the wake of the Asian economic crisis, these countries' primary energy demand rebounded by 1.4% in 1999 and by 1.9% in 2000. The 1998 decline was the first time the region's energy demand had fallen since 1983. Japanese energy consumption has been stagnant in recent years because of low economic growth. The country's total primary energy demand edged up by 0.9% in 1999 and 1.8% in 2000, in line with real GDP growth. Australian energy use also resumed its upward trajectory, which had stalled in 1998.

Table 6.2: **Key Economic and Energy Indicators of Japan, Australia and New Zealand, 2000**

	Japan	Australia	New Zealand	OECD
GDP (in billion 1995 $, PPP)	3,036	484	71	24,559
GDP per capita (in 1995 $, PPP)	23,910	25,246	18,367	21,997
Population (million)	127	19	4	1,116
TPES (Mtoe)	525	110	19	5,291
TPES/GDP*	0.17	0.23	0.26	0.22
Energy production/TPES	0.2	2.1	0.8	0.7
Per capita TPES (toe)	4.1	5.7	4.8	4.7

* Toe per thousand dollars of GDP, measured in PPP terms at 1995 prices.

Current Trends and Key Assumptions

Macroeconomic Context

Recent economic performance and near-term prospects vary markedly among the three countries. The Japanese economy remains very weak, while those of Australia and New Zealand have weathered the global downturn well. The combined GDP of the three countries was stable in 2001, a considerable deceleration from the 2.6% growth registered in 2000. Combined GDP is expected to fall fractionally in 2002.

Japan's economy tipped back into recession in 2001 after a modest acceleration in 2000. GDP fell by 0.4%. Falling worldwide demand for information and communication technology (ICT) goods led to a sharp contraction of industrial production and investment, which then spread to other sectors. Private consumption weakened as a result of rising unemployment, some of it resulting from corporate restructuring. Investment may fall further in response to worsening business and consumer confidence, and deteriorating corporate profits. Retail prices have dropped slightly because of weak consumer demand and increased competition. These trends continue despite repeated attempts by the government to counter deflationary forces with fiscal stimuli and an easing of monetary policy. Recent data suggest that the downturn is finally coming to an end and that the economy will pick up in the second half of 2002, with an improvement in global trade. But the recovery is likely to be weak. There are major uncertainties about the implementation of much-needed structural reforms, the write-off of non-performing bank loans and government's large and growing fiscal deficit.

Following a modest slowdown in late 2000, *Australia*'s economy is growing strongly, led by buoyant consumption and new housing (Box 6.1). Although export demand may suffer if the global economic recovery disappoints, sound finances and easy monetary conditions are expected to underpin domestic demand and support growth. GDP is projected to expand by 3.7% in 2002 compared to 2.4% the year before.

New Zealand's economy also rebounded strongly in 2001, breaking with the trend in the rest of the world. The economy was driven by booming export volumes, exceptionally high commodity export prices and a softer currency. Exports fell sharply at the end of 2001, but accelerating domestic consumption and high investment — the result of softer monetary policy and a fall in the unemployment rate to its lowest level since 1988 — cushioned the impact. The immediate prospects for output remain bright.

Box 6.1: **Australia's Economic Resilience**

In contrast to most of the rest of the world, the Australian economy boomed in 2001 and shows every sign of notching up growth of at least 3% in 2002. GDP expanded by more than 4% in the fourth quarter of 2001, by far the best performance of any industrialised country. And while most world stock markets fell sharply from their peaks in early 2000, share prices in Sydney continued to soar, reaching record highs in early 2002.

The prices of the basic commodities, which Australia exports, have fallen since 2000, and recession has hit both of its two main export markets, Japan and the United States. Nevertheless, the Australian economy has benefited from a weak currency, buoyant private consumption and a housing boom. The weak Australian dollar improved the international competitiveness of Australian manufactured goods and raised the local currency profits of energy and other commodity producers whose exports are priced in US dollars. A surge in housing values supported consumer spending and stimulated construction. Moreover, the low importance of ICT firms in Australia protected the economy from the global slump in that industry. Rising household debt and a widening trade deficit pose threats to long-term growth, but the government's budget surplus and low inflation provide a cushion against varying risks, including a collapse in private consumption, higher interest rates and a recession elsewhere in the world.

GDP for the three countries as a whole is assumed to increase at an average of 1.6% annually over the *Outlook* period, sharply down from the 3.2% achieved per year between 1971 and 2000 (Table 6.3). The OECD foresees economic recovery for Japan after 2002, but with slower growth than in the 1970s and 1980s.[1] The region's economic fortunes will remain highly dependent on trade with South-East Asian countries, whose growth is also expected to be less rapid than in the recent past. The region's overall population is assumed to remain stable over the first half of the projection period, with a decline in Japan offset by continued growth in Australia and

1. OECD (2002).

New Zealand. Population will start to fall after 2015 mainly due to a falling birth rate in Japan (Box 6.2).

Table 6.3: **Reference Scenario Assumptions for Japan, Australia and New Zealand**

	1971	2000	2010	2030	Average annual growth 2000-2030 (%)
GDP (in billion 1995 $, PPP)	1,436	3,590	4,200	5,717	1.6
Population (million)	121	150	153	150	0.0
GDP per capita (in 1995 $, PPP)	11,902	23,939	27,407	38,149	1.6

Box 6.2: **The Impact of Demography on Energy Use in Japan**

Japan is expected to undergo large demographic changes over the projection period, and these will have important yet uncertain effects on energy demand. Japan's population will start declining by around 2010, leading eventually to a fall in the number of households. The fertility rate will remain far below what is required to maintain the current population. The National Institute of Population and Social Security Research of Japan estimates that population could peak as early as 2006. The institute's projections show the share of the population aged 65 and over growing from 17% in 2000 to 28% in 2020 and to almost 30% in 2030, while that of the working-age population (15 to 64) will decline from 68% in 2000 to 60% in 2020 and 59% in 2030.

A dwindling and ageing population will detrimentally affect the Japanese economy. It will reduce labour inputs, a key driver of economic growth. Household saving rates will probably drop, and the government's budget will come under pressure as the social security burden increases. On the other hand, tightening labour markets will provide more job opportunities for women and older people, partly offsetting the labour supply effects. The abundance of capital relative to labour will encourage expansion in capital-intensive sectors.

Technological change could improve labour productivity. The net impact will depend on government policies on social security, on retirement rules and on technological change.

Energy demand in the residential sector will probably be more affected by these demographic changes than in other sectors. One of the main determinants of residential energy demand is the number of households, and this number is expected to start declining around 2015. As a result, demand in that sector will slow in the second half of the *Outlook* period. A decline in the number of people per household will also drive down energy use. And the fuel mix in the residential sector will also change. As the population ages, the use of electric appliances and equipment, especially for heating and cooking, will increase faster than that of appliances powered by other fuels, mainly because electric appliances are safer and easier to operate. Energy demand growth in the transport sector will slow. Passenger-vehicle ownership will reach saturation, while the number of kilometres driven may grow more slowly if there is a marked shift to public transport. The direct impact of population changes on energy demand will also depend on the development of home-care services and on improvements in transport infrastructure.

Energy Liberalisation and Prices

Japan continues to liberalise its energy sector with the aim of improving efficiency and lowering pre-tax prices, which are among the highest in the world:

- The deregulation of the oil sector, now largely complete, has unleashed fierce competition in oil product marketing. The resulting pressure on margins is forcing firms to rationalise their operations and cut costs. Several companies have merged since 2000, and a large part of the country's refining capacity has been shut down.
- The Japanese Diet passed a bill in May 1999 amending the Electric Utilities Industry Law to allow a partial opening of the power sector to competition. About 8,000 large industrial and commercial consumers — accounting for about a third of the total power market — may now choose their electricity suppliers. Regional utilities are obliged to allow power from other suppliers to pass

through their grids to eligible consumers. Independent power producers have already entered the market. The government plans to fully open the power market to competition by 2007. Japanese electricity prices are currently the highest in the OECD.

- Gas reforms, which have partially opened up the market to competition, were launched in 1999. Suppliers can now compete for eligible customers outside their traditional service areas. The electric utilities and oil companies are considering selling gas in competition with the city-gas distribution companies. Competition is most intensive in the Kansai area, where Osaka Gas and Kansai Electric Power Company are seeking to gain market shares in each other's territories.

In *Australia*, market reforms in the electricity and gas sectors were launched in the early 1990s. Competition in the power sector was launched in Victoria in 1994 and in 1996 in New South Wales. A National Electricity Market was set up in 1998, but is not yet strongly integrated. The amount of electricity traded across state borders is still low and prices can differ, especially when there are transmission constraints. Nevertheless, the reforms have led to productivity gains and lower prices (Figure 6.1). Moves to open up the gas market to competition based on third-party access are more recent. The Commonwealth Gas Pipelines Access Act and related state legislation adopted in 1997 and 1998 provide the basis for a competitive national gas market. But delays in approving access regimes at the state level have held up the development of effective competition. Full retail competition, with all customers being able to choose their suppliers, is due to be introduced in both the electricity and gas sectors in 2002.

Electricity market reform in *New Zealand* has brought the separation of generation, transmission and distribution functions and the introduction of a light-handed regulatory regime. It has helped to reduce wholesale prices. New legislation adopted in 2001 is aimed at increasing retail competition and driving down prices to household customers.

The implementation of energy-sector reforms and their ultimate impact on end-user prices is a key source of uncertainty in the energy demand projections for all three countries. The *Outlook* assumes that oil product prices will follow the trend in international crude oil markets.[2] The Japanese import price for LNG, which serves as the marker gas price for all the Asia/Pacific countries, is assumed to fall gradually relative to crude oil

2. See Chapter 2 for international price assumptions.

Figure 6.1: **Average Electricity Prices in Australia**

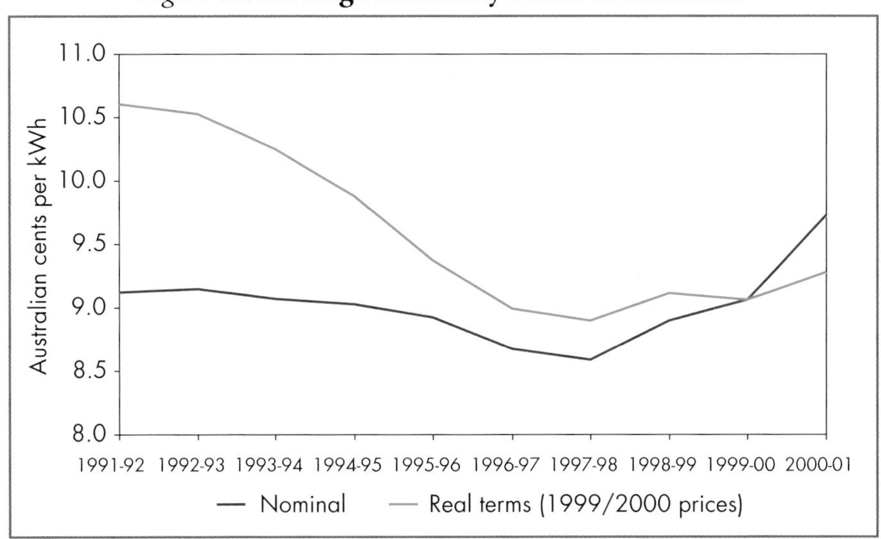

prices. The strong historical link between gas and oil prices, which is embedded in Japan's long-term contracts with suppliers, is expected to weaken as competition intensifies. The expiry of several long-term import contracts over the next few years will provide Japanese buyers with opportunities to press for lower LNG prices in new contracts and to seek out cheaper spot supplies. An increase in short-term trading in LNG, combined with falling shipping costs, will promote a degree of convergence between gas prices in the region, currently the highest in the world, and prices in America and Europe. The price of internationally traded hard coal is assumed to remain flat until 2010 and then rise very slowly through to 2030.

Other Energy and Environmental Policies

Japan is committed under the Kyoto Protocol to reducing its greenhouse gas emissions by 6% below their 1990 level during the period from 2008 to 2012. The government decided to ratify the Protocol in 2002. In 2001, the Japanese Government's Advisory Committee for Natural Resources and Energy completed a review of energy objectives and policies for the next ten years. Recognising the difficulty of increasing Japan's nuclear capacity as fast as originally planned, the review calls for further efforts to save energy, to promote the use of renewables and to encourage switching from coal to natural gas. A package of new or enhanced measures would, it argues, enable Japan to achieve its Kyoto emissions reduction target and to secure stable long-term energy supplies (Table 6.4).

This *Outlook*'s Reference Scenario takes into account measures enacted by mid–2002.[3] These include measures required by the Energy Conservation Law, revised in 1999, the Law on the Rational Use of Energy and the Law on the Promotion of Measures to Cope with Global Warming, both of which were adopted in 1999. The main measures include:

- *Industry*: Voluntary action plans issued by more than 30 industrial associations under the "Keidanren Programme".
- *Residential and commercial sector*: Existing energy efficiency standards for heating and cooling in houses as well as standards for the prevention of heat loss for six kinds of buildings. For electric equipment and appliances such as televisions and refrigerators, Japan has introduced the "top-runner" concept, which sets standards based on the most efficient product available on the market.
- *Transport*: Fuel economy standards for passenger vehicles and small trucks depending on size and fuel. The Reference Scenario assumes that the *average* fuel economy of new cars will rise by 20% in 2010 compared to 1995.

The pace of implementation of the new measures, to be introduced soon under the Global Warming Initiatives programme, as well as their ultimate impact on demand are key uncertainties in the energy outlook for Japan. These factors are analysed in the Alternative Policy Scenario (Chapter 12).

3. Most of the measures correspond to those listed in the base case of the review (Ministry of Energy, Trade and Industry, 2001a).

Table 6.4: **Reduction in Japanese Energy Use in 2010 from Proposed New Measures**

Sector/measure	Base case		Policy case	
	Mtoe	%	Mtoe	%
Industry	**18.59**	**5.4**	**18.96**	**5.5**
Keidanren Voluntary Action Plan	18.59	5.4	18.59	5.4
Promotion of efficient industrial furnaces	0	0	0.37	0.1
Residential/services sector	**12.96**	**3.8**	**17.21**	**5.0**
Efficiency standards for appliances	5	1.5	6.11	1.8
Efficiency standards for buildings	7.96	2.3	7.96	2.3
Promotion of efficient appliances	0	0	0.83	0.2
Promotion of home-energy management system	0	0	0.83	0.2
Promotion of business-energy management	0	0	1.48	0.4
Transport sector	**14.71**	**4.3**	**15.63**	**4.6**
Efficiency standards for cars	5	1.5	5.46	1.6
Promotion of natural gas, hybrid and fuel-cell vehicles	0.74	0.2	1.20	0.4
Promotion of computer-based technology applications	8.97	2.6	8.97	2.6
Cross-sectoral measures	**0**	**0**	**0.92**	**0.3**
Promotion of efficient boilers, lasers and lighting	0	0	0.92	0.3
Total reduction in energy use	**46.26**	**13.5**	**52.72**	**15.4**

Note: Energy savings are expressed as a percentage of total final consumption in 1999. The base case includes measures already introduced. The policy case includes new measures.
Source: Ministry of Economy, Trade and Industry (2001a).

Australia's commitment under the Kyoto Protocol is to hold its greenhouse-gas emissions in 2008-2012 at 8% higher than in 1990. But Australia has not yet ratified the Protocol, and current trends suggest that emissions will turn out much higher than the Kyoto target. The government's response measures include the Greenhouse Gas Abatement Programme, which offers grants for projects that reduce emissions or enhance sinks (forests and other geographical areas that absorb carbon); the Greenhouse Challenge programme, a voluntary energy-efficiency scheme for industry; and mandatory energy-efficiency measures, including standards and labelling. The government has also adopted the Mandatory Renewable Energy Target – a programme designed to raise the share of renewables in electricity generation by 2% by 2010.

Results of the Projections

Overview

Aggregate primary energy demand in Japan, Australia and New Zealand is projected to grow by 0.8% per year from 2000 to 2030 (Table 6.5). Demand growth is fastest in the period to 2010, at 1.2%. It then slows to 0.8% in 2010-2020 and to 0.3% in 2020-2030. These rates are well below the average of 2.4% seen since 1971 and less than those projected for North America. This is because of slower economic expansion, which in turn is partly the result of stagnating population growth in Japan.

Table 6.5: **Primary Energy Demand in Japan, Australia and New Zealand** (Mtoe)

	1971	2000	2010	2030	Average annual growth 2000-2030 (%)
Coal	78	142	147	135	-0.2
Oil	229	308	332	340	0.3
Gas	5	89	113	149	1.7
Nuclear	2	84	105	145	1.8
Hydro	9	11	13	14	0.7
Other renewables	5	19	27	40	2.5
TPES	**329**	**653**	**737**	**823**	**0.8**

Consumption of oil, the main fuel, will continue to grow, but its share in primary energy use will decline steadily, from 47% in 2000 to 41% in 2030 (Figure 6.2). Nuclear energy will grow by 1.8% per year, and its share in the fuel mix will rise from 13% to 18%. Japan will remain the only country of the three with a nuclear power industry. The shares of natural gas and renewable energy grow strongly, reflecting government policies to promote their use.

Primary energy intensity continues its long-term decline, by a projected 0.8% per annum from 2000 to 2030. This is close to the average annual decline since 1971, although energy intensity actually *rose* in the 1990s. The decline will result from a continued shift to low-energy, high value-added manufacturing and services, as well as improvements in energy efficiency. Government climate change policies will also promote efficiency improvements.

Figure 6.2: **Total Primary Energy Demand in Japan, Australia and New Zealand**

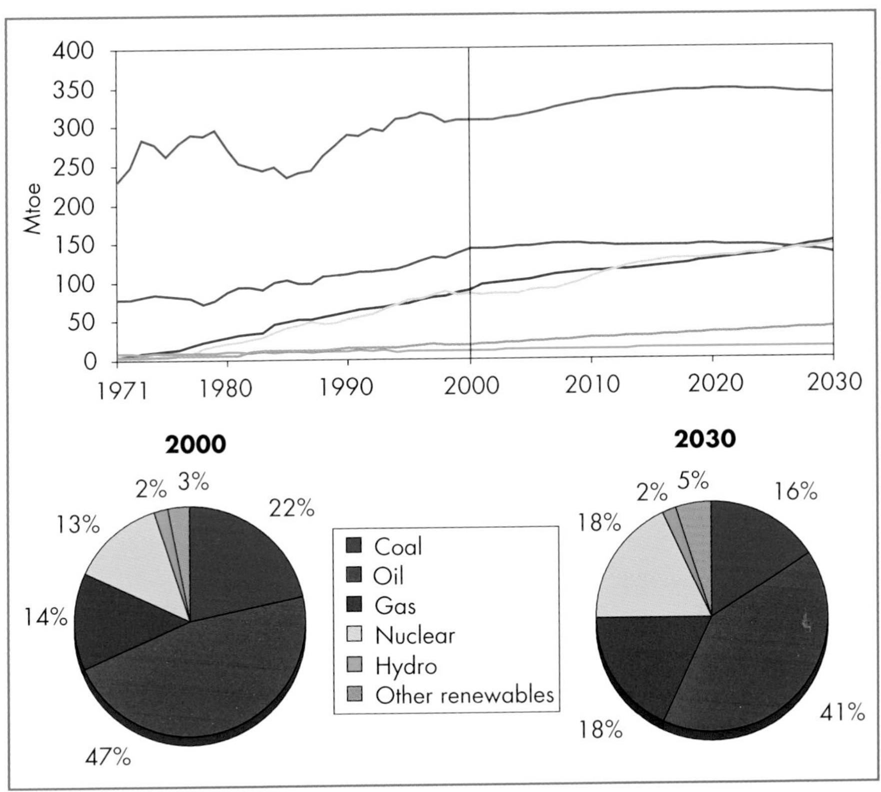

Demand by End-use Sector

Total final energy consumption will grow by 0.7% per year over the *Outlook* period. Oil's share in consumption will continue to decline, and the shares of electricity and to a lesser extent, gas will increase.

Industry remains the largest end-use sector in 2030, but its share of total final consumption declines because demand in the transport, residential and services sectors grows faster (Figure 6.3). Energy use in industry grows modestly, by 0.5% per year from 2000 to 2030. A combination of structural change in Japan's economy and continued efforts to improve energy efficiency will help to restrain growth of industrial energy demand. Heavy industry's energy needs will grow much

Figure 6.3: **Total Final Consumption by Sector in Japan, Australia and New Zealand**

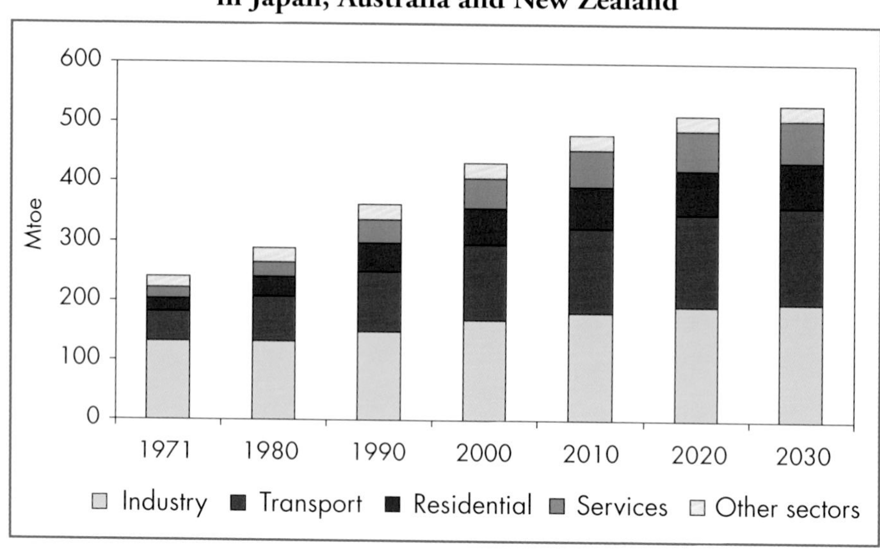

Note: "Other sectors" include agriculture, non-energy use, and non-specified energy use.

more slowly than those of less energy-intensive industries such as machinery, transport equipment and information and communication goods. The fuel mix will continue to shift away from oil and towards gas and electricity, albeit at a slightly slower pace than in recent years.

Demand for transport fuels rises on average by 0.8% per year from 2000 to 2030, but at a decelerating rate. Growth in transport fuel demand will be much slower than it was in the recent past — 2.4% from 1990 to 2000. This decline reflects further improvements in vehicle fuel efficiency (brought about in part by climate change initiatives), saturation in car ownership and increasing oil prices after 2010. Car ownership and use in Japan, Australia and New Zealand is still lower than in other OECD regions. This is due to the well-developed public transport systems, traffic congestion and physical limits on expanding the road network in Japan. These factors, as well as the prospect of an ageing and declining population, will continue to hold back growth in driving.

Energy use by the residential and services sectors (including agriculture) will also grow at an average rate of 0.8% per year over the projection period, slowing progressively because of saturation in demand for space and water heating as the population starts to fall after 2015. More energy-efficient appliances and equipment will also curb residential and services sector use. Electricity consumption grows faster than that of fossil fuels. Electricity will account for half of total energy consumption in the residential and services sectors by 2030, while the share of oil will fall steeply, to 30% in 2030. Demand for renewable energy is also expected to grow rapidly, especially solar power for household water heating in Japan. But the overall share of renewable energy will be only 4% of total energy consumption in the residential and services sectors in 2030.

Oil

Primary oil consumption grows by an average of 0.3% per year, from 6.4 mb/d in 2000 to 7 mb/d in 2030. It will actually decline after 2020, because of falling demand in power generation. Transport accounts for the bulk of the increase in oil demand. Imports, which already meet 87% of the region's needs, will rise even further – most sharply during the first decade of the projection period when production is expected to start declining in Australia (Figure 6.4). Import dependence will stabilise at about 92% between 2010 and 2030. Japan accounts for almost all net oil imports because it is the largest consuming country of the three (86% of total demand in 2000) and because it produces virtually no indigenous oil.

Figure 6.4: **Oil Balance in Japan, Australia and New Zealand**

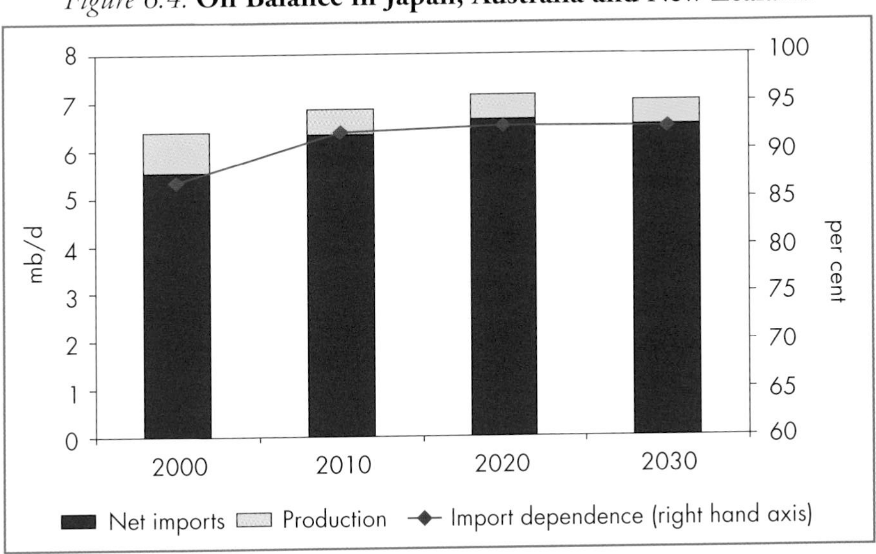

Australia is the only one of the three countries with significant oil resources,[4] but proven reserves – at 2.9 billion barrels at the beginning of 2001[5] – are small relative to current production. Output increased sharply in the last two years as new offshore fields came on stream, reaching 730 kb/d in 2001. Production is nonetheless expected to decline in the coming years, as older fields are depleted and new developments fail to make up the difference. Australia's net imports of oil, which fell to only 100 kb/d in 2001 from 200 kb/d in the late 1990s, will begin to rise again in the near future. New Zealand relies on imports for around 70% of its needs.

Today, over 80% of Japan's imported oil comes from the Middle East, compared to around 65% in the 1980s. Like virtually every importing country, Japan's dependence on Middle East oil is set to rise, as other sources peak and decline. This highlights major concerns about the vulnerability of Japan and other countries to oil supply disruptions and the importance of their being prepared to handle such events.

Natural Gas

Primary gas demand in Japan, Australia and New Zealand will rise from 103 bcm in 2000 to 172 bcm in 2030 – an increase of 1.7% per year. More than 70% of the extra gas will go to power generation. In its recent long-term policy review, the Japanese government increased the priority of promoting natural gas, mainly because of its environmental advantages. Gas consumption will also increase strongly in Australia, where there are low-cost indigenous resources, as well as robust demand in the industrial sector and in power generation. In both countries, gas demand will grow most rapidly in the first decade of the projection period, when gas prices are assumed to decline slightly. Rising prices will dampen demand growth after 2010.

Australia has large and expanding gas reserves and the potential to discover even larger quantities. Proven reserves stand at 3.5 tcm,[6] equivalent to well over a hundred years of production at current rates. Mean undiscovered resources are estimated at 3.1 tcm.[7] Major gas discoveries were made in 1999 and 2000, including those by the West

4. According to the USGS, Australia has undiscovered crude oil resources of 5 billion barrels and undiscovered natural gas liquid resources of 5.8 billion barrels. (USGS, 2000).
5. Radler (2000).
6. Cedigaz (2001).
7. USGS (2000).

Australia Petroleum consortium in the Gorgon area offshore north-western Australia. The Gorgon field and others in the vicinity could contain more than 490 bcm of proven and probable reserves.

Australian production is expected to shift from the Coopers/Eromanga and Gippsland basins, where reserves are declining, to the Carnarvon Basin on the North West Shelf and to the Timor Sea in northern Australia. The latter may provide LNG exports as well as serve local demand, possibly feeding in to a planned pipeline system that could extend to Queensland. LNG exports to other Asia/Pacific countries will underpin the expansion of North West Shelf production (Box 6.3). Total gas production of Japan, Australia and New Zealand is projected to increase threefold from 41 bcm in 2000 to 122 bcm in 2030 (Figure 6.5).[8]

Japan is the only importer of gas among the three countries. Imports are entirely in the form of LNG, coming from Indonesia, Malaysia, Australia, Brunei, the United Arab Emirates, Qatar, Oman and the US state of Alaska. Imports reached 72.3 bcm in 2000, making Japan the world's largest LNG market. LNG is expected to meet most of the country's additional gas needs over the projection period. Japan is already committed to lifting more LNG from Malaysia starting in 2003 and from Australia (NWSP) in 2004. These volumes will partly make up for the

Figure 6.5: **Natural Gas Balance in Japan, Australia and New Zealand**

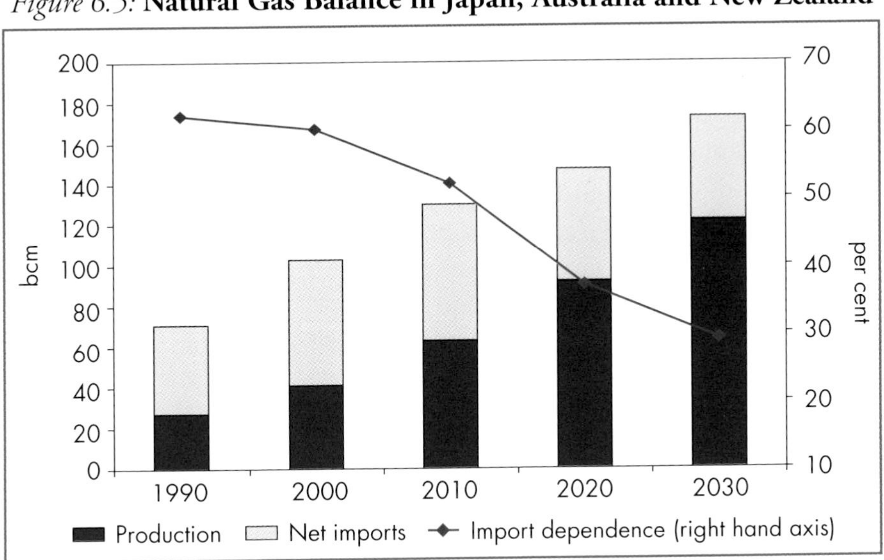

8. This number does not include gas-to-liquid production.

expiry of existing contracts. Buyers are expected to insist on more competitive pricing formulas and more flexible contracts, with shorter terms. A growing proportion of Japan's incremental demand is likely to be met by spot purchases.

Box 6.3: **Australian LNG Export Prospects**

Much of the projected increase in Australian gas output will go to export markets in the form of LNG. Strong growth in demand for gas is expected in Asia/Pacific markets throughout the projection period. Along with Japan, currently the principal market for Australian LNG, China and India are expected to emerge as major potential buyers.

Australia has been exporting LNG from the North West Shelf Project (NWSP) since 1990. Exports reached 10.6 bcm in 2000, with more than 95% going to Japan under long-term contracts. The rest was sold on a spot basis to US and Korean buyers. The partners in the NWSP consortium reached agreement in 2001 on building a fourth train with a capacity of 4.2 Mt/year, the largest ever to be built anywhere. Contracts have been signed with Japanese buyers, and first gas is expected in 2004. In August 2002, Australia secured a contract to export LNG to China's first receiving terminal in Guangdong. The first gas will be delivered in 2005. With this new contract, Australia is considering building a fifth train on the North West Shelf.

Various other LNG projects have been proposed, including: plants based on the Gorgon field some 100 km south of the NWSP; the Scott Reef/Brecknock fields 750 km north-east of the NWSP; the Bonaparte Basin in offshore Northern Territory (the Sunrise Project); the Scarborough field in Western Australia; and the Bayu-Undan field in the Timor Sea. Development of the Bayu-Undan LNG project, with a 3 Mt/year capacity is expected to start in 2002. In March 2002, agreements were signed for the sale of virtually all the gas in the field over 17 years to Tokyo Electric Power Company and Tokyo Gas. The first cargo is scheduled for 2006.

There are two main pipeline projects that could eventually supply the Japanese market with gas from Russia: from the Kovykta field near Irkutsk via China and Korea and from the Sakhalin field via an undersea pipeline. Estimates by the Asia-Pacific Energy Research Center of the cost of transportation suggest that these options will not be competitive with LNG in the near term (Figure 6.6).[9]

Figure 6.6: **Indicative Pipeline and LNG Transportation Costs to Japan**

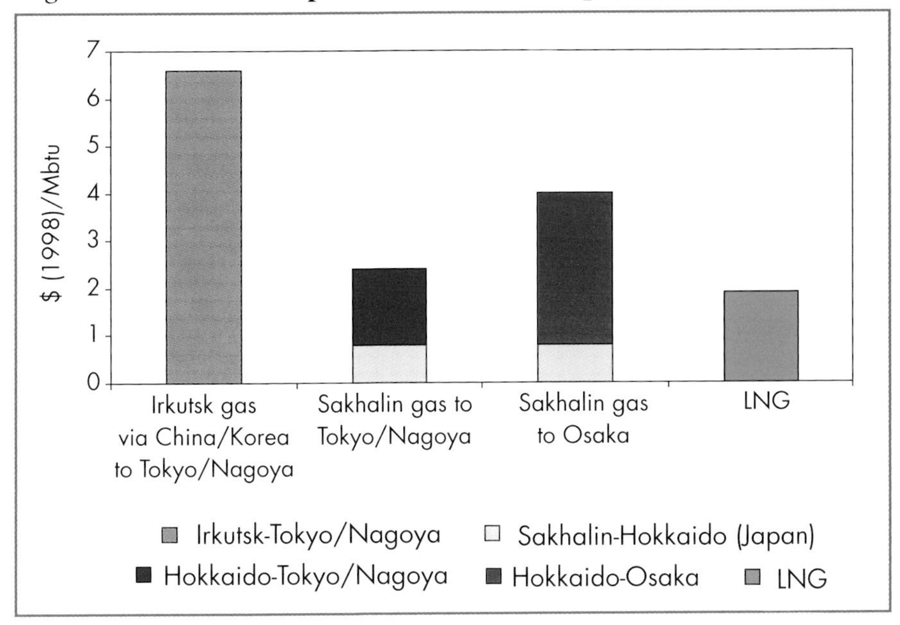

Coal

Primary coal demand in Japan, Australia and New Zealand will decline by 0.2% per year until 2030. Industrial coal use will continue to decline steadily; coal use in power generation will increase slowly to 2020 and then fall.

Australia has virtually all of the region's coal reserves. It is the world's largest exporter and sixth-largest producer of coal. With more than 42.5 billion tonnes of economically recoverable hard coal reserves, it has the potential to expand production greatly. Japan, the world's largest coal importer, will remain the biggest single market for Australian coal, but demand from other markets in Asia is expected to grow more rapidly. The Australian coal industry is undergoing consolidation and rationalisation aimed at improving productivity in the face of intense competitive pressures and changing technology. Increasing exports from China will intensify competition in the regional market.

9. Asia-Pacific Energy Research Center (2000).

Box 6.4: **Gas Hydrates in Japan**

Japan is exploring the potential of gas (methane) hydrates production. The economic extraction of methane from hydrate deposits would dramatically enhance Japan's energy security and bring environmental benefits, if the gas were to replace coal or oil.

Gas hydrates are crystallised ice-like substances, inside which molecules of methane, the chief constituent of natural gas, are trapped. Hydrate deposits in Japan are estimated at about 7.4 tcm of natural gas, 100 times annual domestic consumption now. Developing techniques for safely extracting gas from deep undersea hydrate deposits is a major challenge. The success of a test well in Canada in 2002, which yielded methane from hydrates at more than 1,000 metres underground, has raised hopes that Japan may one day be able to exploit its resources. But considerable reductions in cost will be needed to make gas hydrates competitive with other sources of gas.

Electricity

Electricity demand will rise by 1.1% per year over the projection period. Growth is quicker in the first two decades, rising in line with GDP. Later on, saturation of the market for appliances and energy-using equipment in the residential and services sectors, as well as a stagnating population will cause electricity demand to decelerate. It nonetheless grows faster than coal and oil, so that its share in total final consumption rises from 23% in 2000 to 26% by 2030. It stood at only 14% in 1971.

There are major differences among the three countries in the shares of each fuel used to generate electricity (Table 6.6). Japan generates power mainly from nuclear energy, natural gas and coal. Australia is heavily reliant on coal, while New Zealand depends on hydropower for more than 60% of its electricity.

Table 6.6: **Electricity Generation Mix in Japan, Australia and New Zealand** (TWh)

	2000				2030
	Japan	**Australia**	**New Zealand**	**Total**	**Total**
Coal	254	161	1	416	394
Oil	159	3	0	162	104
Gas	239	26	9	276	405
Hydrogen - Fuel cell	0	0	0	0	105
Nuclear	322	0	0	322	556
Hydro	87	17	25	129	161
Other renewables	20	2	4	25	92
Total	**1,082**	**208**	**39**	**1,329**	**1,818**

The generation mix of the three countries combined is expected to shift towards nuclear energy, which will account for about half of total incremental generation between 2000 and 2030 (Figure 6.7). In 2030, more than 30% of electricity will be generated by nuclear power plants, all of them in Japan. The role of natural gas will also increase. Non-hydro renewables will grow at the fastest rate, by 4.4% per year from 2000 to

Figure 6.7: **Change in Electricity Generation by Fuel in Japan, Australia and New Zealand**

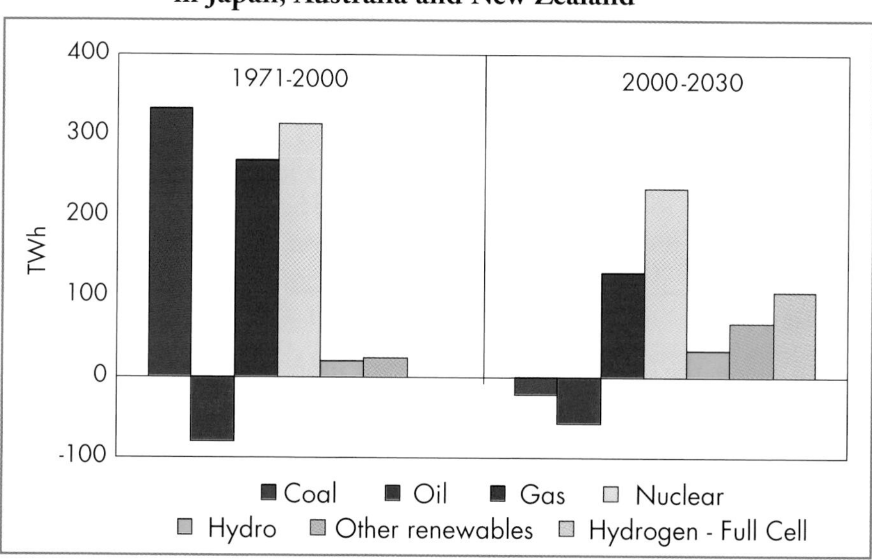

Figure 6.8: **CO_2 Emissions and Intensity of Energy Use in Japan, Australia and New Zealand**

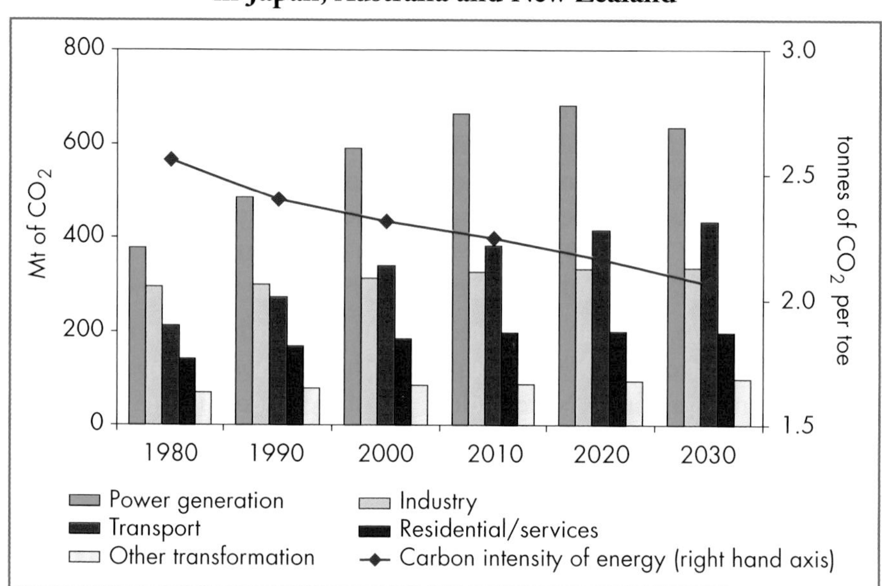

Korea

Energy Market Overview

Korea's economy grew at an impressive 7.4% per year over the past three decades, transforming the country from one of the poorest in the world to one of the richest. The heavy industrial sector, which generated high exports, was the main engine of growth. With a population of 47 million and GDP of $776 billion in the year 2000 (measured in PPP), Korea is the tenth-largest OECD economy.

Primary energy use surged by 8.8% per year between 1971 and 2000 (Table 6.8). Korea now accounts for more than 2% of world demand, up from just 0.3% in 1971. Per capita primary energy demand is approaching that of Japan, Australia and New Zealand.

Korea is highly dependent on imported energy. Imports meet more than 80% of the country's energy demand. Korea is the world's fourth-largest oil importer and the second-largest importer of both coal and LNG. The Middle East supplies about 75% of Korea's oil imports. The country's energy intensity increased in the 1990s, mainly because economic growth in that period was led by energy-intensive industries, such as petro- chemicals, steel and shipbuilding.

Table 6.8: **Key Energy Indicators of Korea**

	1971	1990	2000	Average annual growth 1990-2000 (%)
TPES (Mtoe)	17	93	194	7.7
Coal demand (Mt)	13	45	67	4.1
Oil demand (mb/d)	0.2	1.0	2.1	7.6
Gas demand (bcm)	-	3.2	20	20.0
Nuclear (Mtoe)	-	14	28	7.5
TPES/GDP*	0.17	0.22	0.25	1.4
Energy production/TPES	0.38	0.24	0.17	-3.1
Per capita TPES (toe)	0.5	2.2	4.1	6.6

* Toe per thousand dollars of GDP, measured in PPP terms at 1995 prices.

Oil is Korea's most important fuel with 54% of the primary energy mix in 2000. Coal demand remains strong, especially in power generation, although its share of primary energy has fallen markedly since the 1970s. Low-quality Korean coal has been largely replaced by natural gas in the residential sector. Steam coal for power generation is mostly imported. Natural gas, in the form of LNG, was introduced to Korea only in 1986. Since then, gas consumption has grown quickly, reaching 9% of primary energy demand in 2000. The role of nuclear power has also expanded. It provided 15% of the country's primary energy in 2000.

The Asian economic crisis seriously damaged the Korean economy, causing GDP to fall by 6.7% in 1998. As a result, primary energy demand fell by 7.6%. Oil demand dropped particularly sharply, by 16%, due mainly to weakness in the transport and service sectors.[10] Energy demand resumed its upward trajectory in 1999 with a rapid economic recovery. Demand returned to its 1997 level in 1999 and rose by a further 7% in 2000.

The Korean government has accelerated energy market reforms since the crisis. The oil sector has been largely deregulated, many aspects of the electricity sector are being reformed in phases, gas sector reforms have been discussed and a new regulatory framework is being established for

10. After the crisis, the government raised the tax on petroleum products. This, combined with the dramatic devaluation of the Korean won, resulted in a reduction in oil demand in these two sectors.

electricity and gas. The government plans to restructure and privatise the state-owned energy companies.

Current Trends and Key Assumptions

Macroeconomic Context

The very strong recovery of the Korean economy continued in 2000. GDP grew by 10.7% in 1999 and by 9.3% in 2000. Although falling investment and stagnant exports slowed GDP growth to 3% in 2001, rapid growth resumed in 2002. The recovery was led by strong private demand, especially for motor vehicles and durable goods. Unemployment has fallen to its lowest level since the crisis. In September 2000, the financial authority launched the second stage of its financial sector-restructuring plan, which includes 40 trillion won ($33 billion) of additional public expenditure. The banking sector recorded profits in 2001 for the first time since 1997, and the ratio of non-performing loans to total bank assets fell to 3.4%.

With brisk private consumption and rising exports, economic growth is expected to accelerate in 2003. The *Outlook* assumes that Korea's GDP will grow by 3.6% per year over the projection period. The rate of growth will, however, decrease, from 4.3% in the period 2000 to 2010, to 3.7% between 2010 and 2020 and to 2.9% from 2020 to 2030 (Table 6.9). Although these rates are among the highest in the OECD, they are much lower than Korea's own performance in the 1990s. This is because the economy is maturing and export demand growth is expected to slow. The restructuring of highly indebted companies still poses a risk to economic growth in the first decade of the *Outlook* period.

The population is assumed to expand by 0.4% per year, reaching 53 million in 2030. Korea has the third-youngest population in the OECD after Mexico and Turkey, with an average age of 33 compared to 38 in the OECD region as a whole. But, like Japan, Korea's population is ageing rapidly. Economic growth will slow as workers retire and fewer people enter the workforce. Growing numbers of older people will need expensive pensions and health services.

Table 6.9: **Reference Scenario Assumptions for Korea**

	1971	2000	2010	2030	Average annual growth 2000-2030 (%)
GDP (in billion 1995 $, PPP)	99	776	1,183	2,259	3.6
Population (million)	33	47	50	53	0.4
GDP per capita (in 1995 $, PPP)	3,017	16,433	23,605	42,582	3.2

Energy Liberalisation and Prices

The Korean government has traditionally intervened heavily in the energy sector. It played a major role in managing national energy companies such as Korea Electric Power Corporation (KEPCO) and Korea Gas Corporation (KOGAS). It also regulated the private refinery industry by licensing, by controlling output and by approving exports and imports. Energy reforms were started under the Five-Year Economic Plan in 1993.

Since the economic crisis, Korea has accelerated its progress towards a more market-oriented energy sector. The oil-refining sector was totally deregulated in October 1998, ahead of schedule, in order to attract foreign investment and to introduce competition to the industry. Reform in the electricity sector started in 1999. The government split KEPCO into seven companies in 2001. It plans to partially privatise these companies in stages. In 1999, a plan to privatise KOGAS was also announced, but strong opposition from the labour unions has delayed this move. Two KOGAS subsidiaries are to be sold to private investors by the end of 2002.

For many years, one of Korea's main energy policy objectives was to keep prices down. Low prices were felt to benefit households and keep industry competitive. The government regulated fuel prices through its direct supervision of the state-owned energy companies and through price controls on private companies. Many of these measures have now been lifted:

- Coal prices are now largely determined by the market, but the government still sets a ceiling on the price of domestically produced anthracite, which is mostly consumed by low-income households and by the agricultural sector.

- The prices of crude oil and petroleum products are still regulated, but these controls are to be lifted by 2006. Prices are being gradually adjusted to eliminate distortions (Table 6.10). The large difference between the prices of diesel and kerosene, for example, has encouraged the illegal substitution of cheap kerosene for diesel in motor vehicles.
- The prices of gas and electricity are still subject to government approval. Cross-subsidies are rife, and average prices do not cover the full cost of supply. The government plans to remove all subsidies as part of its restructuring process.

Table 6.10: **Planned Oil-Product Price Adjustments Relative to Gasoline**
(Gasoline price = 100)

	Gasoline	Diesel	LPG	Kerosene
Pre-reform	100	47	26	40
July 2001	100	52	32	43
July 2002	100	56	38	45
July 2003	100	61	43	48
July 2004	100	66	49	50
July 2005	100	70	54	53
July 2006	100	75	60	55

Source: Republic of Korea Ministry of Commerce, Industry and Energy (1999).

The *Outlook* assumes that final energy prices in Korea will rise progressively to international levels by 2010 as the energy reforms proceed. Prices will move thereafter in line with international markets.[11]

Results of the Projections

Overview

Over the *Outlook* period, total primary energy demand in Korea is projected to grow by 2.3% per year on average, reaching 378 Mtoe in 2030 (Table 6.11). Although well above the average for the rest of OECD Pacific and for the other OECD regions, the projected growth in demand is much lower than the 8.8% rate from 1971 to 2000.

11. See Chapter 2 for international price assumptions.

Box 6.5: **Energy Co-operation with North Korea and Other Neighbouring Countries**

In order to enhance Korea's energy security, the government has announced its intention to promote energy co-operation with North Korea and other neighbouring countries. At the request of North Korea, the government in Seoul is examining the feasibility of exporting electricity to the North. The Korea National Oil Corporation (KNOC) is considering joint exploration and development of offshore oil deposits on the North Korean coast. Korea is also considering energy co-operation in North-East Asia with China, Japan, Russia and Mongolia. Joint projects under consideration include a common North-East Asia market in oil and cross-border gas, as well as several power interconnection projects.

Table 6.11: **Primary Energy Demand in Korea** (Mtoe)

	1971	2000	2010	2030	Average annual growth 2000-2030 (%)
Coal	6	42	58	79	2.2
Oil	11	104	126	165	1.6
Gas	-	17	33	61	4.4
Nuclear	-	28	45	65	2.8
Hydro	0	0	0	1	1.7
Other renewables	-	2	3	7	4.0
TPES	17	194	264	378	2.3

Korea's energy intensity, measured by primary energy demand per unit of GDP, is among the highest of OECD countries. It will start to fall in the first decade of the projection period, reversing the upward trend of the last three decades. The rate of decline is projected to average 1.3% per year between 2000 and 2030. This development results from the slower growth in the main energy-consuming industries – iron and steel, cement and petrochemicals. Other factors include higher energy prices and improved energy efficiency.

Figure 6.9: **Total Primary Energy Demand in Korea**

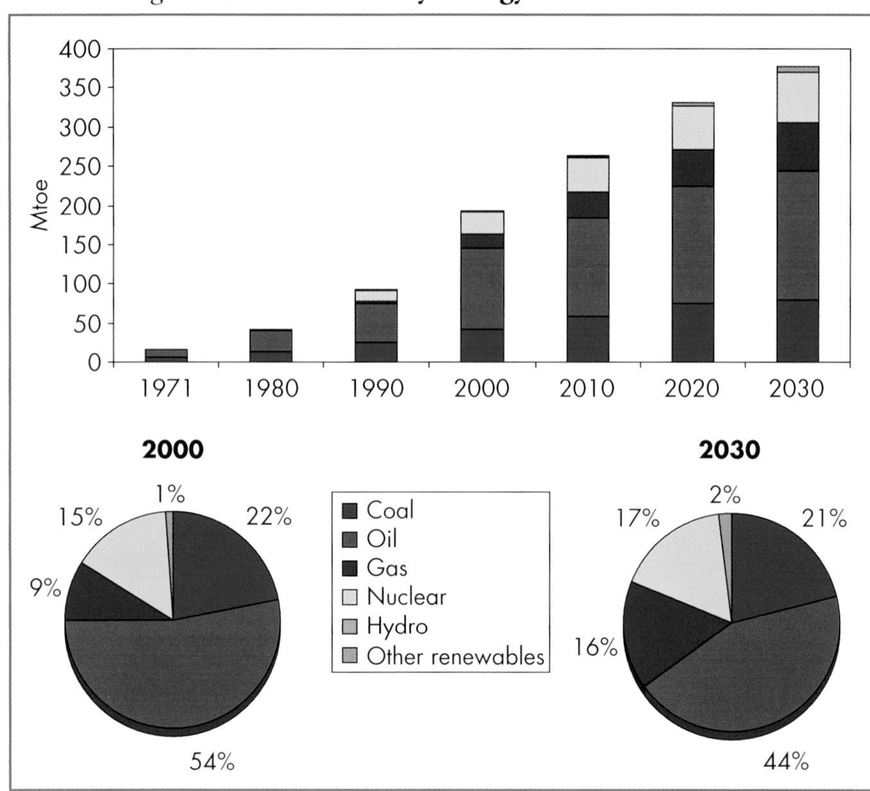

Oil will continue to dominate the primary fuel mix throughout the projection period, but its share will fall sharply, from 54% in 2000 to 44% in 2030 (Figure 6.9). Gas will be the fastest-growing fossil fuel. Gas demand will grow by 4.4% per year, and its share of primary energy demand will rise from 9% to 16%. Nuclear supply will also increase steadily, while coal will lose some of its market share. The supply of non-hydro renewables, mainly wind power, will grow by 4% per year from 2000 to 2030.

Demand by End-use Sector

Korea's final energy consumption is projected to grow by 2.2% per year. Although industry will remain the largest end-use sector in 2030, with 40% of the total final energy demand, energy demand in the transport, residential and services sectors will have increased even more rapidly (Figure 6.10).

Figure 6.10: **Total Final Consumption by Sector in Korea**

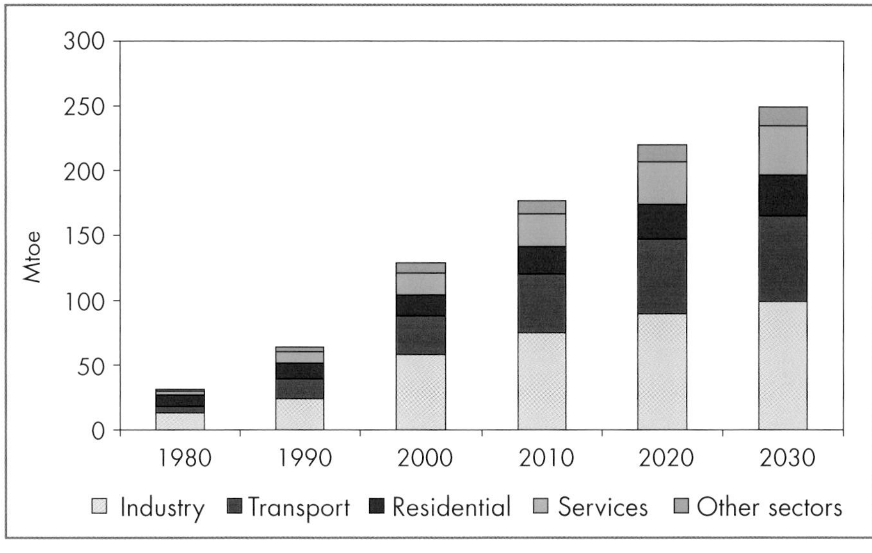

Note: "Other sectors" include agriculture, non-energy use, and non-specified energy use.

Energy use in industry will grow by 1.8% per year from 2000 to 2030, compared to 9.1% per year in the 1990s. Energy demand in heavy manufacturing, which accounted for more than a half of industrial energy consumption in 2000, will grow much more slowly than in lighter industrial sectors. The removal of price subsidies will encourage all industries to use energy more efficiently. Oil will be increasingly replaced by electricity and, to a lesser extent, gas.

Energy use for transport will grow faster than that in any other end-use sector, by 2.7% per year over the projection period. Its share in total final energy consumption will increase by four percentage points to 27% in 2030.

Energy demand in the residential, services and agriculture sectors combined will rise by an average 2.4% per year in 2000-2030. The share of gas will increase further, to 27% in 2030. Electricity demand will grow even more rapidly, with its share reaching 37% by 2030. Rising household incomes will underpin strong sales of electrical appliances and equipment.

Oil

Primary oil demand in Korea will increase from 2.1 mb/d in 2000 to 3.4 mb/d in 2030, an average annual rate of growth of 1.6%. This is much slower than the 8.2% annual rate of the last three decades. Over half of the increase in final oil demand will come from the transport sector with much of the remainder from the industry sector (Figure 6.11).

Figure 6.11: **Change in Oil Demand by Sector in Korea**

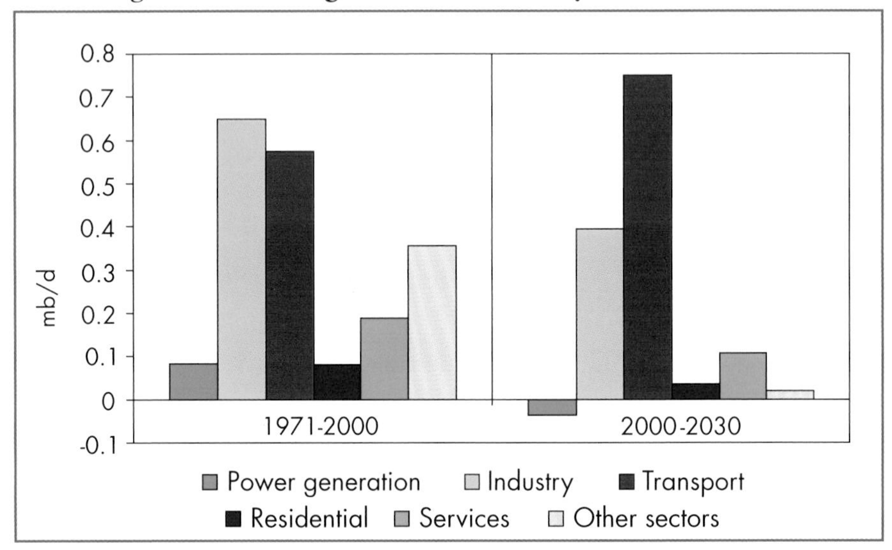

Note: "Other sectors" include other transformations, agriculture, non-energy use and non-specified energy use.

Korea has virtually no domestic oil resources and so must import almost all its needs. The country imports about three-quarters of its oil from the Middle East. Saudi Arabia is the largest supplier, followed by the United Arab Emirates, Iran and Oman. Korea has been trying in recent years to diversify its sources of oil imports. KNOC has been involved in nearly twenty overseas oil exploration and production projects in twelve countries. KNOC's most recent major discovery was in offshore Vietnam, where commercial production is due to begin in 2004. Korea's high dependence on Middle East oil will continue, however, partly because most of its refineries are specially designed to process Middle East grades.

For energy security reasons, the government has tried to maintain 30% more oil refinery capacity than is needed to meet domestic demand. Total capacity of the country's five oil refineries stood at 2.4 mb/d at the beginning of 2000, the fifth-highest in the world. The country has become a major exporter of refined products, with exports of more than 800 kb/d in 2000.

Natural Gas

Primary gas demand will grow from 19.6 bcm in 2000 to 70.2 bcm in 2030, an average increase of 4.4% per year. Just under two-thirds of the incremental gas demand will come from the power generation sector, and half the rest will come from the residential sector (Figure 6.12).

Korea has very limited gas reserves. Demand is currently met entirely by imports, all in the form of LNG. In 2000, 42% of the country's gas imports came from Indonesia, 22% from Qatar and 17% from Malaysia. Korea also imports LNG from Brunei, the United Arab Emirates, Australia and Oman. The share of the Middle East in Korea's LNG imports has been increasingin recent years. All the LNG import contracts have been negotiated by KOGAS. The company operates the LNG receiving terminals, storage facilities and transmission pipelines in the country. It also has a monopoly of gas sales to large industrial customers, power generators and distribution companies. The Donghae-1 field in the south-east of the country could start commercial production in 2003, but

Figure 6.12: **Increase in Natural Gas Demand in Korea by Sector, 2000-2030**

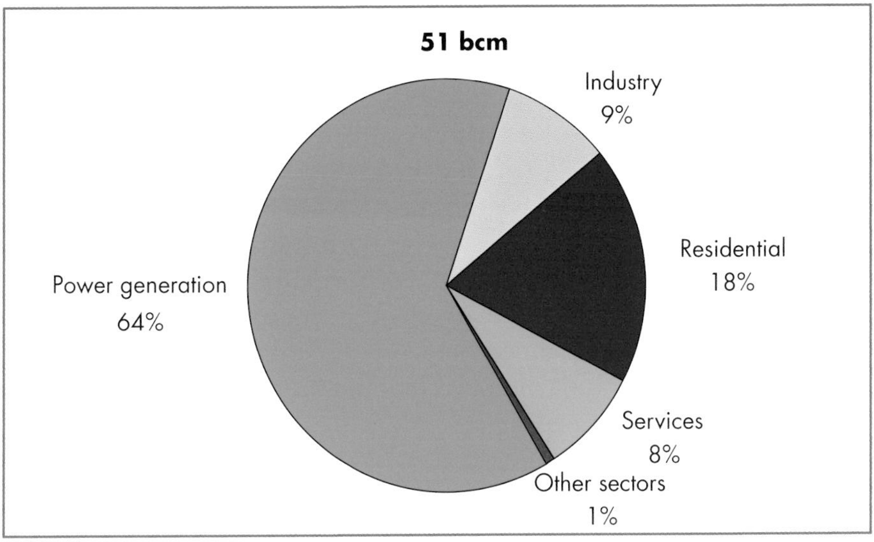

Note: "Other sectors" include other transformation, agriculture, non-energy use and non-specified energy use.

volumes will be small. As with Japan, gas supply by pipeline from Russia is a long-term option for Korea (Box 6.6).

Box 6.6: **The North-East Asia Cross-Border Gas Pipeline Project**

In November 2000, KOGAS, the Russia Petroleum Company and the China National Petroleum Corporation agreed to begin a feasibility study on bringing Russian gas to China and Korea. The Korean government sees the project as a means of diversifying its gas imports and improving supply security.

The gas would come from the Kovykta field in the Irkutsk region of Russia, a field which has proven reserves of from 0.9 tcm to 1.2 tcm. The project will require the construction of a 4,000-kilometre pipeline. There are two possible routes for the line – a direct route traversing Mongolia and a more circuitous one that avoids Mongolia. The preferred option will depend on negotiations with Mongolia over transit fees. The initial cost of a 56-inch pipeline alone would be around $7 billion. Supply could begin as early as 2008, but a number of factors could delay the completion of the project. These include negotiations among China, Russia and Korea over the delivered price of the gas. Another factor is the planned West-East Gas Pipeline in China, which would reduce Chinese demand for Russian gas and raise the unit cost of the Irkutsk-Korea line. Political tension between North and South Korea is an additional risk.

Reform of the gas sector, including the restructuring of KOGAS, is the source of major uncertainty for the gas outlook. A reform plan was announced in November 1999 after the adoption of a similar plan for the electricity industry. The gas plan calls for the privatisation of KOGAS. An initial public offering of 43% of the company's equity was completed in 1999, and the government hopes to sell off its remaining stake by the end of 2002. The gas plan calls for gas-to-gas competition based on open access to the gas-pipeline network and LNG terminals. The following actions will be taken to achieve this goal:

- KOGAS's import and wholesale businesses will be divided into three trading companies, only one of which will become a subsidiary of KOGAS.

- KOGAS's long-term LNG purchasing agreements will be shared out among the three new companies.
- Open access to all terminals and the transmission network will be made mandatory.
- Retail competition will be extended progressively to small consumers by opening up local distribution networks to competing suppliers.
- An independent regulatory body will be set up.

The timetable for completing the privatisation of KOGAS and restructuring the industry is very uncertain. Detailed issues still need to be resolved, including the allocation of the seven existing LNG contracts among the new trading companies and the pricing of network services. Strong resistance from the labour unions, as well as KOGAS's large debts ($4.6 billion in 2001), are holding up full privatisation.

Coal

Korea greatly expanded its coal use after the oil crises in the 1970s, and primary coal consumption has since grown more than sevenfold. Most of this increase has come from industry and power generation. In the residential sector, which used mainly coal in the 1970s, gas and electricity have largely replaced coal. Overall demand for coal in Korea is projected to increase by 2.2% per year from 2000 to 2030.

The power sector's share of coal consumption jumped from 23% in 1990 to 63% in 2000, partly as a result of the low price of Korea's coal relative to natural gas. This change in coal consumption patterns is expected to continue, and 72% of coal will be used in power plants in 2030 (Figure 6.13).

Korea has small reserves of low-quality, high-ash anthracite coal, which will last 19 years at current rates of production. The government plans to continue shutting down inefficient coal mines. Almost all Korea's coal needs are now met by imports, and Korea is the world's second-largest importer of both steam and coking coal. It has benefited from the highly competitive international coal market, switching suppliers according to availability and price. Although the bulk of Korea's steam coal has historically come from Australia, imports from China are increasing rapidly, as are imports from Indonesia and South Africa.

Figure 6.13: **Primary Coal Demand by Sector in Korea**

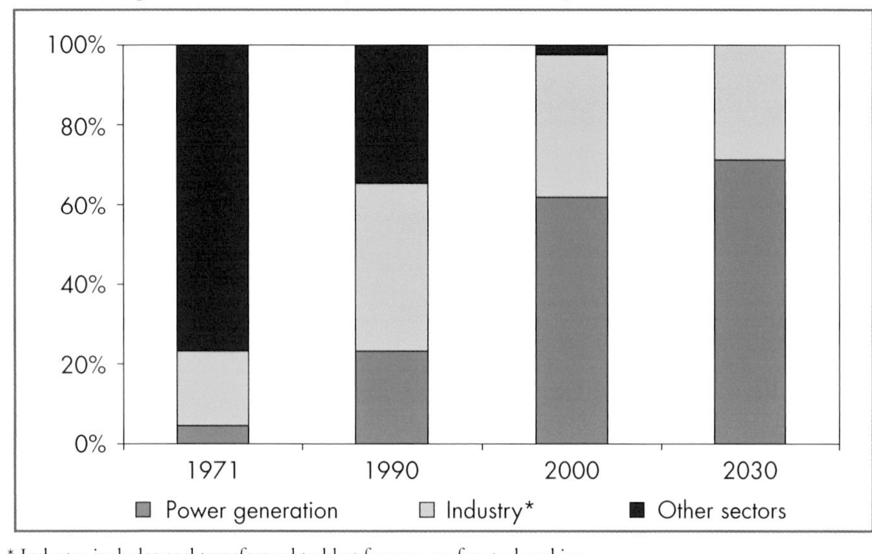

* Industry includes coal transformed to blast furnace gas for steel-making.

Electricity

Final electricity demand will grow by 3% per year over the *Outlook* period (Table 6.12), a sharp deceleration from the 12.1% annual growth rate from 1971 to 2000. Electricity will, nonetheless, remain Korea's fastest growing source of final energy. Electricity's share in total final consumption is projected to increase from 18% in 2000 to 22% in 2030.

Electricity generation reached 292 TWh in 2000. Coal-fired plants produced 43% and nuclear plants 37% (Figure 6.14). The generation fuel mix is now expected to shift towards natural gas, which accounted for less than 10% of total generation in 2000. Gas's share is projected to rise to 17% by 2030. But coal and nuclear energy will nonetheless remain the dominant fuels, accounting for nearly 75% of electricity output in 2030. In order to meet projected electricity demand, Korea needs to build around 100 GW of new generating capacity, at a cost of $88 billion.

Table 6.12: **Electricity Generation Mix in Korea** (TWh)

	2000	**2010**	**2020**	**2030**	**Average annual growth 2000-2030 (%)**
Coal	126	185	253	275	2.6
Oil	25	23	22	18	-1.1
Gas	28	60	95	121	5.0
Hydrogen - Fuel cell	0	0	1	17	–
Nuclear	109	171	210	248	2.8
Hydro	4	5	6	7	1.7
Other renewables	0	2	5	15	12.3
Total	**292**	**446**	**592**	**701**	**3.0**

Korea has 16 nuclear reactors with a total capacity of 14 GW. The current government plan calls for twelve new plants to be built by 2015, including the four already under construction. But difficulties in finding suitable sites, together with strong competition from gas and coal, are expected to slow the rate of construction. The *Outlook* assumes that nuclear capacity will reach 21 GW by 2010 and 31 GW by 2030.

The Korean government has set an ambitious target for renewables, aiming to ensure them a 2% share of primary energy supply by 2006 and 5% by 2010. The share in 2000 was only 1%. The *Outlook* projects that electricity generation based on renewables will grow steadily as they become cheaper. Wind power and biomass are expected to account for most of the projected increase. Wind-power capacity is expected to reach 4 GW in 2030.

Fuel cells using hydrogen, derived largely from natural gas, are expected to contribute to electricity production after 2020. Fuel cell capacity will reach 5 GW in 2030, providing a 2.5% of total output.

Box 6.7: **Reform of Korea's Electricity Supply Industry**[12]

The impact of industry restructuring is a major uncertainty for the projections. In January 1999, the Korean government adopted the

12. For detail, see IEA (2002).

Basic Plan for Restructuring the Electricity-Supply Industry as part of its general regulatory reform process. Key steps in the plan are to:

- unbundle KEPCO, a vertically-integrated monopoly, into generation, transmission, distribution and marketing/supply activities;
- introduce competition in generation and supply by splitting up these two activities into a number of subsidiary companies and privatise them one by one, beginning in 2002;
- open third-party access to all transmission networks;
- create an independent regulator which will initially be housed in the Ministry of Commerce, Industry and Energy (MOCIE), and then become fully independent after the completion of the restructuring;
- introduce customer choice initially to a few large customers and extend it to most other customers by 2009.

Several changes have already been made. KEPCO's non-nuclear generation assets were reallocated to five wholly-owned subsidiary companies in 2001. These companies are to be privatised, starting in 2002. Both foreign and domestic firms will be able to buy shares. Korea's nuclear plants, along with 536 MW of hydroelectric capacity, have been placed in a separate government-owned company. The Electricity Commission, a regulatory body, was set up within MOCIE. But the restructuring programme is now running about a year behind schedule. There are question marks over the terms of privatisation of the generating companies and over the establishment of a wholesale market. Concerns about security of supply and the social impact of planned pricing changes have also emerged.

Energy-related CO_2 Emissions

Korea's energy-related carbon dioxide emissions will increase by 2.3% per year from 2000 to 2030. This is considerably slower than the near 7% rate of increase in the 1990s. Power generation will remain the largest contributor to total emissions. Its share will increase from 35% in 2000 to 41% in 2030. Transport's share will also grow, from 20% in 2000 to 23% in 2030. Rapid increases in the number of vehicles and in vehicle size will also exacerbate local pollution.

Figure 6.14: **Electricity Generation Mix in Korea**

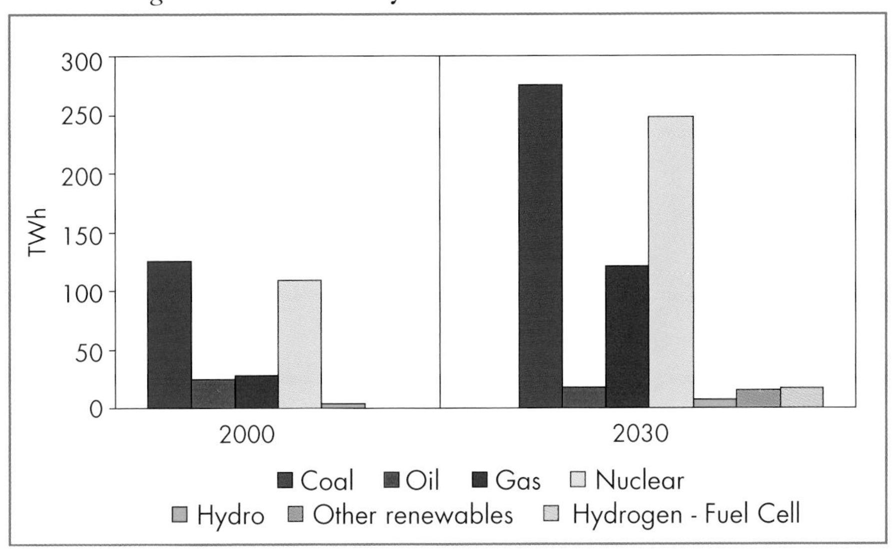

Korea does not have a formal commitment to reduce its CO_2 emissions under the Kyoto Protocol, but it has, nevertheless, formulated a Comprehensive National Action Plan to cut its greenhouse gas emissions. The plan consists of 79 detailed measures, including the promotion of nuclear power and renewable energy.

Per capita CO_2 emissions in Korea were just over 9 tonnes in 2000. This is slightly less than the average of 11 tonnes for the OECD region as a whole. Unless rigorous new measures are adopted under the action plan, Korea's per capita emissions of CO_2 will reach some 16 tonnes in 2030, well above the projected OECD average in that year of 13 tonnes.

CHAPTER 7: CHINA – AN IN-DEPTH STUDY

HIGHLIGHTS

- China, the world's second-largest consumer of primary energy, is a key player in world energy markets, accounting for more than 10% of the world's total primary energy demand. It will continue to be an energy giant in the coming decades as strong economic growth drives up energy demand and imports.
- The Chinese economy is very dependent on coal, of which it has large resources that are cheap to extract. Coal will continue to be a dominant fuel, but the shares of oil, natural gas and nuclear in the primary fuel mix will grow.
- Until the 1990s, the Chinese oil market was largely isolated from the rest of the world, because China produced enough oil to meet its own needs. But oil demand is outstripping production. Imports of crude oil and refined products are growing fast. By 2030, net oil imports are projected to reach almost 10 mb/d – more than 8% of world oil demand. Imports will also have to meet 30% of the country's natural gas needs in 2030. These trends will make China a strategic buyer on world energy markets.
- The investment in energy supply infrastructure needed to meet projected growth in Chinese demand is enormous. Some of the needed funds will come from foreign private investors, but it is not clear whether such capital can be mobilised in a timely fashion. More than $800 billion will be needed for new power generating capacity alone over the next three decades.
- China is already a major contributor to global carbon dioxide emissions. Its share in world emissions is currently 14% and will be even larger by 2030, unless the government takes action. The power sector is responsible for a large part of the increase in emissions, but the share of transport also grows fast.

This chapter analyses in-depth the long-term outlook for China's energy market. The first section, Energy Market Overview, *describes China's growing presence in the world economy and on energy markets, and it examines recent energy trends.* Current Trends and Key Assumptions *reviews recent macroeconomic trends and prospects, including the implications of China's entry into the World Trade Organization, as well as its efforts at energy market reform.* Results of the Projections *presents the outlook for energy demand and supply by sector and by fuel over the period from 2000 to 2030 and the environmental implications of those projections.*

Energy Market Overview

China[1] is the world's most populous country, with about 1.3 billion inhabitants, over 20% of total world population. For the past three decades, it has been one of the fastest growing economies in the world. Its GDP has increased at an average rate of 8.2% per year since 1971 according to official economic statistics. In purchasing power parity terms, its GDP was $4.9 trillion in 2000, making China the second-largest economy in the world after the United States. With its entry into the World Trade Organization (WTO) in 2001, China is likely to become one of the world's top ten *trading* economies in the near future.

China is already a key player in the world energy market. It is the second-largest consumer of primary energy, behind the US, and the third-largest energy producer after the United States and Russia. Its oil imports are growing rapidly. Although the consumption of modern fuels has risen dramatically in recent decades, China remains a big user of biomass for energy purposes, much of it non-commercial. Table 7.1 summarises China's growing importance in the global economy and energy markets.

Table 7.1: **China in the Energy World** (% of world total)

	1971	1990	2000
TPES (excluding biomass)	5	9	11
Coal demand	13	24	28
Oil demand	2	4	7
Final electricity demand	3	5	9
CO_2 emissions	6	11	14
GDP	3	6	12
Population	23	22	21

1. Hong Kong is included with China in the historical data and projections presented here.

China's main fuel is coal. It is the world's largest coal consumer, with 28% of world consumption in 2000. Coal meets nearly 70% of China's primary energy needs. It represents almost 90% of fuel used in the electricity sector. Oil accounts for most of the rest of primary demand, while natural gas contributes a mere 3%.

Because its coal consumption is so large and its coal-burning technologies are so inefficient, China produces a disproportionate 14% of the world's CO_2 emissions, 3 billion tonnes in 2000. This makes it the second-largest CO_2 emitter in the world, after the United States. *Per capita* emissions, on the other hand, are low by international standards: 2.4 tonnes in 2000 compared with the world average of 3.8 tonnes. Local pollution is also a very serious problem in urban areas. China's Tenth Five-Year Plan (2001-2005) calls for reducing dependence on coal by shifting away from energy-intensive manufacturing, by developing oil and gas infrastructure and by promoting energy efficiency and the use of renewables. Natural gas, in particular, is expected to play a major role in meeting this goal.

China has abundant energy resources, including 114 billion tonnes of proven coal reserves –12% of the world total – and 2.3 tcm of proven gas reserves. But these assets are very unevenly distributed. There is plenty of oil, coal and gas in the north and north-west of China, while the main energy-consuming areas are in the eastern and coastal regions. Getting enough resources from the north to market will require massive investment in transport infrastructure.

Primary commercial energy demand, which grew by more than 5% per year from 1990 to 1996, has stagnated in recent years, despite continuing high rates of economic growth. From 1996 to 2000, China's GDP grew by more than 7% per year, according to official government data, while commercial energy consumption grew by only 0.8%. As a result, energy intensity – primary energy use per unit of GDP – fell by 6.4% annually in those four years. In reality, the decline is likely to have been much less, as many studies have revealed that official statistics overstate China's GDP growth rates. If that is the case, then improvements in energy intensity would also have been overstated.[2] Table 7.2 highlights economic and energy indicators compared to global averages.

2. For a discussion of this issue, see IEA (2000).

Table 7.2: **Key Economic and Energy Indicators of China**

	2000		Average annual growth 1990-2000 (%)	
	China	**World**	**China**	**World**
GDP (in billion 1995 $, PPP)	4,861	41,609	9.9	3.0
GDP per capita (in 1995 $, PPP)	3,823	6,908	8.7	1.6
Population (billion)	1.3	6	1.1	1.4
TPES (Mtoe)	950	9,179	3.4	1.5
TPES/GDP*	0.2	0.2	-5.9	-1.5
Energy production/TPES	1	-	-0.7	-
TPES per capita (toe)	0.7	1.5	2.3	0.1
Net oil imports (mb/d)	1.7	-	n.a.	-
CO_2 emissions (million tonnes)	3,052	22,639	2.9	1.2
CO_2 emissions per capita (tonnes)	2.4	3.8	1.8	-0.2

*Toe per thousand dollars of GDP, measured in PPP terms at 1995 prices.
Note: Energy data exclude biomass.

The recent deceleration in China's overall energy demand growth is largely due to a slump in coal demand in industry and, to a lesser extent, in the residential sector (Figure 7.1). There are a number of reasons for this, including the closure of inefficient coal mines, improvements in coal quality, energy conservation and, very likely, statistical error. Oil demand has continued to climb due to increasing motorisation and to switching away from coal and traditional, non-commercial fuels in the residential and services sectors. Primary oil consumption increased at 7.4% per year between 1990 and 2000. Natural gas consumption increased at 6.7% per year over the same period, although it still remains low in absolute terms.

Current Trends and Key Assumptions

The energy projections for China depend on assumptions about a number of key economic and structural factors, for which limited information is available. Acute uncertainties surround China's GDP data, and it is hard to quantify the impact of its recent entry into WTO or the effects of price reform on Chinese energy demand and the fuel mix.

Figure 7.1: **Primary Energy Demand in China**

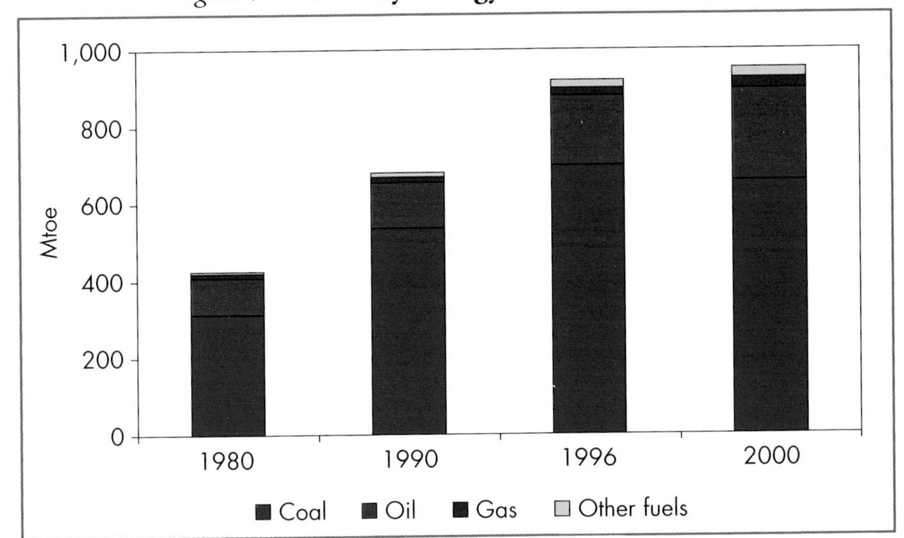

Macroeconomic Context

GDP growth in China accelerated in the early 1990s, reaching a high of 13%, but declined slowly in the second half of the decade. Structural problems, including the inefficient allocation of labour and capital among sectors and among regions, have contributed to this slowdown.[3] China will probably benefit from its WTO membership in the long term, but the adverse effects of trade liberalisation will be challenging, especially in the short term (Box 7.1).

China's economy continues to shift away from agriculture towards industry and services. The share of industry in GDP increased from 42% in 1971 to 49% in 1999 (Table 7.3). Manufacturing makes up most of this amount, at 38% of GDP in 1999, and chemicals are the biggest manufacturing sub-sector, at 13% of GDP. China is the world's largest producer of crude steel and of cement. Output of these two products increased strongly in the 1990s, boosted by a boom in housing and other infrastructure construction. Continued urbanisation and infrastructure development are expected to underpin demand for chemical products, steel and cement, but growing competition with imports could undermine domestic production of these items. The service sector is expanding most rapidly and now accounts for 33% of GDP, while the share of agriculture has dropped to 18%.

3. OECD (2002).

Box 7.1: Impact of World Trade Organisation Membership

Entry into the WTO will inevitably have important economic repercussions for China. Cutting tariffs, lowering non-tariff barriers and opening the domestic economy to foreign participation will produce a more competitive environment, and lead to improvements in efficiency, but these moves could worsen unemployment in the early years. According to an OECD review of various studies on this issue, real GDP growth could be boosted by as much as half a percentage point per year by 2010.[4] The Asian Development Bank (ADB) estimates that the long-term benefits from WTO membership could be much greater, at 1% to 2% of GDP.[5] Labour-intensive sectors such as textiles and clothing, which already account for almost 15% of China's exports, will benefit most from trade liberalisation and so will services. Chinese consumers will gain from lower prices and a wider choice of goods and services.

But these gains will take time to materialise, and structural adjustments across a wide range of sectors could be painful. Lower import duties and foreign competition will increase competitive pressures on domestic producers, particularly those whose activities are land- or capital-intensive. Petrochemical firms and grain and cotton growers may see their situations undermined in the short term. Replacement of domestic agriculture by imports would reduce rural incomes and employment, exacerbating the poverty problem. Industrial sectors that have enjoyed great protection, like automobiles, will also suffer. But the transitional adjustment costs should be manageable, especially if foreign direct investment increases in the service sector. Imports will probably rise, but not dramatically since China has already reduced trade duties, to near-zero for about three-quarters of all imports.

China's accession to the WTO is likely to boost economic growth, but it may also have considerable distributional effects between sectors and regions. There could be a widening of the income divide between coastal areas and the interior and between urban and rural areas. New rifts could also open up between the north and the south and within major metropolitan areas.

4. OECD (2002).
5. ADB (2000).

Table 7.3: **China's Economic Structure**

	Shares in GDP (%)			Average annual growth (%)
	1971	**1990**	**1999**	**1971-1999**
Agriculture	34	27	18	5.7
Industry	42	42	49	8.9
Services	24	31	33	9.5
Total	**100**	**100**	**100**	**8.2**

Source: World Bank (2001).

China's trade with the rest of Asia has been recovering from the 1997 Asian economic crisis. With China's entry into the WTO, imports could well grow faster than exports in the next few years. However, any decline in the current account surplus will probably be offset by increased foreign investment, as the government further opens up the economy under WTO rules (Box 7.2).

Box 7.2: **Foreign Direct Investment in the Chinese Energy Sector**

Foreign direct investment in the Chinese economy has increased rapidly over the last two decades. In 2000, overall foreign investment reached $40 billion, bringing the cumulative total from 1979 to 2000 to $400 billion. China is now the largest recipient of foreign direct investment (FDI) among developing countries.[6] Most foreign investment has gone to coastal provinces, where special economic zones give preferential treatment to foreign-sponsored projects. The manufacturing sector has attracted 60% of total FDI, with virtually all the rest going to the service sector.[7]

But only about $10 billion of foreign investment has gone to the energy sector since 1979. Between 1982 and 2000, the Chinese oil industry signed 140 contracts with 70 foreign partners to explore offshore sites. An estimated $6.5 billion has been invested in these projects.[8] Electricity markets were opened to foreign investors in the

6. Chapter 2 discusses recent trends of FDI in developing countries and transition economies.
7. OECD (2002).
8. This is not recorded as FDI in China's official statistics.

early 1990s, and $2.2 billion in foreign money flowed into 103 power plant projects in 2000.

Liberalisation of foreign investment policies and ongoing reforms in the energy sector will help China to attract more foreign investment, particularly in developing its gas resources in the western provinces and in new power projects. How successful China is in raising FDI, including in the energy sector, will depend on policies to:

- introduce a simpler and more transparent framework for foreign investors;
- remove remaining restrictions on foreign ownership of company shares;
- improve the laws on intellectual property to allow investors to maintain control of the technology they bring to China.

Further work is also needed to improve domestic financial markets in ways that will attract *domestic* institutional investors.

This *Outlook* assumes that China's GDP will more than quadruple to 2030, with annual growth of 4.8% (Table 7.4). This is a marked slowdown from the 1990s, when reported GDP soared by more than 10% a year. The prospect of slower growth reflects both the maturing of the Chinese economy and the likelihood that growth rates in the 1990s were overstated. Moreover, the scope for expansionary fiscal policies, which supported China's growth in recent years, is narrowing as the fiscal deficit rises and the spillover effects of public investment diminish. The economy is assumed to continue its shift away from heavy to light manufacturing industry and towards the service sector. This would have important consequences for energy demand.

Population grew by 1.4% per year between 1971 and 2000. The annual growth rate has been decelerating, largely due to active government policies to reduce fertility rates. This trend is assumed to continue, with average population growth of 0.5% over the *Outlook* period. China's population will still be roughly one-fifth of the world's in 2030. These assumptions imply that per capita income will grow quickly, at 4.3% per year. Per capita income in China will reach one-third of that in the OECD Pacific region in 2030.

Table 7.4: **Macroeconomic and Demographic Assumptions for China**

	1971	2000	2010	2020	2030	Average annual growth 2000-2030 (%)
GDP (in billion 1995 $, PPP)	493	4,861	8,484	13,428	19,753	4.8
Population (million)	845	1,272	1,363	1,442	1,481	0.5
GDP per capita (in 1995 $, PPP)	584	3,823	6,227	9,311	13,338	4.3

Energy Policy Developments

Development of the energy sector is one of the priorities in the Tenth Five-Year Plan for the period 2001-2005. The main objectives are to:

- diversify the energy mix;
- ensure the overall security of energy supply;
- improve energy efficiency;
- protect the environment.

To diversify the energy mix, the government plans to promote the use of gas and renewables. The Tenth Five-Year Plan also calls for the development of nuclear power. It sets numerical targets, including a reduction of coal's share in the primary fuel mix to 64% and an increase in gas's share to 5% by 2005. The government plans to establish a petroleum storage system to provide a cushion in the event of supply disruption. It also plans to establish stronger links with international energy markets and to encourage foreign investment in upstream projects. Energy sector reforms, aimed at boosting competition and attracting more private capital, were launched in 1998. They are to be accelerated. Investment in the energy sector is also intended to contribute to the development of the western provinces. This is part of a larger effort to reduce the disparities in income and development between the provinces in the west and those in the coastal rim. This "Go West" policy, which was officially endorsed at the March 2000 session of the National People's Congress, is expected to have a major impact on the long-term development of the Chinese energy market (Box 7.3).

Box 7.3: Energy Implications of the "Go West" Policy

There are pronounced disparities in income and economic structure among provinces in China. Prosperous city provinces like Beijing and Shanghai have less energy-intensive economies than others, because they focus on lighter manufacturing, services and administrative activities. The north-eastern provinces – the industrial heart of China – are dominated by energy-intensive heavy industries, which accounted for nearly 80% of the north-eastern region's total industrial output in 1999. The coastal provinces, including Shandong and Guangdong, enjoy per capita incomes well above the national average, thanks to the recent explosive growth of light industry and of foreign direct investment.

The central agricultural provinces, including Hunan and Shanxi, and the twelve western provinces, including Xinjiang, Gansu, and Sichuan, are less developed. According to ADB[9], per capita GDP in the western provinces in 1998 was about two-thirds of the national average, about 80% of that of the central provinces, and only a third of that in coastal provinces. Average per capita income in Guizhou is *only* 8% of that in Shanghai. Over half of the 80 million poorest Chinese live in the western provinces.[10]

To compensate the inland provinces for their geographical isolation, the Chinese government seeks to help them make the most of their natural resources. This effort is called the "Go West" policy. There is abundant coal in Shanxi and Inner Mongolia, natural gas in Shanxi, Xinjiang and Sichuan, and hydropower in Sichuan and Guizhou. The government has a vast infrastructure programme to exploit these resources. The 4,000-km West-East gas pipeline project, connecting the Tarim basin in Xinjiang with markets in Shanghai, is a major element of the programme. The central and western provinces are building 150,000 km of new motorways and a 955-km railroad track. In poor agricultural communities, biomass, mainly in the form of agricultural residues, is the primary energy source for heating, cooking and other applications. Economic development in the central

9. ADB (2000).
10. OECD (2001).

and western provinces will determine the pace at which consumers use commercial energy sources.

The government believes the "Go West" policy will generate multiple benefits. Exploration for and production of oil and gas will enhance the country's energy security and bring environmental benefits, by displacing coal. Investment in energy-supply infrastructure will also boost jobs and incomes in the underdeveloped regions. Whether the central and provincial governments can mobilise the required investment is, however, very uncertain. Also, market barriers to inter-provincial trade need to be removed.

Energy Liberalisation and Prices

Energy subsidies have a long tradition in China. Energy price controls and payments to loss-making state-owned enterprises (SOEs) have led to economic inefficiencies throughout the energy sector. Although the pricing system has evolved in recent years to better reflect the underlying costs and prices on international markets, the prices of most fuels sold in China are still set by the authorities:

- *Coal* prices were largely deregulated in 2002. The price of coal from small municipal mines as well as from state and provincial mines is now determined by negotiations between competing producers and industrial end users and distributors.
- Recent reforms have brought *oil* prices more into line with international market levels. Since 1998, domestic crude oil producers and refiners have been free to negotiate the price of crude delivered to refineries. When they cannot reach agreement, the State Development Planning Commission (SDPC) intervenes to set the price. SDPC also sets base prices for retail sales of gasoline and diesel. Since October 2001, these prices have been based on spot prices on the Singapore, New York and Rotterdam markets. Oil companies are allowed to set retail prices within a range of 8% either side of the base prices. All controls over oil pricing are expected to be removed within five years.
- There is a dual-pricing system for *nat*ural gas. Wellhead and retail prices are fixed for gas from projects launched before 1995. Fertilizer producers pay the lowest prices, while commercial and industrial customers pay the highest. In principle, prices for gas from projects begun after 1995 vary from project to project

according to development costs, including a 12% rate of return. These prices are adjusted periodically in line with the prices of competing fuels and inflation. SDPC is considering further reforms to allow prices to reflect more accurately the differences in demand characteristics among buyers, changes in market conditions and project-specific costs.

- Retail *electricity* prices are set by the electricity distributors but must be approved by the government. Up to the early 1980s, they were controlled directly by the government and covered only a small portion of supply costs. Today, they are in line with or higher than long-run marginal supply costs in most cases. However, pricing is still inefficient. Wholesale tariffs paid by distributors to independent generators under long-term agreements take no account of seasonal or time-of-day variations in system load. As a result, generating capacity is not always dispatched in an economically efficient way.[11] Transmission is not priced as a separate service, and costs are not fully reflected in consumer prices. This has restricted power trade among the provinces and encouraged over-building of generating capacity.

This *Outlook* assumes that Chinese energy subsidies will be phased out gradually during the first decade of the *Outlook* period. By 2010, prices will fully reflect the economic cost of supply and will follow trends in international energy prices.[12]

Results of the Projections

Overview

Primary commercial energy demand[13] will grow by 2.7% per year from 2000 to 2030, slowing progressively from 3.2% in 2000-2010 to 2.7% in 2010-2020 and to 2.3% in 2020-2030 (Table 7.5). This growth is much slower than in the past decade, but is still faster than in most other regions and countries. Demand will more than double over the projection period. The projected 1,183 Mtoe increase in demand represents about a

11. World Bank (2001).
12. See Chapter 2 for international price assumptions.
13. The projections presented here refer to commercial energy only. Non-commercial biomass use for energy purposes is discussed towards the end of this chapter and in Chapter 13.

fifth of the total increase in worldwide demand from 2000 to 2030 (Figure 7.2). Primary energy intensity will continue to fall, by 2.1% per year over the *Outlook* period, but that is more slowly than in recent years.[14]

Table 7.5: **Total Primary Energy Demand in China** (Mtoe)

	1971	**2000**	**2010**	**2030**	**Average annual growth 2000-2030 (%)**
Coal	192	659	854	1,278	2.2
Oil	43	236	336	578	3.0
Gas	3	30	57	151	5.5
Nuclear	0	4	23	63	9.3
Hydro	3	19	29	54	3.5
Other renewables	0	1	4	9	6.8
Total primary energy demand	**241**	**950**	**1,302**	**2,133**	**2.7**

The share of coal in China's primary energy demand will drop from 69% in 2000 to 60% in 2030, while that of every other fuel will increase. Coal remains the dominant fuel in power generation, but is increasingly replaced by other fuels in industries and households. Nonetheless, the increase in China's coal demand accounts for almost half of world incremental coal demand between 2000 and 2030. Primary consumption of oil grows steadily, driven mainly by transport demand. Some 16% of the increase in world oil demand comes from China. Natural gas use expands even more rapidly, but from a much smaller base. Although primary gas demand grows at 5.5% per year, it still only meets about 7% of the country's energy needs in 2030. This is below the government's 10% to 15% target for 2020. Nuclear power, which plays a very small role in China's energy supply today, will surge by more than 9% per year, but will account for a mere 3% share of total energy demand in 2030.

14. See IEA (2000) for a discussion of energy intensity trends in China.

Figure 7.2: **Share of China in World Incremental Primary Energy Demand, 2000-2030**

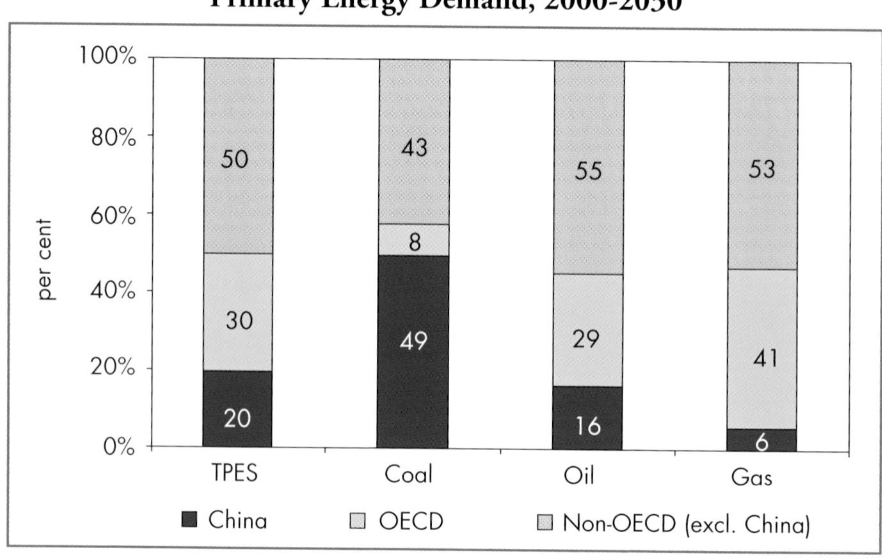

Demand by End-use Sector

Total final energy consumption will grow at 2.7% per year, reaching 1,264 Mtoe in 2030. Coal consumption grows by just 0.7% per year, so that its share in consumption drops dramatically, from 43% in 2000 to 24% in 2030. Oil's share will rise to 40% due to vigorous transport demand. As in all other developing regions, electricity's share will also increase rapidly, by ten percentage points to 26%.

Industrial energy demand accounted for 54% of final consumption in 2000 (Figure 7.3). Its share will fall to 43% in 2030 because of structural changes in the economy. Output of energy-intensive products such as steel and cement is expected to be reined in by competition from imports. Consolidation of small-scale producers will accelerate improvements in their energy efficiency. Less energy-intensive and more labour-intensive manufacturing and service activities are expected to grow faster than heavy industry. Electricity and gas will replace coal, boosting overall energy efficiency and alleviating some environmental problems.

Energy consumption in the *transport* sector will climb by 4.1% per year. Transport's share in total final consumption reaches 23% in 2030, compared to 15% in 2000. Road transport will be the primary driver of demand (Box 7.4), although aviation will be the fastest growing passenger-transport mode. Personal travel in China has soared in the last two decades,

Figure 7.3: **Total Final Consumption by Sector in China**

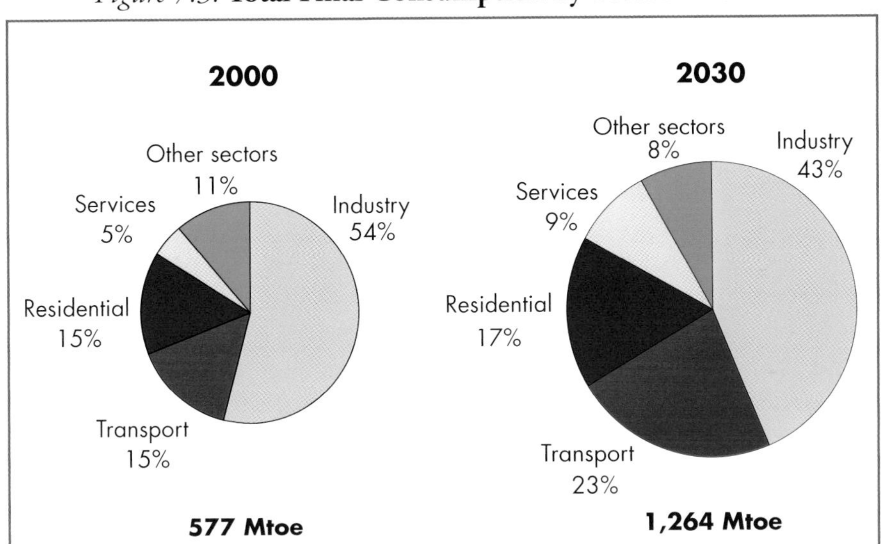

with passenger-kilometres increasing fivefold.[15] Road transport has increased most rapidly, growing eightfold in the last two decades. Passenger-kilometres travelled by air quadrupled in the 1990s. Rising household income and commercial activity will continue to drive strong growth in air travel. Virtually all the increase in transport demand is in the form of oil products. Coal, gas and electricity together will account for a mere 2% of total transport demand in 2030, down from 8% in 2000. About two-thirds of China's incremental final oil demand will come from transport.

Box 7.4: **China's Booming Demand for Road Transport Fuels**

The recent spectacular growth in consumption of road-transport fuel is expected to continue over the *Outlook* period, underpinned by a surge in vehicle ownership. In 2000, vehicle ownership in China was 12 vehicles per 1,000 persons, compared to near 700 in the United States and Canada. GDP per capita recently passed the $3,500 mark, which is seen by the automobile industry as a threshold for rapid motorisation. Vehicle prices are also set to decline with the removal of

15. China State Statistical Bureau.

tariffs on imported cars, and this will further boost ownership and fuel demand. Energy consumption for road transport is projected to grow dramatically, from 54 Mtoe in 2000 to 213 Mtoe in 2030 – an increase of 4.7% per year. Nonetheless, uncertainties surround the pace of road-transport growth, including possible saturation effects as congestion worsens.

There are also uncertainties concerning the structure of demand for road-transport fuel. At present, the government keeps the price of diesel well below that of gasoline. This encourages the use of diesel vehicles, boosts the share of diesel in road-transport energy demand and provokes a mismatch between demand and refinery output of gasoline and diesel. The planned lifting of restrictions on oil product imports and price reforms would help to correct these distortions.

Residential energy use accounted for 15% of final commercial energy consumption in 2000 – a very low share compared to most other countries. Consumption of commercial energy by households even *declined* by 1.4% per year in the 1990s, mainly due to switching from inefficient coal briquettes to gas and electricity for cooking and water heating, and from solid-fuel stoves for central heating. Residential demand is projected to rebound from 84 Mtoe in 2000 to 217 Mtoe in 2030, a rate of increase of 3.2% per year. Rising household incomes will boost the use of electrical appliances. Electricity's share of total residential final consumption will more than double from 18% to 37%, while that of coal will plummet by more than half. There is considerable uncertainty about what is really happening in rural electrification. The grid has been extended to more than 98% of households in China, but the actual usage is still very low, especially in rural areas, partly because electricity services are unreliable.[16] Growth in residential use of gas will depend on the construction of distribution networks and on end-user prices relative to other fuels.

The *services* sector represented only 5% of total final energy consumption in 2000. Commerce has been expanding rapidly, and this trend will continue in the coming decades. As a result, energy consumption in the services sector is projected to surge by 4.5% a year, from 30 Mtoe in 2000 to 111 Mtoe in 2030. Electricity and oil will account for most of this increase.

16. Chapter 13 discusses access to electricity in China and other developing countries.

Oil

The projected 3% annual increase in primary oil demand is driven mainly by transport. Oil consumption, which averaged 5 mb/d in 2001, will more than double by 2030 to 12 mb/d. Most of this additional oil will have to be imported. Net oil imports will rise from 1.7 mb/d in 2001 to 4.2 mb/d in 2010 and 9.8 mb/d in 2030, which is almost equivalent to the net imports of the United States in 2000. China has been a net importer of oil products since 1993 and of crude oil since 1996. The share of imports in total oil demand will reach 82% in 2030 compared to 34% in 2001 (Figure 7.4).

Figure 7.4: **Oil Balance in China**

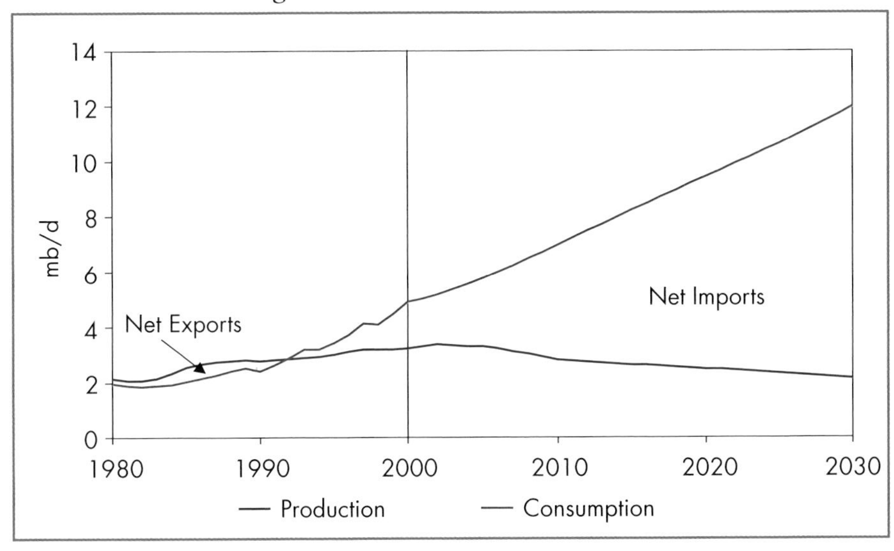

There appears to be little prospect of increasing crude oil production from the ageing fields in China's eastern region, the main source of indigenous supply, but offshore production may increase in the near term. The super-giant Daqing field, the world's fourth-largest producing field, yielded 1.1 mb/d in 2001, roughly a third of total Chinese output. Daqing has been in production since the 1960s and its output is in decline. Recent new finds, many of them offshore, have made up for some of the falling output from Daqing and other mature giant fields, such as Shengli and Liaohe, which are also more than thirty years old. The Chinese National Petroleum Corporation (CNPC) and Philips are jointly developing the 4.3-billion barrel Penglai field in the southern part of the Bohai Sea, where production is due to begin in 2002. Other current projects include exploration of the giant Tahe field in the Tarim basin in western China, a major discovery in the Pearl River Mouth basin of the South China Sea and construction of China's first shallow-water offshore operation in the Shengli oilfield.

CNPC, the Chinese National Offshore Oil Corporation (CNOOC) and the China Petrochemical Corporation (Sinopec), all SOEs, are expected to expand their joint-venture exploration and development projects with foreign companies under production-sharing agreements. But the continuing legal requirement that the Chinese partner must hold a controlling interest in such ventures may discourage investment by foreign companies. To maintain current oil production in the medium term, China will probably focus on maintaining production at Daqing and the other mature fields through enhanced oil-recovery technology. Continued restructuring of the oil and gas industry and reform of the regulatory environment would improve the outlook for indigenous production (Box 7.5).

A large proportion of China's future oil supplies will have to come from overseas, although the rate of increase is a subject of debate. Most imports, largely in the form of crude oil, will undoubtedly come from the Middle East. Russia could also become an important supplier in the coming years. In July 2001, China and Russia agreed to perform a feasibility study on a $1.6 billion, 2,200-km oil pipeline to bring Siberian oil to north-eastern China.[17] Over 25 years, starting in 2005, the pipeline would supply China with 400,000 barrels of oil per day. The capacity would be increased to 600,000 b/d in 2010. If the project were carried through, it would make Russia one of China's largest crude oil suppliers. The Chinese government's support for the project stems from growing concerns about the country's dependence on Middle East oil and its vulnerability to a possible supply disruption. The government is also seeking to secure direct control over foreign oil resources through CNPC, which has acquired interests in exploration and production in Indonesia, Kazakhstan, Venezuela, Sudan, Iraq and Peru.

17. PIRA Energy Group (2002).

Box 7.5: Oil and Gas Reform in China

The oil and gas sector was radically restructured in 1998 with the regrouping of onshore assets into two vertically integrated state companies, CNPC and Sinopec (which also produces petrochemicals). The bulk of offshore assets went to CNOOC. Regulatory functions were taken away from CNPC and Sinopec and transferred to the State Economic and Trade Commission. Most provincial-level companies involved in oil and gas activities were integrated into Sinopec or CNPC. Minority stakes in the operating subsidiaries of these companies were listed on the New York, London and Hong Kong stock markets in 2000.

Oil import restrictions, which have long given these companies a near-monopoly in the downstream market, are being dismantled in the wake of China's entry into the WTO. Until recently, the government has used a quota system to favour imports of crude oil over oil products, in order to maximise domestic refining capacity. Immediately on joining the WTO, the government began issuing crude oil import licences to private trading firms and committed itself to increasing their quotas by 15% per year. Those quotas will be abolished in 2006. Crude oil import tariffs were removed in 2002. Oil product import quotas and tariffs will be removed at the end of 2003. The state companies currently control 90% of oil imports, but this share will fall to 80% by the end of 2002.

China will also open its retail distribution markets fully to foreign oil companies in 2005. Wholesale markets will follow in 2007. Foreign investment to modernise China's run-down refining industry is already being encouraged. Currently, domestic refineries are too small to achieve economies of scale. Many very small refineries operate well under capacity. The competitive pressures that reform is unleashing are forcing the state companies to rationalise their operations, with massive reductions in their workforces, and to introduce modern management practices.

Natural Gas

Consumption of natural gas will increase almost fivefold over the *Outlook* period, from 32 billion cubic metres in 2000 to 61 bcm in 2010 and 162 bcm in 2030. The role of gas in the country's primary energy supply, nonetheless, remains small, at 7% in 2030. These projections are subject to considerable uncertainty, particularly with respect to the cost of supply. Equally uncertain is China's ability to raise funding to develop gas fields, to build LNG regasification terminals and to create an integrated national transportation and distribution network. The competitiveness of gas against coal in power generation will also be a key determinant of gas demand growth.

China's gas resources are uncertain, because of the limited exploration so far carried out. Proven gas reserves were 2.3 tcm at the beginning of 2000, and of this amount 1 tcm is economically recoverable. Total resources, which include mean undiscovered gas, are much larger, at 50.6 tcm, but only 13.3 tcm of these are likely to be recoverable. Roughly 59% of total resources is located in the western and central provinces. Three basins – Tarim, Sichuan and Ordos – hold 52% of national resources. The Sichuan basin accounted for most of the country's gas output of 30.5 bcm in 2000, but output from Tarim, Ordos, Qinghai and offshore fields is growing. The government recently announced a 200 bcm discovery in the northern part of the Tarim basin in Xinjiang province. Connecting the producing fields in the west to the main potential markets in the east will require the construction of long-distance transmission lines and the expansion of distribution networks (Box 7.6).

Box 7.6: **West-East Gas Pipeline Project**

The 4,000 km West-East pipeline to transport gas to Shanghai from the Tarim basin in the west and the Ordos basin in central China is the centrepiece of the government's plan to establish a national gas market. This is one of several major infrastructure projects to promote economic development in the poor central and western provinces. The line, which will have a capacity of 12 bcm per year, is expected to cost $6 billion, but investments in upstream and downstream facilities connected with the pipeline networks could bring the total cost close to $20 billion. PetroChina, the main operating subsidiary of CNPC, and a consortium led by Royal Dutch Shell will build the pipeline.[18] If full-scale construction on the first phase of the project

18. PetroChina will hold a 50 % stake under agreement with the consortium in July 2002. The pipeline concession will last 45 years. A full discussion of the project will be included in IEA (2002, forthcoming).

begins in 2002 as planned, first gas from the Changqing field in the Sichuan basin would reach Shanghai in 2004. The go-ahead will require commitments from very large customers, including power plants and big industrial end users, to lift sufficient gas to make the project economically viable. Technical challenges in building the line across difficult terrain also need to be overcome.[19]

China's gas production will increase from 30 bcm in 2000 to 55 bcm in 2010 and to 115 bcm in 2030. But even if the West-East gas pipeline and other domestic projects go ahead, China will need to import large amounts of gas to meet the projected increase in demand (Figure 7.5). Imports will come from South-East Asia and the Middle East during the early years of the *Outlook* period. Gas will be imported from Russia and possibly Central Asian transition economies after 2015. Those countries will account for nearly a half of China's gas imports by 2030.

Figure 7.5: **Natural Gas Balance in China**

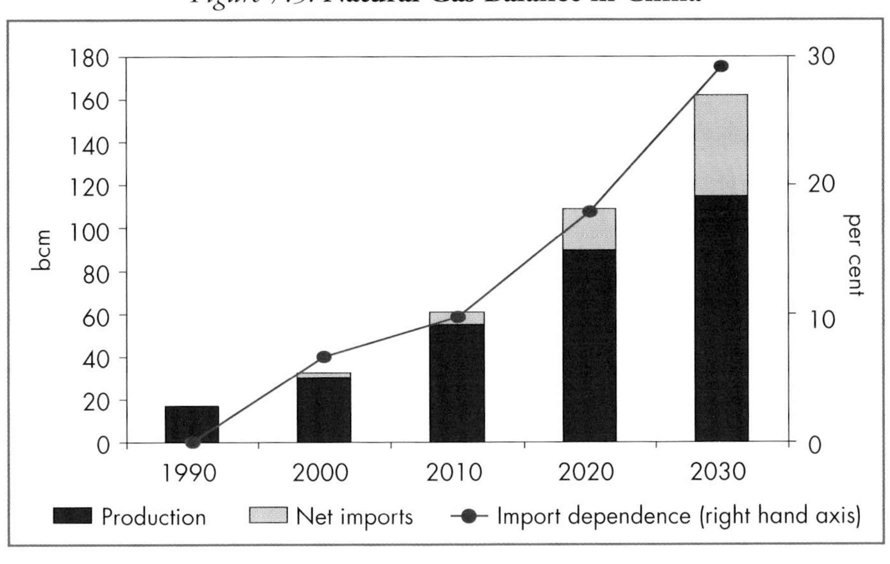

19. PetroChina has so far signed 45 letters of intent to supply customers with 0.8 bcm of gas in 2003, 8 bcm in 2005, and 12 bcm in 2008.

In early 2001, plans to import liquefied natural gas moved forward when the Chinese government approved the construction of the country's first LNG receiving terminal. It will be sited in Guangdong, north of Hong Kong, and will have an initial capacity of 3 Mt/year. In 2002, China and Australia made a contract under which gas will be supplied from Australia's North West Shelf Projects for 25 years. First gas is expected in 2005. A second terminal in Fujian is also planned. In the longer term, China could import gas by pipeline from Russia, most likely from the Kovykta field near Irkutsk in Siberia. The line could run as far as Korea, and possibly on to Japan. China's interest in this project has, however, waned since the late 1990s because of the prospects for a West-East gas pipeline. Doubts about the true size of Irkutsk's reserves and the cost of developing the field and building the line have also dampened Chinese enthusiasm.[20] Still, Irkutsk gas could probably be delivered to Shanghai at lower cost than gas from western China through the West-East pipeline. If the Irkutsk pipeline *were* built, Russia would establish itself as a pivotal supplier to Asian markets, as it already is to European markets. Pipelines from more distant foreign sources, such as Turkmenistan or Kazakhstan, have also been proposed, but they are unlikely to proceed within the next two decades.

Coal

Primary coal demand, which increased by 4.3% per year throughout the last three decades, will continue to rise, but at a much slower rate. It will grow at an average rate of 2.2% from 2000 to 2030, reaching 2 billion tonnes. Most of this increase will be for power generation, but industrial uses, such as coking coal in steel production, will remain important (Figure 7.6). Final coal consumption in the residential and commercial sectors will be broadly flat over the projection period. The recent sudden reversal in the upward trend of Chinese coal consumption heightens the uncertainties surrounding the outlook for demand (Box 7.7).

20. There are two possible routes for the line – a direct route traversing Mongolia and a more circuitous one that avoids it. The preferred option will depend on negotiations with Mongolia over transit fees. The 25-bcm/year pipeline alone would cost around $7 billion.

Box 7.7: **Declining Coal Demand in China**

The decline in China's coal demand since the mid-1990s has attracted a lot of attention and is a major source of uncertainty for China's future energy consumption and CO_2 emissions. Several factors appear to have contributed to this decline:[21]

- The consolidation of small firms has increased economies of scale and cut energy needs. Over capacity in the steel, cement and fertilizer industries, which accounted for nearly 70% of industrial coal consumption in 1996, forced the closure of the most inefficient plants, cutting coal demand. Increasing competition is encouraging state-owned enterprises to use energy more efficiently.
- Improvements in coal quality mean that less coal is needed to provide the same amount of heat. Official data probably do not take adequate account of this factor.
- Energy conservation measures, and to a lesser extent, stricter environmental regulations, have forced inefficient plants to shut down or to use energy more efficiently.
- Residential and commercial consumers in urban areas have been switching to oil products and natural gas.
- Since 1998, the government has forced more than 47,000 small coal mines to close, mainly for safety and environmental reasons. Some private mines have reopened illegally, but their output is no longer included in the official statistics.

The country has an estimated 114 billion tonnes of proven coal reserves, nearly 12% of the world total. Of this 83% is steam coal, and the rest is coking and gas coal (low-quality coal and lignite used to manufacture town gas). Coal output more than doubled between 1980 and 1996, from 620 Mt to 1,402 Mt, but fell back to 1,231 Mt in 2000 because of the closure of small mines and a slump in domestic demand. China remains the world's largest coal producer, with 27% of total world production in 2000.

21. See for example, Sinton and Fridley (2000).

Figure 7.6: **Primary Coal Demand by Sector in China**

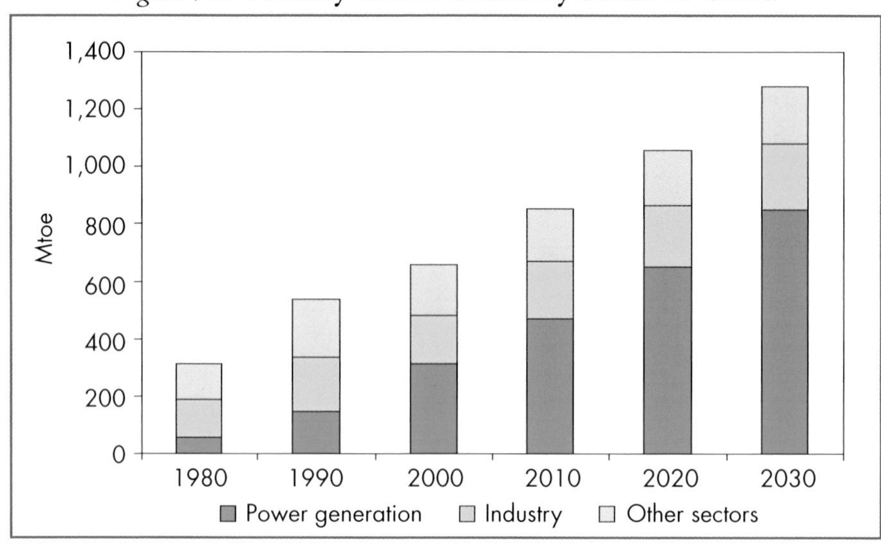

The labour productivity of Chinese coal mining is very low by international standards but is improving as the most inefficient mines are shut and jobs are shed. The government launched a major programme of mine closures in 1998 to correct oversupply and to meet health and safety concerns. At the end of 2000, the government announced that over 47,000 small, loss-making mines, many of which were highly polluting and dangerous, had been closed. Some of these mines are still, in fact, operating.[22] Coal mining and transportation will undoubtedly be rationalised still further.

The proportion of coal reserves available to depths of 150 metres is limited, so future production will focus on deeper underground mining, which is usually more costly. Several large new mines are due to open in the next few years. Pingsu Coal's surface operation in Anjialing will enter full production soon, adding 15 Mt per year to its capacity. Capacity at the Yinzhou Jining III mine will expand by 4 Mt, while new mines at Juye and at Shanxi (Sitaigo) will together add 9 Mt. Expansions at Anhui, Hebei, Shandong and Shanxi are expected to add another 10 Mt.

The government plans to increase coal exports, which have surged in recent years. Exports, mostly of steam coal, jumped from 37 Mt in 1999 to 91 Mt in 2001, helping to dampen prices in the Asia/Pacific market. Japan

22. Local governments often protect illegal small mines both for the income they generate and the employment that they provide. Mine closures have increased social tensions in some regions.

and South Korea take over 60% of China's exports. It is not certain that the Chinese coal industry can meet projected domestic demand and also continue to expand exports. The implementation of announced reforms to open the coal sector to foreign investment in existing large-scale mines and in new projects would boost China's long-term production prospects. Foreign participation would assist the introduction of new technologies, such as coal liquefaction, coal-bed methane production and slurry pipeline transport. These technologies would, in turn, bring environmental benefits and, to the extent that coal-based liquid fuels replace imported oil, they would enhance energy security as well. The government is also planning to restructure the industry along similar lines to the oil and gas sector, creating several large corporations. The newly formed firms could seek foreign capital through international stock offerings, just as CNPC, Sinopec and CNOOC have done. But major changes in property laws, in export rights and in regulations governing the repatriation of profits may be needed to reassure investors.

Hong Kong already imports significant quantities of coal from Australia. In the long term, imported coal could become economic in some coastal regions of China too, as import tariffs are reduced and the remaining price subsidies on domestic coal are removed. The rail network, which is the principal mode of coal transport, is already used to nearly full capacity. This could make seaborne imports more economic than domestic supply in coastal regions.

Biomass

Biomass, which for the most part is not commercially traded and is not included in the *Outlook*'s energy balance, remains an important source of energy in the Chinese residential sector in poor rural areas, although its share in total energy use has been declining. In 2000, biomass use was equivalent to some 18% of total primary energy demand, almost equivalent to primary oil consumption. Rural households consume far more biomass than they do commercial energy. Biomass consists mainly of fuelwood and agricultural residues, burned directly for cooking and hot water and for space heating. Its share in the residential sector's total final consumption was estimated at 72% in 2000.

Biomass use is expected to decline by 0.3% per year on average over the *Outlook* period (Table 7.6). It will remain broadly flat until 2010 and then start to decline at an accelerating rate, as incomes rise and as more people migrate to towns and cities. The estimated share of biomass in residential demand will fall to 49% in 2030.

Table 7.6: **Biomass Energy Demand in China** (Mtoe)

	1971	2000	2010	2030	Average annual growth 2000-2030 (%)
Biomass	163	213	213	195	-0.3
TPES including biomass	404	1,162	1,514	2,326	2.3

The shift away from biomass could aggravate China's greenhouse gas emissions, but it could help to reduce local pollution, especially particulate emissions, with important benefits for public health. Because final consumption of biomass is generally inefficient, its replacement by commercial fuels results in overall gains in energy efficiency and a decline in energy intensity.

Electricity

China's electricity consumption will grow by 4.2% per year on average over the *Outlook* period. The rate of growth will slow slightly as the years pass. Electricity's share of total final consumption will jump from 16% to 26%. Demand will be boosted by growth in personal incomes and by switching from coal and oil, as well as by rising output in industry and the commercial sector.

Electricity generation was 1,387 TWh in 2000, 78% of it from coal-fired plants (Table 7.7). While coal will remain the dominant fuel in the generation mix, its share is projected to drop to 73% in 2030. The shares of nuclear energy and gas will increase. Gas will provide about 7% of total generation and nuclear power 5% in 2030.

China had 319 GW of generating capacity at the end of 2000. In order to meet the rapid growth of electricity demand and to replace plants that are to be retired, China will have to build 800 GW of new capacity by 2030, raising the total to 1,087 GW. More than $800 billion investment will be needed for new generating capacity alone. Figure 7.7 shows capacity additions by fuel in the three decades from 2000 to 2030.

Table 7.7: **Electricity Generation Mix in China** (TWh)

	1971	**2000**	**2010**	**2020**	**2030**
Coal	98	1,081	1,723	2,509	3,503
Oil	16	46	51	53	54
Gas	0	19	74	209	349
Nuclear	0	17	90	163	242
Hydro	30	222	333	511	622
Other renewables	0	2	10	16	42
Total	**144**	**1,387**	**2,282**	**3,461**	**4,813**

The power sector's heavy dependence on coal will not change much over the next three decades. Because China has abundant coal resources and coal-plant construction costs are low, most new generating capacity will be coal-fired. The capacity of coal-fired plants is projected to increase threefold, to reach 696 GW in 2030. The efficiency of China's coal-fired plants, somewhere between 27% and 29%, is very low compared to those in OECD countries. Average efficiency is projected to improve over the projection period, reaching 35% by 2030, because new power plants will be more efficient and many small, inefficient plants will be closed.[23]

Figure 7.7: **Power Generation Capacity Additions in China**

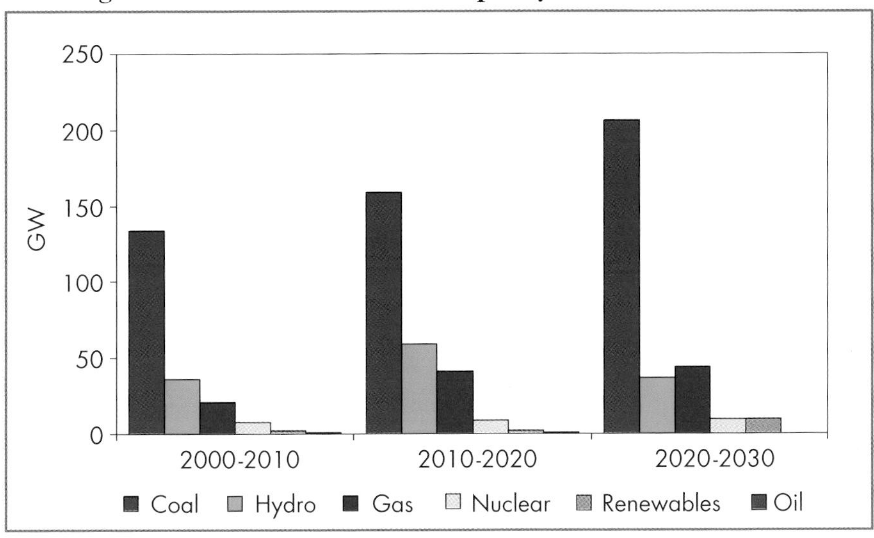

23. Since 1998, construction of small thermal power plants under 135 MW has required official permission from the government. In 2002, the government announced that loans and equipment would no longer be provided to existing small plants.

Concerns about the environmental impact of coal-fired plants have been increasing, and the government plans to introduce more stringent environmental requirements. Liberalisation of trade and investment will open up opportunities for China to import cleaner coal technologies.

Electricity generation based on natural gas is projected to surge, from 19 TWh in 2000 to 349 TWh in 2030. Growth will be particularly strong in coastal regions where much of the liquefied natural gas that will be imported after 2005 will be burned in power plants. Gas-fired capacity, currently very small, is assumed to reach 27 GW in 2010 and 113 GW in 2030. The government is giving priority to expanding the use of gas in the power sector. It will be essential to the economic viability of the West-East gas pipeline and LNG terminals that there be a strong market for gas in power stations. But there is much uncertainty about the development of gas distribution networks and the competitiveness of gas against coal. Increasing use of gas will hold down investment in oil-fired plants. Oil capacity is projected to increase marginally, with the share of oil in total electricity generation dropping to 1% in 2030.

Nuclear power was introduced in China in 1991. Capacity now stands at 2 GW. Nuclear power generated 1% of Chinese electricity in 2000. Given the long lead times and high capital cost of nuclear plants, capacity will reach only 11 GW in 2010, 21 GW in 2020, and 31 GW in 2030. Assuming an average capacity factor of 90%, the share of electricity generated by nuclear plants will reach 5% in 2030. These projections are more conservative than those in China's official long-term plan, which envisages 40 GW of capacity by 2020. Public acceptance may play an important role in the future of nuclear.

China has abundant hydroelectric resources. The Three Gorges Dam will have a total 18 GW when it is completed. The *Outlook* assumes that total hydro capacity will increase from 79 GW in 2000 to 112 GW in 2010 and to 209 GW in 2030. But there are doubts about whether these capacity increases can be achieved in the face of widespread opposition on environmental and social grounds. There are also uncertainties about the cost of building the transmission lines that would be needed to bring electricity from dams in the remote western areas to markets in the east.

Over the next three decades, the cost of renewable energy sources is expected to fall, and they will play a bigger role in electricity generation, particularly in remote, off-grid locations. Wind power is the most promising technology. Its capacity is projected to grow from 0.4 GW in 2000 to 12 GW in 2030. Geothermal, solar and biomass supplies are also expected to grow rapidly, especially after 2010, but their share in total generation will still be small in 2030.

The Chinese government recently launched a new phase of structural reforms in the electricity sector. The reforms are aimed at improving efficiency and service quality, encouraging competition and lower costs and attracting foreign investment. Competitive bidding will be introduced for generators seeking to sell power, and a modern regulatory framework is to be established. Pilot programmes to create a competitive wholesale market were set up in some provinces in 1998. But the government did not consider them a success, mainly because of problems caused by the joint ownership of power plants and transmission assets. A new government plan agreed in February 2002 involves:

- breaking up the dominant State Power Corporation and completely separating the generation and transmission businesses;
- reorganising the transmission network into two national companies covering the south (Southern Power Grid Company) and north (State Power Grid Company); these companies will be allowed to choose the generators from whom they buy electricity;
- sharing out major generation assets among four companies in addition to the State Power Grid Company, which will retain all peak-load plants;
- creating a national regulatory commission under the State Council.

Bottlenecks in transmission and distribution networks are becoming a serious problem in China.[24] There are 14 regional power networks, their areas defined by administrative boundaries. Only a few of them are interconnected. An excessive focus on generation capacity has led to underfunding for transmission and distribution networks. Very large investments are needed to expand and upgrade networks and build more power plants once the existing overcapacity in generation has been absorbed. Capital inflows from developed countries will be needed, as well as more efficient domestic financial markets. Electricity-sector reform is expected to lead eventually to more efficient plant and network operation, more rational pricing and a more efficient allocation of resources between generating plants and transmission and distribution networks.

24. World Bank (2001).

Environmental Issues

Environmental problems caused by the burning of fossil fuels are a growing concern in China. Pollution from energy-related emissions of sulphur oxides, nitrogen oxides, volatile organic compounds and particulates has led to a serious deterioration in air quality in urban areas. Air quality in rural areas has been deteriorating because of the expansion of industrial activities there. According to World Health Organization data for 1998, seven of the world's ten most polluted cities are in China. Areas affected by acid rain make up 30% of the total land area of the country. In 2000, 80% of SO_2 emissions came from the industrial sector.[25] Coal burning is a main contributor to ambient and indoor air pollution, producing 85% of total SO_2 emissions and 28% of total suspended particulate emissions. Pollution problems in many cities have eased a little since the late 1990s (Figure 7.8). This occurred because of a drop in coal consumption and the installation of anti-pollution equipment in power plants and factories. Another positive factor was the closure of inefficient factories, district heating and power plants in some inner-city areas, such as central Beijing. But noxious emissions and carbon monoxide from motor vehicles are growing rapidly and will become a major source of urban air pollution in the coming decades.

The government has taken several actions to curb pollution. In 1996, the Ninth Five-Year Plan for Social and Economic Development included environmental goals for the first time. In 1998, the government upgraded the State Environmental Protection Agency to ministerial status. In 2000, the National People's Congress adopted a revised Law on Air Pollution Prevention and Control, which provides for more detailed and stringent air pollution limits. Provinces, municipalities, and autonomous regions are now required to keep their pollutant emissions within the prescribed limits. The government intends to cut concentrations of SO_2 and other major pollutants by 10% in 2005 compared to 2000. China phased out leaded gasoline for automobiles in 2000.

China is a major contributor to global greenhouse gas emissions. It is the world's second-largest emitter of carbon dioxide. Power generation and industrial activities are the main sources of CO_2 emissions, because of the low energy efficiency of China's power plant and industrial boilers and its heavy reliance on coal. Energy-related CO_2 emissions are projected to

25. Zhou Fengqi (2001).

Figure 7.8: **Local Pollution in Beijing and Shanghai**
(Index, 1996=1)

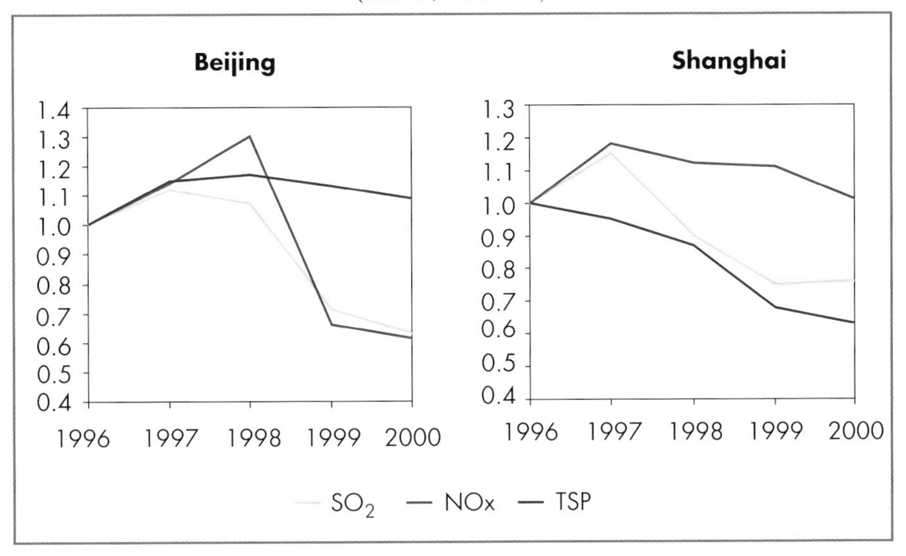

Note: Total suspended particulate (TSP) is atmospheric particulate smaller than 40 micrometers in diameter.
Source: Environment Quality Report of Beijing and Shanghai.

Figure 7.9: **CO_2 Emissions by Sector in China**

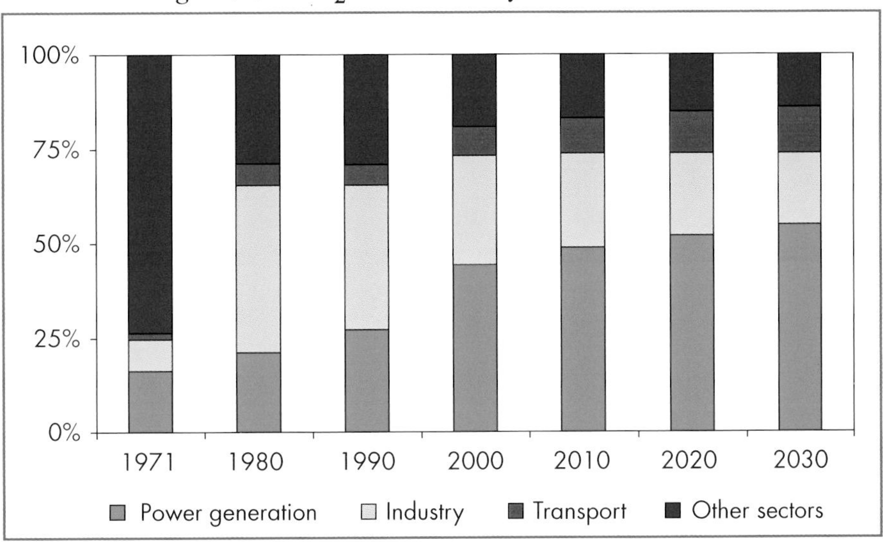

Note: Other sectors include residential, services, agriculture, non-energy use and transformation other than power generation.

increase steadily throughout the projection period, at a rate of 2.7% per year. They will reach 6.7 billion tonnes by 2030, or 18% of world emissions. By comparison, the United States and Canada together will emit 8.3 billion tonnes in that year. The biggest increase will come from the power sector, which will produce more than a half of China's CO_2 emissions in 2030 (Figure 7.9). Transport's share will also increase, as rapid motorisation spurs brisk growth in oil consumption. Emissions from the industrial, residential and commercial sectors will grow much less rapidly than final energy consumption, because coal, the most carbon-intensive fuel, will be increasingly replaced by other fuels.

Liberalisation of trade and investment may alleviate some environmental problems. More efficient and environment-friendly technologies will become available at lower cost. And the phasing-out of subsidies that distort prices and encourage inefficient energy use would also help mitigate pollution, as would stronger measures to promote energy efficiency in the rapidly growing transport, residential and commercial sectors. But these benefits may be offset by an increase in overall economic activity, especially transport.

CHAPTER 8: RUSSIA

HIGHLIGHTS

- Russia will play an increasingly important role in world oil and gas markets over the *Outlook* period. The country is already the world's largest gas exporter and the second-largest exporter of oil and oil products, after Saudi Arabia. Russian exports are set to grow strongly in the next few years.
- The development of Russia's vast resources will be crucial to the energy security of countries within and without the OECD. The current reform process must persist in order for Russia to exploit its huge resource base over the medium term.
- If Russia is to consolidate its role as the largest gas exporter to Europe, it must secure the investment to develop new fields in less accessible areas and to build more pipelines. Russia is also expected to start exporting gas to markets in the Far East, including China.
- The *Outlook* projects that Russia will need to invest some $157 billion in new generating capacity over the next thirty years. The government would like to free up more natural gas for export, but gas has several advantages over other fuels, including higher efficiency, reduced environmental damage and ample supply.
- Russia is the third-largest energy consumer in the world, after the United States and China. Despite a decade of declining energy consumption, Russia's energy intensity is still quite high. Energy efficiency improvements in power generation and end-uses will come about only if current price reforms continue.
- The *Outlook* projects that Russia's energy-related carbon dioxide emissions in 2010 will be 17% below what they were in 1990. If an emissions-trading system is established under the Kyoto Protocol, Russia will be in a position to sell its surplus emissions.

Energy Market Overview

After a decade of severe economic decline, the Russian economy rebounded in the wake of the financial crisis and rouble devaluation of August 1998. High international oil prices and the trade-enhancing effects of its deep rouble devaluation have fuelled steady growth since then. Russia is now negotiating to join the World Trade Organization.

Table 8.1: **Key Energy Indicators for Russia**

	1992	**2000**	**Average annual growth 1992-2000 (%)**
GDP (in million $1995, PPP)	1,370	1,086	-2.9
GDP per capita (in $ 1995, PPP)	9,205	7,465	-2.6
Population (million)	149	145	-0.3
TPES (Mtoe)	776	612	-2.9
TPES/GDP*	0.57	0.56	-0.1
TPES per capita (toe)	5.2	4.2	-2.7
CO_2 emissions (million tonnes)	1,875	1,492	-2.8
CO_2 emissions per capita (tonnes)	12.6	10.3	-2.5

* Toe per $1,000 of GDP, measured in PPP terms at 1995 prices.

Natural gas and electricity occupy a central place in the Russian economy.[1] Lack of transparency and institutionalised secrecy still plague all efforts to regulate Gazprom, the partially state-owned gas company, and the joint stock company, United Energy Systems of Russia (RAO UES), the dominant electricity company. These problems hinder the companies' ability to attract investment.

Russia is one of the most energy-intensive economies in the world. In 2000, it was almost three times as energy-intensive as OECD countries on average – the result of cheap energy and inefficient use. While domestic prices for oil products are now quite close to international levels, domestic gas and electricity prices are still well below them.

1. See IEA (2002) for a detailed discussion of the current energy situation in Russia.

Current Trends and Key Assumptions

Macroeconomic Context

Over the past few years, Russia has achieved impressive economic growth: GDP expanded by over 8% in 2000 and by 5% in 2001.[2] There have also been encouraging signs of macroeconomic stabilisation and improvements in commercial banking procedures. Relations with foreign creditors have improved, and so has political co-operation between the Russian government and the parliament. Sustained high energy prices have been the key factor, but economic policies and institutional change have also contributed to the improved macroeconomic environment. The government has pushed ahead with reforms in a number of areas, including taxation, management of the state budget and business regulation.

If international energy prices hold up, short-term prospects for the Russian economy appear favourable. The medium term, however, is uncertain. Export-driven growth may be unsustainable if energy prices drop or if the rouble appreciates. Further reforms are needed to revive private investment. Substantial capital outflows and a weak small business sector have hindered commerce, competition and investment. A previous lack of investment has led to low reserve replacement and limited turnover in capital stock.

Total foreign investment in Russia declined in the 1990s, especially from countries in the European Union. Investment flows increased slightly in 2001, particularly in the oil and gas sector. Improved energy efficiency would not only help Russia's international competitiveness, but would increase the stock of emissions credits which Russia could sell under the emissions-trading mechanism in the Kyoto Protocol.

The *Outlook* assumes that the pace of economic reform will accelerate. The economy is assumed to expand faster in the second decade of the *Outlook* period, as growth is increasingly driven by private investment and household spending, as well as by buoyant oil and gas exports. Market institutions are expected to become more firmly established. The investment climate in Russia will become more favourable, if Russia creates a stable and attractive fiscal framework for business activities.

GDP is assumed to grow by 3% a year over the full projection period. It is assumed to accelerate from 2.9% a year in the decade to 2010, to 3.5% from 2010 to 2020. It then falls back to 2.6% from 2020 to 2030 as the

2. GDP is expected to grow by over 3% in 2002 (OECD, 2002).

economy matures. In line with past trends, the population is assumed to shrink by 0.6% annually over the *Outlook* period. Real per capita incomes will nearly triple, from $7,465 in 2000 to more than $21,000 by 2030.

Energy Sector Reforms and Prices

One of the most dramatic changes in recent years has been the remonetisation of the gas and electricity sectors. In 1998, the vast majority of domestic sales of gas and electricity were conducted with money surrogates such as barter, debt offsets and bills of exchange. The government effectively banned non-cash transactions in the public sector as from 1 January 2000. Cash sales now predominate, especially for electricity. Arrears in paying energy bills have declined. Progress has been achieved towards normal payment for energy services in cash, despite some remaining policies designed to protect non-paying customers from being cut off.

The Russian government plans to introduce competition into gas and electricity production and supply, based on non-discriminatory third-party access to transmission networks and probably to distribution systems too. In July 2001, the government approved a general reform strategy for the electricity industry. Under the plan, generating firms belonging to RAO UES will be reorganised into independent companies. Vital to the plan's success will be planned increases in electricity prices, so that they cover costs, and the strict enforcement of payments. A similar restructuring of the gas industry is planned. Progress has already been made in the gas sector in raising accounting standards, introducing professional audits and increasing transparency.

The *Outlook* assumes that domestic energy prices will gradually converge with those on international markets.[3]

Results of the Projections

Overview

Russia's total primary energy demand is projected to grow by 1.4% per year over the *Outlook* period, a rate *much* lower than the projected 3% rate for GDP growth. These figures add up to a substantial decline in energy intensity, at an average 1.6% per year. Demand for oil will grow the fastest among fuels, at 1.7% per year, and oil's share in TPES will increase

3. See Chapter 2 for international price assumptions.

Figure 8.1: **Total Primary Energy Demand in Russia**

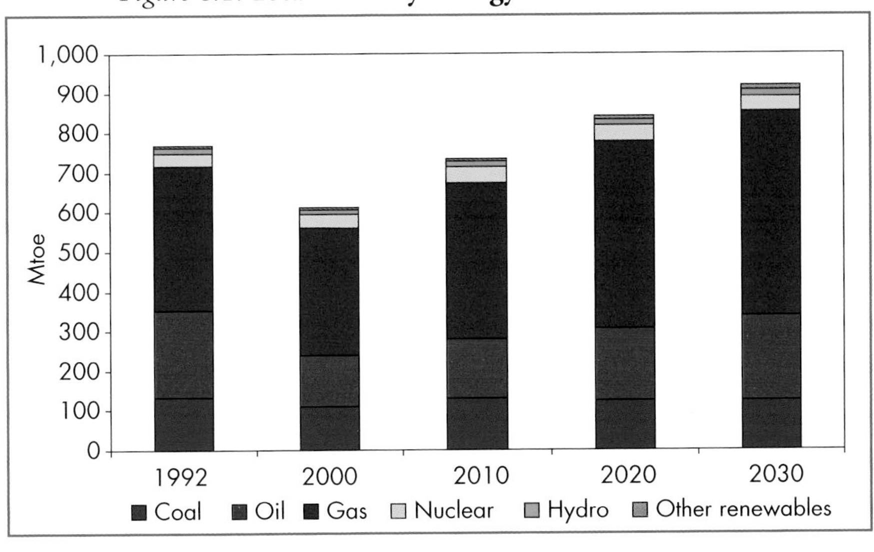

by two percentage points from 2000 to 2030 (Figure 8.1). Gas demand will grow almost as rapidly, at 1.6% a year, and the share of gas in TPES will increase from 52% in 2000 to 56% in 2030. Coal demand will rise by only 0.4% per year. Nuclear power will account for 4% of TPES in 2030, down from almost 6% in 2000. Hydroelectricity's share will fall slightly, to just under 2%.

Total final consumption is expected to grow slightly slower than TPES. It will return to its 1992 level only in the third decade of the projection period. The transport sector is expected to account for nearly 38% of incremental final energy demand and 69% of incremental final oil demand. In both the residential and industry sectors, demand for heat will drop considerably as its price increases and waste is reduced. During the 1990s, electricity and heat prices increased by much less than inflation. This weakened customers' incentive to improve their energy efficiency. As a result, the amount of energy used in space heating and domestic hot water is still some 50% higher in Russia than in OECD countries.

The *Outlook* assumes that efficiency improvements and building stock turnover will, in the longer term, moderate the growth in final energy consumption in commercial and residential buildings. The pace of growth picks up after 2010, by which time prices are assumed fully to reflect costs. Final demand slows down in the last decade of the *Outlook* period, as the economy matures and follows a development path similar to that of OECD countries.

Oil

Russian primary oil consumption will grow throughout the projection period, driven mainly by rising transport demand. Demand for oil will increase, on average, by 1.7% per year. The transport sector's share in total final oil demand will rise from 51% in 2000 to 59% in 2030.

After Saudi Arabia, Russia has the world's second-largest reserves of crude oil, 137 billion barrels (including natural gas liquids), or nearly 14% of total world reserves.[4] In the 1990s, Russia's oil reserves were severely depleted due to a sharp decline in exploration and, more recently, to rising production.

In 2000, Russian oil production grew by 6%, to 6.5 mb/d. Growth was even stronger in 2001, at 8%, with production reaching 7 mb/d. Mature fields in Siberia, primarily in Tyumen Oblast, account for nearly 70% of production. The Volga represents another 14%, while the remainder is produced mainly in the Urals and, to a much lesser extent, in European Russia. Three oil companies, Yukos, Sibneft and Surgutneftgaz, together accounted for 60% of the production increase in 2001. The recent upturn in oil production was largely a result of increased capital spending on fracturing and enhancement, and on putting idle wells back onstream. The increasing spending was, in turn, made possible by high international oil prices in recent years and by the 1998 rouble devaluation, which reduced costs dramatically. The higher prices encouraged partnerships with foreign companies and enabled Russian oil companies to gain access to advanced production technologies, including improved reservoir management methods.[5]

The *Outlook* projects that oil production will reach 8.6 mb/d by 2010 and will continue to rise thereafter, reaching 9.5 mb/d by the end of the projection period.

Russia is the world's second-largest oil exporter. In 2001, net exports of crude oil and refined products came to 4.3 mb/d. The current export boom grows out of high international oil prices and the oil companies' desire to increase their hard currency revenues. Russia is now a major non-OPEC source of oil for Western countries. The largest Western European importers of Russian oil are Germany, Italy, France, Finland and the United Kingdom.

4. United States Geological Survey (2000).
5. Yukos and Sibneft have teamed up with Schlumberger, and TNK with Haliburton.

Box 8.1: **Uncertainty Surrounding Future Russian Oil Production**

It is not certain if the recent growth in Russian oil production will continue at such a strong rate. Many producing fields require modern reservoir management to remedy the damage caused by earlier overproduction. In western Siberia, this involved the use of systematic water injection to raise output as quickly as possible. This practice has resulted in an increasingly large share of water mixed in with the oil extracted. By 1990, the "water cut" was 76% for Russia as a whole, up from about 50% in 1976. The share of oil produced from free-flowing wells dropped from 52% in 1970 to only 12% by 1990 and to just over 8% by 1999. Modern tertiary recovery techniques will be required to maximise oil recovery rates. The age of the Russian basins is reflected in the low average flow rate of its wells: seven tonnes per day in Russia versus 243 tonnes in the Middle East and 143 tonnes in the North Sea.

In the short term, the prospects for Russian oil output hinge essentially on how long western Siberia's current plateau of 4.1 mb/d to 4.6 mb/d can be maintained. Better reservoir management and the development of small and difficult fields could attenuate its depletion. The medium-term outlook will depend on how fast new reserves can be put into production in less mature areas, such as Timan-Pechora and Sakhalin. In the long term, new provinces, such as east Siberia, the Pechora Sea, or the Russian sector of the Caspian, could make sizeable contributions. Russia's ability to attract investment in new projects in those regions will be of critical importance. The revised Energy Strategy of the Russian Federation[6] estimates that an average of $8 to $10 billion per year will be needed to reach Russia's production target of over 9 mb/d in 2020. In 2000, when international oil prices were higher than in the past, upstream investment was less than $5 billion. A production-sharing regime can provide the necessary degree of fiscal and legal predictability for attracting large-scale investment, while the Russian tax code and general legal framework are put in place.

6. The revised Russian Energy Strategy has not yet been approved by the government, but is available on the Ministry of Energy's website : www.mte.gov.ru.

In the first decade of the *Outlook* period, Russia's domestic oil production is expected to grow faster than domestic demand. As a result, net exports will rise to 5.5 mb/d in 2010. Production will then slow, while demand will accelerate towards the end of the projection period. Net exports will fall to about 5.3 mb/d in 2020 and will remain at about that level for the rest of the projection period.

Figure 8.2: **Oil Balance in Russia**

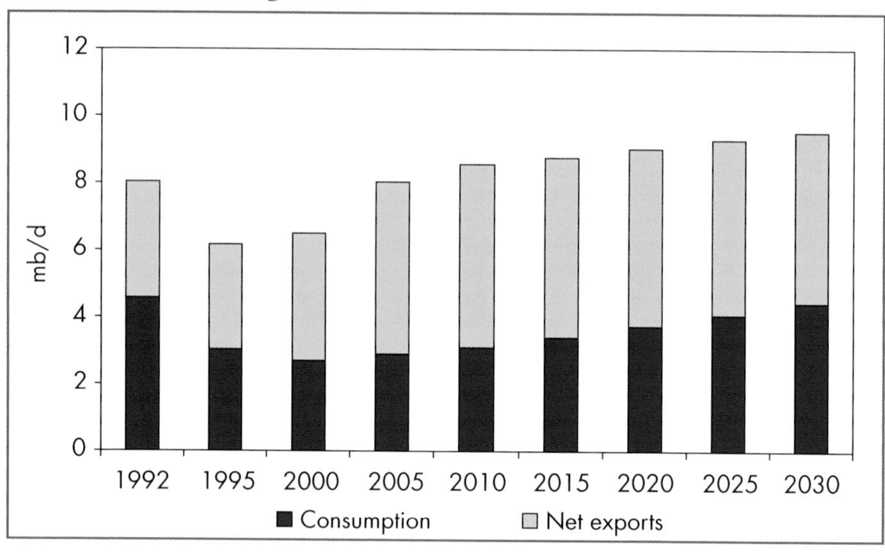

Since the precipitous decline in crude oil production and domestic demand in the early 1990s, total flows through Russian oil pipelines are now much smaller than they once were. But with growing export volumes, bottlenecks exist. Congestion is particularly bad at Novorossisk, Russia's major export port on the Black Sea. In the past, a large portion of Russian crude oil was dispersed to refineries across the FSU, and much of the rest went to Eastern Europe via the Druzhba pipeline. Since the dramatic decline in oil demand in the former Soviet Union and in Eastern Europe, a much larger part of the flow has gone to the country's few seaports. Much of the FSU pipeline system is now redundant as it was designed mainly to move crude oil to internal consuming centres. Existing capacity is probably sufficient to accommodate the projected increase in domestic demand. Some additional export capacity has already been completed, such as the Northern Gateway terminal at Varandey on the Barents Sea. A number of new pipeline projects have been proposed (Table 8.2).

Table 8.2: **Planned Developments in Oil Export Capacity**

Pipeline/route	Capacity (kb/d)	Expected completion	Comments
Caspian Pipeline Consortium	70 (Russian allocation)	2015	Most capacity is for Tengiz production
Baltic Pipeline – pipeline extension, new terminal at Primorsk	600 (pipeline) 240 (terminal)	2003	Diversifies exports, but geography makes it costly
Sakhalin export terminal and pipelines	240 – 300	2005	Pipelines to Japan, Korea and Chinese Taipei across Sakhalin Island
Angarsk (eastern Siberia) to Daqing (China)	400 – 600	2005-2010	Construction to begin in 2003 after route is decided
Druzhba expansion	200	2005	Pipeline to be extended to Omisalj, Croatia on the Adriatic

Source: IEA (2001b).

Natural Gas

The *Outlook* sees natural gas becoming even more dominant in Russia's energy mix than it already is. The share of gas in total primary energy supply is projected to rise from 52% in 2000 to 56% in 2030. Its share of final energy consumption will increase from 27% to 32%. Most of the growth in primary demand for gas will come from the power sector. By 2030, gas will fuel 60% of total electricity generation, compared to 42% in 2000.

At the beginning of 2001, Russia's proven and probable gas reserves stood at 46.6 tcm, according to Cedigaz. Russia holds over 30% of the world's proven natural gas reserves. The west Siberian basin has 37 tcm of reserves, or 79% of the country's total. It also has the largest undiscovered resources. The Continental Shelf, principally the Barents and Kara Seas, and the Sakhalin Shelf, have 4 tcm. Gazprom has licences for the exploitation of 34 tcm of proven and probable reserves, or 73% of the total. Some 60% of the company's reserves are concentrated in a small number of fields in the Nadym-Pur-Taz region of west Siberia. Gas reserves declined somewhat during the 1990s, largely because exploration fell off sharply. Even on conservative estimates, however, Russian gas production can be maintained for more than 40 years at the year 2000 rate.

Gas production fell from 640 bcm in 1990 to 583 bcm in 2000, due to under-investment in the upstream sector and a slump in domestic demand. Russia's main gas-producing fields in west Siberia, which accounted for more than 85% of total Russian gas production in 2000, are about 30% depleted. Output is expected to continue to fall at the region's three super-giant fields, Medvezh'ye, Yamburg and Urengoy. Gazprom hopes to offset this decline by raising production from the Zapolyarnoye gas field, which opened in 2001, in Nadym-Pur-Taz and in fields in new areas. These include the giant Shtokmanovskoye gas field in the Barents Sea and the Yamal Peninsula. Gazprom has already drawn up a plan for developing fields in Yamal at a total cost of $30 billion. Foreign investment and technology will be needed for new field developments, such as Shtokmanovskoye, and for the building of new high-pressure transmission lines, particularly where there are difficult geological and climatic conditions.[7]

Box 8.2: **Competition for Gazprom in the European Gas Market?**

Through its subsidiary company, Gazexport, Gazprom is the sole exporter of Russian gas to Western Europe. Despite production declines in the 1990s, exports to Europe have continued to increase and now account for 37% of total Russian output. In 2001, Gazprom production fell by 2%, following a 5% decline in 2000. Total Russian gas production, however, remained stable due to increasing production from independent gas producers (6% of gas production in 2001) and oil companies (6% of gas production in 2001). Increasingly, these non-Gazprom players will push for more transparency in access to pipelines terms, gas-processing and transportation tariff methodology.

The position of Gazprom in the Russian gas market, however, is so strong that it will be extremely difficult for new companies to compete. Aggressive price reform and market share targets, as well as more favourable access terms guaranteed by the Federal Energy Commission, may help new companies to compete.

7. For more details, see IEA (2001b).

Figure 8.3: **Russian Natural Gas Exports in 2001** (bcm)

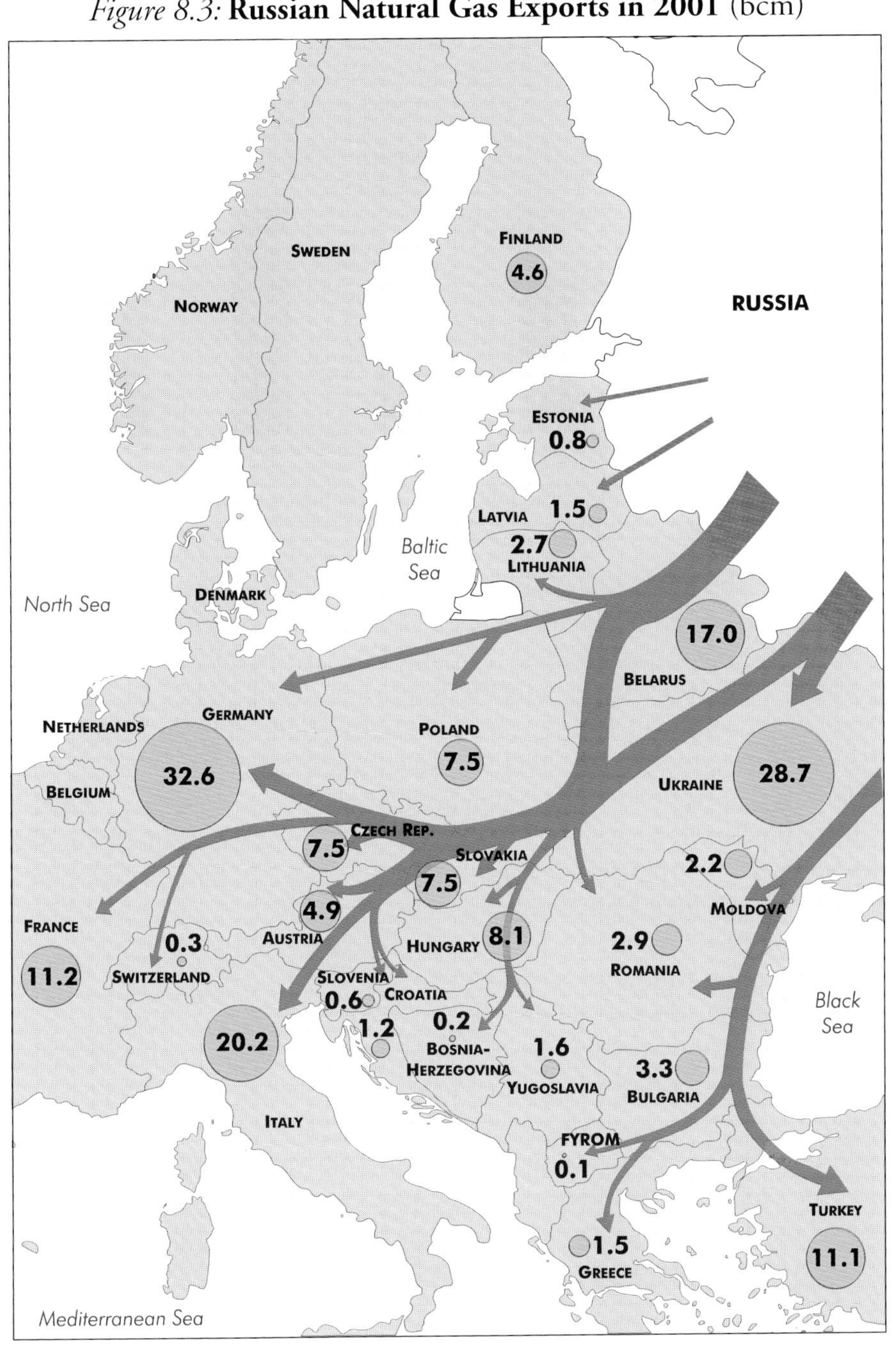

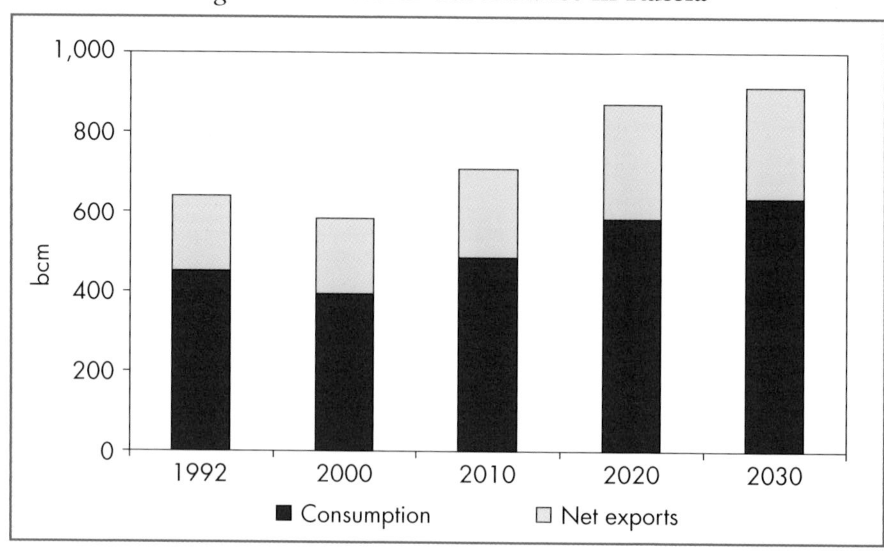

Figure 8.4: **Natural Gas Balance in Russia**

Given the huge resource base, Russian gas production will no doubt increase over the projection period. The *Outlook* expects Russia to satisfy its growing domestic market *and* to remain the largest gas exporter in the world in 2030. Gas exports to OECD Europe totalled 110 bcm in 2001. Western Europe will continue to attract the bulk of exports, but new markets, primarily in Asia, will also be secured. The Sakhalin-2 project plans to export LNG. The Sakhalin-1 project is considering laying a sub-sea pipeline for natural gas exports to the North Island of Japan. The development of a pipeline to China from the Siberian field of Kovykta is another possibility. This line would also enable Russia to export gas to Korea and, possibly, on to Japan. In this *Outlook*, exports to China and the OECD Pacific region are expected to exceed 20 bcm by 2030. This is relatively low compared to the projected 200 bcm of exports to OECD Europe by 2010 (with little increase anticipated to 2030) but it will enable Russia to diversify its potential markets and to a certain extent increase its revenue.

Total Russian gas production is expected to increase continuously over the projection period from 583 bcm in 2000 to 709 bcm in 2010, and to 914 bcm by 2030 (Figure 8.4). Under the Reference Scenario gas price assumptions, total exports will increase from 188 bcm today to 280 bcm in 2030.

Coal

Russian demand for coal fell from 316 million tonnes in 1992 to 219 Mt in 1998. The downward trend reversed itself in 1999, and coal demand has increased in the last two years. While coal's *share* in Russia's total primary energy supply is expected to fall over the next three decades, total primary demand for coal is projected to increase by 0.4% a year. Almost two-thirds of the incremental coal demand will come from power generation. Only in industry will coal demand continue to rise, mainly in the iron and steel sector.

Russia holds 16% of the world's proven recoverable coal reserves.[8] It is the fifth-largest coal producer, accounting for some 5% of total world production in 2000. Russia has a long history of subsidising its coal industry, largely through welfare safety nets for coal workers. Its labour productivity is low by international standards.

In the revised "Energy Strategy of the Russian Federation to 2020", the Russian government lays out plans to increase coal production from the current level of about 270 million tonnes.[9] While the reserve base *does* exist to meet domestic demand over the next few decades, increased coal production will depend on infrastructure development, on attracting new investment, on price reform and on the potential for Russian coal exports.

Coal exports plummeted in the 1990s but recovered to 35 million tonnes in 2000. The location of Russian coal, however, limits export possibilities, as does the coal's low quality. In 2000, some 55% of coal exports were destined for Europe and some 20% for Asia. With its current port facilities, Russia can expand coal exports by no more than 8% to 10% over the next few years. The *Outlook* projects that coal exports will remain flat over the next three decades, given Russia's infrastructure constraints and strong competition from other coal exporters.

Electricity

Electricity demand will grow by 2.3% per year over the *Outlook* period, and electricity will be the fastest growing of all energy sources for final consumption. Growth is expected to accelerate in the second decade

8. World Energy Council (2001).

9. These plans are based on Russia's expectations of high coal demand growth over the next twenty years and an anticipated reduction in the share of gas and heavy fuel oil for heat and power generation. Individual coal companies have already announced new mines and expansion projects at existing mines that will boost hard coal capacity by 25 million tonnes per year by 2010 and by nearly 36 million tonnes by 2020. The government estimates that the investment needed for the development of the coal industry will be close to $20 billion over the next twenty years.

to 3% per year, compared with 2% from 2000 to 2010. Growth will slow after 2020 as saturation effects begin to take hold and economic growth stabilises.

Electricity generation in Russia was 876 TWh in 2000; gas accounted for 42% of this, coal for 20%, hydro for 19%, nuclear energy for 15% and oil for 4% (Table 8.3). The average generating capacity factor was 46%, as against 54% in 1992. The low load factor means there is enough capacity to meet additional demand growth for a few more years. Nevertheless, most plants are old and poorly maintained, and investment is needed to refurbish them so as to continue to operate them. Electricity-sector reform has stalled in Russia — a decision by the Duma on going ahead was still pending in mid-2002.

Fossil-fuel plants generally operate at low load factors in Russia, while the lower running costs of nuclear and hydro plants provide an incentive to operate them as much as possible. Consequently, the share of fossil fuels in electricity output dropped from 71% in 1992 to 66% in 2000. Nuclear-based generation has been increasing since 1994, rising to 131 TWh in 2000.

Table 8.3: **Electricity Generation Mix in Russia** (TWh)

	1992	2000	2010	2020	2030
Coal	154	176	213	216	277
Oil	100	33	33	27	21
Gas	461	370	472	803	1,007
Nuclear	120	131	157	155	143
Hydro	172	164	169	190	196
Other renewables	2	3	8	14	27
Total	**1,008**	**876**	**1,052**	**1,405**	**1,671**

Over the *Outlook* period, gas-based generation is projected to grow to 60% of the total. Over 80% of new generation capacity in the next thirty years will be gas-fired, particularly combined-cycle turbine plants. Higher prices to consumers and the enforcement of payment will be critical in attracting investment in more efficient power technologies. Enormous progress has been made in recent years in improving payment rates and settling arrears, and this trend is expected to continue.

Generation capacity is projected to expand by 205 GW over the next three decades. Nuclear capacity is expected to remain at about 20 GW, and in this *Outlook* nuclear retirements are projected to outweigh additions after 2010. Russia currently has 44 GW of hydroelectric-based generation capacity. A further 10 GW will be added over the projection period. Another 7 GW based on renewable energy, mostly biomass and wind, will also be brought online. Capacity expansions will cost some $157 billion over the thirty-year period. Nearly half of that will be needed in the last decade of the projection period, as existing capacity can meet most of the additional demand to 2010, provided that the necessary upgrading is made.

Environmental Issues

Russia's breakneck industrialisation has left a legacy of pollution and nuclear waste, with which the country now struggles. The cost of environmental clean-up is high and paying it has, so far, not been a top priority of the Russian government. While economic decline in the 1990s brought a steep drop in carbon dioxide emissions, the Russian economy is still very carbon-intensive.

Figure 8.5: **Energy-Related CO_2 Emissions in Russia**

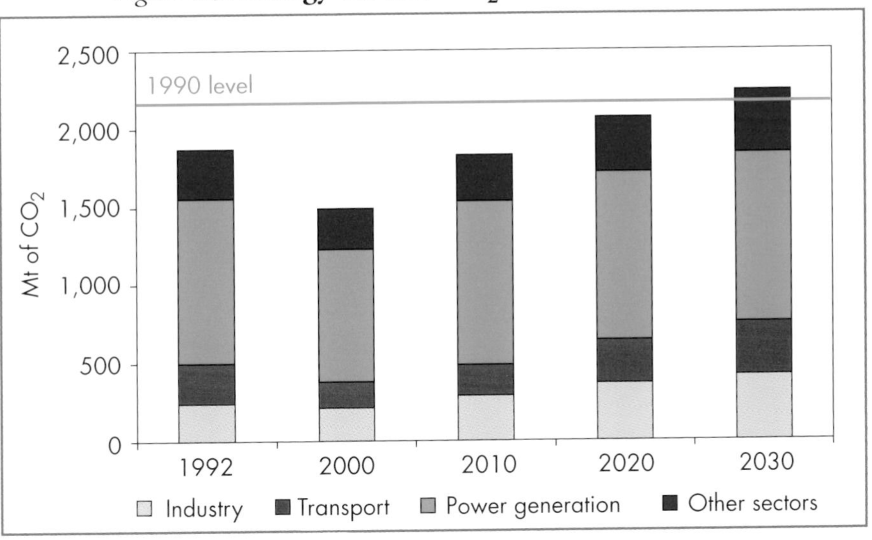

Under the Kyoto Protocol, Russia made a commitment to limit its average annual greenhouse gas emissions in the "performance period", 2008-2012, to their 1990 level. Because of the decline in economic activity over the last decade, emissions will in fact be much lower than that. This provides scope for Russia to sell surplus emissions as part of the emissions-trading system envisaged in the Protocol. This *Outlook* foresees that Russia's energy-related carbon emissions in 2010 will be some 380 million tonnes, or 17%, below 1990 — despite a projected increase of 2.1% per year between 2000 and 2010. Emissions grow more slowly in the last two decades of the projection period (Figure 8.5), in line with projected growth in energy demand.

CHAPTER 9: INDIA

HIGHLIGHTS

- India will become an increasingly important player on world energy markets as continued rapid expansion of the population and strong economic growth drive up energy demand. Primary energy supply will rise by an average 3.1% per year between 2000 and 2030. Final demand for oil, gas and electricity will increase rapidly.
- With limited domestic resources, India will have to import more oil and gas. Coal imports will probably increase too, as demand shifts to higher quality grades that can be acquired more cheaply abroad. The country's oil import dependence will increase sharply, from 65% in 2000 to 94% in 2030.
- The prospects for electricity supply are uncertain, given the industry's severe financial difficulties, the result of decades of underpricing and poor management. Massive investment is needed to boost India's generating capacity, and to improve and expand its transmission and distribution networks to meet growing demand. India's electrification rate is projected to grow, but hundreds of millions of people will still be without electricity in 2030.
- Natural gas could play a much bigger role in India's energy mix in the future. But financial problems in the power sector – the key growth market for gas – will need to be resolved, and financing must be found for LNG and cross-border pipeline projects. Some half of the projected growth in gas demand will be met by imports.
- Further reform of energy pricing is a vital precondition to the development of energy supply infrastructure in India. Foreign investors will have to provide an increasing part of the capital. The cumulative investment needed over the next three decades to meet the projected increase in generating capacity alone is estimated at around $270 billion.

Energy Market Overview

India's primary energy demand has grown spectacularly over the last thirty years, along with rising population and incomes. It now accounts for more than 3% of the world total.[1] Even so, energy intensity – measured both as energy use per unit of GDP and per capita – is still very low. Coal accounts for 55% of total primary energy demand and oil for 34%. Gas, hydropower and nuclear make up the remainder.

Table 9.1: **Key Economic and Energy Indicators of India**

	2000		Average annual growth 1990-2000 (%)	
	India	**World**	**India**	**World**
GDP (in billion 1995 $, PPP)	2,279	41,609	5.6	3.0
GDP per capita (in 1995 $, PPP)	2,247	6,908	3.7	1.6
Population (million)	1,014	6,023	1.8	1.4
TPES (Mtoe)	300	9,039	5.1	1.5
TPES/GDP*	0.13	0.22	-0.5	-1.5
TPES per capita (toe)	0.3	1.5	3.2	0.1
Net oil imports (mb/d)	1.4	-	11.1	-
CO_2 emissions (million tonnes)	937	22,639	4.6	1.2
CO_2 emissions per capita (tonnes)	0.9	3.8	3.0	-0.1

*Toe per thousand dollars of GDP, measured in PPP terms at 1995 prices.
Note: Energy data exclude biomass.

India has limited resources of oil and gas, but plenty of coal, although much of it is of poor quality. Overall, India is a net importer of energy. In 2000, it imported 65% of its crude oil requirements and consumed close to 3% of world oil supply. The country is self-sufficient in natural gas at present, but will soon need to import gas, mainly in the form of liquefied natural gas, if it is to meet the projected surge in demand.

1. The energy demand figures exclude biomass. Biomass demand is discussed before the electricity section in this chapter.

Current Trends and Key Assumptions

Macroeconomic Context

India has the world's fourth-largest economy. Its gross domestic product grew by an average 4.9% per year in the last three decades. Economic growth accelerated in the early 1990s when a major programme of market-oriented fiscal and structural reforms, launched in 1991, began to take effect. Growth averaged 6.7% in the period 1992-1996, and then slowed down in 1997, at the peak of the Asian economic crisis. It picked up in 1998 and averaged 6.1% per year between 1997 and 2000, but it was still slower than in the early part of the decade. This slower pace reflected slower growth across the Asian-Pacific region, volatile oil prices, variable monsoons and the Gujarat earthquake in January 2001. Persistent structural problems include poor infrastructure, pervasive subsidies and high trade barriers. Excessive regulatory constraints in agriculture and industry, a large fiscal deficit, a growing national debt and high interest rates also hold back the Indian economy. The economic reform programme began to stall towards the end of the 1990s, constraining both private and public investment. The government has announced a second phase of fiscal and structural reforms to respond to these problems. This time, the focus will be on farming, international trade, labour markets, social policy and the restructuring and privatisation of publicly-owned enterprises.

India's GDP is assumed to increase at an average of 4.6% per year over the full projection period (Table 9.2). The economy will grow more quickly in the current decade, at 5% per year, and then slow to 4.1% per year from 2020 to 2030. This trend stems in large part from slower population growth and from a maturing of the economy.

Table 9.2: **Reference Scenario Assumptions for India**

	1971	2000	2010	2030	Average annual growth 2000-2030 (%)
GDP (in billion 1995 $, PPP)	570	2,279	3,722	8,787	4.6
Population (million)	560	1,014	1,164	1,409	1.1
GDP per capita (in 1995 $, PPP)	1,017	2,247	3,197	6,236	3.5

Note: All values are in dollars at constant 1995 prices.

India's population is estimated to have passed the one-billion mark in 2000, having grown at an average rate of 2.1% per year over the preceding three decades. But population growth has been slowing and was down to around 1.6% in 2000. The birth rate is assumed to continue declining progressively over the projection period, with population growth averaging 1.1%. At less than $2,300 in purchasing power parity terms, India's average per capita income is well below even the developing world average. The median income is lower still, because a small proportion of the population have incomes well above the average. According to the World Bank, some 44% of the population – close to 450 million people – lived on less than $1 a day in the year 2000.[2] They represent about 45% of the world's extremely poor people. Incomes vary greatly between one part of India and another, and between urban and rural areas. India is less urbanised than most other developing countries in Asia. Approximately 75% of India's poor live in rural areas, although migration from rural to urban areas is increasing.

Energy Prices

The Indian government has taken important steps towards removing price controls on oil and coal and lowering subsidies to energy generally. Coal prices were decontrolled in 2000, and there are no longer any direct subsidies to coal production or consumption. Delivered coal prices, nonetheless, remain below market levels due to continuing subsidies on rail transportation. In April 2002, the government completed the dismantling of the Administered Pricing Mechanism for oil products and natural gas and the removal of all subsidies, except for those on kerosene and LPG used by households. Consumer prices for coal, oil products and gas are assumed to follow international prices over the *Outlook* period.[3]

Indian electricity is still heavily subsidised. In 2000-2001, the average rate of subsidy expressed as a proportion of the estimated full cost of electricity supply was 93% for farmers and 58% for households. Industrial and commercial customers and the railways pay above-cost prices. On average, current retail prices represent 70% of real costs; the figure was 80% in the early 1990s (Figure 9.1). About a half of electricity sales are in fact billed, and only 41% are regularly paid for, partly because of theft and corruption.[4] The central and state governments are attempting to address

2. From World Bank's *World Development Indicators.*
3. See Chapter 2 for international price assumptions.
4. See Chapter 13 for further discussion.

Figure 9.1: **Average Electricity Supply Cost, Revenues and Recovery Rate in India**

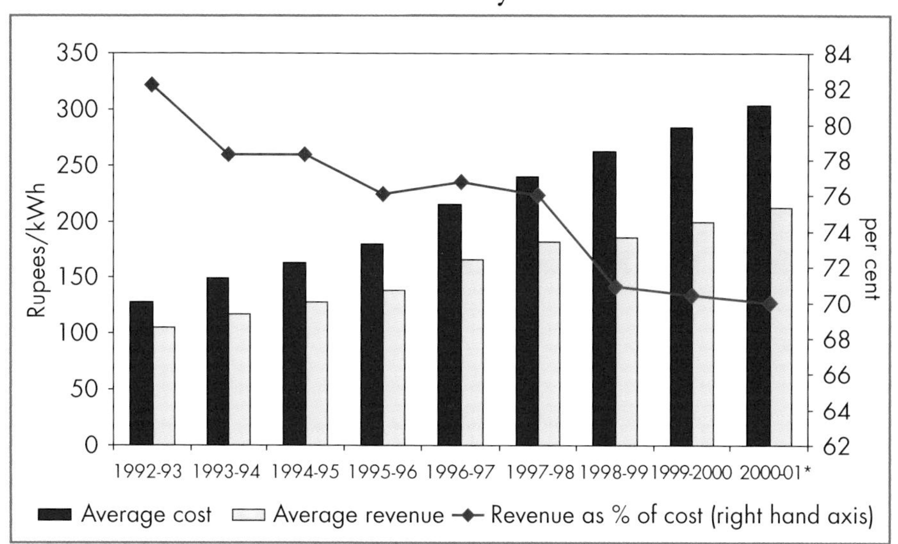

* Annual Plan data.
Source: GOI (2001).

these problems as part of a broader restructuring of the industry, but pricing reform has been blocked or delayed by fierce resistance from some consumers and political leaders. Our projections assume only gradual progress in bringing prices up to cover full costs.

Energy and Environment Policies

Traditionally, the Indian government has intervened heavily in the energy sector, both through state ownership and through regulation, including price controls and subsidies. As part of its economic reform programme, the government has sought in recent years to modify its role in the energy sector. In addition to dismantling price controls and subsidies, it has opened up the energy sector to private and foreign investment, set up independent regulatory commissions in the power and gas sectors and removed restrictions on energy trade.

Among the consequences of India's soaring energy use have been severe air and water pollution, deforestation and rising CO_2 emissions. Air pollution in urban areas, largely the result of fossil-fuel use, contributes to millions of premature deaths each year.[5] Concentrations of sulphur dioxide

5. United Nations Development Program *et al.* (2002).

and airborne particulate in most Indian cities greatly exceed international standards. Major causes include the rapid increase in the use of transport fuels, especially diesel, and the burning of coal in power generation and industry.[6] Widespread use of biomass-fuelled cooking stoves causes indoor pollution. The central and state governments are trying to address these problems. They have tightened quality standards for air and water and are trying to enforce them more strictly.

Results of the Projections

Overview

India's primary energy demand is projected to rise by 3.1% per year between 2000 and 2030 (Figure 9.2) – well below the 5.6% rate between 1971 and 2000.[7] The slowdown reflects, in part, lower GDP growth and population growth over the *Outlook* period. Demand will decelerate gradually over the projection period, in line with the assumed slowdown in economic and population growth, and rising energy prices after 2010. Coal, already the main fuel in the primary energy mix, and oil will account for over 80% of primary energy demand in 2030. Natural gas use will increase rapidly, but from a low base, so its share in total primary energy supply will reach only 13% in 2030 compared to 7% in 2000. Nuclear energy supply grows over the projection period, on the assumption that a small number of new nuclear plants are built.

Primary energy intensity – the amount of commercial energy needed to produce a unit of GDP – will continue to decline in line with the trend since 1995. The economy will continue to shift towards less energy-intensive activities. Meanwhile, there is tremendous scope to improve the efficiency of energy use in power generation and elsewhere. Commercial energy use per person will nonetheless increase slightly over the projection period, because increased commercial activity and higher incomes will continue to boost demand for energy services.

6. Tata Energy Research Institute (2002). An early study by the same institute (1999) quantifies the economic and social costs of environmental degradation.
7. The energy demand projections exclude biomass. Biomass demand is discussed before the electricity section in this chapter.

Figure 9.2: **Total Primary Energy Demand in India**

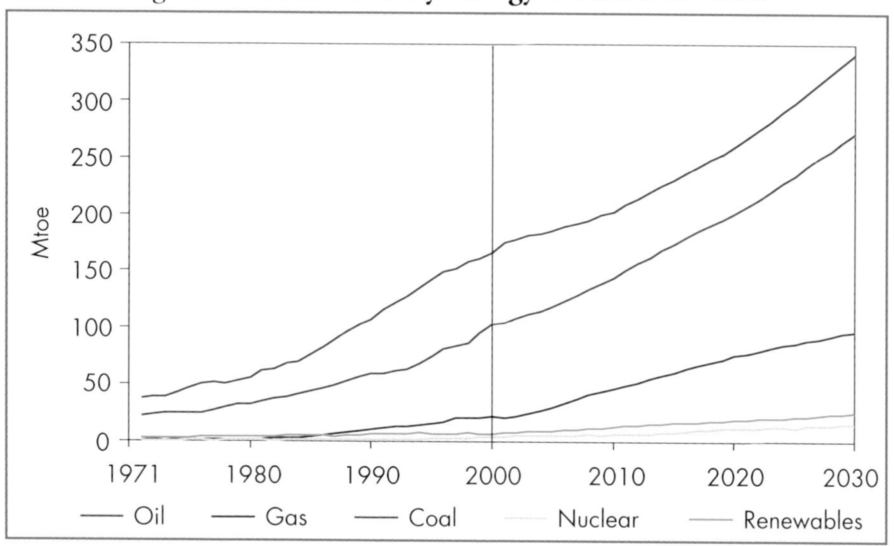

Demand by End-use Sector

Total final energy consumption will increase at 3.5% per annum over the *Outlook* period. This is slightly faster than primary demand, mainly because the thermal efficiency of power stations is projected to increase. The shares of electricity and oil in final consumption are set to rise, mainly at the expense of coal.

The increase in final oil consumption is driven by 4.4% per year growth in transport demand, although this is less than the 6.6% annual rise over the past three decades. Oil's share in final consumption increases from 55% in 2000 to 57% in 2030. Although pump prices rise in line with the assumed increase in international crude oil prices after 2010, this will only slightly dampen demand growth. The projections assume that passenger vehicle ownership grows rapidly and that two-wheeled vehicles still dominate the Indian fleet.

The residential and services sectors currently account for only 22% of final commercial energy use, but this share will increase slightly over the projection period to 23% (Figure 9.3). These sectors' consumption of electricity will grow by 6% per year, as electrification rates rise, but this is still lower than growth rates over the last 30 years, and India's per capita consumption of commercial energy will remain very low by international standards.

Figure 9.3: **Total Final Consumption by Sector in India**

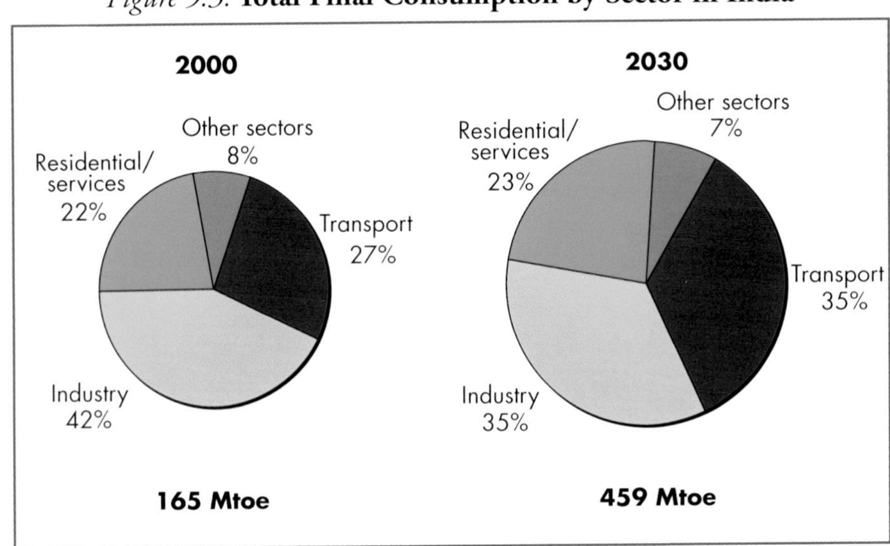

Industry's share of total final consumption will decline from 42% in 2000 to around 35% in 2030. Industrial energy use is concentrated now in a small number of industries; the iron-and-steel and chemicals sectors alone account for 54% of industrial energy consumption (excluding renewables). Electricity and gas will be the most rapidly increasing energy sources in industry. The use of coal and oil in industry will continue to grow as the economy expands, but more slowly than in the past.

Oil

Indian primary oil consumption will grow by an average 3.3% per year, from 2.1 mb/d in 2000 to 3 mb/d in 2010 and 5.6 mb/d in 2030. About 70% of the increase in oil demand will come from transportation. Diesel will remain popular, because it is taxed so much less than gasoline, making it half as expensive at the pump; no change in the tax structure is assumed. A sharp increase in passenger and commercial vehicles, including farm tractors, is the main source of new demand (Figure 9.4). Growth in the vehicle stock slows towards the end of the *Outlook* period as the growth in GDP and population declines.

Oil imports, which meet around 65% of India's current needs, will continue to grow because the indigenous production of crude oil and natural gas liquids is projected to decline. India is trying to limit the country's dependence on oil imports by boosting domestic exploration and

Figure 9.4: **Transport Oil Demand and Passenger Vehicle Ownership in India**

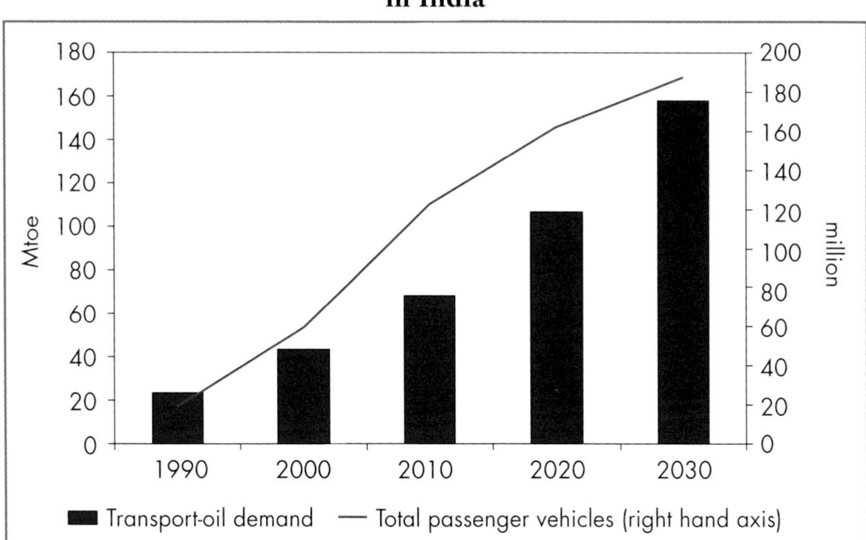

production. In 1997, the government adopted a New Exploration Licensing Policy, which allowed a number of foreign companies to win licences to explore for hydrocarbons, an activity previously reserved for state firms. Foreign investment is expected to raise recovery rates from reservoirs, which currently average around 30%, well below the world average. But prospects for major new oil discoveries are not bright, given the poor results of exploration in recent years. The best chances may lie in deep-water offshore exploration. A find in offshore Gujarat in 2001 is believed to hold about 200 million barrels of recoverable reserves.

Natural Gas

Natural gas demand is projected to jump from 22 bcm in 2000 to 97 bcm in 2030 – an average rate of growth of 5.1% per year. The biggest increase in gas use is expected to come from power generation. By 2030, just over 62% of all gas consumed in India will be used by power stations. Consumption will grow in the residential sector by 5.2% per year, and in industry by 3.5% per year.

India's natural gas production reached 22 bcm in 2000. Gas reserves are mainly in the Bombay High Field. Production is assumed to rise, reaching 58 bcm in 2030, largely through better exploitation of associated gas (gas found in the same fields as oil). In the longer term, the government hopes to tap coal bed methane resources.

Figure 9.5: **LNG Terminal Projects in India**

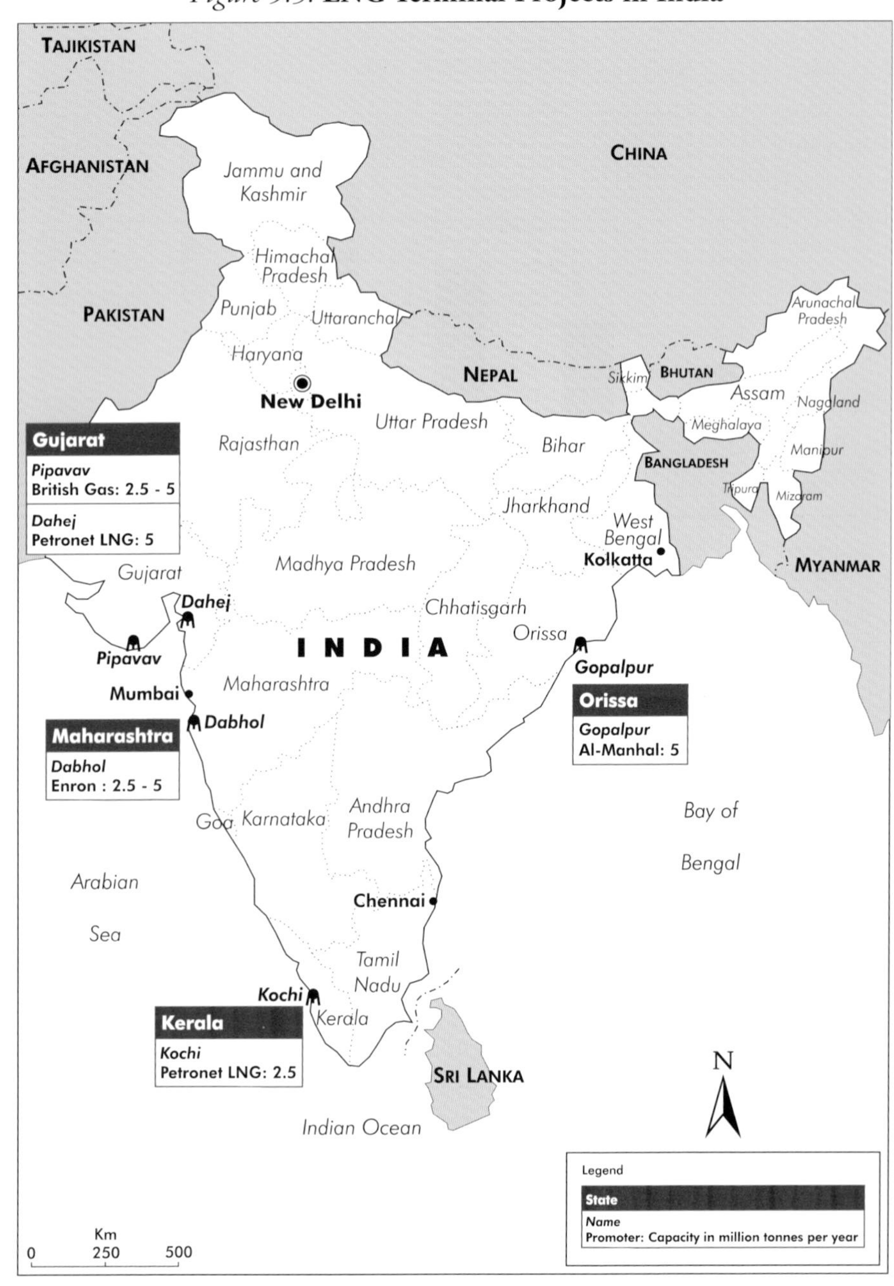

Imports will play a major role in meeting the projected increase in gas demand. Imports are projected reach 9 bcm in 2010 and 38 bcm in 2030. Much of this gas will be imported as LNG. The country's first LNG terminal at Dabhol is due to be commissioned in 2002. It will have a capacity of 5 Mt/year (6.5 bcm). Almost all the LNG delivered to Dabhol is earmarked to supply a power plant already in operation. But a dispute over power tariffs and sales between the owners of the plant, in which the bankrupt Enron company holds a 65% stake, and the Maharashtra State Electricity Board (SEB) has delayed completion of the project. A second 5 Mt per year terminal is to be built by Petronet LNG Ltd at Dahej, with first gas expected in 2004. Several other terminals are planned (Figure 9.5).

Gas could also be imported by pipeline. Bangladesh, which has large proven reserves, is the most likely source, but the Bangladeshi government has not yet authorised exports. In the longer term, India could import gas from Iran, Qatar or Central Asia. But the prospects of any such projects are complicated by political tensions between India and Pakistan, by the need to transit through Afghanistan and by doubts about the availability of gas for export from Iran's massive South Pars field. An offshore line from Iran or Qatar bypassing Pakistani territorial waters has been proposed, but its cost is likely to be prohibitive.

The prospects for both LNG and pipeline imports are very uncertain. How rapidly demand for gas develops will depend on the ability of India's power companies – the main prospective buyers of imported gas – to pay their bills. The Dabhol dispute and the dire financial straits of the state electricity boards have heightened concerns about the risk of non-payment. Moreover, some of the LNG terminals that have been proposed have yet to secure gas supplies or financing for terminal construction and the purchase of ships.

Coal[8]

Primary coal demand in India will increase by 2.4% per annum from 2000 to 2030. Power generation, which already accounts for around 75% of the country's coal use, will remain the main user, while consumption in industry and other sectors will grow more modestly (Figure 9.6). Growth in coal demand will accelerate after 2010. Any tightening of current environmental restrictions on coal use would lead to lower coal demand, although new clean coal technologies, if their cost were to fall sharply,

8. See IEA/CIAB (2002) for a description of the Indian coal sector.

Figure 9.6: **Primary Coal Consumption in India**

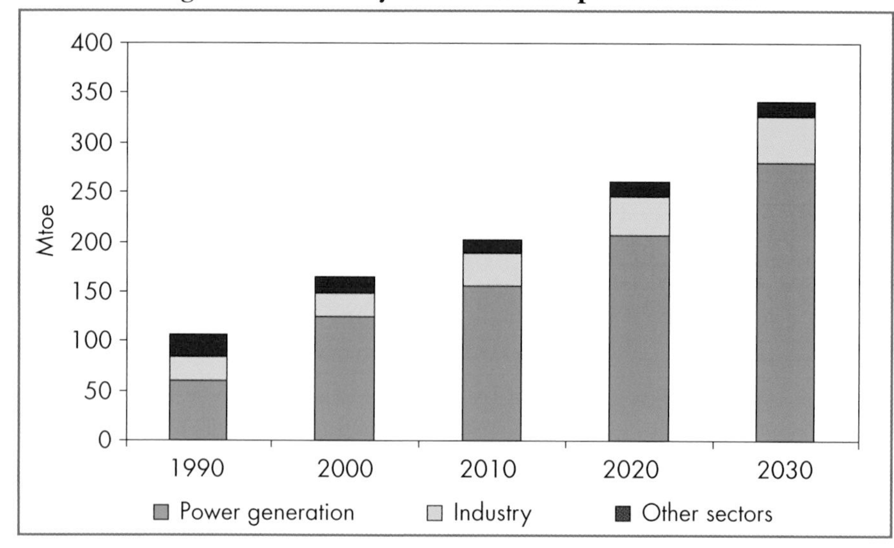

might help to boost demand. This *Outlook* assumes no major changes in India's environmental regulations.

In 2000, India was the world's third-largest coal producer, after China and the United States. Coal production, which surged from 75 Mt in 1971 to about 329 Mt in 2000, is expected to rise, but probably not quickly enough to meet demand. Proven coal reserves are estimated at 82.4 billion tonnes, of which three-quarters are in Bihar, Madhya Pradesh and West Bengal. Although full privatisation of the industry is still deemed unfeasible, the government has taken cautious steps towards liberalising the coal market, including the removal of price controls.

Indian coal is high in ash, low in sulphur and of low calorific value. It needs washing to make it suitable for coke ovens. Mining productivity is low, with mechanisation limited largely to coal cutting. Loading is done mostly by hand. Average production costs, already low by international standards, have been kept stable in real terms by the lower costs of new developments. Costs are likely to rise as opencast mining conditions worsen. About three-quarters of coal production is transported to power plants by rail. The growth in coal consumption projected here implies the need for very substantial investment in new transport capacity. In coastal areas, higher-quality imports are increasingly likely to be preferred to low-quality domestic coal. Australia will probably remain India's main supplier of coking coal, while Indonesia, South Africa and China will probably supply most of the additional steam coal.

Biomass

Biomass, used mainly in the residential sector, is still the largest single source of energy in India. Its consumption reached almost 200 Mtoe in 2000, or 40% of the country's total energy use (Table 9.3). Biomass accounts for roughly 85% of residential energy use. The share is even higher in many rural areas, where modern commercial fuels are either unavailable or unaffordable. Animal waste and crop residues each make up about a quarter of India's total biomass. Consumption in the residential sector is projected to increase by 0.8% from 2000 to 2010, then to slow over the next two decades, as incomes rise and households start to use commercial fuels along with traditional biomass. As in most developing countries, the share of biomass in overall energy use has been declining to be replaced by modern commercial fuels. Demand for biomass is projected to rise by 0.5% a year over the entire *Outlook* period.[9]

Table 9.3: **Biomass Energy Demand in India** (Mtoe)

	2000	**2010**	**2020**	**2030**	**Average annual growth 2000-2030 (%)**
Biomass	*198*	*216*	*224*	*227*	*0.5*
TPES including biomass	**498**	**628**	**788**	**971**	**2.2**

Electricity

Electricity output, which expanded at a brisk 6.5% per year over the past decade, is projected to grow more slowly in the future. Output growth will average a little more than 4% per annum, slowing gradually over the projection period (Table 9.4). The share of coal in total generation will drop from more than three-quarters now to 65% by 2030, while that of gas will increase from 5% to 18%. The share of nuclear energy will also increase, but only marginally. Hydropower is assumed to expand, but its share in total generation will actually fall. New power projects fuelled by imported LNG could be built where coal is expensive to transport, but this would require pricing reform to cover the cost of imported LNG. Large-scale oil-fired generation is marginal, but oil-fired *distributed* generation is

9. Chapter 13 discusses the relationship between poverty and biomass use.

expected to expand, especially in the near term. Problems with the quality of grid-based electricity supply will continue to encourage industries to generate their own power.

Table 9.4: **Electricity Generation Mix in India** (TWh)

	1990	**2000**	**2010**	**2030**	**Average annual growth 2000-2030 (%)**
Coal	193	420	552	1,169	3.5
Oil	8	5	11	15	3.5
Gas	11	25	122	321	9.0
Nuclear	6	17	24	60	4.3
Hydro	72	74	129	208	3.5
Other renewables	0	1	9	31	10.6
Total	**289**	**542**	**848**	**1,804**	**4.1**

The electricity industry faces enormous challenges in providing reliable service and meeting rising demand. A lack of peak-load capacity and the poor performance of the transmission and distribution system cause frequent, widespread blackouts and brownouts. Plant load factors are often low, due to the age of generating units, poor quality coal, defective equipment and insufficient maintenance. The lack of inter-regional grid connections accentuates local power shortages. Power theft, the non-billing of customers and non-payment of bills are common. Some of the state electricity boards have been poorly managed. Pricing policies, which keep tariffs to most customers well below the cost of supply, have starved the SEBs of cash, driven up their debts and discouraged investment. Although the government has encouraged private and foreign investment in new independent power producers, most of the projects proposed have stalled because of financing problems and delays in obtaining regulatory approvals. The dispute over the Dabhol power plant, the largest single foreign investment in India, has drawn attention to the financial risks of independent power producers. Structural and pricing reforms, which are assumed to proceed slowly over the *Outlook* period, constitute a major source of uncertainty for India's electricity supply prospects (Box 9.1). Poor network performance as a result of under-investment and widespread theft have led to high transmission and distribution losses. Losses are assumed to be trimmed from the current level, which is extremely high by world standards, to 20% by 2030. The cumulative investment needed to meet the projected increase in generating capacity is estimated at around $270 billion from 2000 to 2030.

Box 9.1: **Electricity Sector Reforms**

The government has taken a number of steps in recent years to restructure the electricity industry, to reform pricing and to introduce more market-based mechanisms. The 1998 Electricity Regulatory Commissions Act established the Central Electricity Regulatory Commission (CERC) with a mandate to set tariffs for inter-state trade and multi-state generation companies. The act also allowed for the setting-up of state electricity commissions to regulate retail tariffs. In this same year, an amendment to the Electricity Law decreed the separation of generation from transmission functions and gave Powergrid, the newly created central transmission utility, responsibility for inter-state transmission and centralised dispatch. In December 2000, the CERC decreed a change in the way prices were charged to state electricity boards by power plants owned by the central government. The ultimate aim of these actions is to create a competitive wholesale power market.

Still bolder reforms are planned. A new bill, drawn up by the government in 2000, is under discussion. Key measures include:

- easing licensing restrictions for new power projects (other than hydroelectric projects);
- open access in transmission;
- an obligation on states to establish regulatory commissions, which would set retail tariffs on the basis of full costs and promote competition;
- a requirement that any subsidies on electricity retail sales be paid out of state budgets rather than through cross-subsidisation.

Environmental Implications

The projections in this *Outlook* imply a massive increase in pollutant emissions from the burning of fossil fuels, unless technological solutions are implemented to curb this trend. Projected emissions will cause a dramatic deterioration in air quality, especially in cities. India's contribution to global greenhouse gas emissions will also expand, with the rapid growth in energy demand and the continued dominance of coal – the most carbon-intensive fossil fuel. India is now one of the lowest per capita emitters of CO_2, at 0.9 tonnes, or about one-twelfth the OECD average. But the energy sector's carbon intensity is high, and the country's *total* carbon dioxide emissions rank among the world highest. They are projected to reach 2.3 billion tonnes in 2030, up from 937 million tonnes in 2000 (Figure 9.7). Over half of this increase will come from power generation . Carbon intensity will decline steadily, primarily due to structural changes in the Indian economy towards less energy-intensive activities and a small drop in the share of coal in the primary energy mix.

Figure 9.7: **CO_2 Emissions and Intensity in India**

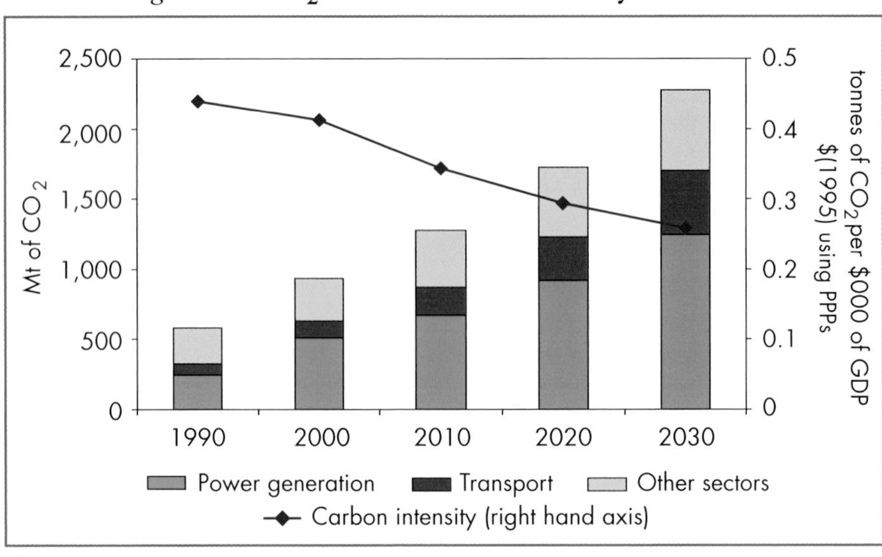

CHAPTER 10: BRAZIL

HIGHLIGHTS

- The *Outlook* projects annual average growth of 3% in primary energy supply over the next three decades in Brazil. Oil and hydropower are expected to remain the key fuels in its energy mix. But gas will make major inroads in power generation, particularly towards the end of the projection period.
- Brazil has large deep-water oil and gas resources. Its ability to exploit them, however, is uncertain, because doing so will be very costly and will require heavy investment. The *Outlook* expects Brazil to become self-sufficient in oil by the second decade of the projection period.
- Over the long term, gas will be increasingly important for Brazilian power generation, partly because new hydro sites are located far from consuming centres. The share of gas in the power generation mix is negligible today, but the *Outlook* expects that, by 2030, it will rise to 35%.
- Gas import dependence will rise rapidly in the first decade of the *Outlook* period. But Brazil is expected to tap its vast gas resources, and import dependence will fall to some 5% by 2030.
- Investment in power projects has so far fallen short of expectations due to Brazil's unstable regulatory regime and unattractive power generation prices. This *Outlook* projects that investment of some $160 billion will be needed over the next three decades to build the necessary additional generating capacity.

Energy Market Overview

Brazil is Latin America's[1] largest energy consumer, accounting for 36% of the region's consumption in 2000 (Table 10.1).[2] Its primary energy

1. The Latin American region as presented here does not include Mexico, which is a Member of the OECD. See Appendix 2 for a list of countries in the Latin American region and Chapter 4 for a detailed analysis of the Mexican energy sector.
2. The energy demand figures exclude biomass. Biomass demand is discussed before the electricity section in this chapter.

mix is dominated by oil (63%), and hydropower (19%). In 2000, hydroelectric generation produced 87% of Brazil's power. Despite its large energy resources, Brazil consumes more energy than it produces. In 2000, gas production was 7 bcm and demand was 9 bcm, while oil production was 1.3 mb/d, and demand was 1.8 mb/d.

Brazil's energy sector is in the midst of a profound restructuring involving privatisation of state-owned assets. Electricity generation and distribution have been opened up to private capital. The monopoly of the state-owned company Petrobras on exploration and production concessions in oil and gas has ended. There are two transnational gas pipelines and several electricity transmission lines linking Brazil with neighbouring countries. Many others are under construction or in the planning stage.

Table 10.1: **Brazil's Weight in Latin America, 2000**

	Level	**Share in region (%)**
GDP (in billion 1995 $, PPP)	1,178	45
GDP per capita (in 1995 $, PPP)	6,923	-
Population (million)	170	41
TPES* (Mtoe)	138	36
Oil production (mb/d)	1.3	19
CO_2 emissions (Mt)	303	35

* Excludes biomass.

Energy demand projections for Brazil are subject to uncertainties about the course of economic growth, the future of reforms in the gas and electricity sectors and the rate of gas penetration in the power sector. Raising the share of gas in Brazil's energy mix in the near term will depend on securing the investment to build transportation and distribution infrastructure. Oil production increases are also contingent on finding investment and on technological advances that would lower the costs of exploration and production. Although Brazil has large oil reserves, they are mainly off-shore and in deep water.

Table 10.2: **Key Energy Indicators of Brazil**

	Brazil 2000	Average annual growth 1990-2000 (%)	Mexico 2000
TPES per capita (toe)	0.8	3.0	1.5
Net oil import dependence (%)	28	-4.7	-77
Electricity consumption per capita (kWh)	1,877	2.8	1,640
TPES/GDP *	0.12	1.7	0.18
CO_2/TPES (tonnes/toe)	2.2	0.2	2.5

* Toe per thousand dollars of GDP, measured in PPP terms at 1995 prices.

Box 10.1: **The Brazilian Electricity Crisis**

In 2001, Brazil suffered a severe electricity shortage, brought on by the worst drought in 70 years and by insufficient investment in electricity generation and transmission capacity in the last 15 years. To avoid rolling blackouts, the government implemented a strict 10-month electricity-rationing programme in the industry, services and residential sectors. This programme has had a profound effect on Brazil's nascent economic recovery. From 2000 to 2001, GDP *contracted* by three percentage points. The rationing, however, forced power plant operators and industry to increase their efficiency, and it raised consumer awareness about energy savings.[3]

Electricity rationing ran from June 2001 to March 2002 and aimed to reduce usage by between 15% and 25%, depending on customer category. At first, the plan applied only to the north-eastern, south-eastern and mid-western regions. In August 2001, the programme was extended to the northern region, and the government sought to reduce electricity demand by 20% nationwide.[4] Power cuts

3. Platts estimates that, under rationing, Brazil used 14% less electricity than in the same months in 2000, yet industrial production declined by only 1.1% (Platts, 2002b).
4. Southern Brazil was unaffected by the drought but transmission from the region was insufficient to benefit the crisis areas.

were most severe in south-eastern Brazil, which has most of the country's population and industry. Electricity prices became extremely volatile.[5] The electricity shortage also dealt a blow to Brazil's liberalisation efforts. Privatisation goals for 2001/2002 were not realised, because political support for the programme declined.

Current Trends and Key Assumptions

Macroeconomic Context and Energy Prices

Over the past three decades, Brazil's economy grew, on average, by some 4% a year. The period included disruptive cycles of contraction and recovery in the 1980s and 1990s and a particularly severe monetary crisis in 1999.[6] In 2000, there were signs that the 1999 devaluation of the Brazilian currency, the *real*, was encouraging domestic and foreign investment in the private sector. The country earned $2.6 billion from foreign trade in 2001, its first surplus since 1994. Growth was about 4.5% in 2000, but slowed to 1.5% in 2001, partly because of electricity rationing and partly because of the global economic downturn. Growth is expected to recover to 2.5% in 2002.[7]

The impact of the 2001 electricity crisis is still the greatest threat to Brazil's economic recovery (Box 10.1). The crisis raised inflation, because the government increased electricity prices in order to restore the financial capacity of power companies. The threat of even higher inflation is exacerbated by high petrol prices, stemming from the removal of subsidies as the market was liberalised. The effects of the crisis in Argentina have also been felt throughout the Mercosur community.[8]

Brazil is one of the leading recipients of net foreign direct investment. From 1967 to 1998, net direct investment flows to Brazil were $101.4 billion. Since the initiation of reforms in the energy sector, the share of foreign capital has increased rapidly. Private investors now own 26% of electricity generation, compared with 0.3% in 1995, and 64% of

5. The spot price for electricity was more than $200 per MWh in July 2001, but fell to less than $4 per MWh in May 2002 (University of Sao Paulo, direct communication).
6. For more information on the economic background of Brazil, see IEA (2000) and OECD (2001).
7. OECD (2002).
8. The Common Market of the South, Mercosur's members are Argentina, Brazil, Paraguay and Uruguay, with Chile and Bolivia as associated members.

distribution, compared with 2.3% in 1995. Competition among Latin American countries for energy-related foreign investment has intensified, however, especially with the fall in private foreign energy investment in the late 1990s.[9]

This *Outlook* assumes that the Brazilian economy will grow by 3% per year from 2000 to 2020, then slow to an annual 2.8% in the final decade of the *Outlook* period. The population will increase by 1% per year, reaching 229 million by 2030 (Table 10.3). Given these population figures and the fact that Brazil's savings rate was 19% of GDP in 2000, growth could be higher than expected.

Table 10.3: **Reference Scenario Assumptions for Brazil**

	1971	2000	2010	2030	Average annual growth 1971-2000 (%)	Average annual growth 2000-2030 (%)
GDP (in billion 1995 $, PPP)	383	1,178	1,577	2,779	3.9	2.9
Population (million)	98	170	192	229	1.9	1.0
GDP per capita (in 1995 $, PPP)	3,897	6,923	8,221	12,135	2.0	1.9

The *Outlook* assumes that energy prices will be increasingly determined by market forces, as reforms proceed and that, as a result, prices of all energy products will follow international price trends more closely.[10]

Results of the Projections

Overview

Over the *Outlook* period, primary energy demand, excluding biomass, will grow by 3% a year on average, compared with 4.9% from 1971 to 2000. Energy demand will be 332 Mtoe in 2030. Per capita consumption will increase from 0.8 toe in 2000 to 1.5 toe in 2030, still low compared

9. Figure 2.10 in Chapter 2 shows past trends in private foreign direct investment in energy.
10. See Chapter 2 for international price assumptions.

with the OECD average of 4.7 toe in 2000. Energy intensity in Brazil increased by 1.7% per year over the last decade. Over the *Outlook* period, energy intensity will begin to decline, as the structure of the Brazilian economy gradually approaches that of OECD countries today.

Gas is expected to be the fastest growing fuel, with demand rising by over 7% per year. By 2030, gas will account for 19% of total energy supply, and 35% of fuel inputs in electricity generation. Oil's share in primary demand will drop by seven percentage points to 56% in 2030. Coal, with 1.9% growth per year, and hydro, with 2.2%, will see their percentage shares of total supply fall by 2030.

Brazil's final energy consumption will increase, on average, by 2.8% per year. Oil, half of which is used in transport, will remain the most important end-use fuel in 2030. Oil demand in the transport sector is projected to increase by 60 Mtoe over the projection period, and will represent over a quarter of the increase in primary oil demand in the whole of Latin America. The share of gas in final consumption will increase from 4% in 2000 to 7% in 2030. Electricity demand will grow by 3% per year over the *Outlook* period. Electricity rationing in 2001/2002 produced short-term energy savings, especially in the industry sector, and many analysts believe that it can lead to long-term savings as well. Growth in electricity demand, however, is still expected to outpace growth in GDP over the next three decades.

Demand for energy in the transport sector will grow by an average of 3.1% per year, making this the fastest growing of all final demand sectors. Transport energy demand will represent 38% of total final consumption in 2030, increasing from 42 Mtoe in 2000 to 104 Mtoe in 2030.[11] A substantial potential remains for more vehicle ownership as incomes rise. Brazil had 140 passenger cars per 1,000 people in 2000, compared with 480 per 1,000 in the United States.[12] Industrial energy use will increase by 2.7% per year. Gas will be the fastest growing fuel in the industry sector, and its share is expected to nearly double. The share of electricity in other sectors will rise from 54% in 2000 to 65% in 2030.[13]

11. Ethanol, which is not included in these demand figures, plays a major role in transport energy use in Brazil (see the section on biomass in this chapter).
12. International Road Federation (2002).
13. "Other sectors" includes the services, residential and agricultural sectors.

Oil

Oil consumption will rise from 1.8 mb/d in 2000 to 3.8 mb/d in 2030, at an annual average rate of 2.5%. The transport sector will account for 62% of incremental oil demand over the next three decades. Oil's share in primary energy demand, however, will fall over the *Outlook* period. After Venezuela, Brazil has the second-largest proven oil reserves in Latin America, at 8.9 billion barrels. Brazil is estimated to have some 47 billion barrels of undiscovered recoverable resources, and 8 billion barrels of undiscovered recoverable NGL.[14] Almost all the oil is in offshore fields, with about 35% in the offshore Campos basin.

In May 2001, Petrobras, the state oil and gas company, increased its five-year budget for exploration and production expenditure by $600 million to $19.2 billion. Petrobras holds the world record for deep-water drilling and production — with production from a well 1,853 metres below the surface. About 23% of Brazilian reserves are found at seawater depths of between 1,000 and 2,000 metres. It is expected that about half the resources yet to be discovered will be found at similar depths.

Brazil produced 1.3 mb/d of oil in 2000, with the Campos basin accounting for roughly 70%. Increases in production are likely to occur with the introduction of private capital, a more competitive environment and an increase in foreign participation in exploration and production. Production is expected to reach 2.3 mb/d in 2010, about 3.2 mb/d in 2020 and some 3.9 mb/d in 2030. Brazil is now an oil importer, taking half a million barrels in 2000, most of it from Venezuela and Argentina. The *Outlook* projects that Brazil will attain oil self-sufficiency some time in the second decade (Figure 10.1).

Brazil has opened up its oil products sector to foreign competition. Nearly all of Brazil's LPG and 90% of its diesel are imported. Foreign and local companies can now import oil products and build new refineries. Brazil currently has some 1.9 mb/d of refinery capacity, over 40% of it in Sao Paulo. Petrobras owns 11 of the country's 13 refineries. In January 2001, Petrobras announced plans to invest $5 billion by 2010 to upgrade its refineries and increase their capacity. An estimated $700 million will be spent at the largest refinery, Replan, which accounts for about 22% of Brazil's total refining output. Another $400 million will be invested in the Mataripe refinery in the north-eastern state of Bahia.[15] The government

14. USGS (2000).
15. DOE/EIA (2001).

Figure 10.1: **Oil Balance in Brazil**

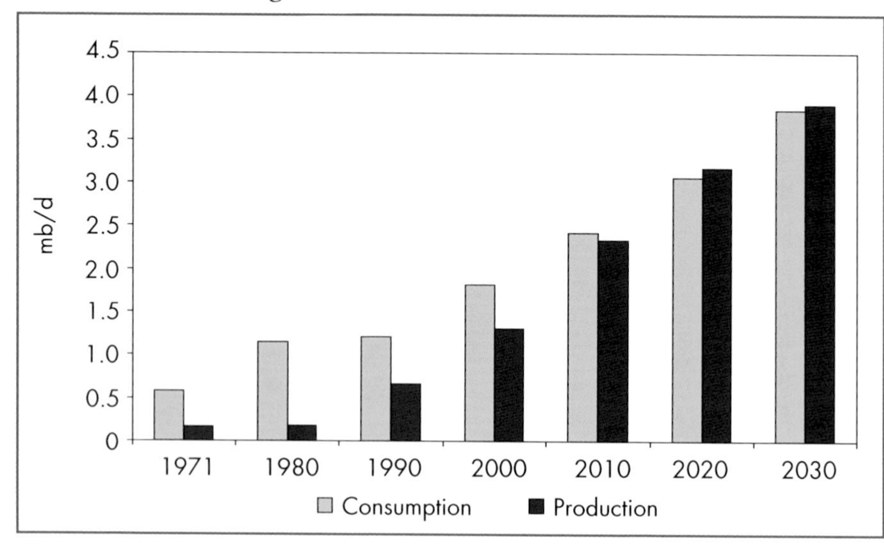

hopes to attract private capital to increase domestic refining capacity, but competition for capital may impede new investment in Brazil.

Foreign companies first entered the exploration and production sector in Brazil in 1997, through joint ventures with Petrobras. The first bidding round open to foreign competition took place in 1999. Since then, there have been four more bidding rounds offering 100 blocks, mostly in offshore deep-water areas. They have so far raised nearly $800 million for the National Petroleum Agency. The agency will continue to offer oil E&P blocks to foreign and local companies in annual auctions.

Natural Gas

Demand for natural gas is expected to rise as Brazil seeks to reduce its reliance on hydroelectricity. Two factors will work in favour of gas for new power generation: Brazil's desire to diversify its energy mix away from hydroelectricity and the great distances between new hydropower sites and major consuming centres. The price of imported gas will also be important. For example, the Bolivian export price for gas doubled from July 1999 to first quarter 2001 in line with the trend in world oil prices.[16] This has made Bolivian gas expensive in Brazil.

16. Enever (2001).

In 2000, gas constituted some 5% of Brazil's total primary energy supply. Over the *Outlook* period, gas demand is expected to grow by 7.3% a year, to 62 Mtoe, or 19% of total demand. Most of the expected growth will occur in power generation, where gas use will reach 39 Mtoe in 2030, 35% of total demand in the power generation sector. The potential for market development is very large, but the investments needed to bring projects to reality are enormous. Prospects for gas will depend crucially on the establishment of a clear and stable fiscal and regulatory environment to win investor confidence. Brazil's regulatory regime for electricity is still very cloudy, and this is hindering the development of gas-fired power plants.

Cedigaz estimates that Brazil has some 240 bcm of proven gas reserves.[17] The Campos and Santos basins hold the largest gas fields. Offshore south-east Brazil is rich in hydrocarbons and still underexplored, with potential for a significant increase in reserves and production. Brazil is estimated to have some 5,505 bcm of undiscovered gas reserves.[18]

Brazil produced 7 bcm of natural gas in 2000.[19] Production will rise to 15 bcm in 2010 and to over 34 bcm in 2020. From 2020 to 2030, production is expected to accelerate, reaching 70 bcm. Gas production will grow by an average annual rate of 8.1% over the entire projection period.

Figure 10.2: **Gas Balance in Brazil**

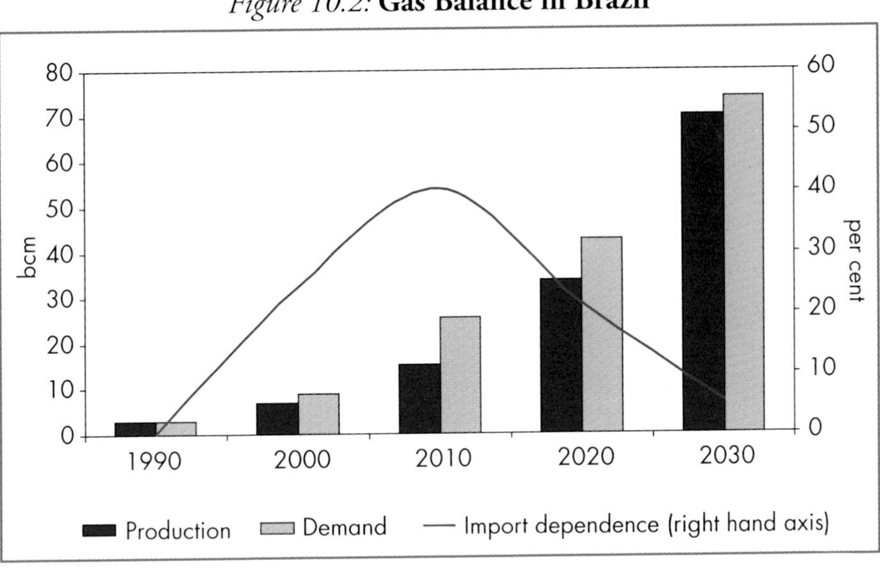

17. Cedigaz (2001).
18. USGS (2001).
19. A roughly equivalent amount of gas was flared or re-injected to enhance oil production.

Figure 10.3: Gas Pipelines in Brazil

In 2000, net imports were 2 bcm. They are projected to peak at some 11 bcm in the second decade of the projection period (Figure 10.2). By 2030, the *Outlook* projects that gas imports will be down to some 4 bcm.

The first pipeline to connect Brazil to a foreign gas source was the Bolivia-to-Brazil pipeline, with a capacity of 11 bcm/year. It was inaugurated in July 1999. It currently supplies 2.9 bcm/year of Bolivian gas to south-eastern and southern Brazil. Supply is expected to increase to at least 5.8 bcm per year by 2006. The second operational pipeline, Transportadora de Gas del Mercosur (TGM), links Argentina to Brazil. It supplies gas to a 600-MW AES power plant in Uruguaiana. Service began in July 2000. An extension of the TGM line, which will connect Uruguaiana to Porte Alegre, is currently under construction. As gas demand grows in Latin America, the Southern Cone market will become more integrated and trading volumes will rise dramatically (Figure 10.3).

Additional Argentina-Brazil pipelines are in various stages of planning, although recent gas discoveries in Bolivia and potential Brazilian discoveries could discourage development of these projects. Argentina-to-Brazil links now under study include the Cruz del Sur, Trans-Iguacu, and Mercosur pipelines. The Cruz del Sur would extend the Argentine-Uruguayan pipeline to Brazil. The Trans-Iguacu would cross from northern Argentina's Noroeste basin into southern Brazil. The Mercosur would link north-western Argentina's Neuquen basin to Curitiba, Brazil, and could extend to Sao Paulo. There are also plans to double the Bolivia-Brazil pipeline.[20]

Coal

Brazil's coal demand is expected to increase, on average, by 1.9% per year over the *Outlook* period, much more slowly than the 6% annual growth from 1971 to 2000. The share of coal in Brazil's primary demand will decline, from 10% in 2000 to 7% in 2030. Coal is used mainly in the iron and steel industry, with a smaller quantity going to electricity generation. In 2000, final consumption of coal was 6 Mtoe in industry and 3 Mtoe in power generation. Fuel substitution and efficiency improvements will reduce the demand for coal over the next three decades.[21]

20. See IEA (forthcoming 2002).
21. In Brazil, the share of electric-arc furnace technology in total steel production is over 20%.

Brazil has proven coal reserves of 12 billion tonnes, mostly steam coal quality.[22] It produced 7 million tonnes of hard coal in 2000 and imported over 13 million tonnes, roughly 40% from North America and another 35% from Australia. Brazil's steel industry will continue to rely on coking coal imports over the next few decades to meet growth in domestic steel demand.

The *Outlook* projects that, by 2030, Brazil will need 7 Mtoe of coal to fuel power plants and 11 Mtoe for industry. Local coal will be promoted as an alternative to imported gas in areas with no gas infrastructure, but major coal reserves are mainly in the southern states, where competition from imported Bolivian and Argentinean gas will be strong.

Biomass

Biomass demand, which is not included in the *Outlook*'s energy balances, represented nearly a quarter of Brazil's primary energy demand in 2000. Unlike in many other developing countries, biomass use in Brazil is not largely confined to firewood and charcoal for cooking and heating.[23] The industry sector accounts for over one-half of biomass use, mainly for food processing.

Table 10.4: **Biomass Energy Demand in Brazil** (Mtoe)

	2000	**2010**	**2020**	**2030**	**Average annual growth 2000-2030 *(%)***
Industry	22	24	27	30	1.0
Transport	6	8	9	10	1.7
Other sectors*	9	6	4	3	-4.0
Final biomass consumption	37	38	40	43	0.5
Primary biomass consumption	43	44	47	50	0.7
TPES including biomass	**179**	**236**	**299**	**380**	**2.5**

* "Other sectors" include the services, residential and agricultural sectors.

22. World Energy Council (2001).
23. See Chapter 13 for a discussion of biomass use in developing countries.

In 2000, biomass met more than a third of energy demand in the industry sector. While industrial demand for biomass will continue to rise over the projection period, its *share* in total industry demand will fall to less than a quarter.

Sugar cane plays a major role in the production of ethanol, for use in the transport sector. The share of alcohol in the fuel mix of the transport sector has been declining for quite some time in Brazil, driven by the decline in the stock of cars running only on alcohol. Demand for *blended* fuel, however, has been increasing. The current mix is some 24% alcohol to 76% gasoline, but the government is considering raising the alcohol content to 26%. In 2000, ethanol consumption was 0.13 mb/d, half of which was blended into gasoline. By 2030, the *Outlook* projects that ethanol consumption will be 0.22 mb/d.

Electricity

Final consumption of electricity is expected to increase by 3% per year over the *Outlook* period, faster than the assumed GDP growth rate. Electricity demand will nearly double by 2020 and reach 2.5 times its current level by 2030. In rural areas in Brazil, some 8 to 10 million people still lack access to electricity, but households are expected to be fully electrified by the end of the *Outlook* period.

Table 10.5: **Electricity Generation Mix in Brazil** (TWh)

	1990	2000	2010	2020	2030
Coal	5	10	15	18	30
Oil	6	17	16	16	12
Gas	0	2	56	104	226
Nuclear	2	6	12	21	21
Hydro	207	305	391	507	589
Other renewables	4	9	13	19	32
Total	**223**	**349**	**505**	**685**	**909**

In 2000, electricity generation in Brazil was 349 TWh. Hydropower represented 87%, oil some 5% and coal 3%. Biomass represented another 3% of generation, mostly in the form of bagasse, the residual product from sugar-cane processing. Nuclear power supplied less than 2% of electricity in 2000.

Generation is projected to grow by 3.2% a year over the next 30 years. Brazil is expected to continue to develop its hydropower resources. Brazil's largest hydropower plant, the 11-GW Belo Monte, is expected to be completed in this decade. Belo Monte will be the first large dam to be built in Amazonia since the Tucurui dam was completed in the early 1980s. But the pace of development of Brazil's hydro resources will gradually slow down, especially from 2020 to 2030, as the best hydro sites will have been exploited and natural gas becomes increasingly available. The *Outlook* sees the share of hydropower in total generation falling to 65% in 2030.

Box 10.2: Recent Electricity Reform Efforts in Brazil

Brazil is creating a new regulatory and market framework which aims to improve market rules and to foster private investment in the electricity sector. In March 2002, the power regulator, ANEEL, published rules for the country's new wholesale electricity market. While the new structure appears to be more transparent than in the past, the new pricing policies are still a cause for concern.

One of the biggest challenges facing the government is how to integrate expensive new thermal plants into a system dominated by older, lower-cost hydro plants — whose costs have already been amortised. Re-regulation of existing hydro plants, owned by Eletrobras, the national electricity company, and the subsidisation of Bolivian gas transported via the Gasbol pipeline are part of the government's plan to revitalise the power sector. The programme's ultimate goal is a substantial increase in the share of gas-fired plants in total generation.

Most new capacity in the near term is expected to be hydroelectric. Many of the gas-fired projects have been halted or had their construction delayed.[24] The government would not only like to increase capacity but also to diversify energy supply. At the same time, however, it would like to avoid undermining previous liberalisation efforts and the future for competition in the electricity market. The privatisation process in the electricity sector has been deferred pending the presidential elections in October 2002.

24. Platts (2000a).

Brazil and neighbouring countries have abundant natural gas reserves, which are expected to be increasingly exploited. As gas pipeline networks in Latin America become more integrated, the gas-fired generation will become cheaper. It will represent 25% of generation in 2030, growing by over 16% a year over the *Outlook* period. Nearly half of new gas-fired electricity generation will be built in the third decade of the projection period.

Brazil has two nuclear power plants, Angra I and Angra II. Angra II was connected to the grid in July 2000. Construction of a third nuclear power plant, Angra III, was halted because of political and economic factors, but may be resumed in the next few years. A decision will be taken by the country's next president. Even if the decision is affirmative, Angra III will not go online before 2010. The *Outlook* assumes that Angra III will add another 1,300 MW of capacity in south-eastern Brazil after 2010.

The Brazilian National Development Bank is working on projects to finance additional power generation from biomass. Electricity generation from biomass is projected to rise to 22 TWh in 2030. Wind power will increase to nearly 10 TWh.

To meet demand growth over the *Outlook* period, Brazil will need to add 123 GW of new capacity by 2030. New gas-fired capacity will make up 40% of the additional capacity and hydropower 46%. Oil-fired capacity will decline modestly. Seven GW of additional capacity from renewable

Figure 10.4: **Power Generation Capacity in Brazil**

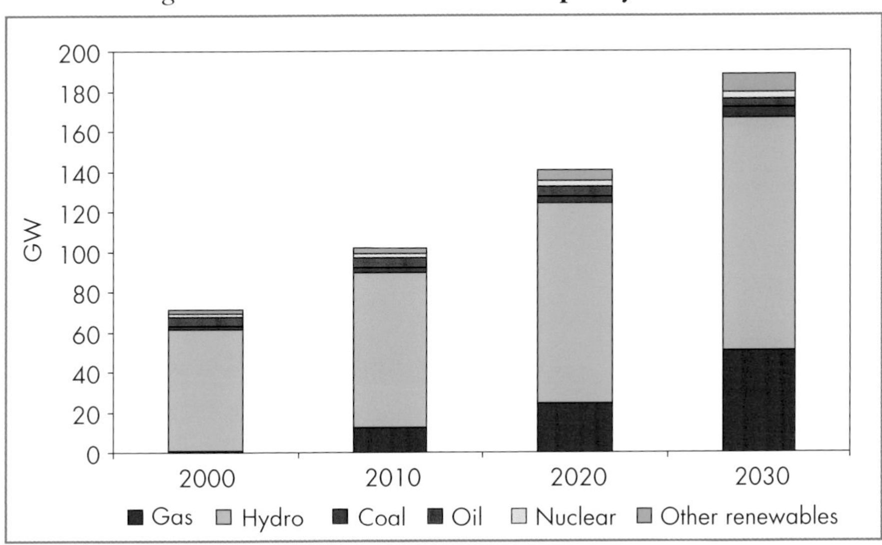

energy sources will come online by 2030, mostly biomass and wind. Brazil's solar resources will be exploited towards the end of the projection period, if the economics of their use improves.

The investment required to build the additional generating capacity over the next three decades in Brazil is enormous, some $160 billion. This is nearly equal to the investment in generating capacity required to meet additional demand in the whole of the rest of Latin America. The Brazilian public sector alone will not be able to provide the finance needed to maintain and to expand the energy infrastructure and supply in Brazil. Private investment will only be forthcoming, however, if Brazil's regulatory regime becomes more transparent and consistent.

Environmental Issues

Brazil's energy-related environmental problems include oil spills, air pollution, flooding and deforestation. Oil spills are frequent and cause severe environmental damage. There were twelve spills in 2000, in part because of weak environmental oversight. Air pollution is mainly due to rapid urbanisation and to industrial activities. Pollution levels have also been fuelled by low gasoline prices and inefficient vehicle engines.

Energy-related CO_2 emissions are expected to reach 760 million tonnes by 2030, up from 303 million tonnes in 2000. Brazil's energy system is one of the least carbon-intensive in the world, because of the wide use of hydropower and active government encouragement of biomass fuels. Even if per capita emissions increase as projected, from 1.8 tonnes to 3.3 tonnes, they will still be far below the expected OECD North American average in 2030, of 16.2 tonnes per capita.

CHAPTER 11: INDONESIA

HIGHLIGHTS

- As a major energy exporter and an increasingly important consumer, Indonesia will continue to play an important role in international energy markets. The country's primary energy demand is projected to grow rapidly in the next three decades, at an average 3.5% per year.
- Now an oil exporter, Indonesia will become a net oil importer in the second decade of the *Outlook* period. Production will continue to decline at existing fields, and domestic demand will rise rapidly, mainly for transport. Oil will still dominate Indonesia's fuel mix in 2030.
- Indonesia is the world's largest exporter of liquefied natural gas. Indonesian gas will increasingly supply growing markets in the Asia-Pacific region, including Japan and Korea. Domestic gas demand will also rise over the *Outlook* period, at an average 3.4% per year.
- Final electricity demand will grow rapidly, by over 5% a year, nearly doubling its share in final consumption by 2030. There could be an electricity shortage in the next few years. Investment in new power projects will be crucial to meeting projected demand. The *Outlook* estimates that $73 billion must be invested in power plants over the projection period.
- Uncertainties surrounding the energy projections for Indonesia are particularly acute. The economy is still reeling from the effects of the 1997 economic crisis. The recent global economic downturn, together with political instability at home, is clouding near-term prospects for sustained economic growth. Investor confidence will be crucial to Indonesia, both for its macroeconomic outlook and for the development of its energy supply projects.

Energy Market Overview

Indonesia, with its abundant fossil fuel resources, is a major energy producer and has a growing domestic market. In 2000, it produced 1.4 mb/d of oil and 69 bcm of natural gas. Indonesia is a major exporter of crude oil and is a member of OPEC. Its oil exports have, however, been dropping since the late 1970s, while its own consumption has surged. Indonesia has some 3.8 tcm of natural gas reserves and is the world's largest exporter of LNG. Its gas exports of 37 bcm represented nearly 6% of global gas trade in 2000. Also, Indonesia has proven coal reserves of 5.4 billion tonnes and is the world's third-largest exporter of hard coal.

Oil dominates energy use in Indonesia.[1] It accounts for more than half of primary consumption, with gas and coal making up most of the rest.

Table 11.1: **Key Economic and Energy Indicators of Indonesia**

	Indonesia 2000	Average annual growth 1990-2000 (%)	World 2000
GDP (in billion 1995 $, PPP)	573	4.2	41,609
GDP per capita (in 1995 $, PPP)	2,731	2.5	6,908
Population (million)	210	1.7	6,023
TPES (Mtoe)	98	6.5	9,039
Oil demand (mb/d)	1.1	4.9	75
Gas demand (bcm)	32	7.4	2,565
Coal demand (Mt)	22	13.3	4,654
TPES/GDP*	0.17	2.2	0.22
Energy production/TPES	1.6	-1.0	-
TPES per capita (toe)	0.5	4.8	1.5
Electricity consumption per capita (kWh)	377	13.3	2,101

* Toe per thousand dollars of GDP, measured in PPP terms at 1995 prices.
Note: Energy data exclude biomass.

Primary energy consumption increased by 6.5% per year in the 1990s, with fossil fuel use growing rapidly (Table 11.1). Energy use fell back in 1998 because of the Asian economic crisis, but recovered strongly in 2000. Although electricity consumption climbed at the annual growth rate of

1. The energy demand figures exclude biomass. Biomass demand is discussed before the electricity section in this chapter.

11% in the 1990s, it is still comparatively low. In 2000, it accounted for only 10% of the country's total final consumption of commercial energy. Per capita electricity consumption in Indonesia was less than 20% of the world average.

Current Trends and Key Assumptions

Macroeconomic Context

Indonesia is a large archipelago. With 210 million inhabitants, it is the fourth most populous nation in the world. The majority live on Java, one of the five main islands. Per capita GDP was only $2,731 in 2000, a fact which puts Indonesia in the bottom quarter of countries ranked by income.

The country has been faced with acute economic and political instability since the late 1990s. Its economy was devastated by the 1997 Asian economic crisis, which led to a collapse in the currency and a run on the banks. It left three-quarters of Indonesian businesses in technical bankruptcy. GDP plummeted by 13% in 1998, forcing the government to turn to the International Monetary Fund for emergency debt relief. The economy grew by 5% in 2000, but has since been hit by a slump in export demand. Political turmoil, including independence movements in several provinces, has added to the country's economic difficulties. The presidency changed hands four times between 1998 and 2001.

Economic growth was 3.6% in 2001, as investment and exports reflected weaker external markets and internal political uncertainty. After a modest weakening in early 2002, GDP growth is likely to pick up slightly in the second half. The near-term economic outlook for Indonesia depends crucially on maintaining the pace of economic reform, increasing domestic security and improving investor confidence. The country suffers from lack of foreign direct investment, because of its unstable political situation and the lack of a clear legal framework. The large government budget deficit and high public and private debt also tend to undermine investment flows. Total external debt is currently over $140 billion, half of which is owed by the government. Subsidies in the energy sector have also contributed to the government's financial problems, but the promised structural and economic reforms are materialising slowly, in the face of popular unrest.

This *Outlook* assumes that the Indonesian economy will grow by 3.9% a year from 2000 to 2030 (Table 11.2). Economic growth is expected to slow down towards the end of the projection period, reflecting the maturing of the economy. Population growth is assumed to average 1% per year, down from 1.7% in the 1990s.

Table 11.2: **Reference Scenario Assumptions for Indonesia**

	1971	2000	2010	2030	Average annual growth 1971-2000 (%)	Average annual growth 2000-2030 (%)
GDP (in billion 1995 $, PPP)	103	573	847	1,792	6.1	3.9
Population (million)	120	210	235	280	1.9	1.0
GDP per capita (in 1995 $, PPP)	853	2,731	3,600	6,409	4.1	2.9

Energy Prices

Energy prices in Indonesia are still heavily subsidised, although recent reforms imposed by the IMF have reduced the subsidies on oil products and electricity.[2] According to the World Bank, the domestic prices of petroleum products, like gasoline, kerosene and industrial diesel oil, were, on average, only 43% of international prices in 2000. They did not cover production costs.[3] In January 2002, an automatic price adjustment system was introduced. Under the system, Pertamina, the state-owned oil company, resets domestic oil products prices at 75% of international prices every month. The price of kerosene for households is an exception; it is set at around 63% of the international price.[4] The cost of cooking with kerosene and LPG is a major expense for many poor Indonesians in urban areas, and previous attempts to raise their prices have met with violent protests. Natural gas prices are also kept below economic costs, although gas is not as heavily subsided as petroleum products. Fuel switching to gas has, thus, been limited.

2. IEA (1999) analyses the possible impacts of energy subsidy removal in eight developing countries, including Indonesia.
3. World Bank (2000).
4. Premium gasoline and aviation fuel prices are already at international levels.

Since the economic crisis, the state-owned electricity company, PLN, has received direct subsidies from the government, and electricity consumers have been paying less than the actual cost of what they use. The government expects the direct financial cost of energy subsidies to drop from 21% of total government spending in 2001 to 14% in 2002. It intends to eliminate subsidies eventually, but, in the face of public unrest, has not yet taken a final decision on the timing. It is assumed in these projections that price reforms will be continued and that energy prices will follow international energy prices by the end of the first decade of the projection period.[5]

Results of the Projections

Overview

Indonesia's primary energy demand (excluding biomass) is projected to grow at an average annual rate of 3.5% from 2000 to 2030 (Table 11.3). Coal demand will grow at an annual average rate of 5.2% over the projection period, and coal's share in the primary fuel mix will increase from 14% in 2000 to 23% in 2030, due to strong demand for power generation (Figure 11.1). Gas demand will grow at an average annual rate of 3.4% over the projection period. Oil will still dominate Indonesia's primary fuel mix, but its share will decline from 54% in 2000 to 43% in 2030.

Primary energy intensity, measured as total primary energy supply (excluding biomass) per unit of GDP, is projected to fall, on average, by 0.3% per year between 2000 and 2030. This would reverse the upward trend between 1990 and 2000, when intensity increased by 2.2% per year. A shift of industrial structure to lighter manufacturing, the use of more efficient equipment in end-use sectors and the rise in energy prices as subsidies are removed will contribute to this trend.

5. See Chapter 2 for international price assumptions.

Table 11.3: **Total Primary Energy Demand in Indonesia** (Mtoe)

	1971	2000	2010	2020	2030	Average annual growth 2000-2030 (%)
Coal	0	14	24	40	63	5.2
Oil	8	53	73	96	118	2.7
Gas	0	28	45	64	78	3.4
Hydro	0	1	2	2	2	3.8
Other renewables	0	2	8	12	16	6.7
TPES*	9	98	152	213	276	3.5

* Excludes biomass.

Figure 11.1: **Total Primary Energy Demand in Indonesia**

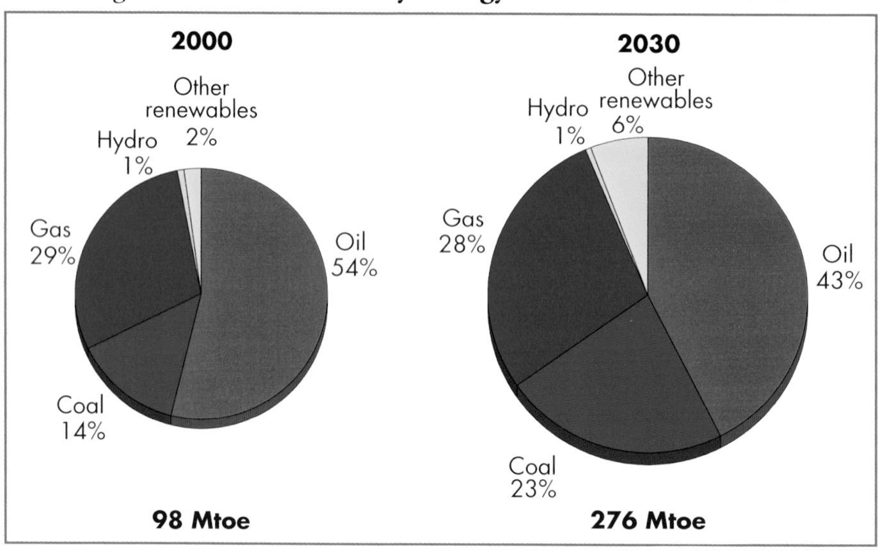

Total final energy consumption will increase, on average, by 3.2% per year from 2000 to 2030. Growth of industrial energy use will continue but slow, increasing most rapidly in the services sector (Figure 11.2). Higher per capita incomes are expected to spur growth in the use of road transport. Transport energy demand will grow at 3.4% per year and incremental oil demand for transport will represent around 60% of the increase in total final oil demand between 2000 and 2030. Commercial energy use in the residential sector will grow at 3.3%, as growth in household incomes stimulates consumers to use oil products and to buy electric appliances. They will, however, continue to use biomass energy *along with* more commercial fuels. Industry's share in final energy consumption will decline to 29% in 2030 from 35% in 2000, while the shares of the transport, residential and service sectors each will increase by 3 percentage points.

Figure 11.2: **Total Final Consumption by Sector in Indonesia**

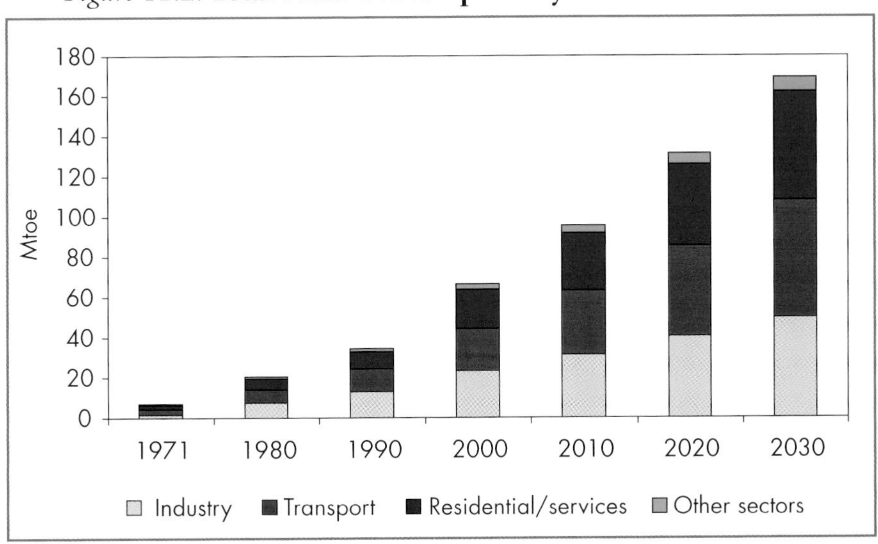

Note: "Other sectors" include agriculture, non-energy and non-specified energy use.

Oil

Primary oil demand will grow by an average 2.7% per year over the projection period, from 1.1 mb/d in 2000 to 2.4 mb/d in 2030. Production, which was running at around 1.4 mb/d in 2000, is expected to rise to 1.5 mb/d in 2010, but will start declining towards the end of the *Outlook* period. Indonesia will become a net importer of oil around 2010, and net imports will reach 0.4 mb/d in 2030 (Figure 11.3). Net exports were 0.3 mb/d in 2000, 90% of which went to Asia-Pacific countries.

Indonesia has proven oil reserves of 5 billion barrels[6], mostly onshore, but there are accessible offshore reserves in north-western Java, east

6. United States Geological Survey (2000).

Figure 11.3: **Oil Balance in Indonesia**

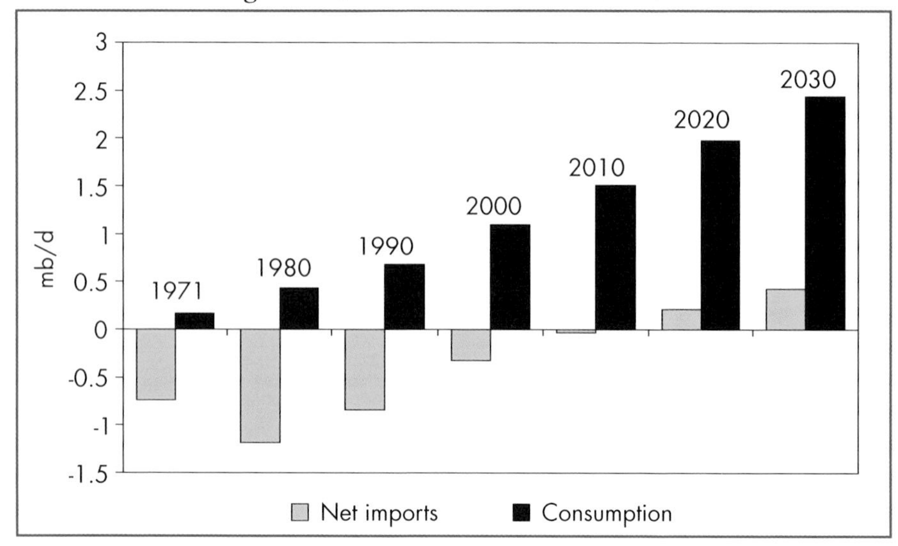

Kalimantan and the Natuna Sea. Central Sumatra is the largest oil-producing province. Estimated reserves have declined by 14% since 1994, as new exploration, in frontier regions of eastern Indonesia, has failed to find enough oil to offset the depletion of producing fields. Indonesia's oil production is subject to OPEC production quotas. A 15% fall in the country's output in the two years to 2001 was entirely due to a cut in OPEC quotas. Indonesia also produces 235,000 b/d of natural gas liquids and condensates, which are not counted in its quota. Indonesia's largest oil producer, PT Caltex, controls over 50% of crude oil production through production-sharing contracts. To increase production, the Indonesian government signed nine new production-sharing contracts in 2001 and plans to offer 17 new blocks for tender in 2002.[7]

Three new oil projects are expected to begin production before 2004. Unocal's West Seno field (offshore east Kalimantan) is expected to produce 60,000 b/d by the end of 2002. Conoco's Belanek project (west Natuna Block B) is planned to produce 100,000 b/d in 2004. And ExxonMobil's Banyu Urip field (Java), with estimated recoverable oil in excess of 250 million barrels, is expected to come onstream in 2003.[8] It is hoped that these projects will compensate for declining output at existing fields, but they are not likely to add significantly to overall production. The impact of

7. DOE/EIA (2002).
8. http : //www2.exxonmobil.com/Corporate/Newsroom/Newsreleases/corp_xom_nr_120401.asp

oil sector reforms on long-term production prospects is a key source of uncertainty (Box 11.1).

Box 11.1: **Oil Sector Reforms**

For the past three decades, Indonesia's oil sector has been vertically integrated and dominated by Pertamina. Pertamina has had a monopoly in the downstream sector and has participated in all exploration and production activities. In October 2001, a law was adopted aimed at stimulating investment in the oil industry. Under this law, Pertamina's monopoly on upstream oil developments will be ended in 2003. An Implementation Agency will be established to award and supervise production-sharing contracts with foreign oil companies, previously Pertamina's sole domain. A second agency will be set up to regulate refining, storage, distribution and marketing activities. The law relaxes requirements on foreign firms in gaining regulatory approvals, and it grants limited powers to regional governments to tax oil company profits. The regional governments have been pressing the central government for some time for a greater share of tax revenues from the oil and gas industry. The oil companies fear that this move will lead to increased taxes overall.

The government is considering allowing foreign investment in its oil and gas sector without the participation of an Indonesian partner. It reckons that around $5 billion of new investment will be needed every year to maintain current oil and gas production.[9] Foreign oil firms invested an estimated $4.5 billion in 2000 and $5.8 billion in 2001.

Natural Gas

Primary consumption of natural gas, including the oil and gas industry's own use, is projected to increase from 32 bcm in 2000 to 89 bcm in 2030, an average annual rate of growth of 3.4%. Power generation, fertilizers and industries, such as minerals processing (including the use of gas as an energy input in LNG processing), are the main users. In 1984, the government charged PTPGN, the national gas company, with the primary responsibility for distributing gas to small and medium-sized industries, to

9. Organization of Petroleum Exporting Countries (2001).

the service sector and to households. Although PT PGN has been responsible for some of the transmission system since 1994, 90% of the domestic gas market is still dominated by Pertamina. The lack of an integrated transmission and distribution network, combined with energy price distortions, constrains gas use.

Indonesia's proven gas reserves at the beginning of 2001 were 3.8 tcm, equivalent to 50 years of production at current rates.[10] The US Geological Survey estimates mean undiscovered resources at just over 3 tcm. More than 71% of proven reserves are offshore, with the largest concentrations off Natuna Island (33%), at east Kalimantan (30%), Irian Jaya (15%), Aceh (7%) and in south Sumatra (6%).

Indigenous production is expected to increase more rapidly than inland demand. Indonesia will increase its exports of LNG to Japan, Korea and Chinese Taipei and of piped gas to neighbouring ASEAN countries (Box 11.2). In 2000, Indonesia exported some 37 bcm of LNG, 25 bcm of it to Japan. Chinese Taipei imported 4 bcm from Indonesia. Korea currently depends on Indonesia for 40% of its gas requirements. In 2001, Indonesia started exporting gas to Singapore from the west Natuna field through an undersea line. Another line from Sumatra to Singapore is planned to supply gas to a power plant. Total exports will rise from 37 bcm in 2000 — 53% of total production — to 95 bcm in 2030 (Figure 11.4).

Figure 11.4: **Natural Gas Balance in Indonesia**

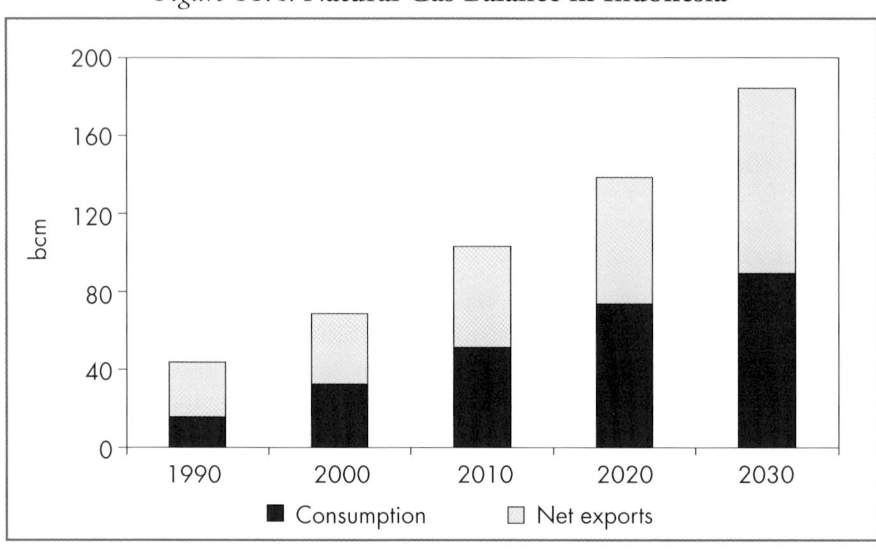

10. Cedigaz (2001).

Box 11.2: **Liquefied Natural Gas Export Prospects**

Increased LNG exports will probably come from capacity expansions at the existing plant at Bontang in east Kalimantan, which has 8 trains and 21.6 Mt per year of capacity, and from new greenfield projects. Production at the other existing plant, at Arun in Aceh province, which has 6 trains and 12.3 Mt per year of capacity, is expected to continue to decline gradually as the fields supplying the plant are depleted. Two trains were shut down in 2000 due to falling production. Civil unrest in Aceh province forced the entire Arun plant to shut for several months in 2001, cutting exports to Japan and Korea.

BP's proposed LNG project, based on gas from the recently discovered Tangguh field in Irian Jaya, is now a top priority for the Indonesian government. The field has proven reserves of at least 420 bcm. There are plans to build two trains there by 2006, with a combined capacity of 6 Mt per year. Pertamina is also studying the feasibility of relocating the idle trains from Arun to Irian Jaya. Significant reserves have also been discovered in east Natuna, but their development remains uncertain due to the cost of processing the gas, which has very high CO_2 content. Indonesia's proximity to East Asian and Pacific markets gives it a cost advantage over competing projects in Australia and the Middle East. But export plans are threatened by political instability, which is undermining investor confidence.

Shell and the Indonesian government are investigating the feasibility of building a gas-to-liquids (GTL) plant in partnership with Pertamina. The plant would exploit gas reserves that are not well located for LNG projects or that are remote from inland markets. The 75,000 b/d plant, which would use Shell's middle-distillate synthesis technology, could start operation in 2005. It would require a gas intake of around 6.2 bcm. Indonesia is expected to become one of the world's major producers of GTL by 2030.

Coal

Primary coal use will grow by 5.2% per year to 2030, driven mainly by the power sector. Industrial demand will grow only very slowly. Exports, which grew tenfold in the 1990s, accounted for more than 70% of production in 2000. Indonesia is the world's third-largest hard coal exporter, behind Australia and South Africa. It exported 55 Mt in 2000. Major destinations include Japan, Korea and Chinese Taipei. Growth in exports will remain the driving force behind coal production over the projection period. Producers have a strong incentive to export coal, due to the large difference between prices received on export markets and prices paid by the State Energy Authority for deliveries to inland power stations.

Indonesia has the coal resources to support a major expansion of exports. Proven reserves total 5.4 billion tonnes, of which 790 million tonnes are hard coal.[11] Indonesia has proven reserves to sustain ten years of hard coal production.[12] Since the country possesses large reserves of sub-bituminous coal and lignite, which could be used for domestic power generation, expansion of proven reserves will involve more pre-mining exploration and planning. Future production increases are likely to be based on very large-area surface mines. Rising exports and inland demand will, nonetheless, require substantial investment in transport infrastructure, including strengthening roads, rivers and port facilities against weather disruptions. Investment has slowed over the past few years because of political and economic instability.

Foreign firms operating in the Indonesian coal-mining sector are obliged to sell majority stakes to Indonesian companies. There has been recent litigation involving a contract between the east Kalimantan government and Kaltim Prima Coal, a 50-50 joint venture between Rio Tinto and BP. The case involves the pace of divestiture and the value of the stake. Uncertainties caused by this dispute and others of its kind, as well as stricter environmental laws, illegal mining and the prospect of more autonomy to the regions, cloud the prospects for Indonesian coal exports. The Australian Bureau of Agricultural and Resource Economics, estimates that these risks could raise coal production costs by 10% and cut exports by 20% in 2010.[13]

Biomass

Biomass, mostly wood, is the primary fuel for cooking and other purposes in rural areas. If it were included in Indonesia's energy balance in 2000, biomass would represent over 70% of total residential demand. Over

11. World Energy Council (2001).
12. IEA (2001a).
13. Australian Bureau of Agricultural and Resource Economics (2002).

the *Outlook* period, the use of biomass by households will decline by 0.7% per year, as increases in per capita income make oil products such as kerosene and electricity more affordable. This trend, combined with rising scarcity of biomass resources in many areas, will result in its share in residential use falling to 45% in 2030.[14]

Table 11.4: **Biomass Energy Demand in Indonesia** (Mtoe)

	2000	2010	2020	2030	Average annual growth 2000-2030 (%)
Biomass	47	50	46	40	-0.5
TPES including biomass	**145**	**201**	**257**	**313**	**2.6**

Electricity

Electricity demand is projected to grow by a strong 5.4% per year from 2000 to 2030, although this is much slower than the 13.7% rate from 1971 to 2000. By 2030, electricity demand will be nearly five times higher than in 2000, and electricity will meet about 19% of Indonesia's final energy needs. Only half of the population is estimated to have access to electricity today, and annual per capita consumption, at 377 kWh, is very low compared with most other countries.

In 2000, electricity generation in Indonesia was 93 TWh, about a third coming from gas-fired plants, another third from coal and about a fifth from oil (Table 11.5). Renewable energy sources, mostly hydro and geothermal, account for the remainder. Electricity generation from coal is projected to increase by 7.3% per year over the *Outlook* period, and its share in the electricity generation mix increases from 31% in 2000 to over 54% in 2030. Gas-fired generation also shows rapid growth, at 5% per year on average over the projection period. Most of the increase in gas-fired generation will occur in the first two decades. As Indonesia becomes an oil importer, reliance on oil for power generation will decline, and its share will fall from 22% in 2000 to less than 5% in 2030. Indonesia is rich in geothermal resources, and geothermal generation will increase at 6% per year, to account for 3.4% of the generation mix in 2030.

14. See Chapter 13 for a discussion of biomass use and its link to poverty in developing countries.

Table 11.5: **Electricity Generation Mix in Indonesia** (TWh)

	1990	2000	2010	2020	2030
Coal	11	29	68	128	238
Oil	17	20	20	23	23
Gas	1	32	63	111	136
Hydro	6	9	18	23	28
Other renewables	1	3	10	15	21
Total	**37**	**93**	**180**	300	**445**

Box 11.3: **Indonesia's Potential Power Shortage and Prospects for New Power Projects**

In the early 1990s, Indonesia's electricity generation sector was opened to independent power producers (IPPs) to meet rapid growth in electricity demand. The independent producers sold power to the national utility, PLN, under long-term contracts denominated in dollars. The recent economic crisis, which led to a collapse of the Indonesian rupiah, squeezed PLN especially hard, since its customers pay in rupiah at regulated prices.

Indonesia faced problems of under-capacity in generation in the mid-1990s. After the Asian economic crisis, the situation changed to one of *over*-capacity. Many independent projects were cancelled in the wake of the crisis, and few new projects have come forward since. With demand once again picking up, capacity shortages have again become a problem. Already regions outside of Java have experienced rotating power blackouts; and the threat of a crisis in Java, the centre of politics and the economy, has compelled the government to rethink reforms in the electricity sector. In June 2002, it said it would allow some of the independent producers to resume their cancelled projects to prevent a power shortage.

The government has drafted a new bill aimed at restructuring the electricity supply industry. Key elements include:

- introducing competition in generation and encouraging private sector participation;
- setting up a regulatory body;
- restoring the financial viability of PLN so that it can be privatised;

- establishing a transparent mechanism for subsidising electricity through a fund and gradually setting retail prices at economic levels by 2005; and
- reforming power purchase agreements.

Figure 11.5: **Power Generation Capacity Additions in Indonesia**

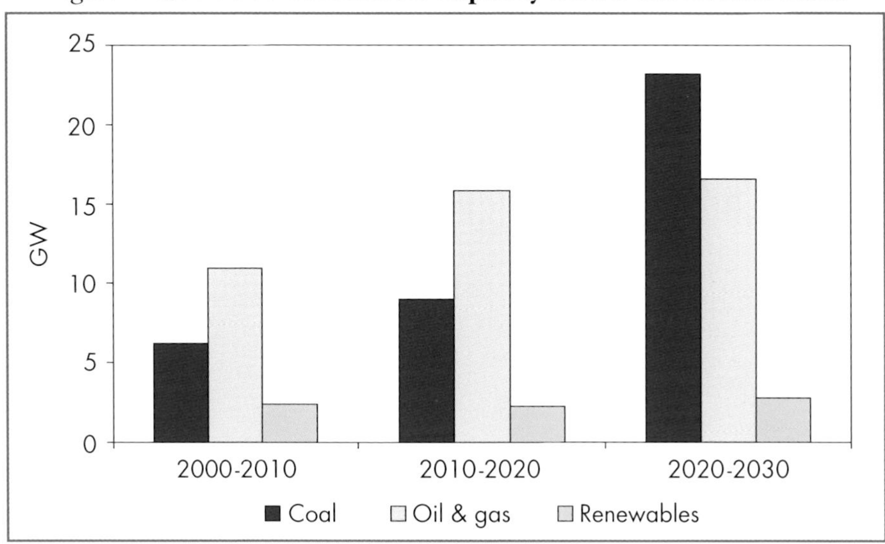

Note: Renewables include hydroelectric plants.

To meet demand growth, some 90 GW of new capacity will have to be installed before 2030 (Figure 11.5). Of the incremental capacity, coal and gas will each account for more than 40%. The *Outlook* estimates that $73 billion must be invested in power plants if supply is to meet electricity demand over the next three decades. A major expansion of transmission and distribution infrastructure will also be required.

Environmental Issues

Indonesia's main environmental challenge is to reduce the degradation of its biomass resources. In the past, the government has failed to curb illegal logging and the excessive conversion of forest areas into agricultural land. More than a million hectares of Indonesian forest disappear every year. Air pollution caused by illegal logging activities and burning for agricultural purposes has led to smog that affects Indonesia and neighbouring countries. Although legislation for environment management is adequate, enforcement capability of the government is still weak.

Carbon dioxide emissions from energy use grew by over 7% per year in the 1990s, reaching some 270 million tonnes in 2000. CO_2 intensity (CO_2 emissions per primary energy consumption) is 2.7 tonnes per toe, higher than the world average of 2.5 tonnes per toe largely because fossil fuels meet 97% of primary energy demand. CO_2 emissions from energy use are projected to increase by 3.4% annually over the *Outlook* period, in line with growth in energy demand. Emissions will grow faster than energy demand between 2020 and 2030, as the country increasingly relies on coal-fired power generation.

PART C

SPECIAL ISSUES ARISING FROM THE *OUTLOOK*

CHAPTER 12: THE OECD ALTERNATIVE POLICY SCENARIO

HIGHLIGHTS

- Implementation of policies currently under consideration in OECD countries would reduce CO_2 emissions by some 2,150 Mt in 2030, or 16%, below the Reference Scenario in 2030. This is roughly equal to the total emissions of Germany, United Kingdom, France and Italy today. Because of the slow pace at which energy capital stock is replaced, CO_2 savings in earlier years are relatively small – only 3% by 2010 and 9% by 2020. Total OECD CO_2 emissions would eventually stabilise, but only towards the end of the *Outlook* period.
- Energy savings, which amount to 9% of the primary energy demand of the Reference Scenario in 2030, are smaller than CO_2 savings, because the latter reflect the benefits of both energy savings and fuel switching to less carbon-intensive fuels.
- The biggest reduction in CO_2 emissions will come from the power generation sector because of rapid growth of renewables and savings in electricity demand. This reflects the emphasis that OECD governments are currently giving to renewables and energy efficiency in their long-term plans for curbing CO_2 emissions and enhancing energy security.
- The reductions in energy demand lower the OECD's dependence on oil and gas imports. In 2030, OECD gas demand will be reduced below the Reference Scenario by 260 bcm, or 13%. The reduction in EU gas imports in 2030 is slightly less than today's imports from Norway and Russia combined. The savings in oil demand, stemming mainly from the transport sector, reach 4.6 mb/d, or 10%.
- Reductions in CO_2 emissions below the Reference Scenario will be largest in the European Union at 19% in 2030, followed by Japan, Australia and New Zealand at 15%, and the United States and Canada at 14%.
- Despite these reductions, the three OECD regions do not individually reach their targets under the Kyoto Protocol. However, if the United States is excluded, their targets could be

met through the savings achieved in this Alternative Policy Scenario and the emissions credits from other Annex-B countries.

- If governments wish to achieve larger or faster savings in energy and CO_2 emissions, they will need to take stronger measures to shape long-term energy and environmental outcomes.

This chapter analyses the impact in OECD countries of additional policies to address climate change and energy security concerns.[1] *It describes the analytical approach taken and details the implications of the additional policies, beyond those taken into account in the Reference Scenario, for energy demand trends, CO_2 emissions, and energy imports. Detailed results are then provided by sector with a discussion of the key assumptions driving the results.*

Background and Approach

The *Reference Scenario* takes into account government policies and measures on climate change and energy security that had been adopted by mid-2002. The Reference Scenario does *not* include policy initiatives that were under serious discussion, but not enacted.

The *Alternative Policy Scenario* analyses the impact on energy markets, fuel consumption and energy-related CO_2 emissions of the policies and measures that OECD countries are currently considering.[2] These include policies aimed at curbing CO_2 emissions, at addressing pollution and reducing energy import dependence.[3] The basic assumptions about macroeconomic conditions and population are the same as in the Reference Scenario, even though there may be some economic feedback from the new policies. Energy prices in the Alternative Policy Scenario adjust to the new energy supply and demand balance.

1. The OECD Alternative Policy Scenario considers new policies in three major regional blocs : the United States and Canada (OECD North America less Mexico); the European Union; and Japan, Australia and New Zealand (OECD Pacific less Korea). The OECD, as defined in this chapter, therefore differs from the Reference Scenario.
2. This analysis does not take into account the impact of OECD policies in the non-OECD regions.
3. The key policies and measures included in the Alternative Policy Scenario are provided in Tables 12.3, 12.5, 12.7 and 12.10.

The Alternative Policy Scenario does not analyse *all* possible, or even likely, policies. It does not, for example, consider single measures with small or locally confined effects, but focuses on policies with the potential to have a major impact on energy use. It does not assess the economic cost-effectiveness of the policies considered, nor does it consider supply-side policies outside the power-generation sector. The key policy assumptions included in this analysis are discussed in each sectoral section. The changes in energy and CO_2 trends are a function of the policies considered and are expressed as percentage changes relative to the Reference Scenario.

The Alternative Policy Scenario makes detailed assumptions on the impact of the policies considered and feeds these into the World Energy Model. Many of the policies considered have effects that operate at a very micro-level in the economy and their impacts cannot be modelled without having a similarly detailed model. For example, the impact of mandatory efficiency standards cannot be estimated from past patterns of energy use since standards impose a new technical standard on the energy system. To meet this challenge, detailed "bottom-up" sub-models of the energy system were incorporated into the World Energy Model, allowing a wide set of policy issues to be analysed within a single modelling framework.[4]

An important element of this approach is the explicit representation of efficiency, the activities that drive energy demand (e.g. kilometres driven) and the physical capital stock important to energy. Slow capital stock turnover affects the rate of energy-efficieny improvements.[5] A power plant has an economic life of forty years or more, while building structures last sixty years, a century or longer. Even cars and major household appliances have average lifetimes of between one and two decades. The very long life of energy capital limits the rate at which more efficient technology can penetrate and reduce energy demand. The detailed capital stock turnover sub-models within the World Energy Model estimate how alternative policies affect energy use.

4. See Appendix 1 for a description of the structure and main characteristics of the World Energy Model.
5. See Figure 2.16.

Key Results

Energy and CO_2 Emissions Savings

In the Alternative Policy Scenario, OECD energy consumption and CO_2 emissions are much lower than in the Reference Scenario in the long term (Table 12.2 and Figure 12.1). In 2010, total primary energy demand is 69 Mtoe below the Reference Scenario, but by 2030 it is 9%, or 529 Mtoe, lower. CO_2 savings are larger than those for energy use because some of the policies considered promote less carbon-intensive energy sources. OECD CO_2 emissions are 3%, or 331 million tonnes, lower in 2010; and 16%, or 2,150 million tonnes, lower in 2030. Significantly, total OECD emissions peak after 2020, at around 10% higher than in 2000, and even decline a little at the end of the *Outlook* period. The total savings of CO_2 in 2030 from the Reference Scenario are nearly the same as today's emissions from France, Germany, Italy and the United Kingdom combined.

Figure 12.1: **Total OECD CO_2 Emissions in the Reference and Alternative Policy Scenarios**

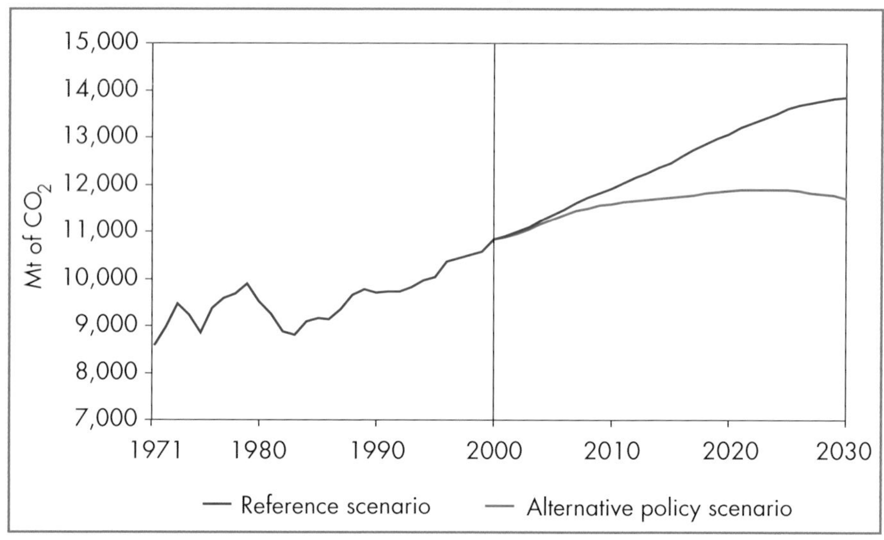

Electricity demand is reduced significantly below the Reference Scenario, with savings reaching 11%, or 107 Mtoe, in 2030. Around 41% of the savings are attributable to the residential sector and two-thirds to the residential and service sectors combined.

As a consequence, in the first decade, the most important reductions in fossil fuels are from natural gas (Figure 12.2). Because the electricity demand reductions below the Reference Scenario up to 2010 mean that a number of new gas-fired generating plants that were in the Reference Scenario will not, in fact, be needed. An increase in renewables-based capacity will also help save gas. In the second decade, gas savings are still larger than other fuels, but savings in coal consumption grow rapidly. The acceleration in coal savings between 2010 and 2020 reflects the reduction in new coal-fired capacity required due to lower electricity demand and higher generation from renewables. By 2030, the savings in coal exceed those for both oil and gas. Around 28% of the savings in gas in 2030 comes from end-use sectors. Oil savings come predominantly from the transport sector and are modest up to 2010, as more efficient vehicles are only then beginning to enter the vehicle fleet.

Figure 12.2: **Fossil Fuel Savings in the Alternative Policy Scenario**

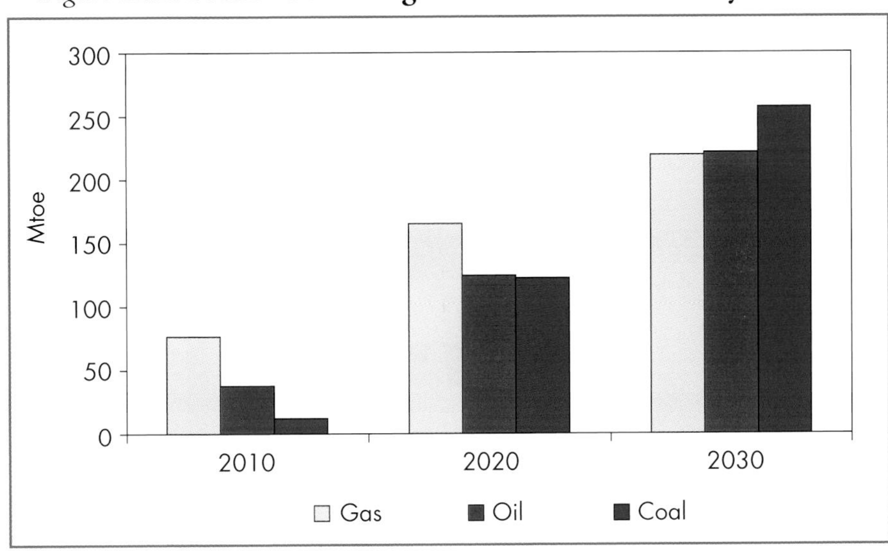

In the Alternative Policy Scenario, the use of renewable energy sources increases much more rapidly than in the Reference Scenario, displacing both gas and coal in the power sector. By 2030, total primary energy demand for non-hydro renewables is 40% higher than in the Reference Scenario.

Implications for CO_2 Emissions

Natural gas accounts for the largest share of energy and CO_2 savings in 2010. The savings in coal consumption have a large impact on CO_2 emissions, because coal has a higher carbon content than gas or oil. Reductions in CO_2 emissions from coal combustion increase rapidly after 2010 and by 2030 coal represents around half of the CO_2 savings in the Alternative Policy Scenario.

The percentage reduction in CO_2 emissions below the Reference Scenario in 2030 is largest in the European Union, at 19%. The European Union is followed by Japan, Australia and New Zealand, at 15%, and the US and Canada at 14% (Figure 12.3 and Table 12.2). In each region, savings in the power sector contribute the largest share of the CO_2 savings, due mainly to policies that promote renewables and reduce electricity demand. The larger percentage reduction in the European Union reflects, in part, more aggressive renewables targets (Figure 12.5). The savings in the United States and Canada are the largest in absolute terms, due to the higher level of emissions in this region.

Figure 12.3: **Reductions in CO_2 Emissions below the Reference Scenario in the Alternative Policy Scenario by Region, 2030**

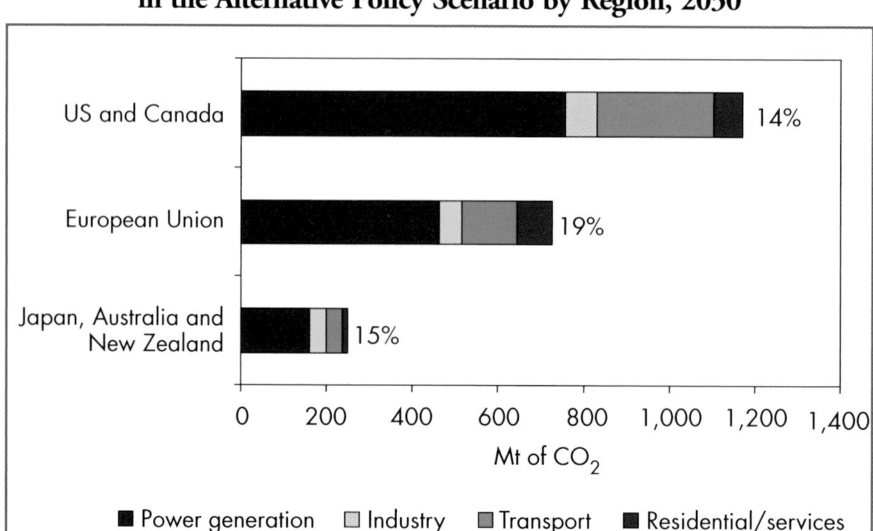

Under the Reference Scenario, *no* OECD region is projected to meet its Kyoto Protocol target with domestic reductions alone. Although the Alternative Policy Scenario yields additional reductions that bring total

OECD emissions down closer to the Kyoto target, it is still the case that no OECD region individually meets its target without recourse to the use of flexibility mechanisms, such as emissions trading. In the Alternative Policy Scenario, the OECD regions considered here are projected to exceed their target by some 28% in total or by 19% if the United States is excluded.

"Hot air"[6] is expected to be available in sufficient quantities to meet the Kyoto Protocol target under the Alternative Policy Scenario *if* the United States is excluded. In this case, emissions would be some 1% below the Kyoto target (Figure 12.4). In the Reference Scenario with or without the United States and in the Alternative Policy Scenario with the United States, the available "hot air" would not be sufficient to fill the gap. In the case where "hot air" is sufficient to meet the Kyoto target, there would, nonetheless, be a cost involved for the countries whose emissions do not reach their targets. In addition to the cost of the policies implemented, they would have to buy emission permits from those countries with "hot air".

Figure 12.4: **Kyoto Target CO_2 Emissions Gap in the Reference and Alternative Policy Scenarios**

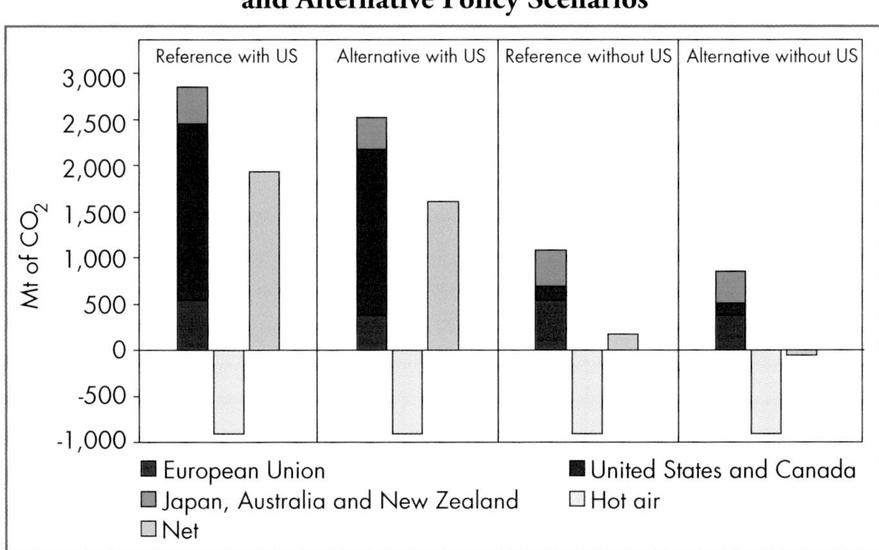

6. "Hot air" refers to the gap between projected emissions in 2010 and those of the Kyoto Protocol targets for individual countries projected to emit *less* than their targets. It does not include projections of emissions credits from possible Joint Implementation or Clean Development Mechanism projects. The amount of available "hot air" is based on the emissions projections given in the Reference Scenario (see Chapter 2).

The maximum price that OECD countries would be prepared to pay for the "hot air" permits would be the marginal cost of abatement in their regions. However, given that the "hot air" available is projected to be sufficient to meet the target under the Alternative Policy Scenario without extra efforts, the traded price of such permits would probably be lower than this.

Implications for Energy Markets

In the Alternative Policy Scenario, the total reduction in demand for fossil fuels below the Reference Scenario would reach around 700 Mtoe in 2030. Oil consumption would be reduced by 4.6 mb/d below the Reference Scenario in 2030, gas by 260 bcm and coal by about 150 Mt in 2030.

The 4.6 mb/d reduction in **oil** demand in the Alternative Policy Scenario would be equivalent to a reduction in the call on OPEC oil of 7% below what is projected in the Reference Scenario for 2030. However, the actual effect on OPEC producers would probably be less than this, because some feedback effects through the oil price would act to reduce non-OPEC oil production and increase world demand.

Table 12.1: **Net Natural Gas Imports by Region, 2030**

	Reference Scenario		Alternative Policy Scenario	
	%*	bcm	%*	bcm
US and Canada	31	371	23	246
European Union	81	632	77	512
Japan, Australia and New Zealand	29	51	22	35

*Per cent of primary gas demand.

In the Alternative Policy Scenario, **natural gas** demand in the OECD is reduced by 12.5% in 2030, or 221 Mtoe. This represents a little more than 5% of world demand. This is likely to have little impact on domestic production in the OECD regions, but will have a significant impact on import requirements (Table 12.1). In the European Union, the projected fall in import needs of around 120 bcm is slightly less than the European Union's current imports from Norway and Russia. The decrease in gas demand and imports would help to ease supply concerns in the projected tight North American gas market. In addition, the projected reduction in the need for gas imports in the United States and Canada, and in Japan, Australia and New Zealand would slow down the expansion of international LNG trade.

The reduction in **coal** consumption represents about 26%, or 257 Mtoe, of projected OECD coal demand in the Reference Scenario in 2030, but is less than 5% of world coal consumption. The impact on world coal trade would be significant, especially in the European Union and in Japan, where most coal is imported.

Figure 12.5: **Share of Renewables in Electricity Generation in the Reference and Alternative Policy Scenarios**

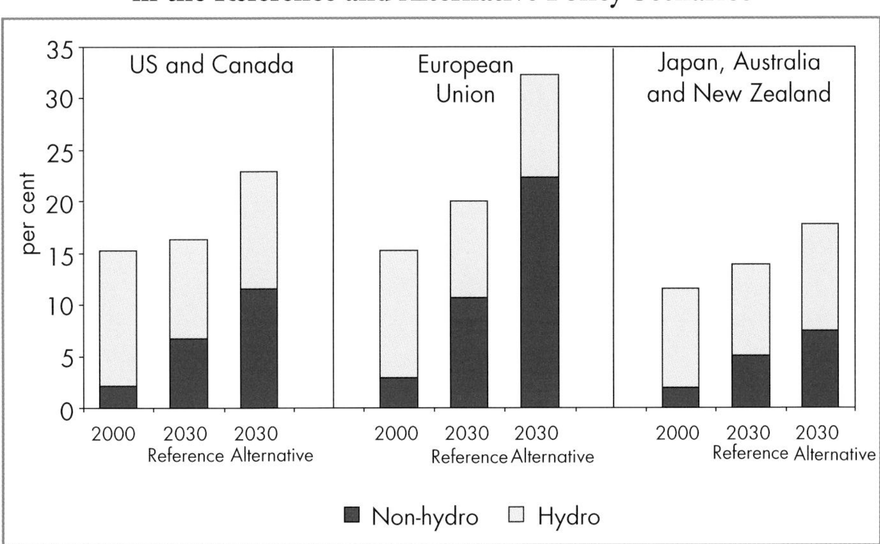

The Alternative Policy Scenario projects higher growth for **renewable energy** than the Reference Scenario, as a result of increased government support. Non-hydro renewables are projected to grow by some 4% per annum between 2000 and 2030 in the Alternative Policy Scenario, compared to 2.7% per annum in the Reference Scenario. This means an additional 166 Mtoe of primary renewable energy, or 40% more than the Reference Scenario, in 2030. The share of renewables, including hydropower, in electricity generation increases substantially, from 14.7% in 2000, to 17.6% in 2010 and to 25.4% in 2030.

Electricity and gas prices respond to the changes in demand and supply from the Reference Scenario. The results presented here take account of the impact of these price changes. The gas price falls in each of the three regions in response to lower demand. On average, gas prices are around 10% lower than in the Reference Scenario in 2030. Electricity prices are higher in the Alternative Policy Scenario than those in the Reference Scenario. This results from the interaction of two opposing forces: the additional cost of policies to encourage renewables and combined heat and power plants on the one hand, and the reduction in generating costs from more efficient technologies and the lower natural gas price on the other hand.

Results by Region

The United States and Canada

In the United States and Canada, total primary energy demand grows by 0.7% per annum in the Alternative Policy Scenario, compared to 1.0% per annum in the Reference Scenario. Total energy demand is 8.5%, or 292 Mtoe, lower in 2030. About half of this reduction comes from policies that reduce electricity demand and, to a lesser extent, those that promote combined heat and power.

Total CO_2 emissions are projected to be 14.1%, or 1,171 million tonnes of CO_2, lower than in the Reference Scenario in 2030. This is twice the level of emissions in Canada today. The power generation sector accounts for around two-thirds of the savings in emissions.

The European Union[7]

Total primary energy demand in the European Union grows by 0.4% per annum in the Alternative Policy Scenario, compared to 0.7% per annum in the Reference Scenario. Energy demand is reduced by 2.1%, or 34 Mtoe, in 2010 and by 9.2%, or 167 Mtoe, in 2030. Natural gas contributes most of the savings, around 100 Mtoe in 2030, with 87% coming from power generation. Oil demand is 86 Mtoe lower than in the Reference Scenario in 2030, around three-quarters of which is attributable to transport. Non-hydro renewables will increase by an extra 73 Mtoe. Some 45% of the energy saved is attributable to power generation, where gas use will be reduced the most.

In the Alternative Policy Scenario, CO_2 emissions in the European Union grow by 0.3% per annum from 2000 to 2010, compared to 0.8% in

7. Although not analysed here, most non-EU members of the OECD in Europe have policies under consideration that are aimed at improving energy efficiency and mitigating the climate change impacts of energy use, notably Norway, Switzerland, the Czech Republic and Hungary.

Table 12.2: **Changes in Total Primary Energy Demand and CO_2 Emissions in the Alternative Policy Scenario**
(percentage change from the Reference Scenario)

	US and Canada			**European Union**			**Japan, Australia and New Zealand**			**Total**		
	2010	**2020**	**2030**	**2010**	**2020**	**2030**	**2010**	**2020**	**2030**	**2010**	**2020**	**2030**
Total primary energy demand by fuel (%)												
Coal	-0.4	-11.3	-23.9	-1.7	-16.6	-35.3	-4.2	-12.7	-24.0	-1.3	-12.5	-26.0
Oil	-0.7	-4.3	-7.9	-3.6	-8.0	-12.9	-2.2	-5.6	-9.0	-1.9	-5.6	-9.5
Gas	-4.3	-9.1	-10.7	-8.8	-13.5	-16.2	-3.2	-6.8	-8.7	-5.7	-10.5	-12.5
Hydro	0.0	0.0	0.0	3.0	4.6	5.0	2.8	3.1	4.4	1.2	1.8	2.1
Other Renewables	18.0	37.6	35.5	36.3	47.6	51.8	2.2	8.4	14.9	22.7	38.4	39.7
Total	-0.7	-4.5	-8.5	-2.1	-5.8	-9.2	-2.2	-5.5	-8.5	-1.3	-5.0	-8.7
CO_2 Emissions (%)												
Total	-1.6	-8.1	-14.1	-4.9	-12.0	-19.0	-3.3	-8.8	-14.8	-2.8	-9.3	-15.5

the Reference Scenario. CO_2 emissions peak around 2010 at around 5% above 1990. In 2030, they will be about 1.4% below 2000 levels and 0.3% below 1990.

Japan, Australia and New Zealand

Primary energy demand in Japan, Australia and New Zealand grows by 0.5% per annum in the Alternative Policy Scenario, compared to 0.8% per annum in the Reference Scenario. In 2010 demand is 2.2%, or 16 Mtoe, lower than in the Reference Scenario, and 8.5%, or 70 Mtoe, lower in 2030. Coal accounts for the largest share, 47%, of total energy savings. Almost half of the energy savings come from power generation, where coal use falls sharply.

In the Alternative Policy Scenario, CO_2 emissions grow by 0.6% per year from 2000 to 2010, compared to 0.9% in the Reference Scenario. Emissions in 2010 are around 23% above 1990, but start to decline around 2010. The average annual rate of decline between 2000 and 2030 is 0.1% per year.

Detailed Results by Sector

Industry

In the Alternative Policy Scenario, industrial energy demand in the OECD is 7.5%, or 85 Mtoe, lower than in the Reference Scenario in 2030. Industrial heat generated from combined heat and power plants *rises* by 19 Mtoe over the *Outlook* period, while the use of all other final energy sources decreases by 103 Mtoe. Savings in electricity demand account for 37 Mtoe of the total fuel savings. Gas accounts for 33 Mtoe.

Assumptions

The Alternative Policy Scenario studies the impact of stronger and broader policies to improve energy efficiency in:

- process heat;
- steam generation;
- motive power;
- buildings.

Policies affecting steam generation and process heat, the two most important energy end-uses, have the potential to reduce industrial energy consumption significantly. Policies on motive power, which is mostly electricity-based, are important to produce significant savings of electricity, and hence the need for power generation.[8]

Table 12.3: **Policies Considered in the Industry Sector for the Alternative Policy Scenario**

Policy category	End uses	Technology impact
Regulations Standards and certification for new motor systems.	Motive power	Improved efficiency of new motor systems.
Voluntary programmes Expansion of existing programmes and establishment of new ones, including: - Information on and assistance in retrofitting, replacing and operating process equipment. - Energy auditing, target setting and monitoring.	Process steam Process heat Motive power Buildings	Improved efficiency of new technologies and accelerated deployment. Improved efficiency of energy use in buildings (building shell and appliances).
Investment programmes Tax incentives and low-interest loans for investment in new efficient technologies.	Process steam Process heat Motive power	Accelerated deployment of new boilers, machine drives, and process-heat equipment.
R&D programmes Increased funding to R&D and demonstration programmes.	Process steam Process heat Motive power	Improved efficiency of new equipment entering the market after 2010-2015.

Estimating the impact of industrial policies is a difficult task due to data limitations and the heterogeneity of the numerous processes and technologies in use. The Alternative Policy Scenario looks at how "bundles" of policies can contribute to the development of more efficient technologies and their increased use in industry. Table 12.3 summarises the four main groups of policies analysed in the Alternative Scenario.

Six industrial sectors are analysed in detail: iron and steel, non-metallic minerals, chemicals, paper and pulp, food and beverages, and "other industry". It is assumed that accelerated technology development will lead to global improvements in new technology that will be shared by industry in all three OECD regions. However, the impact on each region

8. Policies promoting CHP generation have an impact in the industrial and power generation sectors.

will vary. To take just one example, the global efficiency of new equipment for process heat generation in the chemical sector will improve by 45%, compared to 14% in the Reference Scenario, between 2000 and 2030. But the impact on the average efficiency of the capital stock in the chemical sector will vary between regions, from a 31% improvement in Japan, Australia and New Zealand to 35% in the European Union and 54% in the United States and Canada. These variations stem from differences in the efficiency of the existing capital stock in each region and in the different rates at which new technologies penetrate the capital stock.

Results

The impact of the policies considered in the Alternative Policy Scenario results in a reduction in OECD industrial energy demand of 7.5% below the Reference Scenario in 2030. The level of savings is similar in the three regions, varying from 7.1% in the US and Canada to 7.9% in the European Union and Japan, Australia and New Zealand. (Figure 12.6 and Table 12.4).

Table 12.4: **Change in OECD Industrial Energy Consumption in the Alternative Policy Scenario**
(change compared to the Reference Scenario)

	2010	2020	2030
	(%)		
Coal	-3.8	-9.1	-16.2
Oil	-2.3	-4.1	-6.9
Gas	-3.0	-5.8	-10.0
Electricity	-2.5	-6.4	-10.7
Biomass	-1.0	-2.2	-3.3
Purchased heat	21.1	35.0	56.8
Total	**-2.0**	**-4.4**	**-7.5**
	(Mtoe)		
Coal	-2.9	-6.6	-11.2
Oil	-6.8	-12.2	-21.0
Gas	-8.7	-17.8	-32.6
Electricity	-6.9	-20.1	-36.8
Biomass	-0.4	-1.0	-1.6
Purchased heat	5.6	10.4	18.6
Total	**-20.1**	**-47.3**	**-84.6**

Figure 12.6: **Reduction in Industrial Energy Demand by End-use in the Alternative Policy Scenario**
(percentage reduction from the Reference Scenario)

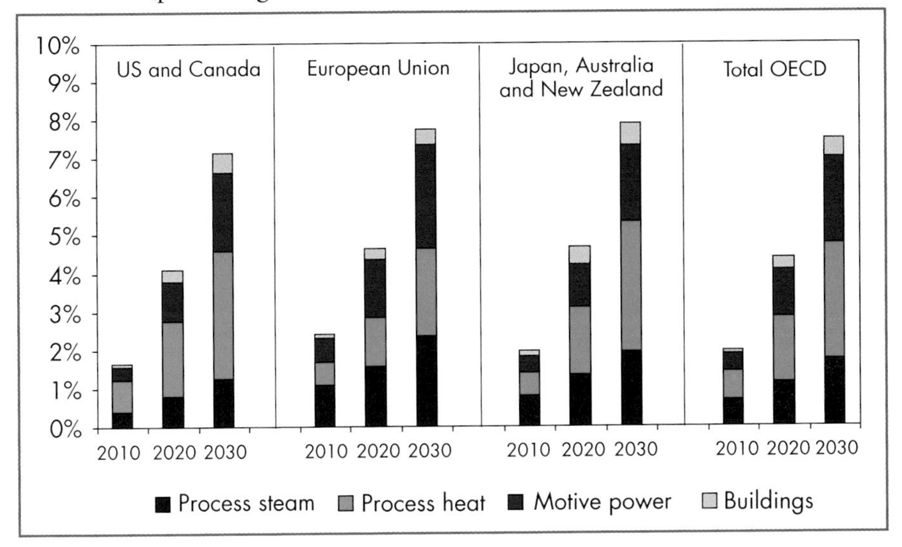

The slow turnover of capital stock in industry limits the impact of new policies in the short to medium term. Even though policies assumed in the Alternative Policy Scenario shorten turnover rates by 5 to 15 years, depending on the type of equipment, the savings achieved by 2010 are only about a quarter of the savings in 2030.

In the US and Canada, and Japan, Australia and New Zealand, efficiency measures affecting process heat demand have the strongest impact on total savings, followed by those affecting machine drives and steam generation. In the European Union measures in each of these three end-uses have about the same effect.

By 2030, OECD energy demand for process heat is around 14% lower than in the Reference Scenario, accounting for about 40% of the total reduction of 7.5% in industry. Savings in motive power, at 13% lower than the Reference Scenario in 2030, account for 30% of the total savings, while more efficient steam generation accounts for 23% of the total 7.5% reduction. The balance of savings, which is attributable to buildings, is modest due to the low share of buildings in total industrial energy demand.

Demand for coal will drop by the largest percentage below the Reference Scenario, but the savings are a modest 11 Mtoe, because of coal's low share in the industrial energy mix. Electricity demand savings represent the largest contribution, at 37 Mtoe, and come primarily from savings in motive power. Gas savings are 33 Mtoe in 2030. Oil savings, at 21 Mtoe, are modest because a large share of the oil in industry in the United States and Canada is used as a feedstock in the chemicals industry, and no policies are considered that would reduce feedstock use. The lower oil savings in the United States and Canada also help to explain why the percentage reduction below the Reference Scenario in this region is less than the other two. Offsetting the savings in energy, as already noted, will be a 19 Mtoe increase in purchased heat consumption due to expanded combined heat and power generation (see power generation).

Industrial savings by fuel differ across regions in the Alternative Scenario, with the largest share of savings for the US and Canada coming from gas saved, primarily in process heat, and for Japan, Australia and New Zealand from oil in process heat and steam. In the European Union, the greatest share of savings comes from electricity, primarily saved in motive power.

Transport

Energy demand for transport, which currently accounts for 35% of total final energy consumption in the OECD, will be 1.5%, or 20 Mtoe, lower in the Alternative Policy Scenario than the Reference Scenario in 2010. The savings accelerate towards the end of the *Outlook* period, and will be 9%, or 144 Mtoe, lower than the Reference Scenario in 2030 (Figure 12.7 and Table 12.6).

In 2030, the largest percentage reduction below the Reference Scenario will have occurred in the United States and Canada, at 9.8%, followed by Japan, Australia and New Zealand, at 8.1% and the European Union at 7.6%. Nine-tenths of the savings will come in road transport in 2010 and 96% in 2030. Virtually all the reductions will be in oil products.

Assumptions

The Alternative Policy Scenario analyses the impact of new policies on fuel consumption by transport mode – road passenger, road freight, rail passenger, rail freight, aviation and navigation. For road transport, the World Energy Model incorporates detailed vehicle fleet and efficiency models in order to track different vehicle types within the total fleet and to assess realistically the impact of the penetration of new, more efficient vehicles. Table 12.5 summarises the policies considered, by type, and region.

In all three OECD regions, fuel-economy programmes now in place are assumed to be tightened and extended after 2010. The Alternative Policy Scenario assumes that the *average* new car in the US and Canada will consume around 34% less fuel in 2030 than the Reference Scenario. This contrasts sharply with the Reference Scenario, which does not assume *any* significant vehicle fuel-economy policies. By the same token, cars in the European Union will consume 35% less fuel in 2030, and those in Japan, Australia and New Zealand about 20% less. The average fuel intensity of the on-road fleet will thus fall relative to the Reference Scenario as new cars replace older ones.

It is assumed that tax measures and regulatory targets will spur increased sales of hybrid-electric and fuel-cell powered vehicles, as well as vehicles running on LPG, natural gas and ethanol, among other alternative fuels.

In the European Union and Japan, the policies analysed reduce the growth in transport activity and encourage the switching of passenger and freight traffic to less energy-intensive modes (from private cars to buses, for example). The policies include improvements in mass transit, charges for vehicles using city streets and more high-speed passenger trains. Various initiatives on freight would promote shifting from road to rail and improving urban logistics for trucks. The net result of all these policies, in 2030, would be a 7% reduction in car traffic below the Reference Scenario in the European Union and an 8% cut in freight truck travel. In Japan, Australia and New Zealand, car use would be reduced by 5% below the Reference Scenario and truck activity by around 10%.

Results

If the new policies considered in the Alternative Policy Scenario are put into effect, OECD transport energy demand is projected to be 9% lower than the Reference Scenario in 2030. These savings grow slowly because of the time required for new, more efficient, vehicles to become a significant share of the total vehicle fleet. By 2030, the Alternative Policy Scenario projects energy consumption by the transport sector to be 31% higher than in 2000, and still growing, albeit at a reduced rate. This is a very different picture from the 44% increase shown in the Reference Scenario.

Table 12.5: **Policies Considered in the Transport Sector for the Alternative Policy Scenario**

Policy aim	Programme/measure	Impacts
Improved vehicle fuel efficiency	Increased CAFE/vehicle efficiency standards (US and Canada) Increased voluntary agreement targets (EU) Top Runner programme and equivalents (Japan, Australia and New Zealand)	New car and light truck efficiency improves.
Increased use of alternative fuels and vehicles	Increased R&D and tax credits (US and Canada) Alternative fuel targets (EU) Green tax for clean fuel vehicles (Japan)	Increases the use of hybrid, gas, and fuel-cell powered vehicles, and of alternative fuels.
Reducing travel demand growth and switching to less energy– intensive modes	Urban road pricing, expansion of high-speed rail and freight initiatives (Japan) White Paper on Transport – package of policies (EU)	Suppress growth in passenger and freight transport and foster modal shift from road and aviation to rail and bus.

Because nearly all the energy savings in transport are of oil, the policies in the Alternative Policy Scenario can have an important impact on the energy security of the oil-importing nations of the OECD. Total OECD oil demand will be half-a-million barrels a day less than in the Reference Scenario in 2010 and that figure rises to 3.6 mb/d in 2030. This represents 6.2% of total primary oil demand in the three regions in the Reference Scenario in 2030. CO_2 emissions are reduced below the Reference Scenario slightly more than energy consumption, by 1.8% in 2010 and by 10.1% in 2030. The slightly higher saving in CO_2 compared to energy in 2030 is due first to a shift from road traffic to rail – which entails a shift from oil products to electricity, and secondly to the increase in the use of alternative, less carbon-intensive fuels.

Sales of hybrid-electric and alternative-fuel vehicles will not be large until after 2010. These technologies will contribute to improved fuel efficiency and will help reduce carbon emissions below the Reference Scenario. In the US and Canada, hybrid and fuel-cell-powered vehicles will make up 18% of the stock of cars and light trucks in 2030. This figure will be influenced by the actual cost of the vehicles, the type of fuel used and the availability of refuelling stations, all of which are highly uncertain.

Table 12.6: **Change in OECD Transport Energy Consumption in the Alternative Policy Scenario**
(change compared to the Reference Scenario)

	2010	2020	2030
		(%)	
Oil	-2.1	-6.8	-11.2
Other fuels	16.3	37.0	58.8
Total	**-1.5**	**-5.5**	**-9.0**
		(Mtoe)	
Oil	-25.7	-95.3	-173.0
Other fuels	6.0	16.0	29.1
Total	**-19.7**	**-79.3**	**-143.9**

The European Union and Japan, Australia and New Zealand benefit from demand-side measures to slow the growth of road traffic, notably shifting some growth from road to rail and from private cars to buses. As a result, the Alternative Policy Scenario shows fuel consumption by railways increasing by 20% above the Reference Scenario in 2030, but this will offset only 4% of the savings achieved in road transport. Energy demand for aviation will decline by 4% in 2030 from the Reference Scenario projection.

Transport energy savings in the Alternative Policy Scenario will be greatest in the US and Canada, at 9.8% below the Reference Scenario in 2030. The European Union will see a reduction of 7.6%, while consumption in Japan, Australia and New Zealand will drop by 8.1% below the Reference Scenario in 2030. This reflects the greater savings potential in the US and Canada, because of the higher current fuel consumption per vehicle, and because in the Reference Scenario for the US and Canada, unlike in the other two regions, no tightening of vehicle efficiency standards has been assumed. The difference between the US and Canada and the other regions would be larger without the additional demand restraint policies in the other two regions.

Figure 12.7: **Reduction in Transport Energy Demand by Mode in the Alternative Policy Scenario**
(change from the Reference Scenario)

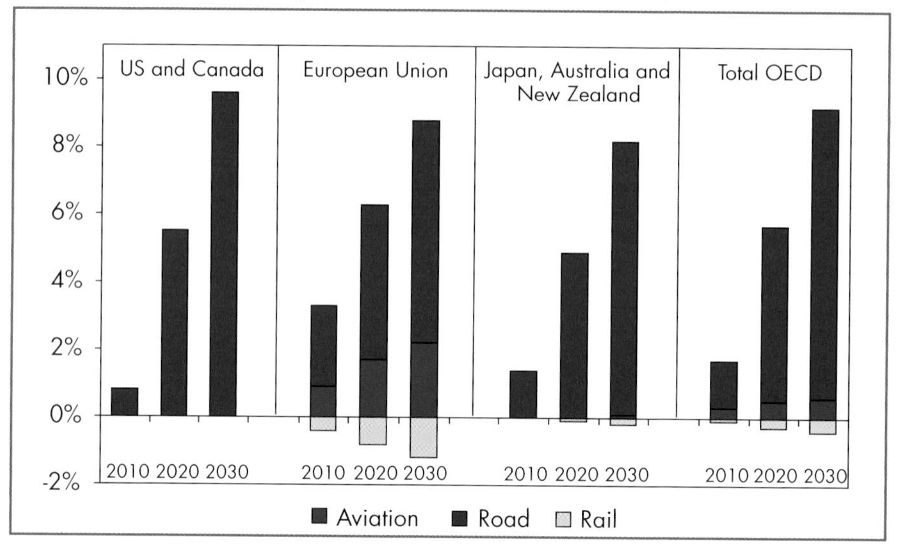

Residential and Services

In the Alternative Policy Scenario, energy demand in the OECD residential sector is projected to be 9.4%, or 75 Mtoe, lower than the Reference Scenario in 2030. In the services, the Alternative Policy Scenario projects savings of 11%, or 57 Mtoe, in 2030. Electricity accounts for 56% of the total savings across both sectors and gas accounts for 34%, corresponding to a 74 Mtoe reduction in electricity demand below the Reference Scenario in 2030 and a 45 Mtoe reduction in gas demand. In the residential sector, nearly half of these savings occur in space heating and a third in household appliances. In services, two-thirds of the savings occur in space and water heating, with the remainder coming from savings in electrical end uses. Savings in Japan, Australia and New Zealand are very different from the average, with the residential sector producing only 5% reduction and services 15% by 2030 (Figure 12.8).

Assumptions

The Alternative Policy Scenario evaluates the impact on energy use in the residential and services sectors of new policies on space heating, water heating, lighting and other electric end-uses. In the residential sector, the other electric end-uses are space cooling and household appliances. In the services sector, other electric end-uses are space cooling, ventilation, office and computer network equipment and a host of miscellaneous end-uses. Table 12.7 summarises the key policies evaluated in the Alternative Policy Scenario for the OECD residential and services sectors by region and policy group.

Table 12.7: **Policies Considered in the Residential and Services Sectors for the Alternative Policy Scenario**

Policy group	Examples of programme/measure	Impacts
	Mandatory policies	
Equipment standards	NAECA standards (US and Canada) Framework Directive (EU) Top Runner and AS/NZS standards (Japan, Australia and New Zealand)	Increased efficiency of new appliances.
Building codes	1993/95 MEC, 1998 IEEC (US and Canada) Buildings Directive (EU) Revised Building Code of Australia	Increased thermal efficiency of buildings.
	Other policies	
Financing	Utility Rebates(US and Canada) Next-Generation Housing loans (Japan) Energy Smart (Australia)	Increased penetration of high efficiency devices.
Endorsement labelling	Energy Star (US and Canada) International Energy Star (Japan)	Increased penetration of high efficiency devices.
Whole-building programmes	Energy Star and Building America (US and Canada) Home/business energy management systems (Japan) Energy Smart (Australia)	Increased efficiency in new and existing buildings.
Accelerated R&D	Federal building science and lighting research (US and Canada) Federal lighting research (Japan)	Increased penetration of emerging super-efficient devices.

NAECA = National Appliance Energy Conservation Act
AS/NZS = Australian Standard/New Zealand Standard
MEC = Model Energy Code
IECC = International Energy Conservation Code

Two main types of *mandatory* policies are considered in the Alternative Policy Scenario: equipment standards and building codes. In some cases, both are accompanied by mandatory labelling schemes. Equipment standards are assumed to reduce the unit energy consumption (UEC) levels of new equipment from 4% to 60% depending on the type of equipment and region. Similarly, building codes are assumed to reduce the average heating and cooling loads of new buildings by between 10% and 40% compared to the Reference Scenario, depending mainly on building type and climate.

Three general types of *voluntary* policies are considered in the Alternative Policy Scenario: financing schemes for efficiency investments, endorsement labelling and "whole-building" programmes. Financing schemes include direct consumer rebates, low-interest loans, and energy-saving performance contracting. Endorsement labelling helps consumers find products that perform well above minimum standards by identifying products that exceed standards. "Whole-building" programmes use systems-optimisation and integrated construction practices to improve energy efficiency. The savings are similar to building codes, but the penetration rates achieved by these voluntary programmes are assumed to be low compared to mandatory policies.

Also considered in the Alternative Policy Scenario are the effects of accelerated research and development (R&D) efforts by governments. More R&D spurs innovation, brings down technology costs and helps to bring high efficiency devices to market more rapidly.

No single regulatory framework for equipment standards and building codes exists in the European Union. Standards have occasionally been created on an *ad hoc* basis in the past, but the current focus is on creating comprehensive regulatory frameworks. The Alternative Policy Scenario includes two important Framework Directives currently being developed by the European Commission. The Framework Directive on end-use equipment standards is expected to be adopted by member states in 2003, with implementing directives to follow by 2005. The Buildings Directive will probably be adopted by the end of 2002, with implementing directives to follow in 2007.

As discussed, policies enacted in 2002 or earlier are already reflected in the Reference Scenario. For the United States, this includes all equipment standards that come into effect by 2007, and for Australia, equipment standards that take effect by 2003. For Japan, this includes the current Top

Runner programme and the 1998 revisions to residential and commercial building codes.

Results

In the Alternative Policy Scenario, total OECD residential energy demand is projected to be 9.4% lower than the Reference Scenario in 2030, while total OECD services energy demand is 11.1% lower (Tables 12.8 and 12.9). In both sectors mandatory policies produce the largest energy savings. Equipment standards produce significant savings by around 2020, but since several key residential and commercial equipment stocks turn over completely before 2030, savings from standards tend to stabilise before the end of *Outlook* period. The savings from building codes accumulate more slowly, due to the very slow turnover in building stock, but are significant between 2020 and 2030 and would continue to grow well beyond 2030.

In the residential sector, policies targeting space heating and appliances produce the largest savings (Figure 12.8). In the European Union, the largest share of savings comes from space heating; in the other regions it is from appliances. In services, policies targeting space heating, space cooling and lighting produce the largest savings. Building codes and standards produce the bulk of these savings in both sectors, but in services, "whole-building" programmes also make a considerable contribution to total savings.

Electricity accounts for 61% of the total savings in the residential sector and gas 24% (Table 12.8). This corresponds to a 15%, or 46 Mtoe, reduction in total OECD residential electricity demand below the Reference Scenario in 2030, and a 6%, or 18 Mtoe, reduction in gas demand. The share of electricity savings is high because residential equipment standards affect mainly electrical equipment.

In the services sector, electricity accounts for 44% of total savings in the Alternative Policy Scenario and natural gas accounts for 41%. For the OECD as a whole, these figures correspond to a 10% reduction in electricity demand in services and a 18% reduction in gas demand in services. Sizeable cuts in gas use in the European Union and the US and Canada, and oil use in Japan, Australia and New Zealand come because these are the principal heating fuels in their respective regions. In the European Union and the US and Canada, the consumption of heat produced from combined-heat-and-power plants grows in response to policies to encourage them in the power sector.

The reduction in residential energy demand below the Reference Scenario in Japan, Australia and New Zealand in 2030 is only 5.7%, compared to 9.0% and 10.4% for the European Union and the United States and Canada, respectively. The lower percentage reduction below the Reference Scenario projected for space heating accounts for most of this difference and partly reflects heating's relatively small share of residential energy demand in Japan, Australia and New Zealand. The low space heating savings also reflect the assumption that Japanese building codes will not be tightened beyond the levels set in 1998.

Table 12.8: **Change in OECD Residential Energy Consumption by Fuel in the Alternative Policy Scenario**

(change relative to the Reference Scenario)

	2010	2020	2030
	(%)		
Oil	-1.6	-6.0	-11.4
Gas	-0.7	-2.9	-5.7
Electricity	-4.1	-11.4	-14.8
Biomass	-0.1	-0.2	-0.3
Purchased heat	-0.1	-0.5	-0.7
Total	**-2.0**	**-6.3**	**-9.4**
	(Mtoe)		
Oil	-1.7	-6.0	-10.6
Gas	-1.9	-8.6	-18.0
Electricity	-9.7	-31.6	-45.9
Biomass	0.0	-0.1	-0.2
Purchased heat	0.0	-0.1	-0.1
Total	**-13.4**	**-46.4**	**-74.8**

In contrast, services energy demand in Japan, Australia and New Zealand is reduced 15.0% below the Reference Scenario in 2030, significantly more than in the European Union (10.2%) and the United States and Canada (10.7%). This difference is primarily due to more aggressive standards, targeting commercial space cooling and lighting, and the assumed high penetration of building energy management systems in the region.

Table 12.9: **Change in OECD Services Energy Consumption by Fuel in the Alternative Policy Scenario**
(change relative to the Reference Scenario)

	2010	**2020**	**2030**
		(%)	
Coal	-0.6	-2.5	-3.9
Oil	-4.7	-13.5	-22.0
Gas	-3.9	-11.2	-18.0
Electricity	-2.6	-7.5	-9.9
Purchased heat	42.3	50.6	75.8
Total	**-2.3%**	**-7.8**	**-11.1**
		(Mtoe)	
Coal	0.0	0.0	-0.1
Oil	-2.4	-6.2	-9.2
Gas	-5.7	-16.6	-26.9
Electricity	-5.8	-19.5	-28.5
Purchased heat	3.6	9.4	7.8
Total	**-10.5**	**-37.6**	**-57.1**

Figure 12.8: **Reduction in Residential and Services Energy Demand by End-Use in the Alternative Policy Scenario**
(percentage change from the Reference Scenario)

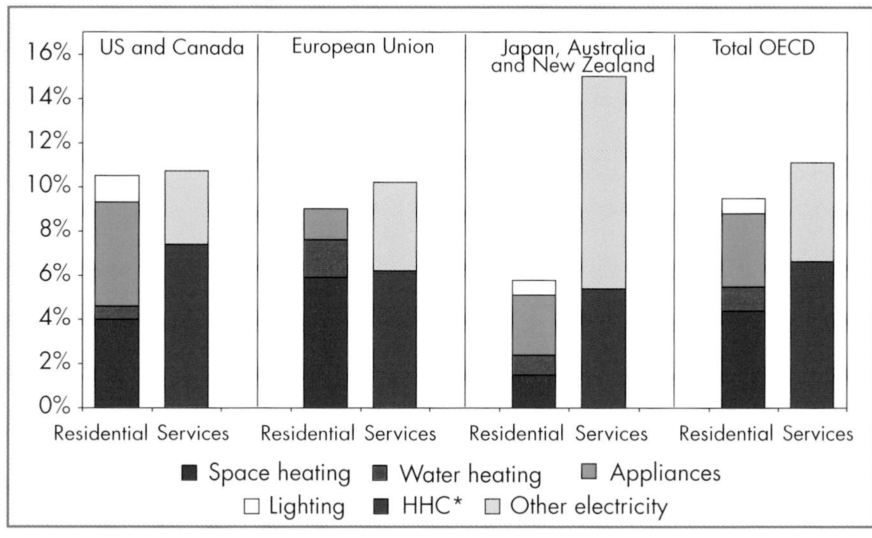

* Heating, hot water and cooking.

Power Generation

In the Reference Scenario, CO_2 emissions from the OECD power sector are expected to rise by a third between now and 2030, while the power-sector *share* of total CO_2 emissions will increase from 38% to 40%. In the Alternative Policy Scenario, the fossil fuel input to electricity generation is around 24% less, or 409 Mtoe, in 2030, compared to the Reference Scenario. Renewable energy sources, especially wind and biomass, increase much faster than in the Reference Scenario.

Assumptions

The Alternative Policy Scenario shows that if existing policies on power generation were enhanced and new policies were adopted to curb emissions and reduce electricity consumption, the savings would be considerable. Policies in the power sector include promoting the use of low-carbon or no-carbon fuels and increasing efficiency (Table 12.10).

Table 12.10: **Policies Considered in the Power Generation Sector for the Alternative Policy Scenario**

Policy type	Programme/measure	Impacts on power generation sector
Increased use of renewables	Renewable Energy Directive (EU) Renewable Portfolio Standard (US and Canada) Renewable Energy Targets (Japan, Australia and New Zealand)	Increases the share of renewables.
Increased CHP	Policies to promote CHP in end-use sectors (All Regions)	Increased share of electricity generation from CHP plants.
Improved efficiency	Various policies and R&D to accelerate the penetration of highly efficiency coal and gas plants and new technologies such as fuel cells (All Regions)	More efficient new gas, coal and fuel cells plants.

For many OECD governments, renewable energy use is the preferred option for reducing CO_2 emissions. Indeed, many of them have set numerical targets for increases in renewables use. In the Alternative Policy Scenario, it is assumed that all the policies needed to meet such targets are indeed adopted. Approaches include guaranteed prices for renewables, mandatory portfolio shares and tax credits for investment in, and production of, renewables.

All European countries have policies to promote renewables. Last year, the European Parliament adopted a Renewable Energy Directive that calls for the share of renewables in gross electricity consumption to increase from 13.9% in 1997 to 22.1% in 2010 (this includes hydro power among renewables). In the Alternative Policy Scenario, it is assumed that the directive is fully carried out by member countries. A Renewable Portfolio Standard (RPS) is under discussion in the United States. This would require a certain share of electricity to be generated from renewables. It is assumed the adoption of an RPS requiring 10% of electricity sales in 2020 to be renewables-based, which corresponds to 8% renewables in generation in the United States and Canada. Japan has set targets for new and renewable energy sources, based largely on more efficient use of wastes, like landfill gas and the installation of 5 GW of photovoltaics by 2010. Japan is assumed to meet its targets.

Combined heat and power generation plants consume less fuel than do separate electricity-only and heat-only plants. The Reference Scenario assumes that CHP's share of electricity generation is constant throughout the projection period. The Alternative Policy Scenario assumes CHP plants provide an increasing share of total electricity generation. This slightly increases the amount of fuel used in the power sector but causes an even larger decrease in industry. The result is a net decline in carbon emissions. Most new CHP plant capacity is likely to be used for on-site generation in industry. Many OECD countries already offer incentives to encourage the use of CHP. It is assumed that, by 2030, the share of electricity produced by CHP plants will be two percentage points higher than in the Reference Scenario.

The Reference Scenario expects cost-effective new energy technologies to come to market only gradually over the *Outlook* period. The Alternative Scenario assumes quicker market penetration for:

- Advanced gas turbines (in simple and combined cycle). These are assumed to be from 2 to 3 percentage points more efficient in the Alternative Policy Scenario.
- Advanced coal technology plants. They are assumed to reach 55% efficiency by 2030, when coal gasification will have become the rule in new coal-fired plants.
- Fuel cells. These are assumed to become economic for power generation around 2015, with fuel-cell technology achieving 70% efficiency by 2030.

The Alternative Policy Scenario assumes no increase in OECD nuclear power capacity beyond the Reference Scenario. Although discussions about the future role of nuclear power have been re-initiated in some countries, there are no concrete policies and measures to support the development of new nuclear plants in OECD countries that could be taken into account in the Alternative Policy Scenario.

Results

In the Alternative Policy Scenario, electricity generation increases by only 0.9% a year to 2030, compared to 1.3% a year in the Reference Scenario. Most of this difference is due to reduced electricity demand as a result of policies described in the end-use sectors. Wind, biomass and gas use increase the most. Fuel cells using natural gas produce 6% of OECD power generation by 2030 (Table 12.11). Coal-fired generation declines to 30% below the Reference Scenario in 2030.

Table 12.11: **OECD Electricity Generation Mix in the Reference and Alternative Policy Scenarios** (TWh)

		2010		2020		2030	
	2000	**Ref.**	**Alt.**	**Ref.**	**Alt.**	**Ref.**	**Alt.**
Coal	3,349	3,315	3,271	3,743	3,178	4,000	2,782
Oil	463	446	438	363	342	254	220
Gas	1,390	2,639	2,188	3,668	2,744	4,126	2,893
Hydrogen-fuel cells	0	0	0	15	81	332	670
Nuclear	2,059	2,174	2,174	1,930	1,930	1,791	1,791
Hydro	1,054	1,118	1,132	1,160	1,180	1,191	1,245
Other renewables	197	391	588	641	1,125	974	1,597
Total	**8,510**	**10,085**	**9,791**	**11,520**	**10,580**	**12,667**	**11,200**

The Alternative Policy Scenario shows a more pronounced trend towards distributed generation compared to the Reference Scenario. Higher efficiencies make diesel generators and gas turbines more attractive for on-site use in the first half of the projection period. In the second half, they are gradually replaced by fuel cells. These distributed sources of electricity, together with photovoltaics in buildings, are expected to meet a substantial part of the incremental need for electricity generation in the Alternative Policy Scenario between 2020 and 2030.

Figure 12.9: **OECD CO_2 Emissions per kWh in the Reference and Alternative Policy Scenarios**

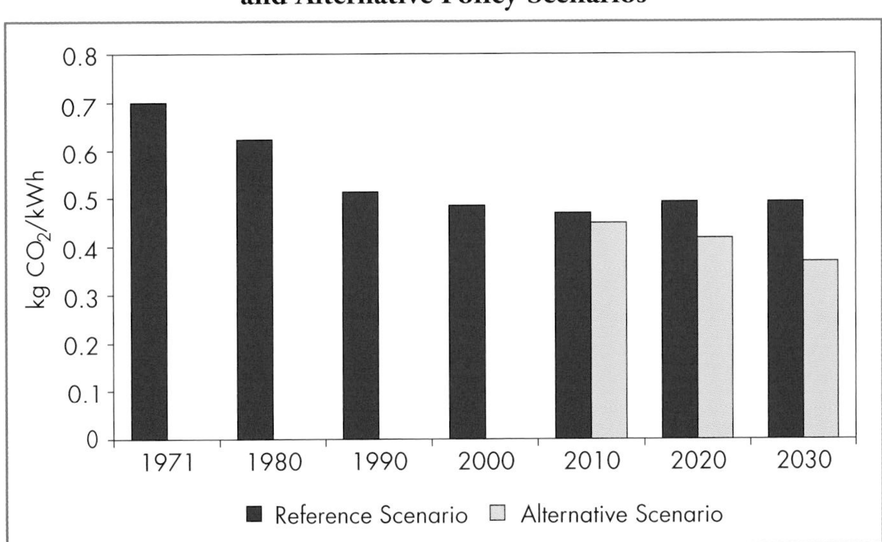

With all the policies and measures considered in the Alternative Policy Scenario, power-sector emissions are projected to be 25% below the Reference Scenario in 2030. Most of the reductions will come from reduced electricity demand and the increased use of renewables. The electricity mix is less carbon-intensive and emissions per kWh are much lower (Figure 12.9). Even so, more than half the electricity generated in 2030 will still be based on fossil fuels.

In the European Union, CO_2 emissions in 2030 are 35% below the Reference Scenario. Gas-fired generation increases much more slowly in the Alternative Scenario, especially in the next ten years. Coal-fired generation declines very rapidly, falling 40% below the Reference Scenario, to account for only 12% of overall power generation in 2030.

In the US and Canada, CO_2 emissions in 2030 are 21% below the Reference Scenario. This limits the increase in CO_2 emissions from power generation to only 16% above 2000 levels, compared to nearly 40% by 2030 in the Reference Scenario.

In the Reference Scenario in Japan, Australia and New Zealand, power sector CO_2 emissions rise only modestly, because of Japan's increasing reliance on nuclear power. But the numbers are even lower in the Alternative Policy Scenario, where emissions are 25% lower than the Reference Scenario in 2030 – and 19% lower than in 2000.

CHAPTER 13: ENERGY AND POVERTY

HIGHLIGHTS

- Some 1.6 billion people – one-quarter of the world population – have no access to electricity. In the absence of vigorous new policies, 1.4 billion people will still lack electricity in 2030.
- Four out of five people without electricity live in rural areas of the developing world, mainly in South Asia and sub-Saharan Africa. But the pattern of electricity deprivation is set to change, because 95% of the increase in population in the next three decades will occur in urban areas.
- Some 2.4 billion people rely on traditional biomass – wood, agricultural residues and dung – for cooking and heating. That number will *increase* to 2.6 billion by 2030. In developing countries, biomass use will still represent over half of residential energy consumption at the end of the *Outlook* period.
- Lack of electricity and heavy reliance on traditional biomass are hallmarks of poverty in developing countries. Lack of electricity exacerbates poverty and contributes to its perpetuation, as it precludes most industrial activities and the jobs they create.
- Investment will need to focus on various energy sources, including biomass, for thermal and mechanical applications to bring productive, income-generating activities to developing countries. Electrification and access to modern energy services do not *per se* guarantee poverty alleviation.
- Renewable energy technologies such as solar, wind and biomass may be cost-effective options for specific off-grid applications, while conventional fuels and established technologies are likely to be preferred for on-grid capacity expansion.

This chapter analyses the relationship between energy use and poverty in developing countries. It describes current patterns of energy use, including rates of electrification. A unique, country-by-country database was especially prepared for this study. This chapter details electricity access and the way households make the transition from traditional fuels to modern forms of energy. It projects biomass use and electricity access rates for the next three decades. It assesses the factors behind these trends, including policies to promote investment in electricity supply and to make electricity more affordable for poor people.

The Link between Energy Use and Poverty

Access to electricity and other modern energy sources is a necessary, but not a sufficient, requirement for economic and social development. The escape from poverty also requires, among other things, clean water, adequate sanitation and health services, a good education system and a communication network. Yet cheap and available energy is indispensable. Electricity provides the best and most efficient form of lighting; household appliances require it. Kerosene and liquefied petroleum gas (LPG) are more energy-efficient cooking fuels than traditional biomass. Diesel, heating and heavy fuel oil are more cost-effective for space heating. Diesel, gasoline and LPG are, and will remain, the primary transport fuels.

Modern energy services enhance the life of the poor in countless ways.[1] Electric light extends the day, providing extra hours for reading and work. Modern cook-stoves save women and children from daily exposure to noxious cooking fumes. Refrigeration allows local clinics to keep needed medicines on hand. And modern energy can *directly* reduce poverty by raising a poor country's productivity and extending the quality and range of its products – thereby putting more wages into the pockets of the deprived.[2]

The extensive use of biomass in traditional and inefficient ways and the limited availability of modern fuels are manifestations of poverty. They also restrain economic and social development:

- *Time spent gathering fuel:* The widespread use of fuelwood and charcoal can result in scarcity of local supplies. This forces people – usually women and children – to spend hours gathering fuelwood and other forms of biomass further afield. In India, two to seven hours each day can be devoted to the collection of fuel for cooking.[3] This reduces the time that people can devote to other productive activities, such as farming and education.
- *Gender:* 70% of all people living in poverty are women.[4] Women place a high value on improved energy services because they are the primary users of household energy. Women are most likely to suffer

1. In September 2000, the World Bank, the International Monetary Fund, members of the Development Assistance Committee of the OECD and many other agencies adopted the Millennium Development Goals. These goals set targets for reductions in poverty, improvements in health and education, and protection of the environment. Improved access to energy services is an underlying component linked to the achievement of these goals. (http: //www.developmentgoals.org).
2. The World Bank Group (2002).
3. United Nations Development Program, United Nations Department of Economic and Social Affairs and World Energy Council (2002).
4. http: //www.undp.org/unifem/ec_pov.htm

the health effects of energy-inefficient appliances. Their exclusion from the decision-making process in many countries has led to the failure of many poverty alleviation programmes.

- *Environment:* Gathering wood for fuel leads to local scarcity and ecological damage in areas of high population density where there is strong demand for wood.
- *Energy efficiency:* In developing countries, biomass fuels are often burned in inefficient stoves. Wood is much less efficient for cooking than modern fuels, such as kerosene and LPG – its net calorific value is four times lower.
- *Health:* The inefficient use of biomass can lead to serious health damage from indoor smoke pollution. Possible effects include respiratory diseases, such as asthma and acute respiratory infections; obstetrical problems, such as stillbirth and low birth-weight; blindness; and heart disease (Box 13.1).
- *Agricultural productivity:* Use of biomass energy reduces agricultural productivity, because agricultural residues and dung are also widely used as fertilizer. The more biomass is put to household use, the less there is available for fertilizer. The dung used as fuel in India would be worth $800 million per year if it were used as fertilizer.[5]

Box 13.1: **Examples of the Impact of Energy Poverty on Health**

The absence of efficient and affordable energy services can severely damage the health of the poor in developing countries.

- In rural sub-Saharan Africa, many women carry 20 kilogrammes of fuel wood an average of five kilometres *every day.*[6] The effort uses up a large share of the calories from their daily meal, which is cooked over an open fire with the collected wood.
- Poor people in the developing world are constantly exposed to indoor particulate and carbon monoxide concentrations many times higher than World Health Organization standards. Traditional stoves using dung and charcoal emit large amounts of CO and other noxious gases. Women and children suffer

5. Tata Energy Research Institute et al. (1999).
6. http: //allafrica.com

most, because they are exposed for the longest periods of time. Acute respiratory illness affects as much as 6% of the world population. The WHO estimates that 2.5 million women and young children in developing countries die prematurely each year from breathing the fumes from indoor biomass stoves.[7]

- A shift from cooking with wood to charcoal reduces the overall health risk by a factor of more than four. A shift to kerosene results in a reduction by a factor of six. Using LPG reduces the overall health risk by a factor of more than 100.[8]
- Often in developing countries, there are no pumps to gather or to purify water. In sub-Saharan Africa, only half of the population has access to an improved water source.[9]
- Lack of refrigeration means that food is spoilt and wasted. Clinics lacking electricity cannot perform such routine functions as sterilising instruments or safely storing medicines.

The share of energy in the total spending of low-income households is high, up to 15% of income (Table 13.1). Energy spending rises with income, but generally at a less than proportional rate.

Table 13.1: **Share of Energy Expenditures in Household Income** (%)

	Uganda	Ethiopia	India	South Africa	United Kingdom
Lowest quintile	15.0	10.0	8.5	7.2	6.6
Highest quintile	9.5	7.0	5.0	5.5	2.0

Sources: African Energy Policy Research Network (Afrepren), direct communication; Tata Energy Research Institute (2001); Davis (1998); Department of Trade and Industry (2002).

The Transition to Modern Fuels

As poor families in developing countries gradually increase their incomes, they can afford more modern appliances, and they demand more and better energy services. But the transition from traditional biomass use to full dependence on modern energy forms is not a straight-line process.

7. United Nations Development Program, United Nations Department of Economic and Social Affairs and World Energy Council (2002).
8. http: //www.ftpp.or.ke/rnews/biomass.htm
9. http: //www.developmentgoals.org/Definitions_Sources.htm

There is a widespread misconception that electricity substitutes for biomass. Poor families use electricity selectively — mostly for lighting and communication devices. They often continue to cook and heat with wood or dung, or with fossil-based fuels like LPG and kerosene.

The three main determinants in the transition from traditional to modern energy use are fuel availability, affordability and cultural preferences. If a modern distribution system is not in place, households cannot obtain access to modern fuels, even if they can afford them. LPG penetration rates are low in many developing countries, partly because distribution infrastructure is lacking.

Even when they *can* afford modern fuels, households may not use them if they are much more expensive than traditional biomass. In rural areas, biomass is often perceived as something that is "free" and readily available. This kind of thinking seriously hampers the switch to modern energy. Even when fuelwood is purchased, it is likely to be cheaper than the cheapest alternative fuel.[10] The affordability of energy-using equipment is just as important as the affordability of fuels. The initial cost of acquiring kerosene and LPG stoves and LPG bottles may discourage some people from switching away from biomass.

In some cases, traditions determine the fuel choice regardless of fuel availability and income. In India, even very rich households keep a biomass stove to prepare their traditional bread.

Figure 13.1 is a static representation of the typical fuel transition in poor households as their income rises. The actual transition is much more dynamic, as nearly all households opt for a combination of fuels.[11] Very poor households can hope to satisfy only their most basic needs: heating, cooking and lighting. Their fuel choices are restricted mainly to different forms of biomass. As their income increases, their fuel choices widen. The incremental energy needs of the highest-income households, whose use of biomass is minimal, tend to be met by electricity. The share of basic needs in total consumption falls off sharply as families grow more prosperous.

Figure 13.1 cannot adequately capture rural-urban differences in fuel choices, nor can it capture fuel switching that takes place within each block. Poor people often switch from one biomass fuel to another when the first becomes scarce. If wood is scarce or labour to collect it is in short

10. See World Bank (1995), which cites the result of a household energy survey in N'Djamena, Chad showing that fuelwood and charcoal are much cheaper than kerosene and LPG, even on the basis of cooking heat delivered.

11. Davis (1998), Masera et al. (2000) and Barnett (2000).

supply, low-income families will use dung or agricultural residues for cooking and heating. In cities, consumption patterns are more likely to be affected by relative fuel prices.

Figure 13.1: **Illustrative Example of Household Fuel Transition**

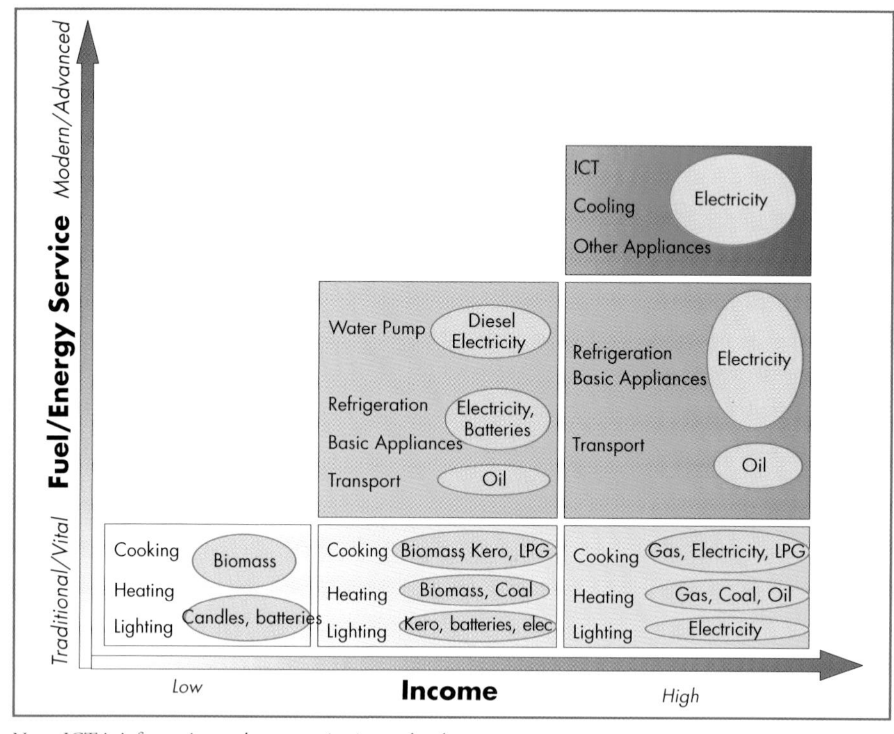

Note: ICT is information and communication technology.
Source: IEA analysis.

Figure 13.2 plots average final energy consumption per capita for 100 developing and transition countries, grouped according to the percentage of their population under the poverty line ($2 a day).[12] In countries where less than 5% of the population is poor, per capita energy consumption is four times higher than in countries where more than 75% of the population lives under the poverty line. Consumption of commercial fuels, especially oil products, is much higher in the

12. World Bank (2001). In this chapter, being below the poverty line is defined as having income of less than $2 per day. People with income of less than $1 per day are categorised as "very poor". Roughly 1 billion come into this category. There is, however, considerable uncertainty over data on the number of people in each of these categories.

Figure 13.2: **Average per Capita Final Energy Consumption and Share of Population Living under Poverty Line, 2000**

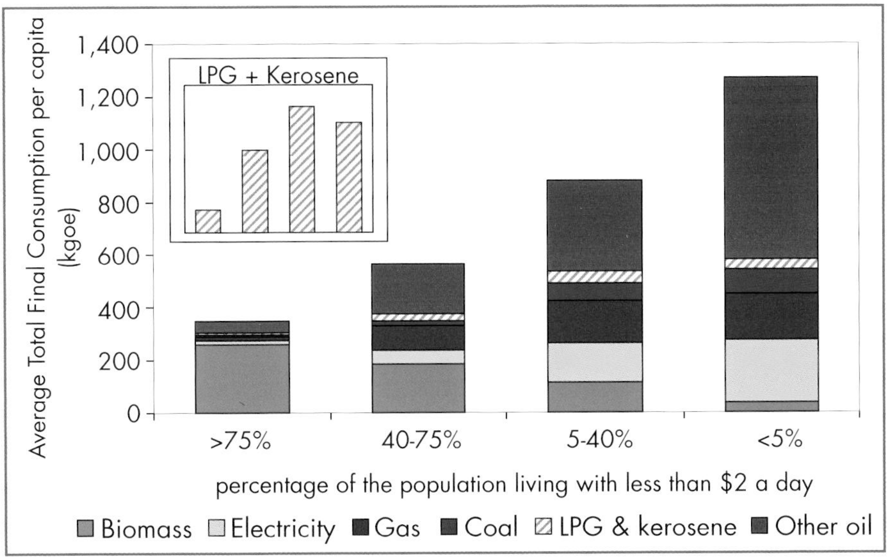

Source: IEA analysis.

richest group of countries, partly because transport demand rises with income. LPG and kerosene are transition fuels in households: their consumption is higher for the intermediate groups, but lower for the richest citizens, who replace them with natural gas and electricity (see insert in Figure 13.2).[13] Electricity consumption is very strongly correlated with wealth. The share of biomass in final energy consumption is lowest in countries where the percentage of poor people is lowest.

Some 2.8 billion of the world's people live on less than \$2 a day – the "poor" as defined in this chapter. Our detailed statistical analysis of energy use in developing countries reveals that 2.4 billion people rely on biomass for cooking and heating, which account for more than 80% of their residential energy needs, and 1.6 billion people use no electricity at all.[14] Most of these people live in South Asia and sub-Saharan Africa (Figure 13.3).

13. LPG accounts for nearly 20% of total residential energy demand in Latin America, but only for some 3% in Asia and Africa.

14. Access to electricity is defined as the number of people with electricity in their homes, either on-grid or off-grid. Our estimate does not include unauthorised connections. See Annex 13.1 for further discussion.

Figure 13.3: **Global Energy Poverty**

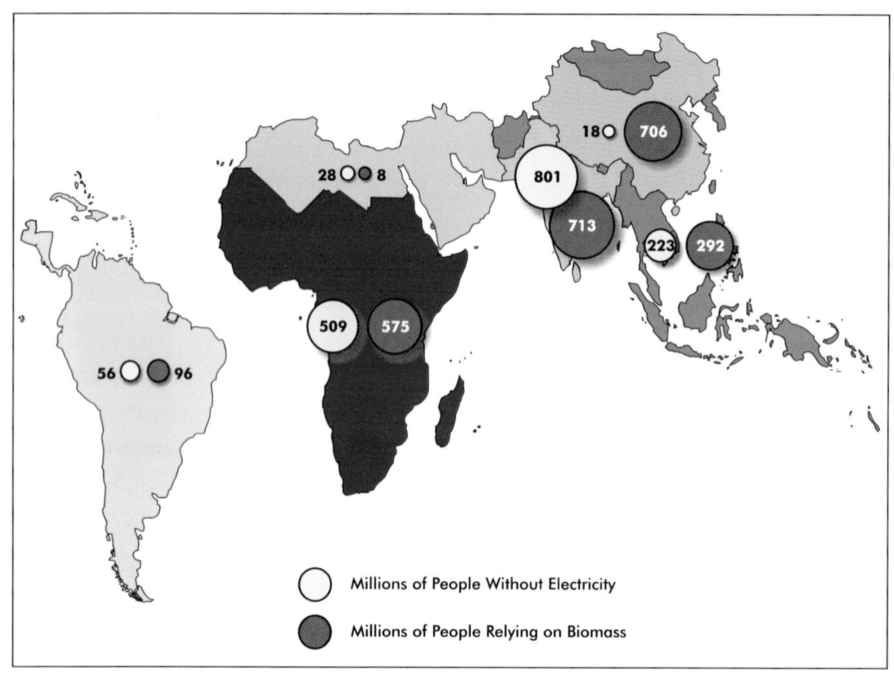

Source: IEA analysis.

***

The transition from energy poverty to relative affluence is a complex and irregular process, varying widely from nation to nation, village to village and family to family. In a general way, it is a journey from nearly exclusive reliance on traditional biomass to the access and use of electricity together with a range of other modern fuels. By 2030, about two billion people will have completed the trip, but more than a billion will still be stranded in primitive energy poverty.

It is a common misconception that electricity simply replaces biomass. In fact, most households use a wide mix of fuels as their income rises, combining biomass with kerosene or LPG to cook or with fuel oil to heat their homes. Nevertheless, traditional biomass and electricity do occupy contrasting positions in the fuel transition, and that is why this chapter concentrates mainly on them. Statistics and analysis of all the other fuels that play a part in the transition from energy poverty to energy affluence are provided in chapters on individual fuels and on regions in the World Energy Outlook.

***

Access to Electricity

To improve our understanding of the electrification process, we have built an extensive database with the best available information for most developing countries on how many people have access to electricity in their homes. The database is broken down by rural and urban areas. Annex 13.1 provides a detailed account of electrification rates for each country covered in the survey.

Aggregate data for 2000 show that the number of people without electricity today is 1.64 billion, or 27% of the world's population. More than 99% of people without electricity live in developing countries, and four out of five live in rural areas. The World Bank estimates that the number of people without electricity has fallen from 1.9 billion in 1970, but not on a straight-line decline, in 1990, the figure was 2 billion.[15] As a proportion of the world's population the number of unelectrified has fallen even more sharply — from 51% in 1970 to 41% in 1990.

The average electrification rate for the OECD, as well as for transition economies, is over 99%. Average electrification rates in the Middle East, North Africa, East Asia/China and Latin America are all above 85%. More than 80% of the people who currently have no access to electricity are located in South Asia and sub-Saharan Africa (Figure 13.4). Lack of electricity is strongly correlated to the number of people living below $2 per day (Figure 13.5). Income, however, is not the only determinant in electricity access. China, with 56% of its people still poor, has managed to supply electricity to more than 98% of its population (Box 13.2).

The rate of improvement in electricity access varies considerably among regions. Rapid electrification programmes in East Asia, especially China, account for most of the progress since 1970. Excluding East Asia/China, the number of people without electricity increased steadily from 1970 to 2000 (Figure 13.6).

Box 13.2: China's Electrification Success Story

China secured electricity access for almost 700 million people in two decades, enabling it to achieve an electrification rate of more than 98% in 2000.[16] From 1985 to 2000, electricity generation in China increased by

15. World Bank (1996).
16. Most sources confirm this electrification rate (See Annex 13.1).

nearly 1,000 TWh, 84% of it coal-fired, most of the rest hydroelectric. The electrification goal was part of China's poverty alleviation campaign in the mid-1980s. The plan focused on building basic infrastructure and on creating local enterprises. China's economy grew by an average annual 9.1% from 1985 to 2000. A key factor in China's successful electrification programme was the central government's determination and its ability to mobilise contributions at the local level. The electrification programme was backed with subsidies and low-interest loans. The programme also benefited from the very cheap domestic production of elements ranging from hydro generators down to light bulbs. China has avoided a trap into which many other nations have fallen: most Chinese customers pay their bills on time. If they do not, their connections are cut off.

This achievement dwarfs the efforts of any other developing country, but it conceals some serious shortcomings. China's transformation and distribution networks still need very large investment to meet modern standards. Electricity services are unreliable and of poor quality. Wiring and meters in homes and offices are undependable, even unsafe. Usage is low, especially in rural areas, where consumers tend to restrict their electricity use to lighting their homes.

Figure 13.4: **Electrification Rates by Region, 2000**

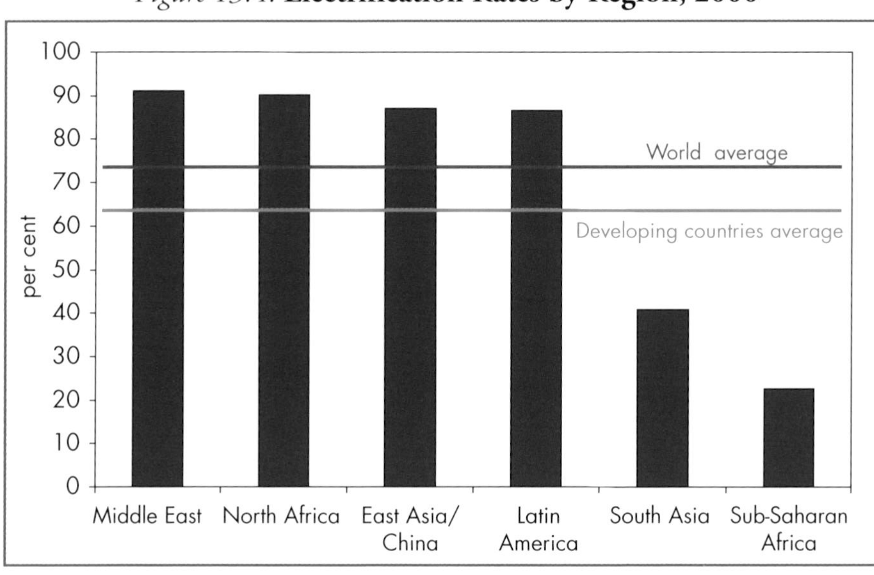

Source: IEA analysis.

Figure 13.5: **The Link between Poverty and Electricity Access**

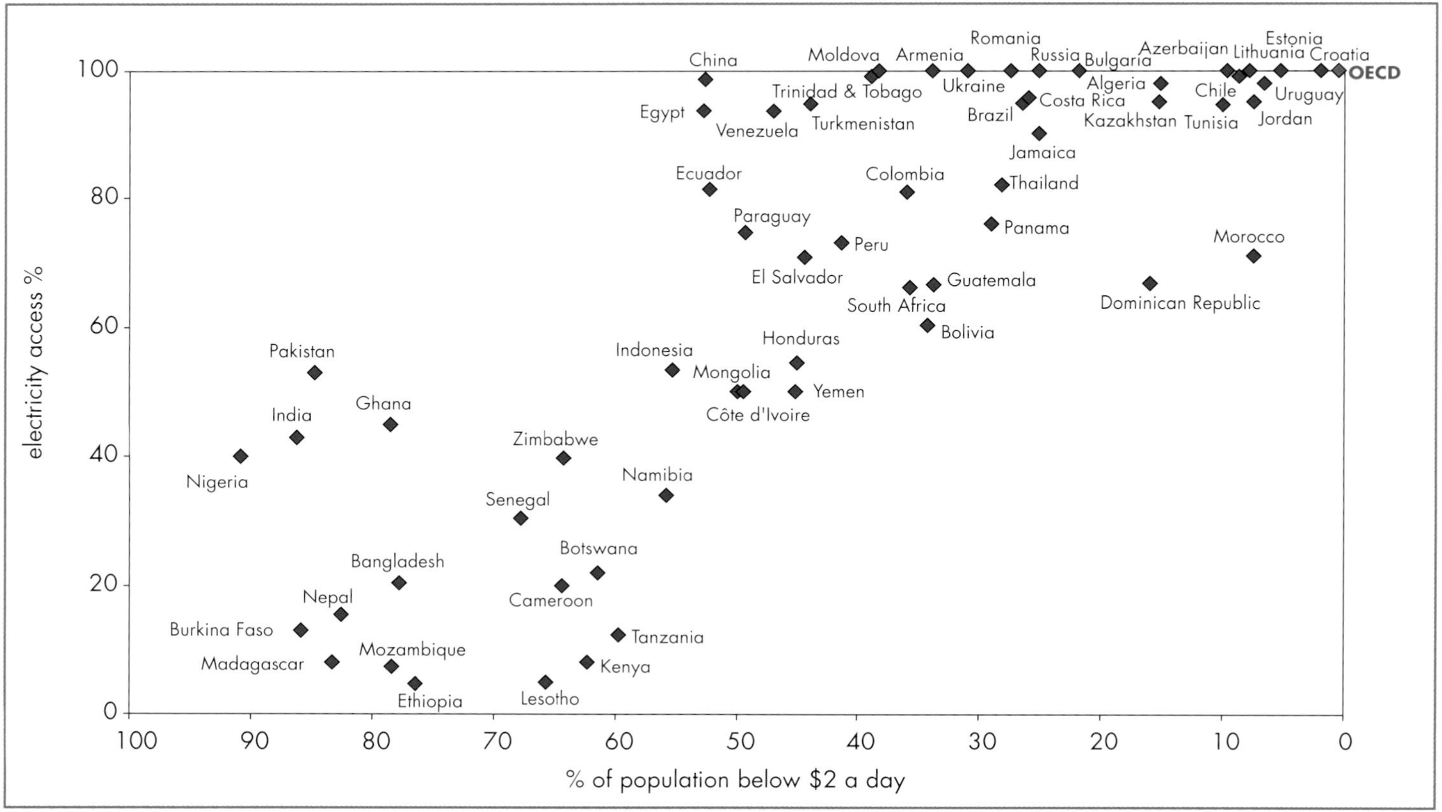

Note: Some transition economies and the OECD average are included for comparison purposes.

Source: IEA analysis; income statistics from the World Bank's *World Development Indicators*, 2001.

Figure 13.6: **Number of People without Electricity, 1970-2000**

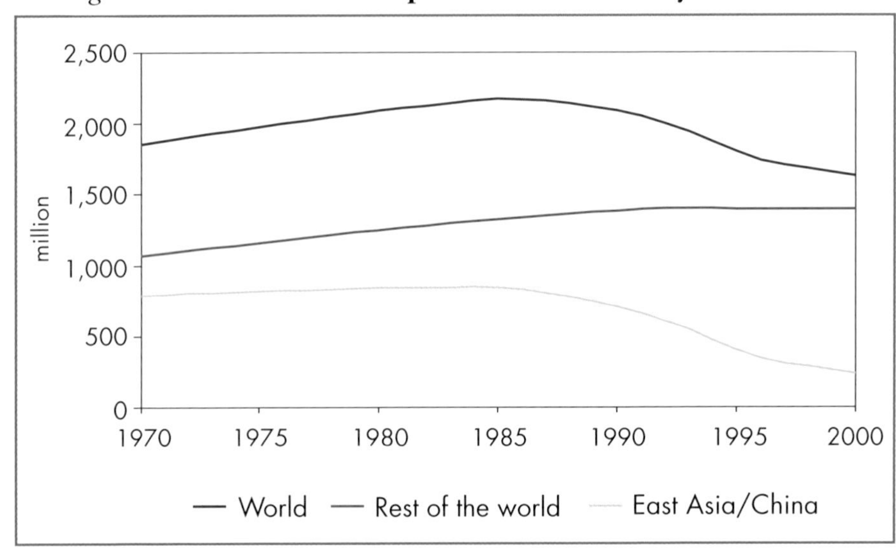

Source: IEA analysis.

With only 23% of its population electrified, sub-Saharan Africa has the lowest electrification rate of any major world region (Table 13.2). More than 500 million Africans are still without access to electricity. The region's poverty is one reason, but so is its low population density, which raises the cost of extending networks.

There are 580 million people lacking electricity in India. Although the electricity network there is technically within reach of 90% of the population, only 43% are actually connected because many poor people cannot afford the cost of connection. Even where incomes are high enough, households are often discouraged from connecting to the grid because of the poor quality of service, including frequent blackouts and brownouts.

Over the past three decades, half the growth in world population occurred in urban areas. Worldwide, electrification has kept pace with urbanisation, maintaining the number of the urban population without electricity at roughly 250 million. Put another way, the urban electrification rate increased from 36% in 1970 to 91% in 2000. The bulk of the urban unelectrified live in Africa and South Asia, where more than 30% of the urban population do not have electricity.

Table 13.2: **Urban and Rural Electrification Rates by Region, 2000** (%)

	Urban	Rural	Total
North Africa	99.3	79.9	90.3
Sub-Sahara	51.3	7.5	22.6
Africa	**63.1**	**16.9**	**34.3**
South Asia	68.2	30.1	40.8
Latin America	98.0	51.5	86.6
East Asia/China	98.5	81.0	86.9
Middle East	98.5	76.6	91.1
Developing countries	**85.6**	**51.1**	**64.2**
World	**91.2**	**56.9**	**72.8**

Four out of five people lacking access to electricity live in rural areas. This ratio has remained constant over the past three decades. The number of the rural unelectrified has fallen by more than 200 million, and rural electrification rose from 12% in 1970 to 57% in 2000.

In Africa, more than 83% of the population in rural areas still lack electricity. In sub-Saharan Africa, more than 92% of the rural population is unelectrified. The number of the people without electricity in this region has doubled in rural areas and tripled in urban areas in the last 30 years. In South Asia, 70% of the rural population has no electricity access.

At the rate of connections of the past decade, it would take more than 40 years to electrify South Asia and almost twice as long for sub-Saharan Africa.

Prospects for Electricity Access in Developing Countries[17]

The *WEO* Reference Scenario projections show that 1.4 billion people will still not have electricity in 2030, some 17% of the world's population, despite assumptions on more widespread prosperity and more advanced technology. The number of unelectrified will be 200 million less than today, in spite of an assumed increase in world population from 6.1 billion in 2000 to 8.2 billion in 2030. Since as much as 95% of population growth will take place in urban areas (Figure 13.7), urban electrification programmes will need to accommodate the swelling mass of the urban poor.

17. A new module has been added to the IEA's World Energy Model to generate projections of electrification rates. The projections are based on many factors, including incomes, fuel prices, demographic trends, technological advances and electricity consumption.

Figure 13.7: **Population Increase in Developing Countries**

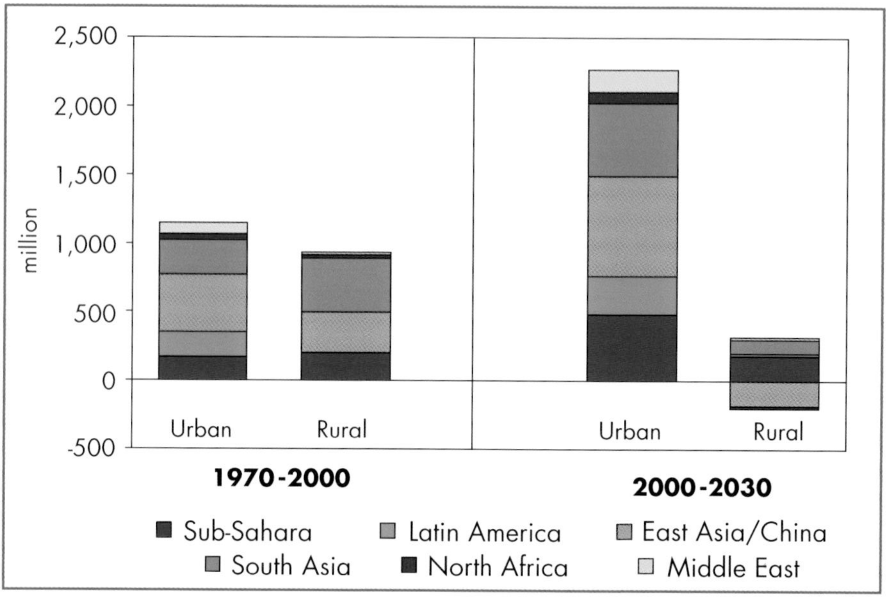

Source: UN Population Statistics for projections, World Bank and OECD for historical data.

Figure 13.8: **Number of People without Electricity, 1970-2030**

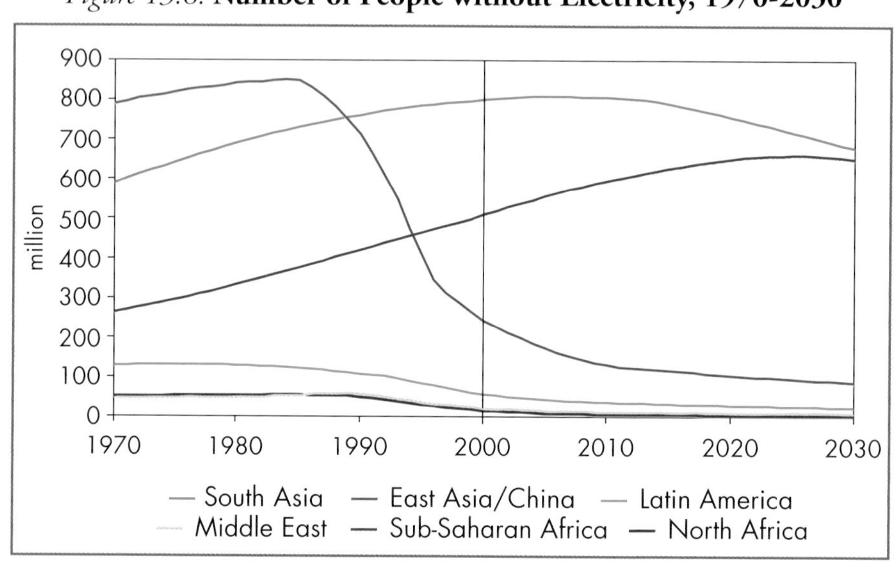

Source: IEA analysis.

Electrification rates and the number of unelectrified people will continue to diverge significantly among regions. Numbers of those without electricity will decline throughout the projection period in China and the rest of East Asia, North Africa, Latin America and the Middle East. They will continue to rise slightly in South Asia, turning down after 2010. They will peak in sub-Saharan Africa around 2025.

Most of the people without access to electricity in 2030 will still be in sub-Saharan Africa (650 million) and South Asia (680 million). In all other regions, including North Africa, the overall electrification rate will exceed 96% (Table 13.3) and will likely be close to 100% in urban areas. On average, 75 million people will gain access to electricity each year for the next 30 years (Figure 13.9).

Figure 13.9: **Annual Average Number of People Gaining Access to Electricity**

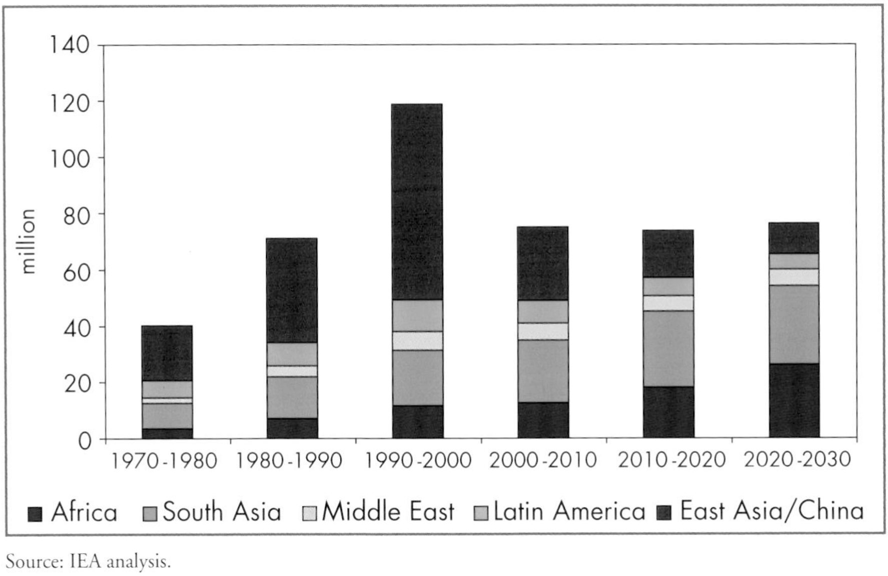

Source: IEA analysis.

In sub-Saharan Africa, the population without electricity will increase steadily until 2025 in line with projected trends in the number of people below the $2 per day poverty line.[18] The number of the unelectrified is expected to stabilise after 2025, mainly because of migration to towns and cities, where access to electricity will be easier. The region's population is projected to

18. See http: //www.developmentgoals.org/Sub-Saharan_Africa.htm for income projections.

double in the next three decades.[19] Three-quarters of the increase will occur in urban areas. By 2030, roughly half the population will have electricity.

The number of the unelectrified in South Asia is expected to peak and level off in this decade and start declining from then on. Affordability and reliability are the main factors in this region. Two out of three people will have access by 2030. If poverty were to be reduced faster than assumed in our analysis, electricity access would expand farther and faster.

These projections are highly dependent on assumptions about incomes and electricity pricing, which together determine the affordability of electric power, and about the rate of investment in expanding electricity supply.

Table 13.3: **Electrification Rates by Region** (%)

	1970	1990	2000	2015	2030
North Africa	34	61	90	98	99
Sub-Sahara	9	16	23	33	49
Africa	**14**	**25**	**34**	**43**	**56**
South Asia	17	32	41	53	66
Latin America	45	70	87	94	96
East Asia/China	30	56	87	94	96
Middle East	36	64	91	97	99
Developing countries	**25**	**46**	**64**	**72**	**78**
World	**49**	**60**	**73**	**78**	**83**

Investment in Electricity Supply Infrastructure

Investment requirements in developing countries fall broadly in these categories:

- additional generation capacity;
- extension of the electricity grid in urban areas;
- mini-grids in medium-sized settlements;
- decentralised installations providing thermal, mechanical and electric power in rural areas;
- maintenance and upgrading of existing infrastructure.

Over the next three decades the investment needed for new power generation capacity in developing countries will amount to $2.1 *trillion* – 10% in Africa and 15% in South Asia.[20] Even if this investment is secured

19. United Nations Population Division (2001).
20. See Chapter 3 for the global and regional electricity generation and investment outlooks.

over the next thirty years, 1.4 billion people will still lack access to electricity in 2030.The average cost of connection varies greatly among countries. In South Africa, the average cost in 2001 was $240, compared to $270 in Sri Lanka and over $1,000 in rural Kenya and Uganda.

Private investment in electricity projects in the developing world increased through most of the 1990s. It fell off, however, after the Asian financial crisis, from $46 billion in 1997 to less than $15 billion in 1999, then recovered to $30 billion in 2000.[21] Private investment in electricity projects in developing countries is concentrated in power generation, rather than transmission and distribution. It is also concentrated in a small number of countries. Only a quarter of *foreign* direct investment in energy goes to South Asia and Africa, while official development aid to Africa from foreign governments has fallen by some 6% a year since 1995.[22]

Since most sub-Saharan African and South Asian countries are strapped for cash, their governments will have to choose which electrification programmes to do first. Meeting the electricity needs of the urban poor costs much less per capita than meeting those of the rural poor. Therefore, providing electricity to the 230 million unelectrified urban people in the two regions is likely to be tackled first.[23] In many countries, strenuous efforts will be made to reduce transmission and distribution losses due to unauthorised connections, non-metering, non-payment of bills and technical malfunctions.

In rural areas, investment is likely to focus on harnessing indigenous energy sources, including fossil fuels, to drive productive, income-generating activities. Smaller, less capital-intensive investments are more likely to be funded and will benefit more poor people. The technology choice should be based on economics and on natural resource availability. Private utilities will not extend networks where it is unprofitable to do so, unless government subsidies make up for financial losses and provide a fair margin of profit. In remote areas, where the distance from the grid renders it too costly to connect communities to the national or regional grid, decentralised micro-projects are an option. (Table 13.4). What is needed is a comprehensive strategy, co-ordinating policies and programmes through which micro-credit, technology uptake and capacity building can take place.

21. These figures do not include investments in oil and upstream gas. (See *Private Participation in Infrastructure Database*, World Bank)
22. OECD (2002).
23. Future electrification investment needs will depend greatly on the level and pattern of rural/urban electrification trends. Detailed projections for rural/urban electrification are in progress and will be made available in the forthcoming study, *WEO 2003 Insights: Global Energy Investment Outlook.*

Table 13.4: **Examples of Off-Grid Power Plant Technologies**

Technology	Applications	Pros	Cons
Diesel engines	Water pumps Mills Refrigeration Lighting and communication	Easy maintenance Continuous energy services (24 hours a day) Allows for income-generating activities	High fuel costs Noxious and CO_2 emissions
Small biomass plants	Water pumps Mills Refrigeration Lighting and communication	Allows for income-generating activities Base load operation, continuous operation possible	Noxious emissions
Mini-hydro	Mills Lighting, communication and other	Long life, high reliability Allows for income-generating activities	Site-specific Intermittent Water availability
Wind	Lighting and communication Mills Pumps	No fuel cost	Expensive batteries Intermittent energy services
PV/Solar	Basic lighting and electronic equipment	No fuel cost	High capital costs High cost of battery replacement Needs further R&D

Frequent electricity blackouts and brownouts (euphemistically known as "load shedding") induce many poor consumers to maintain an alternative energy source. Kerosene is the standard replacement fuel for electricity. In the Indian state of Madhya Pradesh, over 90% of rural electrified households use kerosene as a backup fuel for lighting. In urban areas, "load shedding" is much less frequent and kerosene plays a lesser role. Low-quality electricity service imposes non-negligible costs on consumers, and it can undermine their willingness to pay for electricity. Where it is possible, "scheduled load shedding" is announced to consumers in advance to allow them to adjust and plan for it. Even predicted power outages disrupt activities and increase costs to the end user.

Across India, unauthorised power connections run as high as 20% to 40% of the total.[24] A third of all power produced in India gets stolen. Slums survive on stolen power. Tapped electric lines and poles are a common sight in Indian cities. In Angola, theft is roughly equal to 40% of total electric utility revenue.[25] Poverty drives people to steal electricity and boosts the number of unauthorised grid connections. The expected rise in urban population will exacerbate the problem. In India, half of the electricity supply is estimated to be unmetered. Unpaid bills are also an issue. Unmetered and unauthorised connections lead to very high electricity losses in many developing countries, compared with OECD countries, where losses are technical (Figure 13.10).

Figure 13.10: **Electricity Losses by Country**

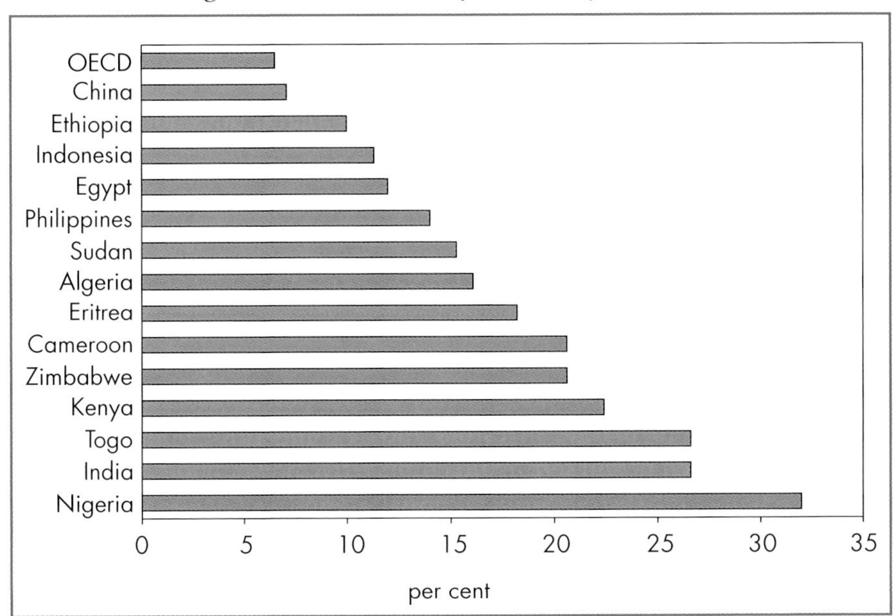

Source: IEA analysis.

Pricing

Even when the infrastructure to supply electricity to the poor exists, they are often unable to pay for it. There are two main barriers:

- the initial cost of connection to the grid and of inside wiring, which can be too high for poor households;

24. Tata Energy Research Institute (2002).
25. Angola News Agency.

- monthly charges for units consumed; since many poor people do not earn a regular wage, it is hard for them to keep up monthly payments for electricity.

As a result, most developing countries subsidise electricity to households, since the benefits subsidies provide are judged to exceed the long-term costs to government. In many countries, the size of the subsidies and the way they are delivered result in heavy losses of economic efficiency, wasteful habits on the part of consumers and adverse environmental effects.[26]

One way to improve this situation is to have the government pay *part of the capital cost of connection,* or have the utility company spread the connection charge out over several months. Another approach is the so-called "lifeline rate," a special subsidy for poor families — with "poverty" defined by both household income and electricity use. The lifeline-rate system avoids a number of the pitfalls of other forms of subsidies, but it is hard to design in such a way that it does not benefit the rich even more than the poor.

Market Reforms

Energy industries in most developing countries are in urgent need of reform. Several countries have begun the process, but in widely varying ways. These include the commercialisation and privatisation of state-owned utilities, unbundling energy production from its distribution, opening markets to private investors and revising price policies. A few countries have even begun setting up competitive power pools. One aim of all these schemes is to attract private capital, in one way or another, into an impoverished energy sector.

Competition is the major theme in OECD reform programmes of the power sector. This may not be the first priority in developing countries. Typically in these countries, the prerequisites for reform are often weak or lacking. Utilities are poor or bankrupt, the institutional framework for investment is non-existent and energy networks are underdeveloped.

Reform strategies in developing countries should address the issue of sustainable financing. Prices charged to consumers must cover the cost of producing and distributing energy. But achieving that end can be very difficult in the world's poorest countries. Subsidies in the form of low-cost energy for consumers are rife. Pilfering, unauthorised grid connections and

26. See IEA (1999).

the non-payment of utility bills can be the rule, not the exception. In India, electricity revenues regularly run at about three-quarters of actual costs.[27]

In these circumstances, innovative thinking and the proper sequencing of events are vital to successful reform. Indeed, there are some cases in which systematic subsidisation for a limited period may be more desirable than attempting to charge full economic prices overnight. As energy industries in the developing world face these challenges, a degree of public support may be essential.

In countries like India, the necessary process of raising prices to market levels is bound to be politically and socially difficult. There is evidence, however, that even very poor families are willing to pay for reliable energy services.

One of the highest barriers to new investment in these countries is the perceived threat of "regulatory risk." If the regulatory framework is perceived as unpredictable or incompetent, investors tend to keep their money in their pockets. A striking example occurred in Argentina earlier this year, when the regulator cut consumer prices after a large foreign investor had built a plant on the basis of the previous tariff lists.

Further complications arise from the lack of domestic capital markets, which can often mean high interest rates, with built-in risk premiums, to offset expected currency volatility and inflation.

The ultimate objective of reform should be an industry structure allowing a sufficient number of players to compete on equal terms, with the monopoly power of established entities truly constrained. This is feasible in most developed countries, but it may be an unrealistic early objective in the developing world. The tendency may be to privatise first and reform industry structures later on. In some cases, there may not be much choice. Most developing countries have inadequate gas and electricity grids, and the deficiency must obviously be made good before real competition can occur. But, in that event, close regulation will be needed to avoid exploitation of customers and to achieve wider electrification.

International lenders and major financial institutions have learned many lessons the hard way. Major lenders have switched away from energy *projects* to energy *programmes* and sector reforms aimed at paving the way for private-sector participation and competition. And there is a new emphasis on the proper *sequencing* of reform steps, which can be expected to vary from country to country.

27. IEA (2002).

Biomass Use

Poor people in developing countries rely heavily on traditional biomass for cooking and heating.[28] Because the *share* of biomass use in final consumption has been decreasing since 1994, the earliest year for which comprehensive IEA data are available[29], it is easy to get the impression that it is being replaced by other fuels and is gradually being phased out. New industries, transport and power generation often *do* use fossil fuels. But wood and other biomass continue to be used in many industries in developing countries and for cooking and heating.[30]

In developing countries, some 2.4 billion people rely on traditional biomass for cooking and heating. The 2000 estimates presented in Table 13.5 are based on an analysis of rural and urban biomass consumption in developing countries. While the people included in the table may or may not have access to electricity for lighting, they generally lack access to modern energy services for cooking and heating.

Over half of all people relying on biomass for cooking and heating live in India and China, but the *proportion* of the population depending on biomass is heaviest in sub-Saharan Africa. Extreme poverty and the lack of access to other fuels mean that 80% of the overall African population relies primarily on biomass to meet its residential needs. In Kenya, Tanzania, Mozambique and Zambia, nearly all rural households use wood for cooking, and over 90% of urban households use charcoal. In Indonesia, nearly all rural households use wood for cooking. In East Asia, the heaviest biomass use occurs in the Philippines, Thailand, Myanmar and Vietnam. In China, the government has discouraged people from using straw and other agricultural waste for fuel. Nevertheless, in rural areas, many families do still rely on biomass. In the past, coal replaced biomass for heating and cooking in China's cities. Poor people in many Central American countries, especially Guatemala, Honduras, Nicaragua and Haiti, rely on wood for cooking and heating.

28. See definition in Annex 13.1. Biomass use here is mostly non-commercial.

29. Despite the importance of biomass in energy demand in developing countries, only recently has it been included in global energy statistics. This is mainly because national biomass statistics have been incomplete or of poor quality. Biomass use can vary widely within a single country, and statistics based on a small number of villages can be misleading when extrapolated to the national level. The IEA has created a specialised database on biomass, with historic data from 1994. Projections for biomass energy demand by region/country are in Annex 13.1.

30. Hulscher (1997) finds that there is no inverse correlation between per capita consumption of biomass and GDP per capita for countries in South and South-East Asia. As GDP per capita rises, people in these countries are likely to go on using biomass, and complement it with other fuels.

Table 13.5: **Number of People Relying on Traditional Biomass for Cooking and Heating in Developing Countries, 2000**

	million	% of total population
China	706	56
Indonesia	155	74
Rest of East Asia	137	37
India	585	58
Rest of South Asia	128	41
Latin America	96	23
North Africa/Middle East	8	0.05
Sub-Saharan Africa	575	89
Developing countries	**2,390**	**52**

Biomass consumption in any country is largely a function of that country's relative poverty (Figure 13.11). But other factors come into play as well, including the country's degree of urbanisation and what fuels are readily available. Biomass use can vary sharply between two countries with similar poverty levels. Costa Rica has a similar number of people living below the poverty line as Thailand, but because of the ample availability of hydroelectric power, many Costa Ricans use electricity for cooking and heating. In Eastern Europe and in Russia, modern fuels are available and a higher percentage of people live in cities. When more than half of the population has an income of less than $2 per day, residential energy demand tends to be dominated by biomass.

In 2000, biomass accounted for over 70% of residential energy consumption in developing countries (Figure 13.12). The amount of biomass consumed and the form it takes differ among regions and even within countries, depending on resource availability, the accessibility of commercial fuels, cultural preferences and incomes. In the majority of sub-Saharan African countries, 80% to 90% of the residential energy needs of low-income households are met by fuelwood or charcoal. In South Asia, dung and agricultural residues account for over half of biomass energy use. In China, agricultural residues alone make up half of biomass energy use. In Latin American countries, biomass use is predominately made up of fuelwood for cooking and heating.

Figure 13.11: **The Link between Poverty and Share of Traditional Biomass in Residential Energy Consumption**

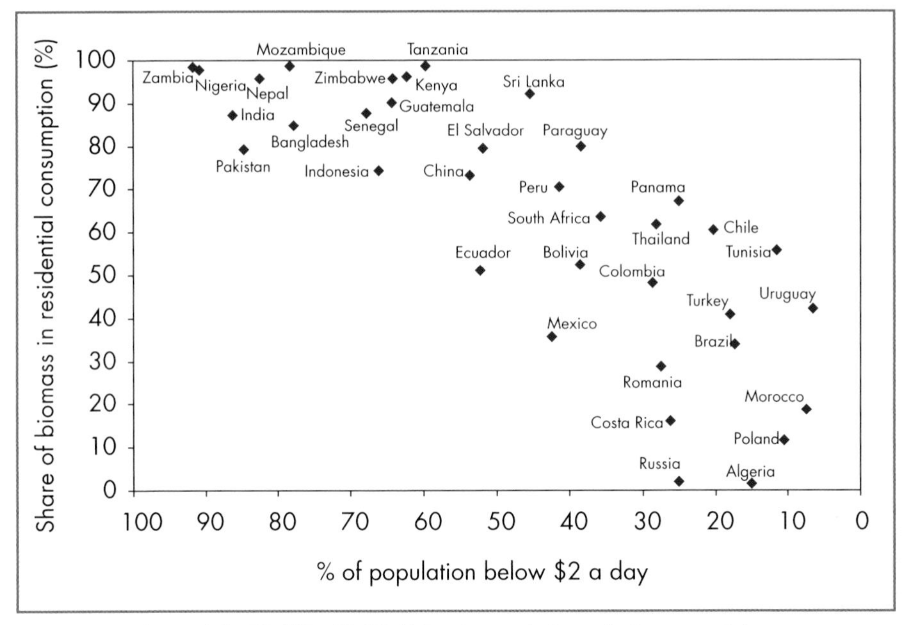

Source: IEA analysis and the World Bank's *World Development Indicators* for income statistics.

Figure 13.12: **Share of Traditional Biomass in Residential Energy Consumption, 2000**

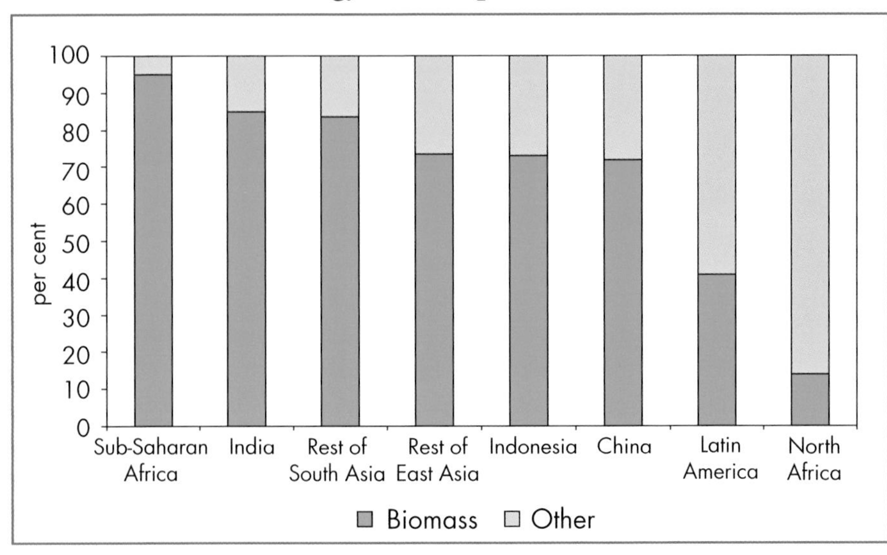

Source: IEA analysis.

Rural and urban consumption patterns also differ. But some general observations can be made:

- Biomass consumption is usually higher in rural areas, where fuelwood is more likely to be gathered than bought commercially.
- In urban areas, biomass competes with commercial energy sources, and consumption patterns are largely determined by relative fuel prices.
- Much of the fuelwood used is collected from the roadside or vacant lots. Very little comes from natural forests.
- Cooking stoves used in rural households are generally of poorer quality than those used in cities.
- In peri-urban areas, poor people are likely to use whatever waste is available as fuel for cooking and heating.

Rural industry in developing countries relies heavily on biomass for fish-smoking, brick-making, tobacco curing, food processing, furniture-making, ceramics and bakeries. These industries employ tens of thousands of people, and the income they produce frees many rural households from poverty. Enormous potential exists to improve the efficiency of industrial ovens, dryers and bakeries that run on biomass.[31]

Prospects for Biomass Use in Developing Countries

Expanded access to electricity, which low-income households use primarily for lighting, is unlikely to reduce the demand for biomass in many countries in Africa and South Asia. In thirty years' time, biomass will also continue to represent the largest share of residential energy demand in some countries in East Asia. In many countries, biomass consumption will continue to increase in absolute terms. The fact that the number of people relying on biomass will remain high over the projection period is not, in itself, a cause for concern. It is the way in which biomass is used, the technologies and applications used in its combustion, which links biomass use to poverty in many developing countries. This makes it urgent to improve the efficiency of biomass use in order to alleviate the adverse impact on health (see Box 13.1) and other damaging effects.

In the *WEO* Reference Scenario, residential biomass demand in developing countries is projected to rise from 723 Mtoe in 2000 to 788 Mtoe in 2030. The main drivers are macroeconomic and demographic

31. See World Energy Council/FAO (1999) and www.rwedp.org for more information on industrial biomass use in developing countries.

variables, commercial energy use, technology and fuel prices. The availability of biomass resources is also taken into account.[32]

In Africa, biomass will still account for 80% of residential energy use in 2030. In East Asian countries, excluding China, it will account for over 50% of residential consumption in 2030. Biomass use in the residential sector in Indonesia will decline in the last two decades of the projection period, accounting for 45% of residential demand in 2030.[33] In South Asia, the share of biomass use will remain high, at nearly 70% in 2030. In developing countries as a whole, biomass will represent 53% of residential energy consumption in 2030, down from 73% in 2000.

The share of the world population relying on biomass for cooking and heating is projected to decline in most developing regions. But the *total number of people* will rise, mainly due to increases in the number of people relying on biomass in South Asia and sub-Saharan Africa. Over 2.6 billion people in developing countries will continue to rely on biomass for cooking and heating at the end of the projection period.[34] That is an increase of 238 million, or 9% (Table 13.6). In China and Indonesia, the number of people using traditional biomass for cooking and heating will decline. Vigorous government measures to encourage the use of agricultural waste for power generation rather than for cooking will lower the share in China.[35] In much of Latin America, rising incomes, improved availability of modern energy and urbanisation will reduce demand for biomass, although it will likely remain the dominant fuel in the poorer countries of Central America.

Because biomass will continue to dominate energy demand in developing countries in the foreseeable future, the development of more efficient biomass technologies is vital for alleviating poverty, creating employment and expanding rural markets. Modern biomass technologies compete with conventional technologies in many applications, and the room for improving the use of biomass in developing countries is immense.[36] Biomass is routinely listed along with other "renewable" energy

32. See Annex 13.1 for regional projections. Regional assumptions regarding wood fuel availability and biomass utilisation issues were made in co-operation with the Food and Agriculture Organization of the United Nations (FAO-UN).
33. See Chapter 11 for more detail on biomass energy demand in Indonesia.
34. The energy demand projections on which these estimates are based include technological parameters that increase the efficiency of biomass use. The estimates of the number of people relying on biomass for cooking and heating are based on the assumption that biomass demand per capita in each region is constant over the *Outlook* period at 2000 levels. The constant per capita use is a conservative assumption, which defines the lower limit of the number of people who rely on biomass for cooking and heating.
35. Ping (2001).
36. See United Nations Development Program (2000) and IEA (2001).

sources, and the impression persists that it is a "free good" that will continue to be available indefinitely. That, however, is not the case. Urbanisation and industrialisation strain the availability of biomass resources. Peri-urban areas, in particular, will come under increasing pressure to meet demands for biomass energy and agricultural land. Biomass scarcity will worsen living conditions in poor neighbourhoods, by forcing residents to use lower-quality waste as cooking fuel. Rising demand for commercially-traded fuelwood in towns and cities will put pressure on supplies in nearby rural areas. As rural supplies become monetised, traditional "free" sources will diminish.

Table 13.6: **Number of People Relying on Biomass for Cooking and Heating in Developing Countries** (million)

	2000	2030*	2000-2030 (%)
China	706	645	-9
Indonesia	155	124	-25
Rest of East Asia	137	145	6
India	585	632	7
Rest of South Asia	128	187	32
Latin America	96	72	-33
Africa	583	823	27
Developing countries	**2,390**	**2,628**	**9**

* Assuming that biomass use per capita is constant, at some 0.3 toe per capita, over the projection period. This figure is an average across all regions and countries. Analysis indicates that average per capita biomass use varies between some 0.24 toe in South Asia to nearly 0.4 toe in many countries in East Asia. See Tata Energy Research Institute et al. (1999) for a similar approach.

In some rural areas, people must go farther and farther afield to gather fuelwood. Radar imagery, which shows deforestation along roads and trails leading to villages, is presented in Figure 13.13. The figure gives a rough idea of how biomass has been depleted around villages in Central Africa.

In many urban and suburban areas of mega-cities in developing countries, there is virtually no wood left to scavenge — or what is left is many miles away. Populations in slums are growing fast because of the influx of people from rural areas attracted by the perception of opportunities that the cities offer. Some of them resort to fuelwood gathering even in the cities. Currently in Africa, there are two cities with more than 10 million people,

Lagos and Cairo. By 2015, Lagos will have a population of over 23 million and Cairo's population may reach 14 million. With an increasing number of people living in marginalised conditions in mega-cities in developing countries, the use of traded wood fuels is bound to increase.

Figure 13.13: **Radar Image of the "Charcoal Web" in Central African Republic**

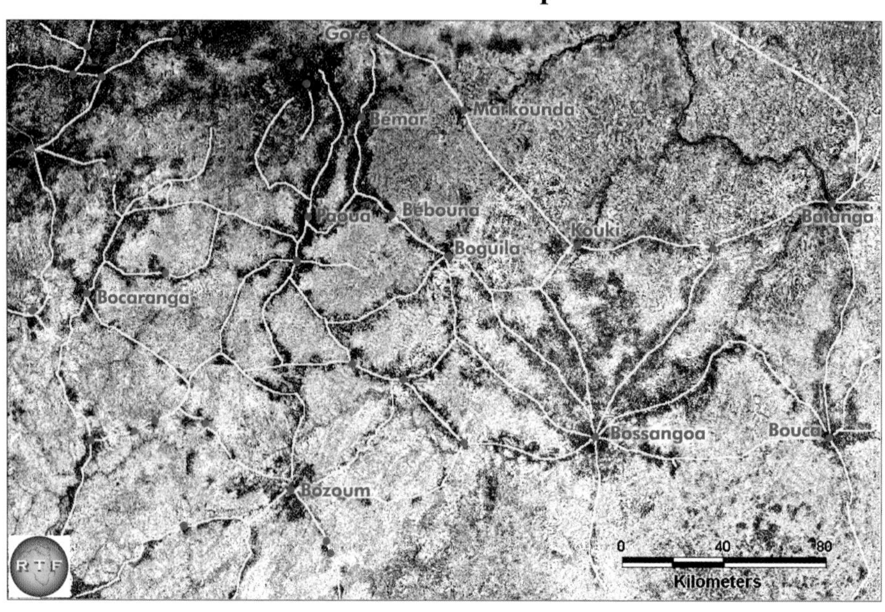

Note: This enhanced radar image shows swathes of deforested land, in darker colour, along the roads and tracks leading out of towns and villages in northwest Bangui in the Central African Republic. The "charcoal web" refers to the deforestation swathes. Most of the wood is used to make charcoal for use by village households.
Source: Courtesy: NASDA; Interpretation: Radar Technologies France.

As a result, poverty proliferates and quality of life erodes even further. One example is Delhi, where sprawling slums have proliferated.[37] When wood becomes scarce, the better-off switch to kerosene. But very poor families have to collect virtually anything lying about — twigs and scraps from construction sites — to heat, cook and light their homes. Together with the low ceilings, cramped quarters and tainted water of the Delhi slums, the noxious fumes from inefficient combustion of this mix of fuels pose a direct threat to the

37. Bhasin (2001).

residents' health. Poverty is also severe in many African cities. In Dar es Salaam, the poor often cannot afford to buy wood for cooking, instead, they collect various sawmill- and agro-residues, twigs from city trees and wastes from public dumps.[38]

Our projections for biomass use take account of expectations for biomass supply. But the link between availability of biomass resources and demand pressures is still not very well understood. The projections are meant to provide a message about possible outcomes of biomass use. The precise numbers are less important than the message they portend.

38. Katyega (2001).

Annex 13.1
EXPLANATORY NOTES

Contents

Definitions

Electricity Access

There is no single internationally accepted definition for electricity access. The definition used here covers electricity access at the household level, i.e. the number of people that have electricity in their home. It comprises commercially sold electricity, both on-grid and off-grid. It also includes self-generated electricity, for those countries where access to electricity has been assessed through surveys by government or government agencies. The data do not capture unauthorised connections.

The main data sources are listed in the tables. Each data point has been validated through a consistency-check process among different data sources and experts. The electrification rate shown in this annex is the number of people with electricity access as a percentage of total population. Rural and urban electrification rates have been collected for most countries. Only the regional averages are shown in this publication. Enquiries about statistics should be addressed to the Economic Analysis Division (http://www.worldenergyoutlook.org).

Secretariat Estimates

Where country data appeared contradictory, outdated or unreliable, the IEA Secretariat made estimates based on cross-country comparisons, earlier surveys, information from other international organisations, annual statistical bulletins, publications and journals.

Population and Urban/Rural Breakdown

Projections are from the *World Urbanisation Prospects – The 2000 Revision 1,* published by the United Nations Population Division. Historical data are from *World Bank Development Indicators, 2001.*

Biomass

Biomass comprises solid fuels (wood, charcoal, wood wastes and agricultural residues and dung), gas (biogas, landfill gas, and other gases from biomass), liquid fuels (alcohols, bio-additives and other liquid fuels) and industrial and municipal waste. Data on biomass consumption are from IEA statistics, *Energy Balances of Non-OECD Countries, 2002 edition.* UN-FAO data are used for information on forest coverage and estimates of biomass supply. In the chapter, traditional biomass refers mainly to non-commercial biomass use, which is largely solid fuels.

Table 13.A1: **Electricity Access in 2000 - Regional Aggregates**

	Electrification rate %	Population without electricity million	Population with electricity million
World	**72.8**	**1644.5**	**4390.4**
Developing countries	**64.2**	**1634.2**	**2930.7**
Africa	34.3	522.3	272.7
Developing Asia	67.3	1041.4	2147.3
Latin America	86.6	55.8	359.9
Middle East	91.1	14.7	150.7
Transition economies	**99.5**	**1.8**	**351.5**
OECD*	**99.2**	**8.5**	**1108.3**

*OECD figures aggregate some important regional variations. The electrification rate for Turkey and Mexico is about 95 %. All other Member countries have 100% electrification.

Table 13.A1: **Electricity Access in 2000 - Africa**

	Electrification rate %	Population without electricity million	Population with electricity million	Sources
Angola	12.0	11.5	1.6	Empresa de Distribuicao de Electricidade Angola EDEL (2001). AFREPREN (2001)
Benin	22.0	4..9	1.4	World Bank (2000)
Botswana	22.0	1.2	0.3	Botswana Power Company (2000), AFREPREN (2001)
Burkina Faso	13.0	10.4	1.6	Société Nationale d'Electricité SONABEL (2001), OECD Development Centre (2001)
Cameroon	20.0	11.9	3.0	Société Nationale d'Electricité SONEL (2001), OECD Development Centre (2001)
Congo	20.9	2.4	0.6	Marchés Tropicaux (2001)
Cote d'Ivoire	50.0	8.0	8.0	Compagnie Ivoirienne d'Electricité CIE (2002), OECD Development Centre (2001)
D. R. of Congo	6.7	47.5	3.4	Marchés Tropicaux (2002)
Eritrea	17.0	3.4	0.7	AFREPREN (2001)
Ethiopia	4.7	61.3	3.0	Ethiopian Electric Power Corporation EEPCO (2001)
Gabon	31.0	0.8	0.4	D. G. de l'Economie - République Gabonaise (1998), OECD Development Centre (2001)
Ghana	45.0	10.6	8.7	Electricity Company of Ghana (2000)
Kenya	7.9	27.7	2.4	Kenya Power & Lighting Company Limited (2001), AFREPREN (2001)
Lesotho	5.0	2.0	0.1	Marchés Tropicaux (2002)
Madagascar	8.0	14.3	1.2	Jiro Sy Rano Malagasy JIRAMA (2000)
Malawi	5.0	9.9	0.5	National Economic Council, Malawi Government (2001), AFREPREN (2001)
Mauritius	100	0.0	1.2	Central Statistics Office of Mauritius CSO (2001), AFREPREN (2001)
Mozambique	7.2	16.4	1.3	AFREPREN (1998)
Namibia	34.0	1.2	0.6	AFREPREN (2001)
Nigeria	40.0	76.1	50.8	National Electric Power Authority NEPA (2001), AFREPREN (2001)

Senegal	30.1	6.6	2.9	Agence Sénégalaise pour l'électrification rurale (2002), OECD Development Centre (2001)
South Africa	66.1	14.5	28.3	National Electricity Regulator (2002), Eskom (2002), Deparment of Minerals and Energy (2001)
Sudan	30.0	21.8	9.3	AFREPREN (2001)
Tanzania	10.5	30.2	3.5	Tanzania Electric Supply Company Limited (2001), Bureau of Statistics Tanzania (2000)
Togo	9.0	4.1	0.4	Compagnie d'énergie électrique du Togo (CEET) (2001), AFREPREN (2001)
Uganda	3.7	22.5	0.9	Uganda Electricity Board (2002), AFREPREN (2001)
Zambia	12.0	8.9	1.2	AFREPREN (2001)
Zimbabwe	39.7	7.6	5.0	AFREPREN (2001)
Other Africa	7.7	71.1	5.9	Secretariat estimate
Sub-Sahara	**22.6**	**508.9**	**148.2**	
Algeria	98.0	0.6	29.8	Société Nationale de l'Électricité et du Gaz (2001)
Egypt	93.8	4.0	60.0	International Private Power Quarterly (2002)
Libya	99.8	0.01	5.3	World Bank (2000)
Morocco	71.1	8.3	20.4	Ministère de l'Industrie et du Commerce, de l'Energie et des Mines (2002)
Tunisia	94.6	0.5	9.1	Ministry of Energy and Mining (2000)
North Africa	**90.3**	**13.4**	**124.6**	
Africa	**34.3**	**522.3**	**272.7**	

Table 13.A1: **Electricity Access in 2000 - Developing Asia**

	Electrification rate %	Population without electricity million	Population with electricity million	Sources
China	98.6	17.6	1244.9	Chinese Energy Research Institute ERI (2002), OECD 2002 (China's Western Development Strategy), USAID
Brunei	99.2	0.003	0.3	APERC (2002)
Cambodia	15.8	10.3	1.9	Ministry of Planning (1998)
Chinese Taipei	98.6	0.3	21.9	Secretariat estimate
DPR of Korea	20.0	17.8	4.5	World Bank (2000)
Indonesia	53.4	98.0	112.4	PLN (2002), ASEAN Center for Energy (2002)
Malaysia	96.9	0.7	22.6	8th Malaysia Plan, Malaysian Government (2000)
Mongolia	90.0	0.3	2.4	World Bank (2000)
Myanmar	5.0	45.3	2.4	World Bank (2000)
Philippines	87.4	9.5	66.1	Rural Electric Cooperatives, Manila Electric Company, DoE Philippines, APERC (2002)
Singapore	100	0.0	4.0	World Bank (2000)
Thailand	82.1	10.9	49.8	Department of Energy Development and Promotion of Thailand, APERC (2002)
Vietnam	75.8	19.0	59.5	APERC (2002)
Other Asia	15.5	10.9	2.0	Secretariat estimate
East Asia/China	**86.9**	**240.7**	**1594.6**	
Afghanistan	2.0	25.4	0.5	Secretariat estimate
Bangladesh	20.4	104.4	26.7	Bangladesh Power Development Board, Agency for International Development (2002)
India	43.0	579.1	436.8	Tata Energy Research Institute (2002), Indian Planning Commission (2002)
Nepal	15.4	19.5	3.5	Nepal Electricity Authority (2000)
Pakistan	52.9	65.0	73.1	Water Power Development Authority of Pakistan (2000)
Sri Lanka	62.0	7.4	12.0	Bureau of Infrastructure Investment (2001)
South Asia	**40.8**	**800.7**	**552.7**	
Developing Asia	**67.3**	**1041.4**	**2147.3**	

Table 13.A1: **Electricity Access in 2000 - Latin America**

	Electrification rate %	Population without electricity million	Population with electricity million	Sources
Argentina	94.6	2.0	35.0	OLADE (1998)
Bolivia	60.4	3.3	5.0	OLADE (1999)
Brazil	94.9	8.7	161.7	OLADE (1999)
Chile	99.0	0.2	15.0	OLADE (1999)
Colombia	81.0	8.0	34.3	OLADE (1999)
Costa Rica	95.7	0.2	3.6	Electricidad y Telecomunicaciones para Costa Rica (2002)
Cuba	97.0	0.3	10.9	Secreteriat estimate based on OLADE
Dominican Republic	66.8	2.8	5.6	World Bank (2000)
Ecuador	80.0	2.5	10.1	National Electricity Council of Ecuador CONELEC (2000)
El Salvador	70.8	1.8	4.5	OLADE (1998)
Guatemala	66.7	3.8	7.6	OLADE (1999)
Haiti	34.0	5.3	2.7	OLADE (1999)
Honduras	54.5	2.9	3.5	OLADE (1998)
Jamaica	90.0	0.3	2.3	Secretariat estimate based on OLADE
Netherlands Antilles	99.0	0.0	0.2	Secretariat estimate
Nicaragua	48.0	2.7	2.4	Nicaraguan National Energy Commission(2000), OLADE (1999)
Panama	76.1	0.7	2.2	Secreteriat estimate based on OLADE
Paraguay	74.7	1.4	4.1	OLADE (1997)
Peru	73.0	6.9	18.8	Estadistica Electrica - PERU Ministry of Energy (2000)
Trinidad and Tobago	99.0	0.01	1.3	Secreteriat estimate based on OLADE
Uruguay	98.0	0.1	3.2	OLADE (1999)
Venezuela	94.0	1.5	22.7	OLADE (1999)
Other Latin America	86.6	0.5	3.1	Secretariat estimate
Latin America	**86.6**	**55.8**	**359.9**	

Table 13.A1: **Electricity Access in 2000 - Middle East**

	Electrification rate %	Population without electricity million	Population with electricity million	Sources
Bahrain	99.4	0.0	0.6	Ministry of Electricity and Water (1999)
Iran	97.9	1.3	62.4	Iran Power Generation & Transmission Company TAVANIR (1999)
Iraq	95.0	1.2	22.1	World Bank (2000)
Israel	100	0.0	6.2	Israel Electric Corporation Ltd. (1997)
Jordan	95.0	0.2	4.7	World Bank (2000)
Kuwait	100	0.0	2.0	Secreteriat estimate
Lebanon	95.0	0.2	4.1	World Bank (2000)
Oman	94.0	0.1	2.3	Secretariat estimate
Qatar	95.0	0.0	0.7	Secretariat estimate
Saudi Arabia	97.7	0.5	20.2	Ministry of Industry and Electricity (1999/2000)
Syria	85.9	2.3	13.9	UNDP (2000)
United Arab Emirates	96.0	0.1	2.8	Secretariat estimate
Yemen	50.0	8.8	8.8	World Bank (2000)
Middle East	**91.1**	**14.7**	**150.7**	

Table 13.A2: **Urban and Rural Electrification Rates, 2000**

	Total population million	Urban population million	Electrification rate %	Urban electrification rate %	Rural electrification rate %
North Africa	138	74	90.3	99.3	79.9
Sub-Sahara	657	226	22.6	51.3	7.5
Africa	**795**	**300**	**34.3**	**63.1**	**16.9**
South Asia and sub-Sahara	**2010**	**608**	**34.9**	**61.9**	**23.2**
Latin America	416	314	86.6	98.0	51.5
East Asia/China	1835	633	86.9	98.5	81.0
South Asia	1353	381	40.8	68.2	30.1
Middle East	165	109	91.1	98.5	76.6
Developing countries	**4565**	**1739**	**64.2**	**85.6**	**51.1**
World	**6035**	**2828**	**72.8**	**91.2**	**56.9**

Table 13.A3: **Projections of Biomass Energy Demand in Developing Countries**

	Levels (Mtoe)				Shares* (%)				Growth rates (% per annum)		
	2000	**2010**	**2020**	**2030**	**2000**	**2010**	**2020**	**2030**	**2000-2010**	**2000-2020**	**2000-2030**
India											
Demand	198	216	224	227	40	34	28	23	0.9	0.6	0.5
of which residential	*175*	*189*	*193*	*190*	*84*	*79*	*74*	*67*	*0.8*	*0.5*	*0.3*
Rest of South Asia											
Demand	43	52	60	65	44	38	32	26	2.0	1.7	1.4
of which residential	*38*	*47*	*53*	*56*	*84*	*78*	*73*	*67*	*2.0*	*1.7*	*1.3*
China											
Demand	213	213	208	195	18	14	11	8	0.0	-0.1	-0.3
of which residential	*212*	*212*	*206*	*193*	*72*	*64*	*56*	*47*	*0.0*	*-0.1*	*-0.3*
Indonesia											
Demand	47	50	46	40	32	25	18	13	0.7	0.0	-0.5
of which residential	*47*	*49*	*44*	*37*	*73*	*66*	*56*	*45*	*0.5*	*-0.2*	*-0.7*
Rest of East Asia											
Demand	62	70	77	83	16	13	11	10	1.2	1.1	1.0
of which residential	*41*	*43*	*44*	*44*	*74*	*65*	*58*	*51*	*0.3*	*0.3*	*0.2*
Africa											
Demand	254	295	321	345	51	46	40	34	1.5	1.2	1.0
of which residential	*181*	*214*	*233*	*247*	*89*	*86*	*81*	*75*	*1.7*	*1.3*	*1.0*
Brazil											
Demand	43	44	47	50	24	19	16	13	0.4	0.5	0.7
of which residential	*7*	*4*	*3*	*2*	*34*	*17*	*10*	*5*	*-5.2*	*-4.7*	*-4.6*
Rest of Latin America											
Demand	37	40	44	50	13	11	9	8	0.7	0.8	1.0
of which residential	*21*	*20*	*20*	*19*	*44*	*37*	*30*	*24*	*-0.3*	*-0.2*	*-0.2*
Developing Countries											
Demand	**898**	**982**	**1030**	**1059**	**25**	**21**	**17**	**14**	**0.9**	**0.7**	**0.6**
of which residential	***723***	***778***	***796***	***788***	***73***	***67***	***60***	***53***	***0.7***	***0.5***	***0.3***

* The share of biomass in total primary energy demand and the share of biomass in total residential energy demand.

PART D

TABLES FOR REFERENCE SCENARIO PROJECTIONS

General Note to the Tables

For OECD countries and for most non-OECD countries, the analysis of energy demand is based on data up to 2000, published in mid-2002 in *Energy Balances of OECD Countries* and in *Energy Balances of Non-OECD Countries.*

The first two tables provide the GDP and population assumptions. The remaining tables show projections of energy demand, electricity generation and capacity, and CO_2 emissions for the following regions:

- World
- OECD
- OECD North America
- United States and Canada
- Mexico
- OECD Europe
- European Union
- OECD Pacific
- Japan, Australia and New Zealand
- Korea
- Transition economies
- Russia
- Developing countries
- China
- East Asia
- Indonesia
- South Asia
- India
- Latin America
- Brazil
- Middle East
- Africa

The definitions for regions, fuels and sectors are in Appendix 2.

The "other renewables" category in primary energy demand includes geothermal, solar, wind, tidal, wave energy and biomass used for power generation. For all OECD countries, except Mexico, this category also includes biomass used for other transformations and for final consumption. Biomass consumed in these sectors is shown separately for all other countries. Projections for traditional biomass use in developing countries are also shown in the Annex to Chapter 13.

Both in the text of this book and in the tables, rounding may cause some differences between the total and the sum of the individual components.

Economic Growth Assumptions
(average annual growth rates, in per cent)

	1971-2000	2000-2030
WORLD	**3.3**	**3.0**
OECD	**3.0**	**2.0**
North America	3.2	2.1
United States and Canada	*3.2*	*2.0*
Mexico	*3.9*	*3.4*
Europe	2.5	2.0
European Union	*2.4*	*1.9*
Pacific	3.7	2.0
Japan, Australia and New Zealand	*3.2*	*1.6*
Korea	*7.4*	*3.6*
Transition Economies	**0.1**	**3.1**
Russia	*-2.9**	*3.0*
Developing Countries	**4.8**	**4.1**
China	8.2	4.8
East Asia	5.6	3.6
Indonesia	*6.1*	*3.9*
South Asia	4.8	4.6
India	*4.9*	*4.6*
Latin America	3.1	3.0
Brazil	*3.9*	*2.9*
Middle East	2.9	2.6
Africa	2.7	3.6

* 1992-2000.
Source: OECD, World Bank, IMF, Asian Development Bank and CEPII.

Population Growth Assumptions
(average annual growth rates, in per cent)

	1971-2000	2000-2030
WORLD	**1.7**	**1.0**
OECD	**0.8**	**0.4**
North America	1.3	0.8
United States and Canada	*1.0*	*0.8*
Mexico	*2.4*	*1.0*
Europe	0.5	0.1
European Union	*0.3*	*-0.1*
Pacific	0.9	0.1
Japan, Australia and New Zealand	*0.8*	*0.0*
Korea	*1.3*	*0.4*
Transition Economies	**0.5**	**-0.3**
Russia	*-0.3**	*-0.6*
Developing Countries	**2.0**	**1.3**
China	1.4	0.5
East Asia	2.0	1.0
Indonesia	*1.9*	*1.0*
South Asia	2.1	1.3
India	*2.1*	*1.1*
Latin America	2.0	1.1
Brazil	*1.9*	*1.0*
Middle East	3.2	2.3
Africa	2.7	2.1

* 1992-2000.
Source: United Nations, OECD, World Bank.

Reference Scenario: World

	Energy Demand (Mtoe)					Shares (%)					Growth Rates (% per annum)			
	1971	2000	2010	2020	2030	1971	2000	2010	2020	2030	1971-2000	2000-2010	2000-2020	2000-2030
Total Primary Energy Supply	**4999**	**9179**	**11132**	**13167**	**15267**	**100**	**100**	**100**	**100**	**100**	**2.1**	**1.9**	**1.8**	**1.7**
Coal	1449	2355	2702	3128	3606	29	26	24	24	24	1.7	1.4	1.4	1.4
Oil	2450	3604	4272	5003	5769	49	39	38	38	38	1.3	1.7	1.7	1.6
of which International Marine Bunkers	*112*	*133*	*145*	*158*	*174*	*2*	*1*	*1*	*1*	*1*	*0.6*	*0.9*	*0.8*	*0.9*
Gas	895	2085	2794	3531	4203	18	23	25	27	28	3.0	3.0	2.7	2.4
Nuclear	29	674	753	719	703	1	7	7	5	5	11.5	1.1	0.3	0.1
Hydro	104	228	274	327	366	2	2	2	2	2	2.7	1.9	1.8	1.6
Other Renewables	73	233	336	457	618	1	3	3	3	4	4.1	3.7	3.4	3.3
Power Generation	**1209**	**3636**	**4608**	**5559**	**6535**	**100**	**100**	**100**	**100**	**100**	**3.9**	**2.4**	**2.1**	**2.0**
Coal	593	1555	1851	2224	2656	49	43	40	40	41	3.4	1.8	1.8	1.8
Oil	270	310	332	329	311	22	9	7	6	5	0.5	0.7	0.3	0.0
Gas	207	725	1170	1631	2032	17	20	25	29	31	4.4	4.9	4.1	3.5
Nuclear	29	674	753	719	703	2	19	16	13	11	11.5	1.1	0.3	0.1
Hydro	104	228	274	327	366	9	6	6	6	6	2.7	1.9	1.8	1.6
Other Renewables	7	144	228	329	466	1	4	5	6	7	11.1	4.7	4.2	4.0
Other Transformation, Own Use & Losses	**565**	**949**	**1106**	**1286**	**1473**						**1.8**	**1.5**	**1.5**	**1.5**
of which electricity	*75*	*235*	*304*	*388*	*476*						*4.0*	*2.6*	*2.5*	*2.4*
Total Final Consumption	**3634**	**6032**	**7254**	**8636**	**10080**	**100**	**100**	**100**	**100**	**100**	**1.8**	**1.9**	**1.8**	**1.7**
Coal	630	554	592	630	664	17	9	8	7	7	-0.4	0.7	0.6	0.6
Oil	1890	2943	3545	4229	4956	52	49	49	49	49	1.5	1.9	1.8	1.8
Gas	604	1112	1333	1566	1790	17	18	18	18	18	2.1	1.8	1.7	1.6
Electricity	377	1088	1419	1812	2235	10	18	20	21	22	3.7	2.7	2.6	2.4
Heat	68	247	260	272	285	2	4	4	3	3	4.5	0.5	0.5	0.5
Renewables	66	88	106	126	150	2	1	1	1	1	1.0	1.8	1.8	1.8

Industry	**1376**	**2102**	**2466**	**2862**	**3241**	**100**	**100**	**100**	**100**	**100**	**1.5**	**1.6**	**1.6**	**1.5**
Coal	279	427	471	510	545	20	20	19	18	17	1.5	1.0	0.9	0.8
Oil	488	593	671	751	822	35	28	27	26	25	0.7	1.2	1.2	1.1
Gas	332	488	593	708	817	24	23	24	25	25	1.3	2.0	1.9	1.7
Electricity	196	458	581	729	879	14	22	24	25	27	3.0	2.4	2.3	2.2
Heat	46	105	111	117	123	3	5	5	4	4	2.9	0.6	0.6	0.5
Renewables	36	31	39	47	55	3	1	2	2	2	-0.5	2.4	2.1	1.9
Transportation	**851**	**1775**	**2220**	**2749**	**3327**	**100**	**100**	**100**	**100**	**100**	**2.6**	**2.3**	**2.2**	**2.1**
Oil	793	1696	2125	2636	3195	93	96	96	96	96	2.7	2.3	2.2	2.1
Other fuels	58	79	94	113	131	7	4	4	4	4	1.1	1.7	1.8	1.7
Other Sectors	**1101**	**1954**	**2334**	**2762**	**3221**	**100**	**100**	**100**	**100**	**100**	**2.0**	**1.8**	**1.7**	**1.7**
Coal	180	108	104	104	105	16	6	4	4	3	-1.8	-0.4	-0.2	-0.1
Oil	457	476	541	611	685	41	24	23	22	21	0.1	1.3	1.3	1.2
Gas	252	571	676	778	879	23	29	29	28	27	2.9	1.7	1.6	1.4
Electricity	160	603	804	1041	1305	15	31	34	38	41	4.7	2.9	2.8	2.6
Heat	22	141	146	152	158	2	7	6	6	5	6.6	0.4	0.4	0.4
Renewables	30	55	64	75	90	3	3	3	3	3	2.1	1.5	1.6	1.7
Non-Energy Use	**306**	**201**	**234**	**263**	**291**						**-1.4**	**1.5**	**1.4**	**1.2**
Electricity Generation (TWh)	**5248**	**15391**	**20037**	**25578**	**31524**	**100**	**100**	**100**	**100**	**100**	**3.8**	**2.7**	**2.6**	**2.4**
Coal	2101	5989	7143	9075	11591	40	39	36	35	37	3.7	1.8	2.1	2.2
Oil	1095	1241	1348	1371	1326	21	8	7	5	4	0.4	0.8	0.5	0.2
Gas	696	2676	4947	7696	9923	13	17	25	30	31	4.8	6.3	5.4	4.5
Hydrogen-Fuel Cell	0	0	0	15	349	0	0	0	0	1	-	-	-	-
Nuclear	111	2586	2889	2758	2697	2	17	14	11	9	11.5	1.1	0.3	0.1
Hydro	1208	2650	3188	3800	4259	23	17	16	15	14	2.7	1.9	1.8	1.6
Other Renewables	36	249	521	863	1381	1	2	3	3	4	6.9	7.7	6.4	5.9

OECD Biomass (included above*)	60	167	212	263	324	1	2	2	2	2	3.6	2.4	2.3	2.2
Non-OECD Biomass (NOT included above)	592	909	986	1024	1035	11	9	8	7	6	1.5	0.8	0.6	0.4
Total Primary Energy Supply (including Biomass)	**5592**	**10089**	**12118**	**14190**	**16302**	**100**	**100**	**100**	**100**	**100**	**2.1**	**1.8**	**1.7**	**1.6**

* Mexico biomass is not included above - for more details see the General Notes to the Tables.

Reference Scenario: World

	Capacity (GW)				Shares (%)				Growth Rates (% per annum)		
	1999	2010	2020	2030	1999	2010	2020	2030	1999-2010	1999-2020	1999-2030
Total Capacity	**3397**	**4408**	**5683**	**7157**	**100**	**100**	**100**	**100**	**2.4**	**2.5**	**2.4**
Coal	1056	1277	1599	2090	31	29	28	29	1.7	2.0	2.2
Oil	495	547	540	507	15	12	10	7	0.9	0.4	0.1
Gas	677	1162	1865	2501	20	26	33	35	5.0	4.9	4.3
Hydrogen-Fuel Cell	0	0	4	100	0	0	0	1	-	-	-
Nuclear	351	379	362	356	10	9	6	5	0.7	0.1	0.0
Hydro	761	911	1080	1205	22	21	19	17	1.6	1.7	1.5
Other renewables	57	133	233	399	2	3	4	6	8.0	6.9	6.5

	Capacity (GW)				Shares (%)				Growth Rates (% per annum)		
	1999	2010	2020	2030	1999	2010	2020	2030	1999-2010	1999-2020	1999-2030
Other Renewables	**57**	**133**	**233**	**399**	**100**	**100**	**100**	**100**	**8.0**	**6.9**	**6.5**
Biomass	35	57	83	118	61	43	36	30	4.6	4.2	4.0
Wind	13	55	112	195	22	41	48	49	14.3	11.0	9.2
Geothermal	8	13	19	27	15	10	8	7	3.9	4.0	3.8
Solar	1	8	18	56	2	6	8	14	21.2	15.1	14.2
Tide/Wave	0	1	1	3	0	0	1	1	8.2	8.5	8.2

	Electricity Generation (TWh)				Shares (%)				Growth Rates (% per annum)		
	2000	2010	2020	2030	2000	2010	2020	2030	2000-2010	2000-2020	2000-2030
Other Renewables	**249**	**521**	**863**	**1381**	**100**	**100**	**100**	**100**	**7.6**	**6.4**	**5.9**
Biomass	167	276	399	568	67	53	46	41	5.1	4.5	4.2
Wind	31	147	307	539	12	28	36	39	16.8	12.1	10.0
Geothermal	49	85	126	174	20	16	15	13	5.6	4.8	4.3
Solar	1	11	27	92	0	2	3	7	25.1	17.1	15.7
Tide/Wave	1	1	4	8	0	0	0	1	9.3	9.5	8.9

Reference Scenario: World

	CO_2 Emissions (Mt)					Shares (%)					Growth Rates (% per annum)			
	1971	2000	2010	2020	2030	1971	2000	2010	2020	2030	1971-2000	2000-2010	2000-2020	2000-2030
Total CO_2 Emissions	**13654**	**22639**	**27453**	**32728**	**38161**	**100**	**100**	**100**	**100**	**100**	**1.8**	**1.9**	**1.9**	**1.8**
change since 1990 (%)		*12.5*	*36.4*	*62.6*	*89.6*									
Coal	5236	8875	10276	11959	13850	38	39	37	37	36	1.8	1.5	1.5	1.5
Oil	6360	9108	10881	12783	14794	47	40	40	39	39	1.2	1.8	1.7	1.6
of which International Marine Bunkers	*362*	*423*	*462*	*501*	*554*									
Gas	2058	4656	6295	7986	9517	15	21	23	24	25	2.9	3.1	2.7	2.4
Power Generation	**3925**	**8958**	**11285**	**13848**	**16457**	**100**	**100**	**100**	**100**	**100**	**2.9**	**2.3**	**2.2**	**2.0**
Coal	2524	6276	7482	8977	10706	64	70	66	65	65	3.2	1.8	1.8	1.8
Oil	862	992	1070	1060	1000	22	11	9	8	6	0.5	0.8	0.3	0.0
Gas	539	1689	2733	3811	4752	14	19	24	28	29	4.0	4.9	4.2	3.5
Transformation, Own Use & Losses	**752**	**1239**	**1436**	**1630**	**1859**	**100**	**100**	**100**	**100**	**100**	**1.7**	**1.5**	**1.4**	**1.4**
Total Final Consumption	**8978**	**12442**	**14732**	**17250**	**19845**	**100**	**100**	**100**	**100**	**100**	**1.1**	**1.7**	**1.6**	**1.6**
Coal	2577	2442	2638	2817	2976	29	20	18	16	15	-0.2	0.8	0.7	0.7
Oil	5104	7474	9075	10900	12852	57	60	62	63	65	1.3	2.0	1.9	1.8
Gas	1296	2526	3020	3532	4017	14	20	20	20	20	2.3	1.8	1.7	1.6
Industry	**3177**	**4230**	**4888**	**5513**	**6079**	**100**	**100**	**100**	**100**	**100**	**1.0**	**1.5**	**1.3**	**1.2**
Coal	1169	1933	2140	2321	2479	37	46	44	42	41	1.8	1.0	0.9	0.8
Oil	1308	1215	1429	1621	1796	41	29	29	29	30	-0.3	1.6	1.5	1.3
Gas	700	1081	1320	1571	1805	22	26	27	28	30	1.5	2.0	1.9	1.7
Transportation	**2320**	**4814**	**6010**	**7449**	**9024**	**100**	**100**	**100**	**100**	**100**	**2.5**	**2.2**	**2.2**	**2.1**
Oil	2160	4666	5842	7245	8785	93	97	97	97	97	2.7	2.3	2.2	2.1
Other fuels	161	148	167	204	238	7	3	3	3	3	-0.3	1.2	1.6	1.6
Other Sectors	**3298**	**3181**	**3589**	**4020**	**4454**	**100**	**100**	**100**	**100**	**100**	**-0.1**	**1.2**	**1.2**	**1.1**
Coal	1265	445	429	429	431	38	14	12	11	10	-3.5	-0.4	-0.2	-0.1
Oil	1480	1415	1605	1811	2027	45	44	45	45	46	-0.2	1.3	1.2	1.2
Gas	554	1320	1556	1780	1996	17	41	43	44	45	3.0	1.7	1.5	1.4
Non-Energy Use	**182**	**218**	**245**	**267**	**287**	**100**	**100**	**100**	**100**	**100**	**0.6**	**1.2**	**1.0**	**0.9**

Reference Scenario: OECD

	Energy Demand (Mtoe)					Shares (%)					Growth Rates (% per annum)			
	1971	2000	2010	2020	2030	1971	2000	2010	2020	2030	1971-2000	2000-2010	2000-2020	2000-2030
Total Primary Energy Supply	**3365**	**5291**	**5994**	**6605**	**7117**	**100**	**100**	**100**	**100**	**100**	**1.6**	**1.3**	**1.1**	**1.0**
Coal	810	1082	1089	1160	1182	24	20	18	18	17	1.0	0.1	0.3	0.3
Oil	1728	2164	2394	2605	2779	51	41	40	39	39	0.8	1.0	0.9	0.8
Gas	653	1143	1477	1774	2012	19	22	25	27	28	1.9	2.6	2.2	1.9
Nuclear	27	581	631	574	538	1	11	11	9	8	11.2	0.8	-0.1	-0.3
Hydro	75	113	122	128	133	2	2	2	2	2	1.4	0.8	0.6	0.6
Other Renewables	72	208	280	364	473	2	4	5	6	7	3.7	3.0	2.8	2.8
Power Generation	**818**	**2148**	**2505**	**2796**	**3028**	**100**	**100**	**100**	**100**	**100**	**3.4**	**1.5**	**1.3**	**1.2**
Coal	407	874	889	963	987	50	41	35	34	33	2.7	0.2	0.5	0.4
Oil	186	134	130	114	90	23	6	5	4	3	-1.1	-0.3	-0.8	-1.3
Gas	117	328	561	781	958	14	15	22	28	32	3.6	5.5	4.4	3.6
Nuclear	27	581	631	574	538	3	27	25	21	18	11.2	0.8	-0.1	-0.3
Hydro	75	113	122	128	133	9	5	5	5	4	1.4	0.8	0.6	0.6
Other Renewables	5	118	172	236	321	1	5	7	8	11	11.2	3.8	3.5	3.4
Other Transformation, Own Use & Losses	**331**	**430**	**455**	**480**	**504**						**0.9**	**0.6**	**0.6**	**0.5**
of which electricity	*51*	*117*	*136*	*152*	*163*						*2.9*	*1.5*	*1.3*	*1.1*
Total Final Consumption	**2561**	**3593**	**4088**	**4547**	**4942**	**100**	**100**	**100**	**100**	**100**	**1.2**	**1.3**	**1.2**	**1.1**
Coal	301	132	122	116	111	12	4	3	3	2	-2.8	-0.8	-0.7	-0.6
Oil	1431	1905	2140	2363	2555	56	53	52	52	52	1.0	1.2	1.1	1.0
Gas	469	704	801	875	933	18	20	20	19	19	1.4	1.3	1.1	0.9
Electricity	277	709	855	997	1117	11	20	21	22	23	3.3	1.9	1.7	1.5
Heat	17	55	63	70	76	1	2	2	2	2	4.1	1.3	1.2	1.1
Renewables	66	88	106	126	150	3	2	3	3	3	1.0	1.8	1.8	1.8

Industry	**963**	**1080**	**1186**	**1285**	**1366**	**100**	**100**	**100**	**100**	**100**	**0.4**	**0.9**	**0.9**	**0.8**
Coal	180	113	107	102	99	19	10	9	8	7	-1.6	-0.5	-0.5	-0.4
Oil	375	346	364	379	388	39	32	31	30	28	-0.3	0.5	0.5	0.4
Gas	227	286	319	349	372	24	26	27	27	27	0.8	1.1	1.0	0.9
Electricity	137	279	328	376	417	14	26	28	29	31	2.5	1.6	1.5	1.3
Heat	10	25	29	32	34	1	2	2	2	3	3.4	1.3	1.1	1.0
Renewables	36	31	39	47	55	4	3	3	4	4	-0.5	2.4	2.1	1.9
Transportation	**642**	**1218**	**1431**	**1638**	**1826**	**100**	**100**	**100**	**100**	**100**	**2.2**	**1.6**	**1.5**	**1.4**
Oil	610	1185	1391	1592	1773	95	97	97	97	97	2.3	1.6	1.5	1.4
Other fuels	31	33	39	46	53	5	3	3	3	3	0.2	1.9	1.8	1.6
Other Sectors	**861**	**1177**	**1335**	**1474**	**1591**	**100**	**100**	**100**	**100**	**100**	**1.1**	**1.3**	**1.1**	**1.0**
Coal	109	18	14	13	12	13	2	1	1	1	-6.0	-2.4	-1.7	-1.4
Oil	354	258	251	246	239	41	22	19	17	15	-1.1	-0.3	-0.2	-0.2
Gas	224	397	457	497	527	26	34	34	34	33	2.0	1.4	1.1	0.9
Electricity	135	420	516	607	685	16	36	39	41	43	4.0	2.1	1.9	1.6
Heat	8	29	32	36	39	1	2	2	2	2	4.7	1.1	1.1	0.9
Renewables	30	55	64	75	90	4	5	5	5	6	2.1	1.5	1.6	1.7
Non-Energy Use	**95**	**118**	**136**	**149**	**159**						**0.8**	**1.4**	**1.2**	**1.0**
Electricity Generation (TWh)	**3810**	**9598**	**11524**	**13358**	**14890**	**100**	**100**	**100**	**100**	**100**	**3.2**	**1.8**	**1.7**	**1.5**
Coal	1506	3733	3775	4283	4620	40	39	33	32	31	3.2	0.1	0.7	0.7
Oil	820	601	595	529	424	22	6	5	4	3	-1.1	-0.1	-0.6	-1.2
Gas	496	1515	2899	4153	4834	13	16	25	31	32	3.9	6.7	5.2	3.9
Hydrogen-Fuel Cell	0	0	0	15	349	0	0	0	0	2	-	-	-	-
Nuclear	104	2230	2421	2202	2064	3	23	21	16	14	11.2	0.8	-0.1	-0.3
Hydro	873	1311	1419	1491	1548	23	14	12	11	10	1.4	0.8	0.6	0.6
Other Renewables	11	208	416	684	1051	0	2	4	5	7	10.5	7.2	6.1	5.5

Biomass (included above*)	76	181	224	273	331	2	3	4	4	5	3.0	2.2	2.1	2.0
Total Primary Energy Supply (including Biomass)	**3373**	**5298**	**6000**	**6610**	**7120**	**100**	**100**	**100**	**100**	**100**	**1.6**	**1.3**	**1.1**	**1.0**

* Mexico biomass is not included above - for more details see the General Notes to the Tables.

Reference Scenario: OECD

	Capacity (GW)				Shares (%)				Growth Rates (% per annum)		
	1999	2010	2020	2030	1999	2010	2020	2030	1999-2010	1999-2020	1999-2030
Total Capacity	**2015**	**2430**	**2847**	**3294**	**100**	**100**	**100**	**100**	**1.7**	**1.7**	**1.6**
Coal	606	608	659	723	30	25	23	22	0.0	0.4	0.6
Oil	271	282	237	173	13	12	8	5	0.3	-0.6	-1.4
Gas	383	674	1006	1238	19	28	35	38	5.3	4.7	3.9
Hydrogen-Fuel Cell	0	0	4	100	0	0	0	3	-	-	-
Nuclear	297	313	284	269	15	13	10	8	0.5	-0.2	-0.3
Hydro-Conventional	326	357	376	391	16	15	13	12	0.8	0.7	0.6
Hydro-Pumped Storage	84	88	92	94	4	4	3	3	0.4	0.4	0.4
Other renewables	47	109	189	306	2	4	7	9	7.9	6.8	6.2

	Capacity (GW)				Shares (%)				Growth Rates (% per annum)		
	1999	2010	2020	2030	1999	2010	2020	2030	1999-2010	1999-2020	1999-2030
Other Renewables	**47**	**109**	**189**	**306**	**100**	**100**	**100**	**100**	**7.9**	**6.8**	**6.2**
Biomass	29	46	65	88	62	43	34	29	4.2	3.8	3.6
Wind	11	47	95	152	24	43	51	50	14.0	10.8	8.8
Geothermal	5	8	11	15	12	7	6	5	3.2	3.4	3.3
Solar	1	7	16	48	2	7	9	16	20.6	14.6	13.6
Tide/Wave	0	1	1	3	1	1	1	1	8.2	8.5	8.2

	Electricity Generation (TWh)				Shares (%)				Growth Rates (% per annum)		
	2000	2010	2020	2030	2000	2010	2020	2030	2000-2010	2000-2020	2000-2030
Other Renewables	**208**	**416**	**684**	**1051**	**100**	**100**	**100**	**100**	**7.1**	**6.1**	**5.5**
Biomass	145	227	321	440	70	55	47	42	4.6	4.0	3.8
Wind	29	128	267	435	14	31	39	41	16.2	11.8	9.5
Geothermal	33	49	69	92	16	12	10	9	4.0	3.8	3.5
Solar	1	10	24	75	1	2	3	7	24.6	16.5	15.1
Tide/Wave	1	1	4	8	0	0	1	1	9.3	9.5	8.9

Reference Scenario: OECD

	CO_2 Emissions (Mt)					Shares (%)					Growth Rates (% per annum)			
	1971	2000	2010	2020	2030	1971	2000	2010	2020	2030	1971-2000	2000-2010	2000-2020	2000-2030
Total CO_2 Emissions	**9355**	**12369**	**13800**	**15311**	**16397**	**100**	**100**	**100**	**100**	**100**	**1.0**	**1.1**	**1.1**	**0.9**
change since 1990 (%)		*12.5*	*25.5*	*39.2*	*49.1*									
Coal	3121	4216	4251	4522	4593	33	34	31	30	28	1.0	0.1	0.4	0.3
Oil	4756	5560	6152	6681	7121	51	45	45	44	43	0.5	1.0	0.9	0.8
Gas	1479	2594	3396	4109	4683	16	21	25	27	29	2.0	2.7	2.3	2.0
Power Generation	**2708**	**4725**	**5327**	**6090**	**6524**	**100**	**100**	**100**	**100**	**100**	**1.9**	**1.2**	**1.3**	**1.1**
Coal	1783	3529	3594	3895	3987	66	75	67	64	61	2.4	0.2	0.5	0.4
Oil	595	431	420	367	293	22	9	8	6	4	-1.1	-0.3	-0.8	-1.3
Gas	331	765	1313	1829	2244	12	16	25	30	34	2.9	5.5	4.5	3.7
Transformation, Own Use & Losses	**557**	**666**	**706**	**738**	**774**	**100**	**100**	**100**	**100**	**100**	**0.6**	**0.6**	**0.5**	**0.5**
Total Final Consumption	**6090**	**6979**	**7767**	**8483**	**9099**	**100**	**100**	**100**	**100**	**100**	**0.5**	**1.1**	**1.0**	**0.9**
Coal	1245	616	595	566	546	20	9	8	7	6	-2.4	-0.3	-0.4	-0.4
Oil	3855	4742	5322	5896	6399	63	68	69	70	70	0.7	1.2	1.1	1.0
Gas	991	1621	1850	2021	2154	16	23	24	24	24	1.7	1.3	1.1	1.0
Industry	**2147**	**1776**	**1870**	**1937**	**1987**	**100**	**100**	**100**	**100**	**100**	**-0.7**	**0.5**	**0.4**	**0.4**
Coal	727	537	525	501	485	34	30	28	26	24	-1.0	-0.2	-0.3	-0.3
Oil	957	590	618	640	653	45	33	33	33	33	-1.7	0.5	0.4	0.3
Gas	464	649	727	796	849	22	37	39	41	43	1.2	1.1	1.0	0.9
Transportation	**1799**	**3326**	**3901**	**4467**	**4980**	**100**	**100**	**100**	**100**	**100**	**2.1**	**1.6**	**1.5**	**1.4**
Oil	1722	3276	3841	4397	4901	96	99	98	98	98	2.2	1.6	1.5	1.4
Other fuels	77	49	60	70	79	4	1	2	2	2	-1.5	1.9	1.7	1.6
Other Sectors	**2054**	**1769**	**1873**	**1946**	**1992**	**100**	**100**	**100**	**100**	**100**	**-0.5**	**0.6**	**0.5**	**0.4**
Coal	468	74	59	54	50	23	4	3	3	2	-6.2	-2.2	-1.6	-1.3
Oil	1100	771	750	737	716	54	44	40	38	36	-1.2	-0.3	-0.2	-0.2
Gas	486	923	1063	1156	1226	24	52	57	59	62	2.2	1.4	1.1	0.9
Non-Energy Use	**91**	**108**	**123**	**133**	**140**	**100**	**100**	**100**	**100**	**100**	**0.6**	**1.3**	**1.0**	**0.9**

Reference Scenario: OECD North America

	Energy Demand (Mtoe)					Shares (%)					Growth Rates (% per annum)			
	1971	2000	2010	2020	2030	1971	2000	2010	2020	2030	1971-2000	2000-2010	2000-2020	2000-2030
Total Primary Energy Supply	**1775**	**2696**	**3040**	**3391**	**3726**	**100**	**100**	**100**	**100**	**100**	**1.5**	**1.2**	**1.2**	**1.1**
Coal	297	579	586	651	685	17	21	19	19	18	2.3	0.1	0.6	0.6
Oil	828	1070	1198	1336	1485	47	40	39	39	40	0.9	1.1	1.1	1.1
Gas	557	653	821	962	1082	31	24	27	28	29	0.5	2.3	2.0	1.7
Nuclear	12	230	233	198	171	1	9	8	6	5	10.8	0.1	-0.7	-1.0
Hydro	38	55	59	62	63	2	2	2	2	2	1.3	0.7	0.6	0.4
Other Renewables	43	110	144	183	241	2	4	5	5	6	3.3	2.7	2.6	2.7
Power Generation	**400**	**1130**	**1297**	**1466**	**1611**	**100**	**100**	**100**	**100**	**100**	**3.6**	**1.4**	**1.3**	**1.2**
Coal	192	530	537	602	634	48	47	41	41	39	3.6	0.1	0.6	0.6
Oil	61	57	58	55	54	15	5	4	4	3	-0.2	0.1	-0.2	-0.2
Gas	97	175	300	405	491	24	16	23	28	30	2.1	5.5	4.3	3.5
Nuclear	12	230	233	198	171	3	20	18	13	11	10.8	0.1	-0.7	-1.0
Hydro	38	55	59	62	63	9	5	5	4	4	1.3	0.7	0.6	0.4
Other Renewables	1	82	111	145	199	0	7	9	10	12	17.5	3.0	2.9	3.0
Other Transformation, Own Use & Losses	**160**	**222**	**242**	**256**	**272**						**1.1**	**0.8**	**0.7**	**0.7**
of which electricity	*26*	*59*	*68*	*76*	*84*						*2.9*	*1.4*	*1.3*	*1.2*
Total Final Consumption	**1383**	**1778**	**2021**	**2272**	**2526**	**100**	**100**	**100**	**100**	**100**	**0.9**	**1.3**	**1.2**	**1.2**
Coal	84	36	34	32	31	6	2	2	1	1	-2.9	-0.7	-0.6	-0.5
Oil	728	953	1073	1212	1358	53	54	53	53	54	0.9	1.2	1.2	1.2
Gas	385	387	431	465	497	28	22	21	20	20	0.0	1.1	0.9	0.8
Electricity	143	355	427	500	569	10	20	21	22	23	3.2	1.9	1.7	1.6
Heat	0	20	24	27	30	0	1	1	1	1	-	1.7	1.4	1.3
Renewables	42	27	32	36	40	3	1	2	2	2	-1.6	1.8	1.5	1.4

Industry	**456**	**461**	**511**	**561**	**610**	**100**	**100**	**100**	**100**	**100**	**0.0**	**1.0**	**1.0**	**0.9**
Coal	67	34	32	30	29	15	7	6	5	5	-2.3	-0.6	-0.6	-0.5
Oil	117	127	134	141	147	26	28	26	25	24	0.3	0.5	0.5	0.5
Gas	182	150	165	179	192	40	33	32	32	31	-0.7	0.9	0.9	0.8
Electricity	59	123	147	173	199	13	27	29	31	33	2.6	1.8	1.7	1.6
Heat	0	18	22	25	28	0	4	4	4	5	-	1.9	1.6	1.4
Renewables	31	9	12	14	16	7	2	2	2	3	-4.3	2.8	2.3	2.0
Transportation	**420**	**701**	**811**	**940**	**1078**	**100**	**100**	**100**	**100**	**100**	**1.8**	**1.5**	**1.5**	**1.4**
Oil	402	678	784	909	1043	96	97	97	97	97	1.8	1.5	1.5	1.4
Other fuels	18	23	27	31	35	4	3	3	3	3	0.8	1.7	1.6	1.4
Other Sectors	**463**	**552**	**622**	**688**	**749**	**100**	**100**	**100**	**100**	**100**	**0.6**	**1.2**	**1.1**	**1.0**
Coal	18	2	2	2	2	4	0	0	0	0	-6.7	-1.8	-1.0	-0.7
Oil	165	84	79	80	80	36	15	13	12	11	-2.3	-0.6	-0.3	-0.2
Gas	186	217	242	259	274	40	39	39	38	37	0.5	1.1	0.9	0.8
Electricity	83	230	279	326	370	18	42	45	47	49	3.6	1.9	1.7	1.6
Heat	0	2	2	2	2	0	0	0	0	0	-	0.1	0.1	0.1
Renewables	11	16	18	20	21	2	3	3	3	3	1.1	1.1	1.0	1.0
Non-Energy Use	**44**	**64**	**76**	**83**	**89**						**1.3**	**1.7**	**1.3**	**1.1**
Electricity Generation (TWh)	**1956**	**4813**	**5758**	**6702**	**7594**	**100**	**100**	**100**	**100**	**100**	**3.2**	**1.8**	**1.7**	**1.5**
Coal	805	2247	2262	2638	2913	41	47	39	39	38	3.6	0.1	0.8	0.9
Oil	253	237	245	240	240	13	5	4	4	3	-0.2	0.3	0.1	0.0
Gas	413	704	1495	2050	2419	21	15	26	31	32	1.9	7.8	5.5	4.2
Hydrogen-Fuel Cell	0	0	0	5	122	0	0	0	0	2	-	-	-	-
Nuclear	45	881	893	758	657	2	18	16	11	9	10.8	0.1	-0.7	-1.0
Hydro	440	640	687	716	728	23	13	12	11	10	1.3	0.7	0.6	0.4
Other Renewables	1	104	178	295	514	0	2	3	4	7	18.0	5.6	5.4	5.5

Biomass (included above*)	51	98	115	131	154	3	4	4	4	4	2.3	1.6	1.5	1.5
Total Primary Energy Supply (including Biomass)	**1783**	**2703**	**3046**	**3396**	**3729**	**100**	**100**	**100**	**100**	**100**	**1.4**	**1.2**	**1.1**	**1.1**

* Mexico biomass is not included above - for more details see the General Notes to the Tables.

Reference Scenario: OECD North America

	Capacity (GW)				Shares (%)				Growth Rates (% per annum)		
	1999	2010	2020	2030	1999	2010	2020	2030	1999-2010	1999-2020	1999-2030
Total Capacity	**961**	**1146**	**1352**	**1595**	**100**	**100**	**100**	**100**	**1.6**	**1.6**	**1.6**
Coal	335	331	383	427	35	29	28	27	-0.1	0.6	0.8
Oil	97	100	84	74	10	9	6	5	0.3	-0.7	-0.9
Gas	222	381	524	635	23	33	39	40	5.0	4.2	3.4
Hydrogen-Fuel Cell	0	0	2	35	0	0	0	2	-	-	-
Nuclear	109	110	96	86	11	10	7	5	0.1	-0.6	-0.8
Hydro-Conventional	157	164	171	174	16	14	13	11	0.4	0.4	0.3
Hydro-Pumped Storage	19	19	19	19	2	2	1	1	0.0	0.0	0.0
Other renewables	22	41	74	144	2	4	5	9	5.7	5.8	6.2

	Capacity (GW)				Shares (%)				Growth Rates (% per annum)		
	1999	2010	2020	2030	1999	2010	2020	2030	1999-2010	1999-2020	1999-2030
Other Renewables	**22**	**41**	**74**	**144**	**100**	**100**	**100**	**100**	**5.7**	**5.8**	**6.2**
Biomass	16	22	28	39	70	52	38	27	3.0	2.8	2.9
Wind	2	12	32	70	10	28	44	48	15.7	13.3	11.6
Geothermal	4	6	9	12	17	14	12	8	3.7	4.0	3.9
Solar	1	2	5	24	2	6	7	16	14.2	10.8	12.8
Tide/Wave	0	0	0	0	0	0	0	0	0.0	0.0	0.0

	Electricity Generation (TWh)				Shares (%)				Growth Rates (% per annum)		
	2000	2010	2020	2030	2000	2010	2020	2030	2000-2010	2000-2020	2000-2030
Other Renewables	**104**	**178**	**295**	**514**	**100**	**100**	**100**	**100**	**5.5**	**5.3**	**5.5**
Biomass	77	104	134	184	74	58	45	36	3.1	2.8	3.0
Wind	6	37	102	220	6	21	35	43	20.1	15.3	12.8
Geothermal	21	33	51	72	20	19	17	14	4.9	4.6	4.2
Solar	1	4	8	38	1	2	3	7	16.1	11.6	13.3
Tide/Wave	0	0	0	0	0	0	0	0	0.0	11.7	9.4

Reference Scenario: OECD North America

	CO_2 Emissions (Mt)					Shares (%)					Growth Rates (% per annum)			
	1971	2000	2010	2020	2030	1971	2000	2010	2020	2030	1971-2000	2000-2010	2000-2020	2000-2030
Total CO_2 Emissions	**4769**	**6534**	**7300**	**8258**	**9075**	**100**	**100**	**100**	**100**	**100**	**1.1**	**1.1**	**1.2**	**1.1**
change since 1990 (%)		*23.2*	*37.6*	*55.7*	*71.1*									
Coal	1146	2260	2276	2523	2644	24	35	31	31	29	2.4	0.1	0.6	0.5
Oil	2346	2802	3138	3504	3901	49	43	43	42	43	0.6	1.1	1.1	1.1
Gas	1278	1472	1886	2232	2530	27	23	26	27	28	0.5	2.5	2.1	1.8
Power Generation	**1364**	**2684**	**3009**	**3503**	**3827**	**100**	**100**	**100**	**100**	**100**	**2.4**	**1.1**	**1.3**	**1.2**
Coal	885	2088	2115	2370	2494	65	78	70	68	65	3.0	0.1	0.6	0.6
Oil	196	187	191	181	176	14	7	6	5	5	-0.2	0.2	-0.2	-0.2
Gas	283	409	704	953	1157	21	15	23	27	30	1.3	5.6	4.3	3.5
Transformation, Own Use & Losses	**349**	**378**	**424**	**449**	**481**	**100**	**100**	**100**	**100**	**100**	**0.3**	**1.2**	**0.9**	**0.8**
Total Final Consumption	**3056**	**3472**	**3867**	**4306**	**4767**	**100**	**100**	**100**	**100**	**100**	**0.4**	**1.1**	**1.1**	**1.1**
Coal	243	168	157	148	144	8	5	4	3	3	-1.3	-0.7	-0.6	-0.5
Oil	1971	2404	2707	3076	3466	64	69	70	71	73	0.7	1.2	1.2	1.2
Gas	842	900	1003	1082	1156	28	26	26	25	24	0.2	1.1	0.9	0.8
Industry	**818**	**689**	**724**	**758**	**793**	**100**	**100**	**100**	**100**	**100**	**-0.6**	**0.5**	**0.5**	**0.5**
Coal	173	158	148	140	136	21	23	21	18	17	-0.3	-0.6	-0.6	-0.5
Oil	277	184	194	205	214	34	27	27	27	27	-1.4	0.5	0.5	0.5
Gas	368	347	381	413	443	45	50	53	55	56	-0.2	0.9	0.9	0.8
Transportation	**1195**	**1964**	**2272**	**2631**	**3016**	**100**	**100**	**100**	**100**	**100**	**1.7**	**1.5**	**1.5**	**1.4**
Oil	1153	1918	2216	2567	2945	97	98	98	98	98	1.8	1.5	1.5	1.4
Other fuels	41	47	56	64	72	3	2	2	2	2	0.4	1.8	1.6	1.4
Other Sectors	**1002**	**763**	**807**	**846**	**881**	**100**	**100**	**100**	**100**	**100**	**-0.9**	**0.6**	**0.5**	**0.5**
Coal	69	9	8	7	7	7	1	1	1	1	-6.8	-1.8	-1.0	-0.7
Oil	499	248	233	234	233	50	32	29	28	26	-2.4	-0.6	-0.3	-0.2
Gas	434	506	567	605	641	43	66	70	72	73	0.5	1.1	0.9	0.8
Non-Energy Use	**41**	**55**	**65**	**71**	**76**	**100**	**100**	**100**	**100**	**100**	**1.0**	**1.6**	**1.3**	**1.1**

Reference Scenario: United States and Canada

	Energy Demand (Mtoe)					Shares (%)					Growth Rates (% per annum)			
	1971	2000	2010	2020	2030	1971	2000	2010	2020	2030	1971-2000	2000-2010	2000-2020	2000-2030
Total Primary Energy Supply	**1736**	**2551**	**2854**	**3152**	**3420**	**100**	**100**	**100**	**100**	**100**	**1.3**	**1.1**	**1.1**	**1.0**
Coal	295	572	578	643	675	17	22	20	20	20	2.3	0.1	0.6	0.6
Oil	801	976	1085	1198	1316	46	38	38	38	38	0.7	1.1	1.0	1.0
Gas	548	620	771	888	979	32	24	27	28	29	0.4	2.2	1.8	1.5
Nuclear	12	227	230	195	169	1	9	8	6	5	10.8	0.1	-0.8	-1.0
Hydro	37	52	55	57	58	2	2	2	2	2	1.2	0.6	0.4	0.3
Other Renewables	43	104	134	171	224	2	4	5	5	7	3.1	2.6	2.5	2.6
Power Generation	**394**	**1080**	**1230**	**1377**	**1495**	**100**	**100**	**100**	**100**	**100**	**3.5**	**1.3**	**1.2**	**1.1**
Coal	192	525	532	596	626	49	49	43	43	42	3.5	0.1	0.6	0.6
Oil	58	34	33	26	22	15	3	3	2	1	-1.8	-0.5	-1.4	-1.5
Gas	95	164	279	371	439	24	15	23	27	29	1.9	5.4	4.2	3.3
Nuclear	12	227	230	195	169	3	21	19	14	11	10.8	0.1	-0.8	-1.0
Hydro	37	52	55	57	58	9	5	5	4	4	1.2	0.6	0.4	0.3
Other Renewables	1	76	101	133	181	0	7	8	10	12	17.2	2.9	2.8	2.9
Other Transformation, Own Use & Losses	**153**	**196**	**211**	**218**	**225**						**0.9**	**0.7**	**0.5**	**0.5**
of which electricity	*25*	*56*	*63*	*69*	*74*						2.8	1.3	1.1	0.9
Total Final Consumption	**1354**	**1692**	**1907**	**2124**	**2334**	**100**	**100**	**100**	**100**	**100**	**0.8**	**1.2**	**1.1**	**1.1**
Coal	83	35	32	30	29	6	2	2	1	1	-2.9	-0.8	-0.7	-0.6
Oil	709	893	996	1116	1237	52	53	52	53	53	0.8	1.1	1.1	1.1
Gas	379	376	416	443	467	28	22	22	21	20	0.0	1.0	0.8	0.7
Electricity	140	341	407	471	530	10	20	21	22	23	3.1	1.8	1.6	1.5
Heat	0	20	24	27	30	0	1	1	1	1	-	1.7	1.4	1.3
Renewables	42	27	32	36	40	10	6	7	7	7	-1.6	1.8	1.5	1.4

Industry	**444**	**430**	**472**	**510**	**546**	**100**	**100**	**100**	**100**	**100**	**-0.1**	**0.9**	**0.9**	**0.8**
Coal	66	32	30	28	27	15	8	6	6	5	-2.4	-0.7	-0.7	-0.6
Oil	114	116	122	128	132	26	27	26	25	24	0.1	0.5	0.5	0.5
Gas	176	140	151	160	168	40	33	32	31	31	-0.8	0.8	0.7	0.6
Electricity	58	115	135	155	175	13	27	29	30	32	2.4	1.6	1.5	1.4
Heat	0	18	22	25	28	0	4	5	5	5	-	1.9	1.6	1.4
Renewables	31	9	12	14	16	7	2	2	3	3	-4.3	2.8	2.3	2.0
Transportation	**410**	**664**	**761**	**873**	**988**	**100**	**100**	**100**	**100**	**100**	**1.7**	**1.4**	**1.4**	**1.3**
Oil	392	641	734	842	953	96	97	96	96	97	1.7	1.4	1.4	1.3
Other fuels	18	23	27	31	35	4	3	4	4	3	0.8	1.7	1.6	1.4
Other Sectors	**458**	**535**	**600**	**660**	**713**	**100**	**100**	**100**	**100**	**100**	**0.5**	**1.2**	**1.1**	**1.0**
Coal	18	2	2	2	2	4	0	0	0	0	-6.7	-1.8	-1.0	-0.7
Oil	161	74	67	66	64	35	14	11	10	9	-2.7	-0.9	-0.6	-0.4
Gas	186	216	240	255	269	41	40	40	39	38	0.5	1.1	0.8	0.7
Electricity	82	225	271	315	354	18	42	45	48	50	3.5	1.9	1.7	1.5
Heat	0	2	2	2	2	0	0	0	0	0	-	0.1	0.1	0.1
Renewables	11	16	18	20	21	3	4	4	4	4	1.1	1.1	1.0	1.0
Non-Energy Use	**43**	**62**	**74**	**81**	**87**						**1.3**	**1.7**	**1.3**	**1.1**
Electricity Generation (TWh)	**1925**	**4609**	**5464**	**6285**	**7016**	**100**	**100**	**100**	**100**	**100**	**3.1**	**1.7**	**1.6**	**1.4**
Coal	805	2228	2240	2612	2882	42	48	41	42	41	3.6	0.1	0.8	0.9
Oil	242	140	136	112	99	13	3	2	2	1	-1.9	-0.3	-1.1	-1.1
Gas	407	664	1398	1871	2123	21	14	26	30	30	1.7	7.7	5.3	4.0
Hydrogen-Fuel Cell	0	0	0	5	122	0	0	0	0	2	-	-	-	-
Nuclear	45	873	882	748	647	2	19	16	12	9	10.8	0.1	-0.8	-1.0
Hydro	426	607	644	663	670	22	13	12	11	10	1.2	0.6	0.4	0.3
Other Renewables	1	98	165	272	473	0	2	3	4	7	17.8	5.4	5.3	5.4

Biomass (included above)	43	91	109	127	151	2	4	4	4	4	2.6	1.9	1.7	1.7
Total Primary Energy Supply (including Biomass)	**1736**	**2551**	**2854**	**3152**	**3420**	**100**	**100**	**100**	**100**	**100**	**1.3**	**1.1**	**1.1**	**1.0**

Reference Scenario: United States and Canada

	Capacity (GW)				Shares (%)				Growth Rates (% per annum)		
	1999	2010	2020	2030	1999	2010	2020	2030	1999-2010	1999-2020	1999-2030
Total Capacity	**922**	**1086**	**1265**	**1473**	**100**	**100**	**100**	**100**	**1.5**	**1.5**	**1.5**
Coal	332	328	379	422	36	30	30	29	-0.1	0.6	0.8
Oil	79	79	58	46	9	7	5	3	0.0	-1.5	-1.7
Gas	216	362	488	577	23	33	39	39	4.8	4.0	3.2
Hydrogen-Fuel Cell	0	0	2	35	0	0	0	2	-	-	-
Nuclear	107	109	95	84	12	10	7	6	0.1	-0.6	-0.8
Hydro-Conventional	147	151	156	157	16	14	12	11	0.2	0.3	0.2
Hydro-Pumped Storage	19	19	19	19	2	2	2	1	0.0	0.0	0.0
Other renewables	21	39	68	133	2	4	5	9	5.5	5.7	6.1

	Capacity (GW)				Shares (%)				Growth Rates (% per annum)		
	1999	2010	2020	2030	1999	2010	2020	2030	1999-2010	1999-2020	1999-2030
Other Renewables	**21**	**39**	**68**	**133**	**100**	**100**	**100**	**100**	**5.5**	**5.7**	**6.1**
Biomass	16	21	27	36	73	55	39	27	2.8	2.6	2.7
Wind	2	11	30	65	11	28	44	49	15.1	13.0	11.3
Geothermal	3	4	7	10	14	11	10	8	3.3	4.1	4.1
Solar	1	2	4	22	3	6	6	17	14.2	10.2	12.7
Tide/Wave	0	0	0	0	0	0	0	0	0.0	0.0	0.0

	Electricity Generation (TWh)				Shares (%)				Growth Rates (% per annum)		
	2000	2010	2020	2030	2000	2010	2020	2030	2000-2010	2000-2020	2000-2030
Other Renewables	**98**	**165**	**272**	**473**	**100**	**100**	**100**	**100**	**5.4**	**5.3**	**5.4**
Biomass	76	102	129	172	78	62	48	36	2.9	2.7	2.7
Wind	6	35	97	207	6	21	36	44	19.6	15.1	12.6
Geothermal	15	24	39	58	15	14	14	12	5.0	5.0	4.7
Solar	1	4	7	36	1	2	3	8	16.3	11.0	13.2
Tide/Wave	0	0	0	0	0	0	0	0	0.0	0.0	0.0

Reference Scenario: United States and Canada

	CO_2 Emissions (Mt)					Shares (%)					Growth Rates (% per annum)			
	1971	2000	2010	2020	2030	1971	2000	2010	2020	2030	1971-2000	2000-2010	2000-2020	2000-2030
Total CO_2 Emissions	**4672**	**6175**	**6840**	**7670**	**8327**	**100**	**100**	**100**	**100**	**100**	**1.0**	**1.0**	**1.1**	**1.0**
change since 1990 (%)		*16.4*	*29.0*	*44.6*	*57.0*									
Coal	1140	2233	2247	2489	2605	24	36	33	32	31	2.3	0.1	0.5	0.5
Oil	2274	2546	2829	3125	3440	49	41	41	41	41	0.4	1.1	1.0	1.0
Gas	1257	1396	1765	2055	2282	27	23	26	27	27	0.4	2.4	2.0	1.7
Power Generation	**1351**	**2563**	**2851**	**3295**	**3561**	**100**	**100**	**100**	**100**	**100**	**2.2**	**1.1**	**1.3**	**1.1**
Coal	885	2067	2092	2344	2464	65	81	73	71	69	3.0	0.1	0.6	0.6
Oil	187	112	107	84	71	14	4	4	3	2	-1.8	-0.5	-1.4	-1.5
Gas	280	384	652	867	1026	21	15	23	26	29	1.1	5.4	4.2	3.3
Transformation, Own Use & Losses	**338**	**327**	**364**	**379**	**394**	**100**	**100**	**100**	**100**	**100**	**-0.1**	**1.1**	**0.7**	**0.6**
Total Final Consumption	**2982**	**3285**	**3625**	**3996**	**4372**	**100**	**100**	**100**	**100**	**100**	**0.3**	**1.0**	**1.0**	**1.0**
Coal	238	162	150	141	137	8	5	4	4	3	-1.3	-0.7	-0.7	-0.6
Oil	1916	2247	2508	2823	3147	64	68	69	71	72	0.6	1.1	1.1	1.1
Gas	829	876	967	1032	1089	28	27	27	26	25	0.2	1.0	0.8	0.7
Industry	**789**	**637**	**661**	**681**	**702**	**100**	**100**	**100**	**100**	**100**	**-0.7**	**0.4**	**0.3**	**0.3**
Coal	168	152	142	132	128	21	24	21	19	18	-0.4	-0.7	-0.7	-0.6
Oil	266	161	169	178	184	34	25	26	26	26	-1.7	0.5	0.5	0.5
Gas	355	325	350	371	389	45	51	53	54	55	-0.3	0.8	0.7	0.6
Transportation	**1164**	**1864**	**2136**	**2449**	**2773**	**100**	**100**	**100**	**100**	**100**	**1.6**	**1.4**	**1.4**	**1.3**
Oil	1123	1817	2080	2385	2702	96	97	97	97	97	1.7	1.4	1.4	1.3
Other fuels	41	47	56	64	72	4	3	3	3	3	-	-	-	-
Other Sectors	**990**	**733**	**768**	**800**	**827**	**100**	**100**	**100**	**100**	**100**	**-1.0**	**0.5**	**0.4**	**0.4**
Coal	69	9	8	7	7	7	1	1	1	1	-6.8	-1.8	-1.0	-0.7
Oil	488	219	199	196	191	49	30	26	24	23	-2.7	-0.9	-0.6	-0.4
Gas	433	505	561	597	628	44	69	73	75	76	-	1.1	0.8	0.7
Non-Energy Use	**38**	**51**	**60**	**66**	**70**	**100**	**100**	**100**	**100**	**100**	**1.0**	**1.7**	**1.3**	**1.1**

Reference Scenario: Mexico

	Energy Demand (Mtoe)					Shares (%)					Growth Rates (% per annum)			
	1971	2000	2010	2020	2030	1971	2000	2010	2020	2030	1971-2000	2000-2010	2000-2020	2000-2030
Total Primary Energy Supply	**39**	**144**	**187**	**239**	**306**	**100**	**100**	**100**	**100**	**100**	**4.6**	**2.6**	**2.6**	**2.5**
Coal	1	7	8	8	10	4	5	4	4	3	5.3	1.4	1.2	1.3
Oil	27	94	113	138	168	70	65	60	58	55	4.3	1.9	1.9	2.0
Gas	9	33	51	74	103	23	23	27	31	34	4.7	4.3	4.1	3.9
Nuclear	0	2	3	3	3	0	1	1	1	1	-	2.3	1.1	0.8
Hydro	1	3	4	5	5	3	2	2	2	2	2.9	2.6	2.3	1.9
Other Renewables	0	6	9	12	18	0	4	5	5	6	-	4.5	3.7	3.7
Power Generation	**6**	**50**	**67**	**89**	**117**	**100**	**100**	**100**	**100**	**100**	**7.8**	**3.0**	**2.9**	**2.9**
Coal	0	5	5	6	7	0	10	8	7	6	19.4	1.3	1.3	1.4
Oil	3	23	25	29	32	49	47	38	33	27	7.6	0.8	1.1	1.0
Gas	2	11	20	34	52	29	22	31	38	45	6.7	6.6	5.9	5.4
Nuclear	0	2	3	3	3	0	4	4	3	2	-	2.3	1.1	0.8
Hydro	1	3	4	5	5	22	6	6	5	4	2.9	2.6	2.3	1.9
Other Renewables	0	6	9	12	18	0	12	14	14	15	-	4.5	3.7	3.7
Other Transformation, Own Use & Losses	**7**	**26**	**31**	**38**	**47**						**4.6**	**1.9**	**1.9**	**2.0**
of which electricity	*0*	*4*	*5*	*7*	*10*						*7.4*	*3.8*	*3.7*	*3.7*
Total Final Consumption	**29**	**86**	**114**	**148**	**192**	**100**	**100**	**100**	**100**	**100**	**3.8**	**2.8**	**2.7**	**2.7**
Coal	1	1	1	1	2	4	2	1	1	1	0.8	0.5	0.4	0.3
Oil	19	61	76	96	121	67	70	67	65	63	4.0	2.4	2.3	2.3
Gas	6	10	16	22	30	21	12	14	15	15	1.9	4.2	3.8	3.5
Electricity	2	14	20	29	40	8	16	18	19	21	6.5	3.7	3.6	3.5
Heat	0	0	0	0	0	0	0	0	0	0	-	-	-	-
Renewables	0	0	0	0	0	0	0	0	0	0	-	-	-	-

Industry	**12**	**31**	**40**	**51**	**64**	**100**	**100**	**100**	**100**	**100**	**3.3**	**2.5**	**2.5**	**2.4**
Coal	1	1	1	1	2	9	4	4	3	2	0.8	0.5	0.4	0.3
Oil	4	11	12	13	14	31	36	31	26	22	3.9	0.7	0.8	0.8
Gas	6	10	14	18	24	49	32	34	37	38	1.8	3.3	3.3	3.0
Electricity	1	9	12	18	24	11	28	32	35	38	6.7	3.8	3.7	3.5
Heat	0	0	0	0	0	0	0	0	0	0	-	-	-	-
Renewables	0	0	0	0	0	0	0	0	0	0	-	-	-	-
Transportation	**11**	**37**	**50**	**68**	**90**	**100**	**100**	**100**	**100**	**100**	**4.4**	**3.1**	**3.0**	**3.0**
Oil	11	37	50	67	90	100	100	100	100	100	4.4	3.1	3.0	3.0
Other fuels	0.0	0.1	0.2	0.2	0.3	0	0	0	0	0	5.5	5.5	4.5	3.9
Other Sectors	**5**	**17**	**22**	**28**	**36**	**100**	**100**	**100**	**100**	**100**	**4.1**	**2.8**	**2.7**	**2.6**
Coal	0	0	0	0	0	0	0	0	0	0	-	-	-	-
Oil	4	11	12	14	15	77	64	56	49	42	3.5	1.5	1.3	1.2
Gas	0	1	2	4	6	6	4	10	13	15	2.9	12.1	8.6	7.2
Electricity	1	5	8	11	15	18	32	34	38	43	6.2	3.5	3.7	3.6
Heat	0	0	0	0	0	0	0	0	0	0	-	-	-	-
Renewables	0	0	0	0	0	0	0	0	0	0	-	-	-	-
Non-Energy Use	**1**	**1**	**2**	**2**	**2**						**0.9**	**1.2**	**0.9**	**0.7**
Electricity Generation (TWh)	**31**	**204**	**294**	**417**	**578**	**100**	**100**	**100**	**100**	**100**	**6.7**	**3.7**	**3.6**	**3.5**
Coal	0	19	22	26	32	1	9	8	6	5	17.9	1.5	1.5	1.7
Oil	11	97	109	128	141	36	48	37	31	24	7.8	1.1	1.4	1.2
Gas	5	40	97	179	296	18	20	33	43	51	7.1	9.2	7.7	6.9
Hydrogen-Fuel Cell	0	0	0	0	0	0	0	0	0	0	-	-	-	-
Nuclear	0	8	10	10	10	0	4	4	2	2	-	2.3	1.1	0.8
Hydro	14	33	43	52	58	46	16	15	13	10	2.9	2.6	2.3	1.9
Other Renewables	0	6	13	22	41	0	3	4	5	7	-	8.1	6.8	6.7

Biomass (NOT included above)	8	7	6	5	3	17	5	3	2	1	-0.4	-1.5	-2.1	-2.7
Total Primary Energy Supply (including Biomass)	**47**	**152**	**193**	**244**	**309**	**100**	**100**	**100**	**100**	**100**	**4.1**	**2.4**	**2.4**	**2.4**

Reference Scenario: Mexico

	Capacity (GW)				Shares (%)				Growth Rates (% per annum)		
	1999	2010	2020	2030	1999	2010	2020	2030	1999-2010	1999-2020	1999-2030
Total Capacity	**39**	**60**	**87**	**121**	**100**	**100**	**100**	**100**	**4.0**	**3.9**	**3.7**
Coal	3	3	4	5	8	6	5	4	1.4	1.5	1.7
Oil	18	22	25	28	46	36	29	23	1.7	1.7	1.5
Gas	6	19	35	58	16	31	41	48	10.6	8.6	7.5
Hydrogen-Fuel Cell	0	0	0	0	0	0	0	0	-	-	-
Nuclear	1	1	1	1	3	2	2	1	0.0	0.0	0.0
Hydro-Conventional	10	13	15	17	25	21	18	14	2.5	2.3	1.9
Hydro-Pumped Storage	0	0	0	0	0	0	0	0	-	-	-
Other renewables	1	2	5	11	2	4	6	9	9.7	8.8	8.5

	Capacity (GW)				Shares (%)				Growth Rates (% per annum)		
	1999	2010	2020	2030	1999	2010	2020	2030	1999-2010	1999-2020	1999-2030
Other Renewables	**1**	**2**	**5**	**11**	**100**	**100**	**100**	**100**	**9.7**	**8.8**	**8.5**
Biomass	0	1	1	3	13	20	19	27	14.1	10.8	11.0
Wind	0	1	2	5	0	24	38	45	61.9	36.3	27.0
Geothermal	1	1	2	2	85	53	31	17	5.1	3.7	3.0
Solar	0	0	1	1	1	2	12	10	13.9	20.1	15.6
Tide/Wave	0	0	0	0	0	0	0	0	-	-	-

	Electricity Generation (TWh)				Shares (%)				Growth Rates (% per annum)		
	2000	2010	2020	2030	2000	2010	2020	2030	2000-2010	2000-2020	2000-2030
Other Renewables	**6**	**13**	**22**	**41**	**100**	**100**	**100**	**100**	**7.2**	**6.4**	**6.4**
Biomass	0	2	4	12	7	16	19	30	16.5	11.9	11.7
Wind	0	1	5	12	0	11	22	30	60.1	34.5	25.7
Geothermal	6	9	12	14	92	73	53	34	4.7	3.5	2.9
Solar	0	0	1	2	0	0	5	5	5.6	19.6	15.4
Tide/Wave	0	0	0	0	0	0	1	1	-	-	-

Reference Scenario: Mexico

	CO$_2$ Emissions (Mt)					Shares (%)					Growth Rates (% per annum)			
	1971	2000	2010	2020	2030	1971	2000	2010	2020	2030	1971-2000	2000-2010	2000-2020	2000-2030
Total CO$_2$ Emissions	**97**	**360**	**460**	**588**	**748**	**100**	**100**	**100**	**100**	**100**	**4.6**	**2.5**	**2.5**	**2.5**
change since 1990 (%)		*23.1*	*57.5*	*101.5*	*156.0*									
Coal	5	27	30	33	39	5	8	6	6	5	5.9	0.9	1.0	1.2
Oil	72	256	310	378	461	74	71	67	64	62	4.5	1.9	2.0	2.0
Gas	20	77	121	177	248	21	21	26	30	33	4.7	4.7	4.3	4.0
Power Generation	**13**	**121**	**158**	**208**	**266**	**100**	**100**	**100**	**100**	**100**	**8.0**	**2.7**	**2.7**	**2.7**
Coal	0	21	23	26	31	1	17	14	12	12	19.7	0.9	1.1	1.3
Oil	9	75	84	97	105	69	62	53	46	39	7.6	1.1	1.3	1.1
Gas	4	25	51	86	131	30	21	33	41	49	6.7	7.4	6.3	5.6
Transformation, Own Use & Losses	**11**	**52**	**60**	**70**	**86**	**100**	**100**	**100**	**100**	**100**	**5.5**	**1.5**	**1.6**	**1.7**
Total Final Consumption	**74**	**187**	**242**	**310**	**395**	**100**	**100**	**100**	**100**	**100**	**3.3**	**2.6**	**2.6**	**2.5**
Coal	5	6	7	7	8	7	3	3	2	2	0.8	1.0	0.9	0.8
Oil	55	157	200	252	320	74	84	82	81	81	3.7	2.4	2.4	2.4
Gas	14	24	36	50	67	19	13	15	16	17	1.8	4.2	3.8	3.5
Industry	**29**	**52**	**63**	**77**	**92**	**100**	**100**	**100**	**100**	**100**	**2.1**	**1.9**	**2.0**	**1.9**
Coal	5	6	7	7	8	17	12	11	9	8	0.8	1.0	0.9	0.8
Oil	11	23	25	27	29	37	45	40	36	32	2.8	0.7	0.8	0.8
Gas	13	22	31	42	54	46	43	49	55	59	1.8	3.3	3.3	3.0
Transportation	**30**	**100**	**136**	**182**	**243**	**100**	**100**	**100**	**100**	**100**	**4.2**	**3.1**	**3.0**	**3.0**
Oil	30	100	136	182	243	100	100	100	100	100	4.2	3.1	3.0	3.0
Other fuels	0	0	0	0	0	0	0	0	0	0	0.0	0.0	0.0	0.0
Other Sectors	**12**	**31**	**39**	**46**	**54**	**100**	**100**	**100**	**100**	**100**	**3.3**	**2.4**	**2.1**	**2.0**
Coal	0	0	0	0	0	0	0	0	0	0	0.0	0.0	0.0	0.0
Oil	11	29	33	38	42	94	95	87	82	76	3.3	1.5	1.3	1.2
Gas	1	2	5	8	13	6	5	13	18	24	2.9	12.1	8.6	7.2
Non-Energy Use	**3**	**5**	**5**	**5**	**6**	**100**	**100**	**100**	**100**	**100**	**1.6**	**1.2**	**0.9**	**0.7**

Reference Scenario: OECD Europe

	Energy Demand (Mtoe)					Shares (%)					Growth Rates (% per annum)			
	1971	2000	2010	2020	2030	1971	2000	2010	2020	2030	1971-2000	2000-2010	2000-2020	2000-2030
Total Primary Energy Supply	**1244**	**1748**	**1953**	**2087**	**2191**	**100**	**100**	**100**	**100**	**100**	**1.2**	**1.1**	**0.9**	**0.8**
Coal	429	319	298	287	283	34	18	15	14	13	-1.0	-0.7	-0.5	-0.4
Oil	660	683	738	773	790	53	39	38	37	36	0.1	0.8	0.6	0.5
Gas	91	384	511	638	721	7	22	26	31	33	5.1	2.9	2.6	2.1
Nuclear	13	239	249	192	157	1	14	13	9	7	10.5	0.4	-1.1	-1.4
Hydro	28	46	50	53	56	2	3	3	3	3	1.8	0.8	0.7	0.6
Other Renewables	24	77	107	144	185	2	4	5	7	8	4.1	3.4	3.2	3.0
Power Generation	**326**	**676**	**774**	**824**	**870**	**100**	**100**	**100**	**100**	**100**	**2.6**	**1.4**	**1.0**	**0.8**
Coal	189	230	219	215	214	58	34	28	26	25	0.7	-0.5	-0.3	-0.2
Oil	74	40	32	20	13	23	6	4	2	1	-2.1	-2.3	-3.3	-3.7
Gas	19	96	182	278	339	6	14	23	34	39	5.8	6.5	5.4	4.3
Nuclear	13	239	249	192	157	4	35	32	23	18	10.5	0.4	-1.1	-1.4
Hydro	28	46	50	53	56	9	7	6	6	6	1.8	0.8	0.7	0.6
Other Renewables	4	24	43	67	90	1	4	6	8	10	6.8	5.9	5.2	4.5
Other Transformation, Own Use & Losses	**132**	**122**	**126**	**132**	**136**						**-0.3**	**0.3**	**0.4**	**0.4**
of which electricity	*21*	*42*	*49*	*54*	*58*						*2.4*	*1.6*	*1.3*	*1.1*
Total Final Consumption	**924**	**1254**	**1410**	**1541**	**1635**	**100**	**100**	**100**	**100**	**100**	**1.1**	**1.2**	**1.0**	**0.9**
Coal	187	62	55	49	46	20	5	4	3	3	-3.7	-1.3	-1.1	-1.0
Oil	525	609	676	722	746	57	49	48	47	46	0.5	1.0	0.9	0.7
Gas	77	268	307	337	356	8	21	22	22	22	4.4	1.4	1.1	0.9
Electricity	99	230	275	319	353	11	18	19	21	22	3.0	1.8	1.6	1.4
Heat	17	32	35	37	39	2	3	2	2	2	2.2	0.8	0.8	0.7
Renewables	20	52	64	77	95	2	4	5	5	6	3.4	2.0	2.0	2.0

Industry	**372**	**393**	**418**	**443**	**461**	**100**	**100**	**100**	**100**	**100**	**0.2**	**0.6**	**0.6**	**0.5**
Coal	91	48	43	39	37	24	12	10	9	8	-2.2	-1.1	-1.0	-0.9
Oil	171	110	110	109	107	46	28	26	25	23	-1.5	0.0	0.0	-0.1
Gas	42	113	125	138	145	11	29	30	31	32	3.5	1.0	1.0	0.8
Electricity	54	99	112	125	134	15	25	27	28	29	2.1	1.3	1.2	1.0
Heat	10	7	7	6	6	3	2	2	1	1	-1.2	-0.3	-0.3	-0.2
Renewables	3	16	21	26	31	1	4	5	6	7	5.7	2.6	2.3	2.2
Transportation	**169**	**360**	**432**	**485**	**520**	**100**	**100**	**100**	**100**	**100**	**2.6**	**1.8**	**1.5**	**1.2**
Oil	157	352	422	473	505	93	98	98	98	97	2.8	1.8	1.5	1.2
Other fuels	12	7	10	12	15	7	2	2	2	3	-1.6	2.6	2.5	2.4
Other Sectors	**349**	**464**	**520**	**568**	**609**	**100**	**100**	**100**	**100**	**100**	**1.0**	**1.1**	**1.0**	**0.9**
Coal	84	14	11	10	9	24	3	2	2	1	-6.0	-2.3	-1.8	-1.5
Oil	165	110	103	96	88	47	24	20	17	14	-1.4	-0.6	-0.7	-0.7
Gas	34	154	180	197	209	10	33	35	35	34	5.3	1.6	1.2	1.0
Electricity	41	125	155	184	208	12	27	30	32	34	3.9	2.1	1.9	1.7
Heat	8	25	28	31	33	2	5	5	5	5	4.2	1.1	1.0	0.9
Renewables	17	36	42	51	62	5	8	8	9	10	2.6	1.7	1.8	1.9
Non-Energy Use	**35**	**36**	**41**	**44**	**46**						**0.1**	**1.1**	**1.0**	**0.8**
Electricity Generation (TWh)	**1392**	**3164**	**3763**	**4339**	**4777**	**100**	**100**	**100**	**100**	**100**	**2.9**	**1.7**	**1.6**	**1.4**
Coal	617	943	892	948	1037	44	30	24	22	22	1.5	-0.6	0.0	0.3
Oil	316	178	143	93	62	23	6	4	2	1	-2.0	-2.2	-3.2	-3.5
Gas	76	508	1007	1625	1889	5	16	27	37	40	6.8	7.1	6.0	4.5
Hydrogen-Fuel Cell	0	0	0	5	105	0	0	0	0	2	-	-	-	-
Nuclear	51	918	954	736	603	4	29	25	17	13	10.5	0.4	-1.1	-1.4
Hydro	322	538	581	616	652	23	17	15	14	14	1.8	0.8	0.7	0.6
Other Renewables	9	79	186	315	429	1	2	5	7	9	7.8	8.9	7.2	5.8

Biomass (included above)	21	69	90	118	148	2	4	5	6	7	4.1	2.7	2.7	2.6
Total Primary Energy Supply (including Biomass)	**1244**	**1748**	**1953**	**2087**	**2191**	**100**	**100**	**100**	**100**	**100**	**1.2**	**1.1**	**0.9**	**0.8**

Reference Scenario: OECD Europe

	Capacity (GW)				Shares (%)				Growth Rates (% per annum)		
	1999	2010	2020	2030	1999	2010	2020	2030	1999-2010	1999-2020	1999-2030
Total Capacity	**698**	**831**	**973**	**1109**	**100**	**100**	**100**	**100**	**1.6**	**1.6**	**1.5**
Coal	193	180	170	194	28	22	17	17	-0.6	-0.6	0.0
Oil	84	84	61	37	12	10	6	3	0.0	-1.5	-2.6
Gas	98	196	353	434	14	24	36	39	6.5	6.3	4.9
Hydrogen-Fuel Cell	0	0	1	30	0	0	0	3	-	-	-
Nuclear	131	126	95	78	19	15	10	7	-0.3	-1.5	-1.7
Hydro-Conventional	135	154	163	172	19	19	17	16	1.2	0.9	0.8
Hydro-Pumped Storage	37	38	38	39	5	5	4	4	0.2	0.1	0.1
Other renewables	19	53	92	126	3	6	9	11	9.7	7.8	6.3

	Capacity (GW)				Shares (%)				Growth Rates (% per annum)		
	1999	2010	2020	2030	1999	2010	2020	2030	1999-2010	1999-2020	1999-2030
Other Renewables	**19**	**53**	**92**	**126**	**100**	**100**	**100**	**100**	**9.7**	**7.8**	**6.3**
Biomass	9	16	26	37	48	31	28	29	5.4	5.1	4.6
Wind	9	33	59	73	46	64	65	58	13.0	9.5	7.1
Geothermal	1	1	1	1	4	2	1	1	1.1	1.1	1.0
Solar	0	2	4	13	1	3	4	10	24.9	17.5	15.7
Tide/Wave	0	1	1	2	1	1	2	2	8.6	8.9	7.5

	Electricity Generation (TWh)				Shares (%)				Growth Rates (% per annum)		
	2000	2010	2020	2030	2000	2010	2020	2030	2000-2010	2000-2020	2000-2030
Other Renewables	**79**	**186**	**315**	**429**	**100**	**100**	**100**	**100**	**9.0**	**7.2**	**5.8**
Biomass	50	89	144	204	63	48	46	47	6.1	5.5	4.8
Wind	22	87	155	194	29	47	49	45	14.5	10.1	7.4
Geothermal	6	6	7	8	8	3	2	2	0.6	0.8	0.9
Solar	0	1	6	19	0	1	2	4	29.7	21.9	18.6
Tide/Wave	1	1	3	5	1	1	1	1	9.7	9.4	7.8

Reference Scenario: OECD Europe

	CO$_2$ Emissions (Mt)					Shares (%)					Growth Rates (% per annum)			
	1971	2000	2010	2020	2030	1971	2000	2010	2020	2030	1971-2000	2000-2010	2000-2020	2000-2030
Total CO$_2$ Emissions	**3635**	**3890**	**4249**	**4573**	**4778**	**100**	**100**	**100**	**100**	**100**	**0.2**	**0.9**	**0.8**	**0.7**
change since 1990 (%)		*0.4*	*9.6*	*18.0*	*23.3*									
Coal	1682	1251	1186	1140	1122	46	32	28	25	23	-1.0	-0.5	-0.5	-0.4
Oil	1764	1763	1889	1962	1994	49	45	44	43	42	0.0	0.7	0.5	0.4
Gas	188	876	1174	1471	1662	5	23	28	32	35	5.4	3.0	2.6	2.2
Power Generation	**1054**	**1301**	**1429**	**1597**	**1715**	**100**	**100**	**100**	**100**	**100**	**0.7**	**0.9**	**1.0**	**0.9**
Coal	774	948	903	884	883	73	73	63	55	52	0.7	-0.5	-0.3	-0.2
Oil	236	128	101	65	41	22	10	7	4	2	-2.1	-2.3	-3.3	-3.7
Gas	43	224	424	648	790	4	17	30	41	46	5.8	6.6	5.5	4.3
Transformation, Own Use & Losses	**164**	**182**	**172**	**169**	**167**	**100**	**100**	**100**	**100**	**100**	**0.3**	**-0.5**	**-0.4**	**-0.3**
Total Final Consumption	**2417**	**2407**	**2648**	**2807**	**2895**	**100**	**100**	**100**	**100**	**100**	**0.0**	**1.0**	**0.8**	**0.6**
Coal	848	277	267	244	229	35	12	10	9	8	-3.8	-0.4	-0.6	-0.6
Oil	1429	1514	1672	1785	1843	59	63	63	64	64	0.2	1.0	0.8	0.7
Gas	140	616	709	778	824	6	26	27	28	28	5.2	1.4	1.2	1.0
Industry	**1003**	**690**	**715**	**723**	**725**	**100**	**100**	**100**	**100**	**100**	**-1.3**	**0.4**	**0.2**	**0.2**
Coal	440	220	214	196	184	44	32	30	27	25	-2.4	-0.3	-0.6	-0.6
Oil	473	216	216	214	211	47	31	30	30	29	-2.7	0.0	0.0	-0.1
Gas	90	254	285	313	330	9	37	40	43	46	3.7	1.1	1.0	0.9
Transportation	**457**	**934**	**1116**	**1253**	**1337**	**100**	**100**	**100**	**100**	**100**	**2.5**	**1.8**	**1.5**	**1.2**
Oil	424	932	1113	1248	1331	93	100	100	100	100	2.8	1.8	1.5	1.2
Other fuels	33	2	3	4	6	7	0	0	0	0	-9.7	5.7	4.8	4.4
Other Sectors	**916**	**747**	**776**	**787**	**788**	**100**	**100**	**100**	**100**	**100**	**-0.7**	**0.4**	**0.3**	**0.2**
Coal	362	54	43	38	35	39	7	6	5	4	-6.3	-2.3	-1.8	-1.5
Oil	505	332	312	288	265	55	44	40	37	34	-1.4	-0.6	-0.7	-0.7
Gas	50	360	421	461	487	5	48	54	59	62	7.1	1.6	1.2	1.0
Non-Energy Use	**41**	**37**	**41**	**44**	**46**	**100**	**100**	**100**	**100**	**100**	**-0.4**	**1.0**	**0.9**	**0.7**

Reference Scenario: European Union

	Energy Demand (Mtoe)					Shares (%)					Growth Rates (% per annum)			
	1971	2000	2010	2020	2030	1971	2000	2010	2020	2030	1971-2000	2000-2010	2000-2020	2000-2030
Total Primary Energy Supply	**1041**	**1456**	**1625**	**1729**	**1811**	**100**	**100**	**100**	**100**	**100**	**1.2**	**1.1**	**0.9**	**0.7**
Coal	307	212	191	186	180	29	15	12	11	10	-1.3	-1.0	-0.7	-0.6
Oil	606	593	635	659	670	58	41	39	38	37	-0.1	0.7	0.5	0.4
Gas	82	339	453	556	620	8	23	28	32	34	5.0	2.9	2.5	2.0
Nuclear	13	225	232	179	153	1	15	14	10	8	10.4	0.3	-1.2	-1.3
Hydro	19	27	28	30	31	2	2	2	2	2	1.2	0.3	0.4	0.4
Other Renewables	16	60	87	120	157	1	4	5	7	9	4.8	3.8	3.6	3.3
Power Generation	**263**	**555**	**638**	**677**	**714**	**100**	**100**	**100**	**100**	**100**	**2.6**	**1.4**	**1.0**	**0.8**
Coal	142	162	147	147	144	54	29	23	22	20	0.5	-1.0	-0.5	-0.4
Oil	69	35	27	16	10	26	6	4	2	1	-2.3	-2.6	-3.8	-4.0
Gas	16	84	165	244	292	6	15	26	36	41	5.8	7.0	5.5	4.2
Nuclear	13	225	232	179	153	5	41	36	26	21	10.4	0.3	-1.2	-1.3
Hydro	19	27	28	30	31	7	5	4	4	4	1.2	0.3	0.4	0.4
Other Renewables	3	21	39	61	83	1	4	6	9	12	6.6	6.2	5.4	4.7
Other Transformation, Own Use & Losses	**112**	**90**	**93**	**96**	**97**						**-0.7**	**0.3**	**0.3**	**0.2**
of which electricity	*17*	*29*	*35*	*39*	*40*						*1.8*	*2.0*	*1.5*	*1.1*
Total Final Consumption	**772**	**1052**	**1181**	**1286**	**1359**	**100**	**100**	**100**	**100**	**100**	**1.1**	**1.2**	**1.0**	**0.9**
Coal	127	32	28	24	21	16	3	2	2	2	-4.7	-1.3	-1.4	-1.3
Oil	478	529	581	617	633	62	50	49	48	47	0.4	0.9	0.8	0.6
Gas	68	240	272	296	312	9	23	23	23	23	4.5	1.2	1.0	0.9
Electricity	83	192	228	263	289	11	18	19	20	21	2.9	1.7	1.6	1.4
Heat	5	20	24	27	30	1	2	2	2	2	5.1	1.6	1.5	1.3
Renewables	12	38	48	59	74	2	4	4	5	5	4.0	2.2	2.2	2.2

Industry	**309**	**325**	**346**	**367**	**380**	**100**	**100**	**100**	**100**	**100**	**0.2**	**0.6**	**0.6**	**0.5**
Coal	65	26	23	20	18	21	8	7	5	5	-3.1	-1.3	-1.3	-1.3
Oil	159	96	96	94	92	51	30	28	26	24	-1.7	0.0	-0.1	-0.1
Gas	36	102	111	122	129	12	31	32	33	34	3.6	0.9	0.9	0.8
Electricity	45	83	93	104	111	15	25	27	28	29	2.1	1.2	1.2	1.0
Heat	1	4	4	5	5	0	1	1	1	1	3.9	0.7	0.7	0.7
Renewables	3	14	18	22	26	1	4	5	6	7	5.3	2.5	2.2	2.0
Transportation	**145**	**317**	**377**	**421**	**448**	**100**	**100**	**100**	**100**	**100**	**2.7**	**1.7**	**1.4**	**1.2**
Oil	140	311	369	411	436	96	98	98	98	97	2.8	1.7	1.4	1.1
Other fuels	5	6	8	10	12	4	2	2	2	3	0.3	2.5	2.4	2.3
Other Sectors	**288**	**380**	**426**	**463**	**493**	**100**	**100**	**100**	**100**	**100**	**1.0**	**1.1**	**1.0**	**0.9**
Coal	57	5	4	3	3	20	1	1	1	1	-8.1	-1.6	-1.7	-1.7
Oil	151	92	84	76	67	53	24	20	16	14	-1.7	-0.9	-1.0	-1.0
Gas	31	138	159	173	181	11	36	37	37	37	5.2	1.5	1.1	0.9
Electricity	36	105	129	152	169	12	28	30	33	34	3.8	2.1	1.9	1.6
Heat	4	17	20	23	25	1	4	5	5	5	5.4	1.8	1.6	1.4
Renewables	9	24	29	37	47	3	6	7	8	9	3.4	2.0	2.1	2.2
Non-Energy Use	**30**	**30**	**33**	**36**	**37**						**0.0**	**0.9**	**0.9**	**0.7**
Electricity Generation (TWh)	**1165**	**2572**	**3064**	**3511**	**3834**	**100**	**100**	**100**	**100**	**100**	**2.8**	**1.8**	**1.6**	**1.3**
Coal	509	704	640	687	724	44	27	21	20	19	1.1	-1.0	-0.1	0.1
Oil	300	161	126	77	51	26	6	4	2	1	-2.1	-2.4	-3.6	-3.8
Gas	72	450	905	1414	1598	6	18	30	40	42	6.5	7.2	5.9	4.3
Hydrogen-Fuel Cell	0	0	0	5	105	0	0	0	0	3	-	-	-	-
Nuclear	49	864	889	685	588	4	34	29	20	15	10.4	0.3	-1.2	-1.3
Hydro	225	319	328	343	360	19	12	11	10	9	1.2	0.3	0.4	0.4
Other Renewables	9	74	176	300	409	1	3	6	9	11	7.7	9.1	7.3	5.9

Biomass (included above)	14	54	73	98	124	1	4	4	6	7	4.9	3.0	3.0	2.8
Total Primary Energy Supply (including Biomass)	**1041**	**1456**	**1625**	**1729**	**1811**	**100**	**100**	**100**	**100**	**100**	**1.2**	**1.1**	**0.9**	**0.7**

Reference Scenario: European Union

	Capacity (GW)				Shares (%)				Growth Rates (% per annum)		
	1999	2010	2020	2030	1999	2010	2020	2030	1999-2010	1999-2020	1999-2030
Total Capacity	**573**	**679**	**792**	**901**	**100**	**100**	**100**	**100**	**1.6**	**1.6**	**1.5**
Coal	146	134	122	136	25	20	15	15	-0.8	-0.9	-0.2
Oil	78	77	55	33	14	11	7	4	-0.1	-1.7	-2.7
Gas	90	176	310	372	16	26	39	41	6.3	6.1	4.7
Hydrogen-Fuel Cell	0	0	1	30	0	0	0	3	-	-	-
Nuclear	124	118	88	76	22	17	11	8	-0.5	-1.6	-1.6
Hydro-Conventional	84	91	95	100	15	13	12	11	0.7	0.6	0.5
Hydro-Pumped Storage	33	33	33	34	6	5	4	4	0.1	0.1	0.1
Other renewables	18	50	87	120	3	7	11	13	9.8	7.8	6.3

	Capacity (GW)				Shares (%)				Growth Rates (% per annum)		
	1999	2010	2020	2030	1999	2010	2020	2030	1999-2010	1999-2020	1999-2030
Other Renewables	**18**	**50**	**87**	**120**	**100**	**100**	**100**	**100**	**9.8**	**7.8**	**6.3**
Biomass	8	15	23	33	46	29	27	28	5.3	5.1	4.6
Wind	9	33	57	71	48	65	66	59	12.8	9.4	7.0
Geothermal	1	1	1	1	3	1	1	1	1.0	1.0	1.0
Solar	0	2	4	13	1	3	5	10	25.9	18.1	16.1
Tide/Wave	0	1	1	2	1	1	2	2	8.6	8.9	7.5

	Electricity Generation (TWh)				Shares (%)				Growth Rates (% per annum)		
	2000	2010	2020	2030	2000	2010	2020	2030	2000-2010	2000-2020	2000-2030
Other Renewables	**74**	**176**	**300**	**409**	**100**	**100**	**100**	**100**	**9.1**	**7.3**	**5.9**
Biomass	46	83	134	190	62	47	45	46	6.1	5.5	4.8
Wind	22	85	151	189	30	48	50	46	14.3	10.0	7.4
Geothermal	5	5	6	6	6	3	2	1	0.4	0.7	0.8
Solar	0	1	6	18	0	1	2	5	30.7	22.4	19.0
Tide/Wave	1	1	3	5	1	1	1	1	9.7	9.4	7.8

Reference Scenario: European Union

	CO_2 Emissions (Mt)					Shares (%)					Growth Rates (% per annum)			
	1971	2000	2010	2020	2030	1971	2000	2010	2020	2030	1971-2000	2000-2010	2000-2020	2000-2030
Total CO_2 Emissions	**3015**	**3146**	**3422**	**3689**	**3829**	**100**	**100**	**100**	**100**	**100**	**0.1**	**0.8**	**0.8**	**0.7**
change since 1990 (%)		*1.1*	*10.0*	*18.6*	*23.1*									
Coal	1228	840	757	736	709	41	27	22	20	19	-1.3	-1.0	-0.7	-0.6
Oil	1618	1532	1626	1673	1689	54	49	48	45	44	-0.2	0.6	0.4	0.3
Gas	168	774	1039	1280	1431	6	25	30	35	37	5.4	3.0	2.6	2.1
Power Generation	**844**	**981**	**1084**	**1232**	**1315**	**100**	**100**	**100**	**100**	**100**	**0.5**	**1.0**	**1.1**	**1.0**
Coal	586	673	611	610	598	69	69	56	49	46	0.5	-1.0	-0.5	-0.4
Oil	221	112	86	52	33	26	11	8	4	3	-2.3	-2.6	-3.8	-4.0
Gas	38	196	387	571	683	5	20	36	46	52	5.8	7.0	5.5	4.3
Transformation, Own Use & Losses	**149**	**150**	**139**	**134**	**130**	**100**	**100**	**100**	**100**	**100**	**0.0**	**-0.8**	**-0.6**	**-0.5**
Total Final Consumption	**2021**	**2014**	**2199**	**2323**	**2385**	**100**	**100**	**100**	**100**	**100**	**0.0**	**0.9**	**0.7**	**0.6**
Coal	594	150	132	114	101	29	7	6	5	4	-4.6	-1.3	-1.4	-1.3
Oil	1301	1312	1442	1528	1566	64	65	66	66	66	0.0	1.0	0.8	0.6
Gas	126	553	625	681	718	6	27	28	29	30	5.2	1.2	1.0	0.9
Industry	**836**	**541**	**547**	**553**	**553**	**100**	**100**	**100**	**100**	**100**	**-1.5**	**0.1**	**0.1**	**0.1**
Coal	323	128	113	98	87	39	24	21	18	16	-3.1	-1.3	-1.3	-1.3
Oil	435	183	182	180	176	52	34	33	33	32	-2.9	0.0	-0.1	-0.1
Gas	78	230	251	275	290	9	42	46	50	52	3.8	0.9	0.9	0.8
Transportation	**389**	**823**	**976**	**1089**	**1156**	**100**	**100**	**100**	**100**	**100**	**2.6**	**1.7**	**1.4**	**1.1**
Oil	377	822	974	1086	1151	97	100	100	100	100	2.7	1.7	1.4	1.1
Other fuels	12	1	2	3	5	3	0	0	0	0	-7.1	4.4	4.3	4.2
Other Sectors	**760**	**620**	**643**	**645**	**638**	**100**	**100**	**100**	**100**	**100**	**-0.7**	**0.4**	**0.2**	**0.1**
Coal	249	20	17	14	12	33	3	3	2	2	-8.4	-1.6	-1.7	-1.7
Oil	464	278	255	228	203	61	45	40	35	32	-1.7	-0.9	-1.0	-1.0
Gas	47	322	372	403	423	6	52	58	62	66	6.8	1.5	1.1	0.9
Non-Energy Use	**36**	**31**	**33**	**36**	**37**	**100**	**100**	**100**	**100**	**100**	**-0.6**	**0.8**	**0.8**	**0.7**

Reference Scenario: OECD Pacific

	Energy Demand (Mtoe)					Shares (%)					Growth Rates (% per annum)			
	1971	2000	2010	2020	2030	1971	2000	2010	2020	2030	1971-2000	2000-2010	2000-2020	2000-2030
Total Primary Energy Supply	**346**	**847**	**1001**	**1127**	**1200**	**100**	**100**	**100**	**100**	**100**	**3.1**	**1.7**	**1.4**	**1.2**
Coal	84	184	205	221	215	24	22	20	20	18	2.7	1.1	0.9	0.5
Oil	240	412	458	496	505	69	49	46	44	42	1.9	1.1	0.9	0.7
Gas	5	106	145	174	210	2	13	15	15	18	11.0	3.2	2.5	2.3
Nuclear	2	112	150	184	210	1	13	15	16	17	14.7	2.9	2.5	2.1
Hydro	9	11	13	14	14	3	1	1	1	1	0.6	1.3	0.9	0.8
Other Renewables	5	21	29	37	47	1	3	3	3	4	5.4	3.3	2.9	2.7
Power Generation	**92**	**342**	**434**	**506**	**547**	**100**	**100**	**100**	**100**	**100**	**4.6**	**2.4**	**2.0**	**1.6**
Coal	27	113	132	147	139	29	33	30	29	25	5.1	1.6	1.3	0.7
Oil	51	37	40	38	24	56	11	9	8	4	-1.2	1.0	0.2	-1.4
Gas	2	57	80	98	128	2	17	18	19	23	13.0	3.5	2.8	2.7
Nuclear	2	112	150	184	210	2	33	34	36	38	14.7	2.9	2.5	2.1
Hydro	9	11	13	14	14	10	3	3	3	3	0.6	1.3	0.9	0.8
Other Renewables	1	12	18	24	32	1	3	4	5	6	8.4	4.8	3.8	3.4
Other Transformation, Own Use & Losses	**39**	**86**	**86**	**93**	**96**						**2.7**	**0.1**	**0.4**	**0.4**
of which electricity	*5*	*16*	*19*	*21*	*22*						*4.3*	*1.7*	*1.4*	*1.0*
Total Final Consumption	**254**	**561**	**657**	**733**	**781**	**100**	**100**	**100**	**100**	**100**	**2.8**	**1.6**	**1.3**	**1.1**
Coal	30	33	34	34	34	12	6	5	5	4	0.4	0.1	0.1	0.1
Oil	178	343	392	429	451	70	61	60	59	58	2.3	1.3	1.1	0.9
Gas	8	48	63	73	79	3	9	10	10	10	6.6	2.7	2.1	1.7
Electricity	35	124	154	178	195	14	22	23	24	25	4.4	2.2	1.9	1.5
Heat	0	3	4	5	6	0	0	1	1	1	-	3.7	3.2	2.8
Renewables	4	10	11	13	15	1	2	2	2	2	3.5	1.3	1.5	1.5

Industry	**136**	**226**	**256**	**281**	**296**	**100**	**100**	**100**	**100**	**100**	**1.8**	**1.3**	**1.1**	**0.9**
Coal	22	32	33	33	33	16	14	13	12	11	1.3	0.3	0.2	0.2
Oil	86	109	120	129	134	63	48	47	46	45	0.8	1.0	0.9	0.7
Gas	3	22	28	32	35	2	10	11	12	12	7.0	2.5	1.9	1.5
Electricity	23	57	68	78	85	17	25	27	28	29	3.1	1.8	1.6	1.3
Heat	0	0	0	0	0	0	0	0	0	0	-	-	-	-
Renewables	2	6	7	8	9	1	3	3	3	3	5.0	1.0	1.2	1.2
Transportation	**53**	**157**	**188**	**213**	**228**	**100**	**100**	**100**	**100**	**100**	**3.8**	**1.8**	**1.5**	**1.3**
Oil	51	155	185	209	225	97	98	98	98	99	3.9	1.8	1.5	1.3
Other fuels	2	3	3	3	3	3	2	2	2	1	1.3	1.2	1.1	1.0
Other Sectors	**49**	**160**	**193**	**218**	**233**	**100**	**100**	**100**	**100**	**100**	**4.2**	**1.9**	**1.5**	**1.3**
Coal	8	2	1	1	1	15	1	1	0	0	-5.1	-4.2	-2.4	-1.9
Oil	25	63	68	71	71	50	40	35	33	31	3.3	0.7	0.5	0.4
Gas	4	26	34	41	44	9	16	18	19	19	6.3	2.9	2.3	1.8
Electricity	11	64	83	97	107	21	40	43	45	46	6.4	2.6	2.1	1.7
Heat	0	2	2	3	3	0	1	1	1	1	-	2.8	2.6	2.3
Renewables	2	3	4	5	6	4	2	2	2	3	1.8	1.9	2.0	1.9
Non-Energy Use	**16**	**18**	**20**	**22**	**24**						**0.3**	**1.2**	**1.1**	**0.9**
Electricity Generation (TWh)	**462**	**1622**	**2003**	**2317**	**2519**	**100**	**100**	**100**	**100**	**100**	**4.4**	**2.1**	**1.8**	**1.5**
Coal	85	542	621	697	670	18	33	31	30	27	6.6	1.4	1.3	0.7
Oil	250	186	207	196	122	54	11	10	8	5	-1.0	1.1	0.3	-1.4
Gas	7	304	398	478	526	2	19	20	21	21	13.7	2.7	2.3	1.8
Hydrogen-Fuel Cell	0	0	0	5	122	0	0	0	0	5	-	-	-	-
Nuclear	8	431	574	707	804	2	27	29	31	32	14.7	2.9	2.5	2.1
Hydro	110	133	151	159	168	24	8	8	7	7	0.6	1.3	0.9	0.8
Other Renewables	2	26	52	74	108	0	2	3	3	4	10.2	7.3	5.4	4.9

Biomass (included above)	4	15	19	24	29	1	2	2	2	2	5.0	2.8	2.4	2.3
Total Primary Energy Supply (including Biomass)	**346**	**847**	**1001**	**1127**	**1200**	**100**	**100**	**100**	**100**	**100**	**3.1**	**1.7**	**1.4**	**1.2**

Reference Scenario: OECD Pacific

	Capacity (GW)				Shares (%)				Growth Rates (% per annum)		
	1999	2010	2020	2030	1999	2010	2020	2030	1999-2010	1999-2020	1999-2030
Total Capacity	**356**	**453**	**523**	**591**	**100**	**100**	**100**	**100**	**2.2**	**1.9**	**1.6**
Coal	78	97	107	103	22	21	20	17	2.0	1.5	0.9
Oil	90	97	92	62	25	21	18	10	0.7	0.1	-1.2
Gas	63	97	130	169	18	21	25	29	4.0	3.5	3.3
Hydrogen-Fuel Cell	0	0	2	35	0	0	0	6	-	-	-
Nuclear	57	77	94	105	16	17	18	18	2.7	2.4	2.0
Hydr Conventional	34	40	42	44	10	9	8	7	1.3	1.0	0.8
Hydro-Pumped Storage	27	31	34	36	8	7	7	6	1.0	1.1	0.9
Other renewables	6	15	23	36	2	3	4	6	8.9	6.9	6.1

	Capacity (GW)				Shares (%)				Growth Rates (% per annum)		
	1999	2010	2020	2030	1999	2010	2020	2030	1999-2010	1999-2020	1999-2030
Other Renewables	**6**	**15**	**23**	**36**	**100**	**100**	**100**	**100**	**8.9**	**6.9**	**6.1**
Biomass	5	9	11	13	79	58	46	36	5.9	4.2	3.5
Wind	0	2	4	9	2	12	17	24	31.5	20.0	16.0
Geothermal	1	1	2	2	16	8	6	5	2.6	2.5	2.2
Solar	0	3	7	12	4	22	31	32	27.4	17.9	13.6
Tide/Wave	0	0	0	1	0	0	0	2	-	-	-

	Electricity Generation (TWh)				Shares (%)				Growth Rates (% per annum)		
	2000	2010	2020	2030	2000	2010	2020	2030	2000-2010	2000-2020	2000-2030
Other Renewables	**26**	**52**	**74**	**108**	**100**	**100**	**100**	**100**	**7.3**	**5.4**	**4.9**
Biomass	19	34	43	53	75	66	58	49	5.9	4.1	3.4
Wind	0	4	10	22	1	8	13	20	30.5	19.0	15.3
Geothermal	6	9	11	13	24	17	15	12	3.9	3.1	2.6
Solar	0	5	10	18	0	9	13	17	47.6	26.4	19.3
Tide/Wave	0	0	0	2	0	0	0	2	-	-	-

Reference Scenario: OECD Pacific

	CO$_2$ Emissions (Mt)					Shares (%)					Growth Rates (% per annum)			
	1971	2000	2010	2020	2030	1971	2000	2010	2020	2030	1971-2000	2000-2010	2000-2020	2000-2030
Total CO$_2$ Emissions	**951**	**1945**	**2251**	**2479**	**2545**	**100**	**100**	**100**	**100**	**100**	**2.5**	**1.5**	**1.2**	**0.9**
change since 1990 (%)		*27.5*	*47.5*	*62.5*	*66.8*									
Coal	293	705	790	859	828	31	36	35	35	33	3.1	1.1	1.0	0.5
Oil	646	995	1125	1215	1227	68	51	50	49	48	1.5	1.2	1.0	0.7
Gas	13	245	336	405	490	1	13	15	16	19	10.7	3.2	2.5	2.3
Power Generation	**290**	**740**	**890**	**990**	**982**	**100**	**100**	**100**	**100**	**100**	**3.3**	**1.9**	**1.5**	**0.9**
Coal	123	492	576	641	609	43	67	65	65	62	4.9	1.6	1.3	0.7
Oil	163	116	128	121	76	56	16	14	12	8	-1.2	1.0	0.2	-1.4
Gas	4	132	186	228	297	1	18	21	23	30	13.0	3.5	2.8	2.7
Transformation, Own Use & Losses	**43**	**105**	**109**	**120**	**126**	**100**	**100**	**100**	**100**	**100**	**3.1**	**0.4**	**0.7**	**0.6**
Total Final Consumption	**618**	**1099**	**1252**	**1369**	**1437**	**100**	**100**	**100**	**100**	**100**	**2.0**	**1.3**	**1.1**	**0.9**
Coal	154	171	171	173	172	25	16	14	13	12	0.4	0.0	0.1	0.0
Oil	455	824	943	1035	1090	74	75	75	76	76	2.1	1.4	1.1	0.9
Gas	8	105	138	161	174	1	10	11	12	12	9.2	2.7	2.2	1.7
Industry	**327**	**397**	**431**	**456**	**469**	**100**	**100**	**100**	**100**	**100**	**0.7**	**0.8**	**0.7**	**0.6**
Coal	114	160	163	165	165	35	40	38	36	35	1.2	0.2	0.2	0.1
Oil	207	190	208	221	228	63	48	48	48	49	-0.3	0.9	0.8	0.6
Gas	6	47	61	70	75	2	12	14	15	16	7.6	2.6	1.9	1.6
Transportation	**147**	**428**	**513**	**583**	**627**	**100**	**100**	**100**	**100**	**100**	**3.7**	**1.8**	**1.6**	**1.3**
Oil	145	427	512	582	626	98	100	100	100	100	3.8	1.8	1.6	1.3
Other fuels	3	1	1	1	1	2	0	0	0	0	-3.5	-1.4	0.5	0.8
Other Sectors	**136**	**259**	**290**	**313**	**323**	**100**	**100**	**100**	**100**	**100**	**2.3**	**1.2**	**0.9**	**0.7**
Coal	37	11	8	8	7	27	4	3	3	2	-4.2	-2.4	-1.5	-1.2
Oil	96	191	206	215	218	71	74	71	69	68	2.4	0.8	0.6	0.4
Gas	3	57	76	90	98	2	22	26	29	30	11.3	2.9	2.3	1.8
Non-Energy Use	**8**	**16**	**17**	**18**	**18**	**100**	**100**	**100**	**100**	**100**	**2.4**	**0.5**	**0.5**	**0.4**

Reference Scenario: Japan, Australia and New Zealand

	Energy Demand (Mtoe)					Shares (%)					Growth Rates (% per annum)			
	1971	2000	2010	2020	2030	1971	2000	2010	2020	2030	1971-2000	2000-2010	2000-2020	2000-2030
Total Primary Energy Supply	**329**	**653**	**737**	**796**	**823**	**100**	**100**	**100**	**100**	**100**	**2.4**	**1.2**	**1.0**	**0.8**
Coal	78	142	147	147	135	24	22	20	18	16	2.1	0.4	0.2	-0.2
Oil	229	308	332	346	340	70	47	45	43	41	1.0	0.8	0.6	0.3
Gas	5	89	113	127	149	2	14	15	16	18	10.3	2.4	1.8	1.7
Nuclear	2	84	105	130	145	1	13	14	16	18	13.6	2.3	2.2	1.8
Hydro	9	11	13	13	14	3	2	2	2	2	0.6	1.3	0.9	0.7
Other Renewables	5	19	27	33	40	1	3	4	4	5	5.0	3.5	2.8	2.5
Power Generation	**90**	**274**	**328**	**365**	**380**	**100**	**100**	**100**	**100**	**100**	**3.9**	**1.8**	**1.4**	**1.1**
Coal	26	87	93	94	82	29	32	29	26	22	4.2	0.7	0.4	-0.2
Oil	49	31	35	33	20	55	11	11	9	5	-1.6	1.4	0.4	-1.4
Gas	2	51	66	75	94	2	19	20	20	25	12.6	2.5	1.9	2.0
Nuclear	2	84	105	130	145	2	31	32	36	38	13.6	2.3	2.2	1.8
Hydro	9	11	13	13	14	10	4	4	4	4	0.6	1.3	0.9	0.7
Other Renewables	1	10	16	21	25	1	3	5	6	7	7.7	5.3	3.9	3.3
Other Transformation, Own Use & Losses	**37**	**63**	**64**	**67**	**68**						**1.8**	**0.2**	**0.3**	**0.3**
of which electricity	*5*	*13*	*15*	*16*	*16*						*3.7*	*1.4*	*1.0*	*0.7*
Total Final Consumption	**241**	**432**	**480**	**514**	**532**	**100**	**100**	**100**	**100**	**100**	**2.0**	**1.0**	**0.9**	**0.7**
Coal	25	26	26	25	25	10	6	5	5	5	0.2	-0.2	-0.2	-0.2
Oil	170	257	278	293	299	71	60	58	57	56	1.4	0.8	0.6	0.5
Gas	8	37	45	50	53	3	9	9	10	10	5.7	2.0	1.5	1.2
Electricity	34	101	119	132	140	14	23	25	26	26	3.8	1.6	1.3	1.1
Heat	0	1	1	1	1	0	0	0	0	0	-	3.4	3.0	2.4
Renewables	4	9	11	13	14	1	2	2	2	3	3.4	1.3	1.5	1.4

Industry	**130**	**168**	**181**	**192**	**197**	**100**	**100**	**100**	**100**	**100**	**0.9**	**0.8**	**0.7**	**0.5**
Coal	22	25	25	24	24	17	15	14	13	12	0.6	-0.2	-0.2	-0.2
Oil	81	73	76	78	79	62	43	42	41	40	-0.4	0.5	0.4	0.3
Gas	3	20	24	27	28	2	12	13	14	14	6.5	2.0	1.5	1.2
Electricity	23	44	50	55	57	18	26	27	28	29	2.3	1.2	1.1	0.9
Heat	0	0	0	0	0	0	0	0	0	0	-	-	-	-
Renewables	2	6	7	8	9	1	4	4	4	4	5.0	1.0	1.2	1.2
Transportation	**51**	**127**	**143**	**155**	**162**	**100**	**100**	**100**	**100**	**100**	**3.2**	**1.2**	**1.0**	**0.8**
Oil	49	125	140	152	159	97	98	98	98	98	3.3	1.2	1.0	0.8
Other fuels	2	2	3	3	3	3	2	2	2	2	1.1	1.1	1.0	0.9
Other Sectors	**44**	**124**	**142**	**153**	**159**	**100**	**100**	**100**	**100**	**100**	**3.6**	**1.4**	**1.1**	**0.8**
Coal	3	1	1	1	1	6	1	1	1	1	-3.3	-0.2	-0.4	-0.5
Oil	24	47	48	48	47	56	38	34	31	30	2.3	0.3	0.1	0.0
Gas	4	18	21	23	24	10	14	15	15	15	4.9	2.0	1.4	1.0
Electricity	10	55	67	75	80	23	44	47	49	50	5.9	2.0	1.6	1.3
Heat	0	1	1	1	1	0	0	1	1	1	-	3.4	3.0	2.4
Renewables	2	3	4	5	6	5	3	3	3	4	1.6	1.9	2.0	1.9
Non-Energy Use	**16**	**13**	**14**	**14**	**14**						**-0.6**	**0.3**	**0.3**	**0.3**
Electricity Generation (TWh)	**451**	**1329**	**1557**	**1725**	**1818**	**100**	**100**	**100**	**100**	**100**	**3.8**	**1.6**	**1.3**	**1.0**
Coal	84	416	435	444	394	19	31	28	26	22	5.7	0.5	0.3	-0.2
Oil	242	162	184	174	104	54	12	12	10	6	-1.4	1.3	0.4	-1.5
Gas	7	276	337	383	405	2	21	22	22	22	13.3	2.0	1.7	1.3
Hydrogen-Fuel Cell	0	0	0	5	105	0	0	0	0	6	-	-	-	-
Nuclear	8	322	403	497	556	2	24	26	29	31	13.6	2.3	2.2	1.8
Hydro	109	129	146	153	161	24	10	9	9	9	0.6	1.3	0.9	0.7
Other Renewables	2	25	50	69	92	0	2	3	4	5	10.2	7.2	5.2	4.4

Biomass (included above)	4	12	16	19	22	1	2	2	2	3	4.4	2.9	2.3	2.0
Total Primary Energy Supply (including Biomass)	**329**	**653**	**737**	**796**	**823**	**100**	**100**	**100**	**100**	**100**	**2.4**	**1.2**	**1.0**	**0.8**

Reference Scenario: Japan, Australia and New Zealand

	Capacity (GW)				Shares (%)				Growth Rates (% per annum)		
	1999	2010	2020	2030	1999	2010	2020	2030	1999-2010	1999-2020	1999-2030
Total Capacity	**304**	**367**	**407**	**447**	**100**	**100**	**100**	**100**	**1.7**	**1.4**	**1.3**
Coal	64	73	74	67	21	20	18	15	1.1	0.7	0.1
Oil	80	87	82	54	26	24	20	12	0.7	0.1	-1.3
Gas	52	71	89	116	17	19	22	26	2.8	2.6	2.6
Hydrogen-Fuel Cell	0	0	1	30	0	0	0	7	-	-	-
Nuclear	44	55	67	75	14	15	17	17	2.2	2.1	1.8
Hydro-Conventional	33	38	40	42	11	10	10	9	1.3	0.9	0.8
Hydro-Pumped Storage	26	29	32	34	8	8	8	8	1.0	1.0	0.9
Other renewables	6	14	22	31	2	4	5	7	8.8	6.6	5.6

	Capacity (GW)				Shares (%)				Growth Rates (% per annum)		
	1999	2010	2020	2030	1999	2010	2020	2030	1999-2010	1999-2020	1999-2030
Other Renewables	**6**	**14**	**22**	**31**	**100**	**100**	**100**	**100**	**8.8**	**6.6**	**5.6**
Biomass	4	8	10	12	79	58	46	38	5.8	4.0	3.2
Wind	0	2	3	5	1	11	14	16	31.1	18.9	14.2
Geothermal	1	1	2	2	16	8	7	6	2.6	2.5	2.2
Solar	0	3	7	11	4	22	32	37	28.3	18.4	13.8
Tide/Wave	0	0	0	1	0	0	0	2	-	-	-

	Electricity Generation (TWh)				Shares (%)				Growth Rates (% per annum)		
	2000	2010	2020	2030	2000	2010	2020	2030	2000-2010	2000-2020	2000-2030
Other Renewables	**25**	**50**	**69**	**92**	**100**	**100**	**100**	**100**	**7.2**	**5.2**	**4.4**
Biomass	19	33	40	47	75	66	59	51	5.8	3.9	3.1
Wind	0	4	8	12	1	8	11	13	29.8	17.7	13.3
Geothermal	6	9	11	13	24	18	16	14	3.9	3.1	2.6
Solar	0	4	10	18	0	9	14	19	64.6	33.5	23.8
Tide/Wave	0	0	0	2	0	0	0	2	-	-	-

Reference Scenario: Japan, Australia and New Zealand

	CO_2 Emissions (Mt)					Shares (%)					Growth Rates (% per annum)			
	1971	2000	2010	2020	2030	1971	2000	2010	2020	2030	1971-2000	2000-2010	2000-2020	2000-2030
Total CO_2 Emissions	**900**	**1513**	**1657**	**1723**	**1701**	**100**	**100**	**100**	**100**	**100**	**1.8**	**0.9**	**0.7**	**0.4**
change since 1990 (%)		*16.4*	*27.5*	*32.6*	*30.9*									
Coal	271	536	558	557	504	30	35	34	32	30	2.4	0.4	0.2	-0.2
Oil	616	771	838	871	847	68	51	51	51	50	0.8	0.8	0.6	0.3
Gas	13	207	261	296	350	1	14	16	17	21	10.0	2.4	1.8	1.8
Power Generation	**283**	**590**	**665**	**680**	**635**	**100**	**100**	**100**	**100**	**100**	**2.6**	**1.2**	**0.7**	**0.2**
Coal	122	374	401	403	354	43	63	60	59	56	3.9	0.7	0.4	-0.2
Oil	157	97	111	104	63	55	16	17	15	10	-1.6	1.4	0.4	-1.4
Gas	4	119	152	173	219	1	20	23	25	34	12.6	2.5	1.9	2.1
Transformation, Own Use & Losses	**41**	**70**	**72**	**78**	**82**	**100**	**100**	**100**	**100**	**100**	**1.8**	**0.2**	**0.5**	**0.5**
Total Final Consumption	**576**	**853**	**920**	**965**	**984**	**100**	**100**	**100**	**100**	**100**	**1.4**	**0.8**	**0.6**	**0.5**
Coal	134	140	137	134	131	23	16	15	14	13	0.2	-0.2	-0.2	-0.2
Oil	434	634	687	724	741	75	74	75	75	75	1.3	0.8	0.7	0.5
Gas	8	79	96	107	112	1	9	10	11	11	8.1	2.0	1.5	1.2
Industry	**311**	**313**	**326**	**334**	**336**	**100**	**100**	**100**	**100**	**100**	**0.0**	**0.4**	**0.3**	**0.2**
Coal	113	132	129	126	124	36	42	39	38	37	0.5	-0.2	-0.2	-0.2
Oil	193	140	147	152	153	62	45	45	45	45	-1.1	0.5	0.4	0.3
Gas	6	41	50	56	60	2	13	15	17	18	7.1	2.0	1.5	1.2
Transportation	**141**	**341**	**383**	**416**	**434**	**100**	**100**	**100**	**100**	**100**	**3.1**	**1.2**	**1.0**	**0.8**
Oil	138	340	382	415	433	98	100	100	100	100	3.2	1.2	1.0	0.8
Other fuels	3	1	1	1	1	2	0	0	0	0	-	-	-	-
Other Sectors	**116**	**184**	**196**	**199**	**197**	**100**	**100**	**100**	**100**	**100**	**1.6**	**0.6**	**0.4**	**0.2**
Coal	19	9	8	8	7	16	5	4	4	4	-2.7	-0.2	-0.4	-0.5
Oil	94	138	142	141	139	82	75	73	71	71	1.3	0.3	0.1	0.0
Gas	3	37	45	50	51	2	20	23	25	26	-	2.0	1.4	1.0
Non-Energy Use	**8**	**15**	**16**	**16**	**17**	**100**	**100**	**100**	**100**	**100**	**2.2**	**0.3**	**0.3**	**0.3**

Reference Scenario: Korea

	Energy Demand (Mtoe)					Shares (%)					Growth Rates (% per annum)			
	1971	2000	2010	2020	2030	1971	2000	2010	2020	2030	1971-2000	2000-2010	2000-2020	2000-2030
Total Primary Energy Supply	**17**	**194**	**264**	**331**	**378**	**100**	**100**	**100**	**100**	**100**	**8.8**	**3.2**	**2.7**	**2.3**
Coal	6	42	58	74	79	37	22	22	22	21	6.8	3.3	2.9	2.2
Oil	11	104	126	150	165	63	54	48	45	44	8.2	2.0	1.9	1.6
Gas	0	17	33	47	61	0	9	12	14	16	-	6.7	5.3	4.4
Nuclear	0	28	45	55	65	0	15	17	17	17	-	4.6	3.3	2.8
Hydro	0	0	0	1	1	1	0	0	0	0	3.9	2.0	2.0	1.7
Other Renewables	0	2	3	4	7	0	1	1	1	2	-	1.9	3.4	4.0
Power Generation	**2**	**69**	**106**	**141**	**166**	**100**	**100**	**100**	**100**	**100**	**12.3**	**4.5**	**3.7**	**3.0**
Coal	0	26	39	53	57	11	38	37	38	34	17.1	4.0	3.6	2.6
Oil	2	6	5	5	4	84	9	5	4	2	3.8	-1.0	-0.7	-1.2
Gas	0	6	14	24	33	0	8	14	17	20	-	10.0	7.5	6.1
Nuclear	0	28	45	55	65	0	41	42	39	39	-	4.6	3.3	2.8
Hydro	0	0	0	1	1	5	1	0	0	0	3.9	2.0	2.0	1.7
Other Renewables	0	2	2	4	7	0	3	2	3	4	-	1.9	3.4	4.1
Other Transformation, Own Use & Losses	**2**	**23**	**22**	**26**	**28**						**8.6**	**-0.4**	**0.5**	**0.6**
of which electricity	*0*	*2*	*3*	*4*	*5*						*11.2*	*3.2*	*2.9*	*2.5*
Total Final Consumption	**13**	**129**	**177**	**220**	**249**	**100**	**100**	**100**	**100**	**100**	**8.1**	**3.2**	**2.7**	**2.2**
Coal	5	7	8	9	10	38	5	4	4	4	1.1	1.1	1.2	1.0
Oil	7	86	113	137	152	56	67	64	62	61	8.8	2.8	2.3	1.9
Gas	0	11	18	23	27	0	9	10	10	11	-	4.8	3.8	3.0
Electricity	1	23	35	46	55	6	18	20	21	22	12.1	4.4	3.7	3.0
Heat	0	2	3	4	5	0	2	2	2	2	-	3.7	3.3	2.9
Renewables	0	0	0	0	0	0	0	0	0	0	-	2.1	2.1	2.1

Industry	**6**	**58**	**75**	**89**	**99**	**100**	**100**	**100**	**100**	**100**	**8.4**	**2.6**	**2.2**	**1.8**
Coal	0	7	8	9	10	7	11	11	10	10	10.3	1.9	1.7	1.3
Oil	5	36	44	51	55	84	62	59	57	56	7.3	2.0	1.7	1.4
Gas	0	3	5	6	7	0	5	6	7	7	-	5.9	4.1	3.1
Electricity	1	13	19	24	27	10	22	25	26	28	11.6	3.6	3.0	2.5
Heat	0	0	0	0	0	0	0	0	0	0	-	-	-	-
Renewables	0	0	0	0	0	0	0	0	0	0	-	-	-	-
Transportation	**2**	**30**	**45**	**57**	**66**	**100**	**100**	**100**	**100**	**100**	**9.5**	**4.1**	**3.3**	**2.7**
Oil	2	30	45	57	66	99	99	99	100	100	9.6	4.1	3.3	2.7
Other fuels	0	0	0	0	0	1	1	1	0	0	6.4	2.7	2.2	1.8
Other Sectors	**5**	**36**	**51**	**65**	**74**	**100**	**100**	**100**	**100**	**100**	**6.8**	**3.4**	**2.9**	**2.4**
Coal	5	1	0	0	0	89	2	0	0	0	-7.1	-29.7	-21.0	-15.4
Oil	0	16	20	23	24	6	45	39	35	33	14.4	1.9	1.6	1.3
Gas	0	8	13	17	20	0	23	26	27	27	-	4.5	3.7	3.0
Electricity	0	10	16	23	27	5	26	32	35	37	13.3	5.5	4.4	3.6
Heat	0	1	1	2	2	0	3	3	3	3	-	2.5	2.5	2.3
Renewables	0	0	0	0	0	0	0	0	0	0	-	2.1	2.1	2.1
Non-Energy Use	**0**	**5**	**7**	**8**	**9**						**9.5**	**3.2**	**2.8**	**2.3**
Electricity Generation (TWh)	**11**	**292**	**446**	**592**	**701**	**100**	**100**	**100**	**100**	**100**	**12.1**	**4.3**	**3.6**	**3.0**
Coal	1	126	185	253	275	7	43	42	43	39	19.5	3.9	3.5	2.6
Oil	8	25	23	22	18	81	8	5	4	3	3.7	-0.8	-0.6	-1.1
Gas	0	28	60	95	121	0	10	14	16	17	-	7.9	6.2	5.0
Hydrogen-Fuel Cell	0	0	0	1	17	0	0	0	0	2	-	-	-	-
Nuclear	0	109	171	210	248	0	37	38	35	35	-	4.6	3.3	2.8
Hydro	1	4	5	6	7	13	1	1	1	1	3.9	2.0	2.0	1.7
Other Renewables	0	0	2	5	15	0	0	0	1	2	-	13.1	12.6	12.3

Biomass (included above)	0	2	3	4	6	0	1	1	1	2	-	1.8	3.2	3.7
Total Primary Energy Supply (including Biomass)	**17**	**194**	**264**	**331**	**378**	**100**	**100**	**100**	**100**	**100**	**8.8**	**3.2**	**2.7**	**2.3**

Reference Scenario: Korea

	Capacity (GW)				Shares (%)				Growth Rates (% per annum)		
	1999	2010	2020	2030	1999	2010	2020	2030	1999-2010	1999-2020	1999-2030
Total Capacity	**52**	**86**	**116**	**144**	**100**	**100**	**100**	**100**	**4.8**	**3.9**	**3.4**
Coal	14	24	33	36	27	28	28	25	5.1	4.1	3.1
Oil	10	11	10	8	20	12	9	6	0.3	-0.1	-0.8
Gas	11	26	41	53	20	30	35	37	8.6	6.6	5.4
Hydrogen-Fuel Cell	0	0	0	5	0	0	0	3	-	-	-
Nuclear	14	21	26	31	27	25	23	21	4.1	3.1	2.6
Hydr Conventional	2	2	2	3	3	2	2	2	1.8	1.9	1.6
Hydro-Pumped Storage	2	2	2	3	3	2	2	2	1.8	1.9	1.6
Other renewables	0	0	2	5	0	1	1	4	13.6	13.3	13.0

	Capacity (GW)				Shares (%)				Growth Rates (% per annum)		
	1999	2010	2020	2030	1999	2010	2020	2030	1999-2010	1999-2020	1999-2030
Other Renewables	**0**	**0**	**2**	**5**	**100**	**100**	**100**	**100**	**13.6**	**13.3**	**13.0**
Biomass	0	0	1	1	78	51	38	26	9.4	9.4	9.0
Wind	0	0	1	4	6	38	56	71	35.0	26.3	22.6
Geothermal	0	0	0	0	0	0	0	0	-	-	-
Solar	0	0	0	0	17	11	6	4	9.4	8.1	7.6
Tide/Wave	0	0	0	0	0	0	0	0	-	-	-

	Electricity Generation (TWh)				Shares (%)				Growth Rates (% per annum)		
	2000	2010	2020	2030	2000	2010	2020	2030	2000-2010	2000-2020	2000-2030
Other Renewables	**0**	**2**	**5**	**15**	**100**	**100**	**100**	**100**	**13.1**	**12.6**	**12.3**
Biomass	0	1	3	6	83	63	49	36	10.0	9.7	9.2
Wind	0	0	2	9	4	28	45	61	39.1	27.8	23.4
Geothermal	0	0	0	0	0	0	0	0	-	-	-
Solar	0	0	0	0	13	9	5	3	8.7	7.7	7.0
Tide/Wave	0	0	0	0	0	0	0	0	-	-	-

Reference Scenario: Korea

	CO_2 Emissions (Mt)					Shares (%)					Growth Rates (% per annum)			
	1971	2000	2010	2020	2030	1971	2000	2010	2020	2030	1971-2000	2000-2010	2000-2020	2000-2030
Total CO_2 Emissions	**51**	**432**	**594**	**756**	**844**	**100**	**100**	**100**	**100**	**100**	**7.6**	**3.2**	**2.8**	**2.3**
change since 1990 (%)		*90.9*	*162.7*	*234.2*	*273.1*									
Coal	21	169	232	302	323	42	39	39	40	38	7.4	3.2	2.9	2.2
Oil	30	224	287	345	380	58	52	48	46	45	7.2	2.5	2.2	1.8
Gas	0	39	75	109	141	0	9	13	14	17	-	6.8	5.3	4.4
Power Generation	**7**	**150**	**226**	**309**	**346**	**100**	**100**	**100**	**100**	**100**	**10.9**	**4.1**	**3.7**	**2.8**
Coal	1	118	175	238	255	14	79	77	77	74	17.7	4.0	3.6	2.6
Oil	6	19	17	17	13	86	13	8	5	4	3.8	-1.0	-0.7	-1.2
Gas	0	13	34	55	78	0	9	15	18	23	-	10.0	7.5	6.1
Transformation, Own Use & Losses	**2**	**35**	**37**	**42**	**45**	**100**	**100**	**100**	**100**	**100**	**10.6**	**0.6**	**0.9**	**0.8**
Total Final Consumption	**42**	**246**	**331**	**405**	**453**	**100**	**100**	**100**	**100**	**100**	**6.3**	**3.0**	**2.5**	**2.1**
Coal	20	30	34	39	41	48	12	10	10	9	1.4	1.2	1.3	1.1
Oil	22	190	256	312	349	52	77	77	77	77	7.8	3.0	2.5	2.0
Gas	0	26	41	54	62	0	10	12	13	14	-	4.8	3.8	3.0
Industry	**15**	**84**	**105**	**122**	**133**	**100**	**100**	**100**	**100**	**100**	**6.1**	**2.3**	**1.9**	**1.5**
Coal	1	28	34	39	41	10	34	32	32	31	10.7	1.9	1.7	1.3
Oil	14	49	60	70	76	90	59	57	57	57	4.5	2.0	1.7	1.4
Gas	0	6	11	14	16	0	7	10	11	12	-	5.9	4.1	3.1
Transportation	**6**	**87**	**130**	**167**	**193**	**100**	**100**	**100**	**100**	**100**	**9.4**	**4.1**	**3.3**	**2.7**
Oil	6	87	130	167	193	98	100	100	100	100	9.5	4.1	3.3	2.7
Other fuels	0	0	0	0	0	2	0	0	0	0	-	-	-	-
Other Sectors	**20**	**75**	**94**	**114**	**126**	**100**	**100**	**100**	**100**	**100**	**4.7**	**2.4**	**2.1**	**1.8**
Coal	19	2	0	0	0	93	3	0	0	0	-7.1	-29.7	-21.0	-15.4
Oil	1	53	64	74	79	7	71	68	65	63	13.1	1.9	1.7	1.3
Gas	0	20	30	40	47	0	26	32	35	37	-	4.5	3.7	3.0
Non-Energy Use	**0**	**1**	**1**	**1**	**1**	**100**	**100**	**100**	**100**	**100**	**12.2**	**3.5**	**2.8**	**2.1**

Reference Scenario: Transition Economies

	Energy Demand (Mtoe)					Shares (%)					Growth Rates (% per annum)			
	1971	2000	2010	2020	2030	1971	2000	2010	2020	2030	1971-2000	2000-2010	2000-2020	2000-2030
Total Primary Energy Supply	**865**	**1024**	**1220**	**1373**	**1488**	**100**	**100**	**100**	**100**	**100**	**0.6**	**1.8**	**1.5**	**1.3**
Coal	337	213	252	248	260	39	21	21	18	17	-1.6	1.7	0.7	0.7
Oil	312	222	260	303	343	36	22	21	22	23	-1.2	1.6	1.6	1.5
Gas	202	492	604	708	763	23	48	49	52	51	3.1	2.1	1.8	1.5
Nuclear	2	67	66	62	54	0	7	5	4	4	13.8	-0.1	-0.4	-0.7
Hydro	13	25	28	32	34	1	2	2	2	2	2.3	0.9	1.1	1.0
Other Renewables	0	5	10	22	34	0	0	1	2	2	-	8.1	7.9	6.8
Power Generation	**265**	**516**	**629**	**666**	**673**	**100**	**100**	**100**	**100**	**100**	**2.3**	**2.0**	**1.3**	**0.9**
Coal	122	133	170	157	162	46	26	27	24	24	0.3	2.5	0.8	0.7
Oil	48	41	42	32	19	18	8	7	5	3	-0.5	0.2	-1.2	-2.4
Gas	82	245	313	362	369	31	47	50	54	55	3.9	2.5	2.0	1.4
Nuclear	2	67	66	62	54	1	13	11	9	8	13.8	-0.1	-0.4	-0.7
Hydro	13	25	28	32	34	5	5	4	5	5	2.3	0.9	1.1	1.0
Other Renewables	0	5	10	22	34	0	1	2	3	5	-	8.1	7.9	6.8
Other Transformation, Own Use & Losses	**148**	**119**	**134**	**156**	**174**						**-0.8**	**1.3**	**1.4**	**1.3**
of which electricity	*15*	*37*	*43*	*54*	*62*						*3.1*	*1.5*	*1.9*	*1.8*
Total Final Consumption	**580**	**683**	**769**	**901**	**1019**	**100**	**100**	**100**	**100**	**100**	**0.6**	**1.2**	**1.4**	**1.3**
Coal	117	51	54	61	68	20	7	7	7	7	-2.9	0.6	1.0	1.0
Oil	234	157	188	236	284	40	23	24	26	28	-1.4	1.8	2.1	2.0
Gas	116	218	257	308	352	20	32	33	34	35	2.2	1.7	1.7	1.6
Electricity	62	91	109	139	163	11	13	14	15	16	1.3	1.9	2.2	2.0
Heat	51	166	161	157	153	9	24	21	17	15	4.2	-0.3	-0.3	-0.3
Renewables	0	0	0	0	0	0	0	0	0	0	-	-	-	-

Industry	**278**	**249**	**281**	**333**	**370**	**100**	**100**	**100**	**100**	**100**	**-0.4**	**1.2**	**1.5**	**1.3**
Coal	46	33	36	43	49	16	13	13	13	13	-1.2	1.1	1.4	1.3
Oil	61	36	42	53	61	22	14	15	16	16	-1.8	1.8	2.0	1.8
Gas	92	78	94	117	134	33	31	34	35	36	-0.5	1.9	2.0	1.8
Electricity	44	42	51	66	75	16	17	18	20	20	-0.1	1.9	2.2	1.9
Heat	37	61	57	54	52	13	24	20	16	14	1.8	-0.6	-0.5	-0.5
Renewables	0	0	0	0	0	0	0	0	0	0	-	-	-	-
Transportation	**101**	**113**	**137**	**178**	**219**	**100**	**100**	**100**	**100**	**100**	**0.4**	**2.0**	**2.3**	**2.2**
Oil	86	77	97	129	163	86	69	70	72	74	-0.4	2.2	2.6	2.5
Other fuels	14	35	41	49	56	14	31	30	28	26	3.2	1.5	1.7	1.6
Other Sectors	**175**	**309**	**337**	**375**	**413**	**100**	**100**	**100**	**100**	**100**	**2.0**	**0.9**	**1.0**	**1.0**
Coal	60	17	16	17	18	34	6	5	5	4	-4.3	-0.4	0.1	0.2
Oil	63	32	36	40	43	36	10	11	11	11	-2.3	1.1	1.1	1.0
Gas	24	112	129	149	170	14	36	38	40	41	5.4	1.5	1.5	1.4
Electricity	14	42	51	66	80	8	14	15	18	19	3.9	2.1	2.3	2.2
Heat	14	106	104	103	101	8	34	31	27	25	7.1	-0.2	-0.2	-0.1
Renewables	0	0	0	0	0	0	0	0	0	0	-	-	-	-
Non-Energy Use	**25**	**13**	**14**	**15**	**17**						**-2.3**	**0.9**	**0.9**	**1.0**
Electricity Generation (TWh)	**903**	**1484**	**1765**	**2238**	**2623**	**100**	**100**	**100**	**100**	**100**	**1.7**	**1.8**	**2.1**	**1.9**
Coal	395	340	429	451	579	44	23	24	20	22	-0.5	2.3	1.4	1.8
Oil	154	82	91	74	46	17	6	5	3	2	-2.1	1.0	-0.6	-2.0
Gas	176	509	658	1088	1350	19	34	37	49	51	3.7	2.6	3.9	3.3
Hydrogen-Fuel Cell	0	0	0	0	0	0	0	0	0	0	-	-	-	-
Nuclear	6	255	253	235	207	1	17	14	10	8	13.7	-0.1	-0.4	-0.7
Hydro	150	295	324	367	396	17	20	18	16	15	2.3	0.9	1.1	1.0
Other Renewables	23	3	11	23	46	3	0	1	1	2	-6.7	14.1	10.6	9.4
Biomass (NOT included above)	24	11	11	12	13	3	1	1	1	1	-2.8	0.6	0.7	0.8
Total Primary Energy Supply (including Biomass)	**889**	**1034**	**1231**	**1385**	**1501**	**100**	**100**	**100**	**100**	**100**	**0.5**	**1.8**	**1.5**	**1.2**

Reference Scenario: Transition Economies

	Capacity (GW)				Shares (%)				Growth Rates (% per annum)		
	1999	2010	2020	2030	1999	2010	2020	2030	1999-2010	1999-2020	1999-2030
Total Capacity	**404**	**421**	**526**	**624**	**100**	**100**	**100**	**100**	**0.4**	**1.3**	**1.4**
Coal	112	110	110	142	28	26	21	23	-0.2	-0.1	0.8
Oil	38	39	33	19	9	9	6	3	0.2	-0.7	-2.2
Gas	125	134	232	300	31	32	44	48	0.7	3.0	2.9
Hydrogen-Fuel Cell	0	0	0	0	0	0	0	0	-	-	-
Nuclear	41	38	35	31	10	9	7	5	-0.7	-0.7	-0.9
Hydro	87	97	110	119	22	23	21	19	1.0	1.1	1.0
Other renewables	1	3	6	13	0	1	1	2	15.6	11.7	10.8

	Capacity (GW)				Shares (%)				Growth Rates (% per annum)		
	1999	2010	2020	2030	1999	2010	2020	2030	1999-2010	1999-2020	1999-2030
Other Renewables	**1**	**3**	**6**	**13**	**100**	**100**	**100**	**100**	**15.6**	**11.7**	**10.8**
Biomass	1	2	3	5	94	74	58	36	13.1	9.1	7.4
Wind	0	0	2	7	3	18	29	50	37.4	25.0	21.7
Geothermal	0	0	1	1	3	8	13	10	25.3	19.4	15.0
Solar	0	0	0	1	0	0	1	5	-	-	-
Tide/Wave	0	0	0	0	0	0	0	0	-	-	-

	Electricity Generation (TWh)				Shares (%)				Growth Rates (% per annum)		
	2000	2010	2020	2030	2000	2010	2020	2030	2000-2010	2000-2020	2000-2030
Other Renewables	**3**	**11**	**23**	**46**	**100**	**100**	**100**	**100**	**14.1**	**10.6**	**9.4**
Biomass	3	9	14	21	94	77	62	45	11.8	8.3	6.8
Wind	0	1	4	15	4	10	17	33	26.8	19.4	17.9
Geothermal	0	1	5	9	3	13	22	20	33.7	23.0	17.2
Solar	0	0	0	1	0	0	0	2	-	-	-
Tide/Wave	0	0	0	0	0	0	0	0	-	-	-

Reference Scenario: Transition Economies

	CO$_2$ Emissions (Mt)					Shares (%)					Growth Rates (% per annum)			
	1971	2000	2010	2020	2030	1971	2000	2010	2020	2030	1971-2000	2000-2010	2000-2020	2000-2030
Total CO$_2$ Emissions	**2281**	**2488**	**3041**	**3374**	**3646**	**100**	**100**	**100**	**100**	**100**	**0.3**	**2.0**	**1.5**	**1.3**
change since 1990 (%)		*-34.3*	*-19.7*	*-10.9*	*-3.7*									
Coal	991	808	976	965	1020	43	32	32	29	28	-0.7	1.9	0.9	0.8
Oil	801	575	709	821	916	35	23	23	24	25	-1.1	2.1	1.8	1.6
Gas	490	1104	1356	1588	1710	21	44	45	47	47	2.8	2.1	1.8	1.5
Power Generation	**834**	**1255**	**1570**	**1599**	**1596**	**100**	**100**	**100**	**100**	**100**	**1.4**	**2.3**	**1.2**	**0.8**
Coal	490	552	703	649	670	59	44	45	41	42	0.4	2.4	0.8	0.6
Oil	153	131	133	104	62	18	10	8	6	4	-0.5	0.2	-1.2	-2.4
Gas	191	573	733	846	864	23	46	47	53	54	3.9	2.5	2.0	1.4
Transformation, Own Use & Losses	**96**	**85**	**108**	**123**	**142**	**100**	**100**	**100**	**100**	**100**	**-0.4**	**2.4**	**1.9**	**1.7**
Total Final Consumption	**1352**	**1147**	**1363**	**1652**	**1908**	**100**	**100**	**100**	**100**	**100**	**-0.6**	**1.7**	**1.8**	**1.7**
Coal	464	251	270	314	349	34	22	20	19	18	-2.1	0.7	1.1	1.1
Oil	623	401	513	645	767	46	35	38	39	40	-1.5	2.5	2.4	2.1
Gas	265	495	580	693	792	20	43	43	42	42	2.2	1.6	1.7	1.6
Industry	**580**	**449**	**564**	**693**	**789**	**100**	**100**	**100**	**100**	**100**	**-0.9**	**2.3**	**2.2**	**1.9**
Coal	187	181	200	241	271	32	40	36	35	34	-0.1	1.0	1.4	1.4
Oil	185	100	155	194	222	32	22	27	28	28	-2.1	4.4	3.4	2.7
Gas	207	167	209	259	296	36	37	37	37	37	-0.7	2.2	2.2	1.9
Transportation	**224**	**261**	**312**	**408**	**503**	**100**	**100**	**100**	**100**	**100**	**0.5**	**1.8**	**2.3**	**2.2**
Oil	186	194	242	323	404	83	74	78	79	80	0.1	2.2	2.6	2.5
Other fuels	38	67	70	86	100	17	26	22	21	20	2.0	0.4	1.2	1.3
Other Sectors	**479**	**424**	**474**	**536**	**598**	**100**	**100**	**100**	**100**	**100**	**-0.4**	**1.1**	**1.2**	**1.2**
Coal	233	67	64	68	71	49	16	14	13	12	-4.2	-0.4	0.1	0.2
Oil	189	96	108	119	130	40	23	23	22	22	-2.3	1.1	1.1	1.0
Gas	56	261	302	349	397	12	62	64	65	66	5.4	1.5	1.5	1.4
Non-Energy Use	**69**	**13**	**14**	**15**	**17**	**100**	**100**	**100**	**100**	**100**	**-5.5**	**0.7**	**0.6**	**0.7**

Reference Scenario: Russia

	Energy Demand (Mtoe)					Shares (%)					Growth Rates (% per annum)			
	1971	2000	2010	2020	2030	1971	2000	2010	2020	2030	1971-2000	2000-2010	2000-2020	2000-2030
Total Primary Energy Supply	**n.a.**	**612**	**733**	**841**	**918**	**n.a.**	**100**	**100**	**100**	**100**	**n.a.**	**1.8**	**1.6**	**1.4**
Coal	n.a.	111	131	125	125	n.a.	18	18	15	14	n.a.	1.7	0.6	0.4
Oil	n.a.	130	150	180	214	n.a.	21	20	21	23	n.a.	1.4	1.7	1.7
Gas	n.a.	319	392	471	512	n.a.	52	54	56	56	n.a.	2.1	2.0	1.6
Nuclear	n.a.	34	41	41	38	n.a.	6	6	5	4	n.a.	1.9	0.8	0.3
Hydro	n.a.	14	15	16	17	n.a.	2	2	2	2	n.a.	0.3	0.7	0.6
Other Renewables	n.a.	4	4	7	12	n.a.	1	1	1	1	n.a.	0.4	3.1	3.8
Power Generation	**n.a.**	**342**	**417**	**446**	**452**	**n.a.**	**100**	**100**	**100**	**100**	**n.a.**	**2.0**	**1.3**	**0.9**
Coal	n.a.	80	100	91	89	n.a.	23	24	20	20	n.a.	2.3	0.6	0.4
Oil	n.a.	23	21	17	12	n.a.	7	5	4	3	n.a.	-1.0	-1.7	-2.2
Gas	n.a.	186	235	275	284	n.a.	54	56	62	63	n.a.	2.4	2.0	1.4
Nuclear	n.a.	34	41	41	38	n.a.	10	10	9	8	n.a.	1.9	0.8	0.3
Hydro	n.a.	14	15	16	17	n.a.	4	3	4	4	n.a.	0.3	0.7	0.6
Other Renewables	n.a.	4	4	7	12	n.a.	1	1	2	3	n.a.	0.4	3.1	3.8
Other Transformation, Own Use & Losses	**n.a.**	**65**	**75**	**90**	**105**						**n.a.**	**1.5**	**1.7**	**1.6**
of which electricity	*n.a.*	*23*	*27*	*35*	*41*						*n.a.*	*1.5*	*2.1*	*2.0*
Total Final Consumption	**n.a.**	**417**	**462**	**548**	**622**	**n.a.**	**100**	**100**	**100**	**100**	**n.a.**	**1.0**	**1.4**	**1.3**
Coal	n.a.	21	21	24	26	n.a.	5	4	4	4	n.a.	0.0	0.8	0.8
Oil	n.a.	93	111	143	176	n.a.	22	24	26	28	n.a.	1.8	2.2	2.1
Gas	n.a.	115	136	172	201	n.a.	27	30	31	32	n.a.	1.7	2.0	1.9
Electricity	n.a.	52	64	86	102	n.a.	13	14	16	16	n.a.	2.0	2.5	2.3
Heat	n.a.	137	130	123	117	n.a.	33	28	22	19	n.a.	-0.5	-0.5	-0.5
Renewables	n.a.	0	0	0	0	n.a.	0	0	0	0	n.a.	-	-	-

Industry	**n.a.**	**148**	**165**	**199**	**217**	**n.a.**	**100**	**100**	**100**	**100**	**n.a.**	**1.1**	**1.5**	**1.3**
Coal	n.a.	10	11	14	15	n.a.	7	7	7	7	n.a.	0.8	1.6	1.3
Oil	n.a.	21	25	32	36	n.a.	14	15	16	17	n.a.	2.0	2.2	1.9
Gas	n.a.	40	49	65	74	n.a.	27	30	33	34	n.a.	2.0	2.4	2.1
Electricity	n.a.	27	33	44	50	n.a.	18	20	22	23	n.a.	2.0	2.5	2.1
Heat	n.a.	50	47	44	41	n.a.	34	29	22	19	n.a.	-0.7	-0.7	-0.6
Renewables	n.a.	0	0	0	0	n.a.	0	0	0	0	n.a.	-	-	-
Transportation	**n.a.**	**79**	**95**	**125**	**156**	**n.a.**	**100**	**100**	**100**	**100**	**n.a.**	**1.9**	**2.3**	**2.3**
Oil	n.a.	47	58	80	104	n.a.	59	61	64	67	n.a.	2.1	2.7	2.7
Other fuels	n.a.	32	37	45	52	n.a.	41	39	36	33	n.a.	1.5	1.7	1.6
Other Sectors	**n.a.**	**182**	**193**	**215**	**238**	**n.a.**	**100**	**100**	**100**	**100**	**n.a.**	**0.6**	**0.8**	**0.9**
Coal	n.a.	9	9	9	10	n.a.	5	4	4	4	n.a.	-1.0	-0.2	0.1
Oil	n.a.	18	20	23	26	n.a.	10	11	11	11	n.a.	1.1	1.2	1.1
Gas	n.a.	47	55	68	81	n.a.	26	29	32	34	n.a.	1.6	1.8	1.8
Electricity	n.a.	20	26	36	46	n.a.	11	13	17	19	n.a.	2.5	2.9	2.8
Heat	n.a.	87	83	79	76	n.a.	48	43	37	32	n.a.	-0.4	-0.4	-0.4
Renewables	n.a.	0	0	0	0	n.a.	0	0	0	0	n.a.	-	-	-
Non-Energy Use	**n.a.**	**8**	**9**	**9**	**10**						**n.a.**	**0.5**	**0.6**	**0.8**
Electricity Generation (TWh)	**n.a.**	**876**	**1052**	**1405**	**1671**	**n.a.**	**100**	**100**	**100**	**100**	**n.a.**	**1.8**	**2.4**	**2.2**
Coal	n.a.	176	213	216	277	n.a.	20	20	15	17	n.a.	1.9	1.0	1.5
Oil	n.a.	33	33	27	21	n.a.	4	3	2	1	n.a.	0.0	-1.0	-1.5
Gas	n.a.	370	472	803	1007	n.a.	42	45	57	60	n.a.	2.5	3.9	3.4
Hydrogen-Fuel Cell	n.a.	0	0	0	0	n.a.	0	0	0	0	n.a.	-	-	-
Nuclear	n.a.	131	157	155	143	n.a.	15	15	11	9	n.a.	1.9	0.8	0.3
Hydro	n.a.	164	169	190	196	n.a.	19	16	14	12	n.a.	0.3	0.7	0.6
Other Renewables	n.a.	3	8	14	27	n.a.	0	1	1	2	n.a.	11.9	8.9	8.1

Biomass (NOT included above)	n.a.	3	3	4	4	n.a.	1	0	0	0	n.a.	0.6	0.6	0.7
Total Primary Energy Supply (including Biomass)	**n.a.**	**615**	**737**	**844**	**922**	**n.a.**	**100**	**100**	**100**	**100**	**n.a.**	**1.8**	**1.6**	**1.4**

Reference Scenario: Russia

	Capacity (GW)				Shares (%)				Growth Rates (% per annum)		
	1999	2010	2020	2030	1999	2010	2020	2030	1999-2010	1999-2020	1999-2030
Total Capacity	**216**	**225**	**298**	**360**	**100**	**100**	**100**	**100**	**0.3**	**1.5**	**1.7**
Coal	53	52	50	64	24	23	17	18	-0.2	-0.2	0.6
Oil	12	12	11	9	5	6	4	2	0.5	-0.1	-0.9
Gas	88	90	159	205	40	40	53	57	0.2	2.9	2.8
Hydrogen-Fuel Cell	0	0	0	0	0	0	0	0	-	-	-
Nuclear	20	23	22	21	9	10	7	6	1.2	0.6	0.1
Hydro	44	47	53	54	20	21	18	15	0.5	0.8	0.6
Other renewables	0	2	3	7	0	1	1	2	12.9	9.4	9.1

	Capacity (GW)				Shares (%)				Growth Rates (% per annum)		
	1999	2010	2020	2030	1999	2010	2020	2030	1999-2010	1999-2020	1999-2030
Other Renewables	**0**	**2**	**3**	**7**	**100**	**100**	**100**	**100**	**12.9**	**9.4**	**9.1**
Biomass	0	2	2	3	97	87	69	41	11.7	7.6	6.1
Wind	0	0	1	3	0	8	16	41	47.2	30.1	26.6
Geothermal	0	0	1	1	2	6	16	14	22.5	19.9	15.7
Solar	0	0	0	0	0	0	0	4	-	-	-
Tide/Wave	0	0	0	0	0	0	0	0	-	-	-

	Electricity Generation (TWh)				Shares (%)				Growth Rates (% per annum)		
	2000	2010	2020	2030	2000	2010	2020	2030	2000-2010	2000-2020	2000-2030
Other Renewables	**3**	**8**	**14**	**27**	**100**	**100**	**100**	**100**	**11.9**	**8.9**	**8.1**
Biomass	3	7	10	13	98	87	68	49	10.7	6.9	5.6
Wind	0	0	1	7	0	4	8	24	65.4	37.1	31.0
Geothermal	0	1	4	7	2	9	25	26	28.7	22.8	17.3
Solar	0	0	0	0	0	0	0	1	-	-	-
Tide/Wave	0	0	0	0	0	0	0	0	-	-	-

Reference Scenario: Russia

	CO_2 Emissions (Mt)					Shares (%)					Growth Rates (% per annum)			
	1971	2000	2010	2020	2030	1971	2000	2010	2020	2030	1971-2000	2000-2010	2000-2020	2000-2030
Total CO_2 Emissions	**n.a.**	**1492**	**1829**	**2068**	**2241**	**n.a.**	**100**	**100**	**100**	**100**	**n.a.**	**2.1**	**1.6**	**1.4**
change since 1990 (%)		*-32.6*	*-17.3*	*-6.5*	*1.3*									
Coal	n.a.	450	536	517	520	n.a.	30	29	25	23	n.a.	1.8	0.7	0.5
Oil	n.a.	328	415	497	578	n.a.	22	23	24	26	n.a.	2.4	2.1	1.9
Gas	n.a.	714	878	1054	1144	n.a.	48	48	51	51	n.a.	2.1	2.0	1.6
Power Generation	n.a.	**848**	**1042**	**1079**	**1079**	**n.a.**	**100**	**100**	**100**	**100**	**n.a.**	**2.1**	**1.2**	**0.8**
Coal	n.a.	338	424	383	376	n.a.	40	41	36	35	n.a.	2.3	0.6	0.4
Oil	n.a.	75	68	53	38	n.a.	9	6	5	4	n.a.	-1.0	-1.7	-2.2
Gas	n.a.	436	550	642	665	n.a.	51	53	60	62	n.a.	2.4	2.0	1.4
Transformation, Own Use & Losses	**n.a.**	**50**	**68**	**79**	**96**	**n.a.**	**100**	**100**	**100**	**100**	**n.a.**	**3.0**	**2.3**	**2.2**
Total Final Consumption	**n.a.**	**594**	**720**	**910**	**1066**	**n.a.**	**100**	**100**	**100**	**100**	**n.a.**	**1.9**	**2.2**	**2.0**
Coal	n.a.	110	112	134	145	n.a.	19	16	15	14	n.a.	0.2	1.0	0.9
Oil	n.a.	227	305	394	474	n.a.	38	42	43	44	n.a.	3.0	2.8	2.5
Gas	n.a.	257	303	382	447	n.a.	43	42	42	43	n.a.	1.7	2.0	1.9
Industry	**n.a.**	**210**	**287**	**369**	**417**	**n.a.**	**100**	**100**	**100**	**100**	**n.a.**	**3.2**	**2.9**	**2.3**
Coal	n.a.	70	74	93	101	n.a.	33	26	25	24	n.a.	0.6	1.5	1.2
Oil	n.a.	57	105	133	152	n.a.	27	37	36	37	n.a.	6.3	4.3	3.3
Gas	n.a.	83	108	143	164	n.a.	40	38	39	39	n.a.	2.7	2.8	2.3
Transportation	**n.a.**	**173**	**201**	**270**	**337**	**n.a.**	**100**	**100**	**100**	**100**	**n.a.**	**1.5**	**2.2**	**2.2**
Oil	n.a.	110	135	190	243	n.a.	64	67	70	72	n.a.	2.1	2.7	2.7
Other fuels	n.a.	63	66	81	94	n.a.	36	33	30	28	n.a.	0.4	1.2	1.3
Other Sectors	**n.a.**	**203**	**224**	**263**	**304**	**n.a.**	**100**	**100**	**100**	**100**	**n.a.**	**1.0**	**1.3**	**1.4**
Coal	n.a.	37	34	36	39	n.a.	18	15	14	13	n.a.	-1.0	-0.2	0.1
Oil	n.a.	55	61	69	76	n.a.	27	27	26	25	n.a.	1.1	1.2	1.1
Gas	n.a.	111	130	158	189	n.a.	55	58	60	62	n.a.	1.6	1.8	1.8
Non-Energy Use	**n.a.**	**8**	**8**	**8**	**8**	**n.a.**	**100**	**100**	**100**	**100**	**n.a.**	**0.0**	**0.1**	**0.3**

Reference Scenario: Developing Countries

	Energy Demand (Mtoe)					Shares (%)					Growth Rates (% per annum)			
	1971	2000	2010	2020	2030	1971	2000	2010	2020	2030	1971-2000	2000-2010	2000-2020	2000-2030
Total Primary Energy Supply	**657**	**2732**	**3773**	**5031**	**6487**	**100**	**100**	**100**	**100**	**100**	**5.0**	**3.3**	**3.1**	**2.9**
Coal	302	1060	1361	1721	2165	46	39	36	34	33	4.4	2.5	2.5	2.4
Oil	297	1085	1472	1938	2473	45	40	39	39	38	4.6	3.1	2.9	2.8
Gas	40	449	713	1050	1428	6	16	19	21	22	8.7	4.7	4.3	3.9
Nuclear	0	26	56	84	111	0	1	1	2	2	16.2	8.0	6.0	4.9
Hydro	16	90	124	167	199	2	3	3	3	3	6.1	3.3	3.1	2.7
Other Renewables	1	21	46	72	112	0	1	1	1	2	9.8	8.1	6.3	5.7
Power Generation	**126**	**972**	**1474**	**2097**	**2834**	**100**	**100**	**100**	**100**	**100**	**7.3**	**4.3**	**3.9**	**3.6**
Coal	64	548	792	1103	1507	51	56	54	53	53	7.7	3.8	3.6	3.4
Oil	36	135	160	183	201	29	14	11	9	7	4.7	1.7	1.5	1.3
Gas	8	152	295	488	704	6	16	20	23	25	10.8	6.9	6.0	5.2
Nuclear	0	26	56	84	111	0	3	4	4	4	16.2	8.0	6.0	4.9
Hydro	16	90	124	167	199	13	9	8	8	7	6.1	3.3	3.1	2.7
Other Renewables	1	21	46	72	112	1	2	3	3	4	9.8	8.1	6.3	5.7
Other Transformation, Own Use & Losses	**86**	**400**	**517**	**650**	**796**						**5.5**	**2.6**	**2.5**	**2.3**
of which electricity	*8*	*82*	*126*	*182*	*251*						*8.5*	*4.4*	*4.1*	*3.8*
Total Final Consumption	**493**	**1756**	**2398**	**3189**	**4119**	**100**	**100**	**100**	**100**	**100**	**4.5**	**3.2**	**3.0**	**2.9**
Coal	211	372	416	453	485	43	21	17	14	12	2.0	1.1	1.0	0.9
Oil	225	880	1217	1630	2117	46	50	51	51	51	4.8	3.3	3.1	3.0
Gas	18	190	275	383	506	4	11	11	12	12	8.4	3.8	3.6	3.3
Electricity	39	289	454	676	954	8	16	19	21	23	7.2	4.6	4.3	4.1
Heat	0	25	36	46	56	0	1	1	1	1	-	3.4	3.0	2.7
Renewables	0	0	0	0	0	0	0	0	0	0	-	-	-	-

Industry	**134**	**773**	**1000**	**1244**	**1506**	**100**	**100**	**100**	**100**	**100**	**6.2**	**2.6**	**2.4**	**2.2**
Coal	53	281	327	365	398	40	36	33	29	26	5.9	1.5	1.3	1.2
Oil	52	211	264	318	373	39	27	26	26	25	4.9	2.3	2.1	1.9
Gas	13	124	180	243	311	10	16	18	19	21	8.1	3.8	3.4	3.1
Electricity	16	137	203	287	387	12	18	20	23	26	7.7	4.0	3.8	3.5
Heat	0	19	25	31	37	0	2	3	3	2	-	3.0	2.6	2.3
Renewables	0	0	0	0	0	0	0	0	0	0	-	-	-	-
Transportation	**109**	**445**	**652**	**933**	**1281**	**100**	**100**	**100**	**100**	**100**	**5.0**	**3.9**	**3.8**	**3.6**
Oil	96	434	638	915	1260	89	98	98	98	98	5.3	3.9	3.8	3.6
Other fuels	12	11	14	18	22	11	2	2	2	2	-0.4	2.1	2.3	2.2
Other Sectors	**64**	**469**	**663**	**913**	**1217**	**100**	**100**	**100**	**100**	**100**	**7.1**	**3.5**	**3.4**	**3.2**
Coal	11	73	73	74	75	17	16	11	8	6	6.9	0.1	0.1	0.1
Oil	39	186	254	325	402	61	40	38	36	33	5.5	3.2	2.8	2.6
Gas	4	62	89	132	182	5	13	13	14	15	10.4	3.7	3.8	3.6
Electricity	11	141	236	368	540	17	30	36	40	44	9.3	5.3	4.9	4.6
Heat	0	6	10	14	18	0	1	1	2	2	-	4.7	4.2	3.8
Renewables	0	0	0	0	0	0	0	0	0	0	-	-	-	-
Non-Energy Use	**186**	**70**	**84**	**99**	**115**						**-3.3**	**1.8**	**1.8**	**1.7**
Electricity Generation (TWh)	**534**	**4308**	**6747**	**9983**	**14012**	**100**	**100**	**100**	**100**	**100**	**7.5**	**4.6**	**4.3**	**4.0**
Coal	200	1916	2939	4340	6391	37	44	44	43	46	8.1	4.4	4.2	4.1
Oil	122	557	662	768	857	23	13	10	8	6	5.4	1.7	1.6	1.4
Gas	24	652	1391	2455	3738	5	15	21	25	27	12.0	7.9	6.9	6.0
Hydrogen-Fuel Cell	0	0	0	0	0	0	0	0	0	0	-	-	-	-
Nuclear	1	100	216	321	426	0	2	3	3	3	16.2	8.0	6.0	4.9
Hydro	185	1044	1446	1942	2315	35	24	21	19	17	6.1	3.3	3.1	2.7
Other Renewables	2	38	94	156	284	0	1	1	2	2	10.1	9.5	7.4	7.0

Biomass (NOT included above)	560	891	968	1007	1019	46	25	20	17	14	1.6	0.8	0.6	0.4
Total Primary Energy Supply (including Biomass)	**1217**	**3623**	**4741**	**6038**	**7506**	**100**	**100**	**100**	**100**	**100**	**3.8**	**2.7**	**2.6**	**2.5**

Reference Scenario: Developing Countries

	Capacity (GW)				Shares (%)				Growth Rates (% per annum)		
	1999	2010	2020	2030	1999	2010	2020	2030	1999-2010	1999-2020	1999-2030
Total Capacity	**979**	**1558**	**2311**	**3238**	**100**	**100**	**100**	**100**	**4.3**	**4.2**	**3.9**
Coal	338	559	830	1224	34	36	36	38	4.7	4.4	4.2
Oil	186	226	271	315	19	15	12	10	1.8	1.8	1.7
Gas	169	354	627	963	17	23	27	30	6.9	6.4	5.8
Hydrogen-Fuel Cell	0	0	0	0	0	0	0	0	-	-	-
Nuclear	13	29	42	56	1	2	2	2	7.5	5.8	4.8
Hydro	263	368	502	601	27	24	22	19	3.1	3.1	2.7
Other renewables	10	22	39	79	1	1	2	2	7.7	6.8	7.0

	Capacity (GW)				Shares (%)				Growth Rates (% per annum)		
	1999	2010	2020	2030	1999	2010	2020	2030	1999-2010	1999-2020	1999-2030
Other Renewables	**10**	**22**	**39**	**79**	**100**	**100**	**100**	**100**	**7.7**	**6.8**	**7.0**
Biomass	5	9	15	25	55	42	39	31	5.0	5.0	5.1
Wind	1	8	15	36	15	35	38	46	16.0	11.6	10.9
Geothermal	3	5	7	10	30	22	19	13	4.8	4.6	4.2
Solar	0	0	2	8	0	2	4	10	85.2	47.6	37.0
Tide/Wave	0	0	0	0	0	0	0	0	-	-	-

	Electricity Generation (TWh)				Shares (%)				Growth Rates (% per annum)		
	2000	2010	2020	2030	2000	2010	2020	2030	2000-2010	2000-2020	2000-2030
Other Renewables	**38**	**94**	**156**	**284**	**100**	**100**	**100**	**100**	**9.5**	**7.4**	**7.0**
Biomass	19	40	64	107	50	42	41	38	7.8	6.4	6.0
Wind	2	18	36	89	6	19	23	31	22.8	14.7	12.9
Geothermal	17	35	53	73	44	38	34	26	7.7	5.9	5.0
Solar	0	1	3	16	0	1	2	6	34.7	24.7	22.0
Tide/Wave	0	0	0	0	0	0	0	0	-	-	-

Reference Scenario: Developing Countries

	CO_2 Emissions (Mt)					Shares (%)					Growth Rates (% per annum)			
	1971	2000	2010	2020	2030	1971	2000	2010	2020	2030	1971-2000	2000-2010	2000-2020	2000-2030
Total CO_2 Emissions	**2018**	**7782**	**10612**	**14042**	**18118**	**100**	**100**	**100**	**100**	**100**	**4.8**	**3.2**	**3.0**	**2.9**
change since 1990 (%)		*45.6*	*98.6*	*162.8*	*239.1*									
Coal	1124	3851	5049	6472	8237	56	49	48	46	45	4.3	2.7	2.6	2.6
Oil	804	2973	4020	5281	6756	40	38	38	38	37	4.6	3.1	2.9	2.8
Gas	90	958	1543	2289	3125	4	12	15	16	17	8.5	4.9	4.5	4.0
Power Generation	**383**	**2978**	**4387**	**6159**	**8337**	**100**	**100**	**100**	**100**	**100**	**7.3**	**4.0**	**3.7**	**3.5**
Coal	251	2195	3185	4434	6049	65	74	73	72	73	7.8	3.8	3.6	3.4
Oil	114	431	516	590	645	30	14	12	10	8	4.7	1.8	1.6	1.4
Gas	18	351	686	1136	1643	5	12	16	18	20	10.8	6.9	6.0	5.3
Transformation, Own Use & Losses	**99**	**488**	**622**	**768**	**942**	**100**	**100**	**100**	**100**	**100**	**5.6**	**2.5**	**2.3**	**2.2**
Total Final Consumption	**1535**	**4316**	**5602**	**7115**	**8838**	**100**	**100**	**100**	**100**	**100**	**3.6**	**2.6**	**2.5**	**2.4**
Coal	868	1575	1773	1938	2081	57	36	32	27	24	2.1	1.2	1.0	0.9
Oil	627	2331	3240	4359	5687	41	54	58	61	64	4.6	3.3	3.2	3.0
Gas	40	410	590	818	1071	3	9	11	11	12	8.3	3.7	3.5	3.3
Industry	**450**	**2005**	**2454**	**2883**	**3303**	**100**	**100**	**100**	**100**	**100**	**5.3**	**2.0**	**1.8**	**1.7**
Coal	255	1215	1414	1579	1722	57	61	58	55	52	5.5	1.5	1.3	1.2
Oil	166	525	656	787	921	37	26	27	27	28	4.1	2.2	2.0	1.9
Gas	29	265	384	516	660	6	13	16	18	20	7.9	3.8	3.4	3.1
Transportation	**298**	**1227**	**1797**	**2575**	**3540**	**100**	**100**	**100**	**100**	**100**	**5.0**	**3.9**	**3.8**	**3.6**
Oil	251	1195	1759	2526	3480	84	97	98	98	98	5.5	3.9	3.8	3.6
Other fuels	47	31	37	49	60	16	3	2	2	2	-1.4	1.8	2.2	2.2
Other Sectors	**766**	**988**	**1242**	**1539**	**1864**	**100**	**100**	**100**	**100**	**100**	**0.9**	**2.3**	**2.2**	**2.1**
Coal	564	304	305	308	310	74	31	25	20	17	-2.1	0.0	0.1	0.1
Oil	190	548	747	955	1180	25	55	60	62	63	3.7	3.2	2.8	2.6
Gas	11	136	190	275	373	1	14	15	18	20	9.0	3.4	3.6	3.4
Non-Energy Use	**22**	**97**	**108**	**119**	**131**	**100**	**100**	**100**	**100**	**100**	**5.2**	**1.1**	**1.0**	**1.0**

Reference Scenario: China

	Energy Demand (Mtoe)					Shares (%)					Growth Rates (% per annum)			
	1971	2000	2010	2020	2030	1971	2000	2010	2020	2030	1971-2000	2000-2010	2000-2020	2000-2030
Total Primary Energy Supply	**241**	**950**	**1302**	**1707**	**2133**	**100**	**100**	**100**	**100**	**100**	**4.8**	**3.2**	**3.0**	**2.7**
Coal	192	659	854	1059	1278	80	69	66	62	60	4.3	2.6	2.4	2.2
Oil	43	236	336	455	578	18	25	26	27	27	6.0	3.6	3.3	3.0
Gas	3	30	57	102	151	1	3	4	6	7	8.1	6.5	6.3	5.5
Nuclear	0	4	23	43	63	0	0	2	2	3	-	18.3	12.1	9.3
Hydro	3	19	29	44	54	1	2	2	3	3	7.2	4.1	4.2	3.5
Other Renewables	0	1	4	5	9	0	0	0	0	0	-	10.7	7.0	6.8
Power Generation	**37**	**360**	**561**	**799**	**1058**	**100**	**100**	**100**	**100**	**100**	**8.1**	**4.5**	**4.1**	**3.7**
Coal	30	314	472	649	849	82	87	84	81	80	8.4	4.1	3.7	3.4
Oil	4	16	18	18	18	11	4	3	2	2	4.7	1.2	0.7	0.5
Gas	0	5	16	41	65	0	1	3	5	6	-	12.3	11.0	8.9
Nuclear	0	4	23	43	63	0	1	4	5	6	-	18.3	12.1	9.3
Hydro	3	19	29	44	54	7	5	5	5	5	7.2	4.1	4.2	3.5
Other Renewables	0	1	4	5	9	0	0	1	1	1	-	10.7	7.0	6.8
Other Transformation, Own Use & Losses	**27**	**159**	**199**	**240**	**281**						**6.3**	**2.3**	**2.1**	**1.9**
of which electricity	*2*	*26*	*44*	*66*	*91*						*9.1*	*5.2*	*4.7*	*4.2*
Total Final Consumption	**189**	**577**	**774**	**1011**	**1264**	**100**	**100**	**100**	**100**	**100**	**3.9**	**3.0**	**2.8**	**2.7**
Coal	140	247	273	291	305	74	43	35	29	24	2.0	1.0	0.8	0.7
Oil	38	192	283	394	510	20	33	37	39	40	5.8	3.9	3.7	3.3
Gas	1	18	30	47	69	1	3	4	5	5	9.2	5.0	4.9	4.5
Electricity	10	93	153	232	323	5	16	20	23	26	7.9	5.1	4.7	4.2
Heat	0	25	36	46	56	0	4	5	5	4	-	3.4	3.0	2.7
Renewables	0	0	0	0	0	0	0	0	0	0	-	-	-	-

Industry	**14**	**314**	**396**	**477**	**553**	**100**	**100**	**100**	**100**	**100**	**11.4**	**2.3**	**2.1**	**1.9**
Coal	6	169	197	217	233	43	54	50	45	42	12.2	1.5	1.2	1.1
Oil	8	57	71	84	94	55	18	18	18	17	7.2	2.3	2.0	1.7
Gas	0	12	17	24	32	0	4	4	5	6	34.9	3.9	3.7	3.4
Electricity	0	57	85	120	157	1	18	21	25	28	21.4	4.0	3.8	3.4
Heat	0	19	25	31	37	0	6	6	7	7	-	3.0	2.6	2.3
Renewables	0	0	0	0	0	0	0	0	0	0	-	-	-	-
Transportation	**6**	**85**	**136**	**206**	**286**	**100**	**100**	**100**	**100**	**100**	**9.8**	**4.9**	**4.6**	**4.1**
Oil	6	77	130	200	279	100	92	95	97	98	9.4	5.3	4.9	4.4
Other fuels	0	7	7	7	7	0	8	5	3	2	-	-0.5	-0.4	-0.2
Other Sectors	**1**	**146**	**204**	**283**	**374**	**100**	**100**	**100**	**100**	**100**	**21.1**	**3.4**	**3.4**	**3.2**
Coal	0	60	60	60	60	0	41	30	21	16	-	0.0	0.0	-0.0
Oil	0	44	65	90	114	53	30	32	32	30	18.7	4.0	3.6	3.2
Gas	0	6	12	22	37	4	4	6	8	10	21.9	6.7	6.6	6.0
Electricity	0	29	57	96	145	43	20	28	34	39	17.9	7.0	6.2	5.5
Heat	0	6	10	14	18	0	4	5	5	5	-	4.7	4.2	3.8
Renewables	0	0	0	0	0	0	0	0	0	0	-	-	-	-
Non-Energy Use	**169**	**32**	**38**	**44**	**50**						**-5.5**	**1.5**	**1.6**	**1.5**
Electricity Generation (TWh)	**144**	**1387**	**2282**	**3461**	**4813**	**100**	**100**	**100**	**100**	**100**	**8.1**	**5.1**	**4.7**	**4.2**
Coal	98	1081	1723	2509	3503	68	78	75	72	73	8.6	4.8	4.3	4.0
Oil	16	46	51	53	54	11	3	2	2	1	3.6	1.1	0.7	0.5
Gas	0	19	74	209	349	0	1	3	6	7	-	14.9	12.8	10.3
Hydrogen - Fuel Cell	0	0	0	0	0	0	0	0	0	0	-	-	-	-
Nuclear	0	17	90	163	242	0	1	4	5	5	-	18.3	12.1	9.3
Hydro	30	222	333	511	622	21	16	15	15	13	7.2	4.1	4.2	3.5
Other Renewables	0	2	10	16	42	0	0	0	0	1	-	18.0	11.1	10.8

Biomass (NOT included above)	163	212	212	206	193	40	18	14	11	8	0.9	0.0	-0.1	-0.3
Total Primary Energy Supply (including Biomass)	**404**	**1162**	**1514**	**1913**	**2326**	**100**	**100**	**100**	**100**	**100**	**3.7**	**2.7**	**2.5**	**2.3**

Reference Scenario: China

	Capacity (GW)				Shares (%)				Growth Rates (% per annum)		
	1999	2010	2020	2030	1999	2010	2020	2030	1999-2010	1999-2020	1999-2030
Total Capacity	**300**	**517**	**787**	**1087**	**100**	**100**	**100**	**100**	**5.1**	**4.7**	**4.2**
Coal	199	342	499	696	66	66	63	64	5.0	4.5	4.1
Oil	21	22	22	23	7	4	3	2	0.4	0.4	0.3
Gas	4	27	69	113	1	5	9	10	18.6	14.2	11.2
Hydrogen - Fuel Cell	0	0	0	0	0	0	0	0	-	-	-
Nuclear	2	11	21	31	1	2	3	3	16.6	11.5	9.0
Hydro	73	112	171	209	24	22	22	19	4.0	4.2	3.5
Other renewables	1	3	5	15	0	1	1	1	14.1	9.5	10.2

	Capacity (GW)				Shares (%)				Growth Rates (% per annum)		
	1999	2010	2020	2030	1999	2010	2020	2030	1999-2010	1999-2020	1999-2030
Other Renewables	**1**	**3**	**5**	**15**	**100**	**100**	**100**	**100**	**14.1**	**9.5**	**10.2**
Biomass	0	1	1	1	60	19	16	7	3.0	3.0	3.0
Wind	0	2	4	12	40	71	73	76	20.1	12.7	12.5
Geothermal	0	0	0	1	0	8	8	5	-	-	-
Solar	0	0	0	2	0	2	2	12	-	-	-
Tide/Wave	0	0	0	0	0	0	0	0	-	-	-

	Electricity Generation (TWh)				Shares (%)				Growth Rates (% per annum)		
	2000	2010	2020	2030	2000	2010	2020	2030	2000-2010	2000-2020	2000-2030
Other Renewables	**2**	**10**	**16**	**42**	**100**	**100**	**100**	**100**	**18.0**	**11.1**	**10.8**
Biomass	2	3	4	5	100	26	23	12	3.3	3.2	3.1
Wind	0	6	9	29	0	54	57	68	-	-	-
Geothermal	0	2	3	5	0	19	19	13	-	-	-
Solar	0	0	0	3	0	1	1	7	-	-	-
Tide/Wave	0	0	0	0	0	0	0	0	-	-	-

Reference Scenario: China

	CO_2 Emissions (Mt)					Shares (%)					Growth Rates (% per annum)			
	1971	2000	2010	2020	2030	1971	2000	2010	2020	2030	1971-2000	2000-2010	2000-2020	2000-2030
Total CO_2 Emissions	**812**	**3052**	**4155**	**5393**	**6718**	**100**	**100**	**100**	**100**	**100**	**4.7**	**3.1**	**2.9**	**2.7**
change since 1990 (%)		*33.3*	*81.5*	*135.5*	*193.4*									
Coal	681	2399	3169	3985	4869	84	79	76	74	72	4.4	2.8	2.6	2.4
Oil	124	594	862	1184	1518	15	19	21	22	23	5.5	3.8	3.5	3.2
Gas	7	59	124	224	330	1	2	3	4	5	7.4	7.7	6.9	5.9
Power Generation	**132**	**1349**	**2024**	**2810**	**3683**	**100**	**100**	**100**	**100**	**100**	**8.4**	**4.1**	**3.7**	**3.4**
Coal	118	1282	1923	2648	3462	90	95	95	94	94	8.6	4.1	3.7	3.4
Oil	13	55	62	63	64	10	4	3	2	2	5.0	1.2	0.7	0.5
Gas	0	12	39	98	157	0	1	2	3	4	-	12.3	11.0	8.9
Transformation, Own Use & Losses	**6**	**138**	**186**	**223**	**257**	**100**	**100**	**100**	**100**	**100**	**11.5**	**3.0**	**2.4**	**2.1**
Total Final Consumption	**675**	**1565**	**1946**	**2360**	**2778**	**100**	**100**	**100**	**100**	**100**	**2.9**	**2.2**	**2.1**	**1.9**
Coal	562	1047	1165	1248	1312	83	67	60	53	47	2.2	1.1	0.9	0.8
Oil	109	486	731	1034	1353	16	31	38	44	49	5.3	4.2	3.8	3.5
Gas	3	31	50	78	112	0	2	3	3	4	8.1	4.8	4.7	4.3
Industry	**67**	**881**	**1044**	**1171**	**1274**	**100**	**100**	**100**	**100**	**100**	**9.3**	**1.7**	**1.4**	**1.2**
Coal	44	739	861	947	1014	66	84	82	81	80	10.2	1.5	1.2	1.1
Oil	23	120	150	178	200	34	14	14	15	16	5.8	2.3	2.0	1.7
Gas	0	22	33	46	60	0	3	3	4	5	-	3.9	3.7	3.4
Transportation	**15**	**242**	**391**	**591**	**818**	**100**	**100**	**100**	**100**	**100**	**10.1**	**4.9**	**4.6**	**4.1**
Oil	15	220	368	567	793	100	91	94	96	97	9.7	5.3	4.9	4.4
Other fuels	0	23	23	24	24	0	9	6	4	3	-	0.1	0.2	0.2
Other Sectors	**590**	**392**	**460**	**547**	**636**	**100**	**100**	**100**	**100**	**100**	**-1.4**	**1.6**	**1.7**	**1.6**
Coal	518	253	252	252	252	88	65	55	46	40	-2.4	-0.1	0.0	0.0
Oil	68	130	192	265	334	12	33	42	48	53	2.3	4.0	3.6	3.2
Gas	3	9	16	31	50	1	2	4	6	8	3.4	6.7	6.6	6.0
Non-Energy Use	**3**	**50**	**50**	**51**	**51**	**100**	**100**	**100**	**100**	**100**	**10.5**	**0.1**	**0.1**	**0.1**

Reference Scenario: East Asia

	Energy Demand (Mtoe)					Shares (%)					Growth Rates (% per annum)			
	1971	2000	2010	2020	2030	1971	2000	2010	2020	2030	1971-2000	2000-2010	2000-2020	2000-2030
Total Primary Energy Supply	**79**	**422**	**618**	**836**	**1063**	**100**	**100**	**100**	**100**	**100**	**6.0**	**3.9**	**3.5**	**3.1**
Coal	29	114	160	221	308	36	27	26	26	29	4.8	3.5	3.4	3.4
Oil	47	207	284	372	456	59	49	46	44	43	5.3	3.2	3.0	2.7
Gas	2	72	121	174	216	2	17	20	21	20	14.3	5.3	4.5	3.7
Nuclear	0	10	15	17	19	0	2	2	2	2	-	4.4	2.8	2.2
Hydro	2	7	11	16	19	3	2	2	2	2	4.4	4.5	4.2	3.4
Other Renewables	0	13	27	35	45	0	3	4	4	4	-	7.6	5.2	4.3
Power Generation	**13**	**139**	**229**	**331**	**444**	**100**	**100**	**100**	**100**	**100**	**8.7**	**5.1**	**4.4**	**3.9**
Coal	3	45	82	136	217	20	32	36	41	49	10.4	6.2	5.7	5.4
Oil	8	29	31	32	28	62	21	14	10	6	4.7	0.6	0.6	-0.1
Gas	0	35	62	94	115	2	25	27	28	26	18.1	5.8	5.0	4.0
Nuclear	0	10	15	17	19	0	7	7	5	4	-	4.4	2.8	2.2
Hydro	2	7	11	16	19	16	5	5	5	4	4.4	4.5	4.2	3.4
Other Renewables	0	13	27	35	45	0	9	12	11	10	-	7.6	5.2	4.3
Other Transformation, Own Use & Losses	**8**	**54**	**81**	**104**	**125**						**6.9**	**4.2**	**3.4**	**2.9**
of which electricity	*1*	*9*	*15*	*23*	*31*						*9.0*	*5.1*	*4.5*	*4.1*
Total Final Consumption	**64**	**280**	**392**	**526**	**668**	**100**	**100**	**100**	**100**	**100**	**5.2**	**3.4**	**3.2**	**2.9**
Coal	24	62	69	75	80	38	22	17	14	12	3.3	1.0	0.9	0.8
Oil	34	162	233	316	402	54	58	59	60	60	5.5	3.7	3.4	3.1
Gas	1	15	23	33	43	2	5	6	6	6	9.9	4.4	3.9	3.6
Electricity	4	41	68	103	143	7	15	17	20	21	7.9	5.2	4.7	4.3
Heat	0	0	0	0	0	0	0	0	0	0	-	-	-	-
Renewables	0	0	0	0	0	0	0	0	0	0	-	-	-	-

Industry	**34**	**131**	**168**	**207**	**246**	**100**	**100**	**100**	**100**	**100**	**4.8**	**2.5**	**2.3**	**2.1**
Coal	23	59	65	71	75	67	45	39	34	30	3.4	0.9	0.9	0.8
Oil	8	40	54	65	76	24	31	32	32	31	5.8	2.9	2.4	2.1
Gas	1	13	19	27	35	2	10	11	13	14	10.3	4.3	3.8	3.4
Electricity	2	19	30	44	60	7	14	18	21	24	7.6	4.8	4.4	4.0
Heat	0	0	0	0	0	0	0	0	0	0	-	-	-	-
Renewables	0	0	0	0	0	0	0	0	0	0	-	-	-	-
Transportation	**16**	**89**	**136**	**197**	**262**	**100**	**100**	**100**	**100**	**100**	**6.1**	**4.3**	**4.0**	**3.7**
Oil	16	89	136	197	261	99	100	100	100	100	6.1	4.3	4.0	3.7
Other fuels	0	0	0	0	0	1	0	0	0	0	-0.2	3.8	3.3	2.8
Other Sectors	**7**	**52**	**78**	**110**	**145**	**100**	**100**	**100**	**100**	**100**	**7.2**	**4.2**	**3.8**	**3.5**
Coal	0	3	3	4	4	2	6	4	4	3	11.9	1.7	1.7	1.5
Oil	6	27	36	44	53	86	52	46	41	37	5.4	2.9	2.6	2.3
Gas	0	2	4	6	8	1	4	5	5	6	13.8	5.5	5.0	4.4
Electricity	1	20	35	55	79	12	39	45	51	55	11.7	5.8	5.2	4.7
Heat	0	0	0	0	0	0	0	0	0	0	-	-	-	-
Renewables	0	0	0	0	0	0	0	0	0	0	-	-	-	-
Non-Energy Use	**7**	**8**	**10**	**13**	**16**						**0.2**	**2.6**	**2.5**	**2.3**
Electricity Generation (TWh)	**62**	**585**	**970**	**1461**	**2029**	**100**	**100**	**100**	**100**	**100**	**8.1**	**5.2**	**4.7**	**4.2**
Coal	9	151	312	548	958	15	26	32	38	47	10.1	7.5	6.6	6.3
Oil	29	126	134	142	125	47	22	14	10	6	5.2	0.6	0.6	0.0
Gas	0	171	299	464	571	0	29	31	32	28	24.7	5.7	5.1	4.1
Hydrogen - Fuel Cells	0	0	0	0	0	0	0	0	0	0	-	-	-	-
Nuclear	0	38	59	66	74	0	7	6	5	4	-	4.4	2.8	2.2
Hydro	23	82	128	187	224	37	14	13	13	11	4.4	4.5	4.2	3.4
Other Renewables	0	16	38	54	77	0	3	4	4	4	-	9.1	6.3	5.4
Biomass (NOT included above)	70	108	117	118	115	47	20	16	12	10	1.5	0.7	0.4	0.2
Total Primary Energy Supply (including Biomass)	**149**	**531**	**735**	**953**	**1178**	**100**	**100**	**100**	**100**	**100**	**4.5**	**3.3**	**3.0**	**2.7**

Reference Scenario: East Asia

	Capacity (GW)				Shares (%)				Growth Rates (% per annum)		
	1999	2010	2020	2030	1999	2010	2020	2030	1999-2010	1999-2020	1999-2030
Total Capacity	**159**	**250**	**375**	**524**	**100**	**100**	**100**	**100**	**4.2**	**4.2**	**3.9**
Coal	28	64	113	201	18	26	30	38	7.7	6.9	6.6
Oil	43	50	53	51	27	20	14	10	1.3	1.0	0.5
Gas	51	87	139	185	32	35	37	35	5.0	4.9	4.3
Hydrogen - Fuel Cell	0	0	0	0	0	0	0	0	-	-	-
Nuclear	5	7	8	9	3	3	2	2	4.0	2.6	2.1
Hydro	28	36	52	63	18	14	14	12	2.2	3.0	2.6
Other renewables	4	6	9	15	2	2	2	3	4.3	4.2	4.5

	Capacity (GW)				Shares (%)				Growth Rates (% per annum)		
	1999	2010	2020	2030	1999	2010	2020	2030	1999-2010	1999-2020	1999-2030
Other Renewables	**4**	**6**	**9**	**15**	**100**	**100**	**100**	**100**	**4.3**	**4.2**	**4.5**
Biomass	1	3	4	6	37	41	46	39	5.2	5.3	4.7
Wind	0	0	0	3	0	2	3	22	36.2	23.9	24.9
Geothermal	2	4	5	6	63	58	50	36	3.5	3.0	2.7
Solar	0	0	0	0	0	0	1	3	38.9	43.1	36.3
Tide/Wave	0	0	0	0	0	0	0	0	-	-	-

	Electricity Generation (TWh)				Shares (%)				Growth Rates (% per annum)		
	2000	2010	2020	2030	2000	2010	2020	2030	2000-2010	2000-2020	2000-2030
Other Renewables	**16**	**38**	**54**	**77**	**100**	**100**	**100**	**100**	**9.1**	**6.3**	**5.4**
Biomass	2	11	18	26	10	29	34	34	21.4	13.1	9.8
Wind	0	0	1	8	0	1	1	10	23.5	17.4	20.5
Geothermal	14	27	35	42	90	71	64	55	6.5	4.5	3.6
Solar	0	0	0	1	0	0	0	1	24.1	35.5	31.2
Tide/Wave	0	0	0	0	0	0	0	0	-	-	-

Reference Scenario: East Asia

	CO_2 Emissions (Mt)					Shares (%)					Growth Rates (% per annum)			
	1971	2000	2010	2020	2030	1971	2000	2010	2020	2030	1971-2000	2000-2010	2000-2020	2000-2030
Total CO_2 Emissions	**232**	**1129**	**1594**	**2168**	**2805**	**100**	**100**	**100**	**100**	**100**	**5.6**	**3.5**	**3.3**	**3.1**
change since 1990 (%)		*81.0*	*163.4*	*265.8*	*379.2*									
Coal	108	431	605	842	1181	47	38	38	39	42	4.9	3.4	3.4	3.4
Oil	121	542	735	959	1171	52	48	46	44	42	5.3	3.1	2.9	2.6
Gas	3	156	254	367	453	1	14	16	17	16	14.8	5.0	4.4	3.6
Power Generation	**35**	**352**	**568**	**859**	**1215**	**100**	**100**	**100**	**100**	**100**	**8.3**	**4.9**	**4.6**	**4.2**
Coal	10	177	324	536	855	28	50	57	62	70	10.5	6.2	5.7	5.4
Oil	25	92	98	103	90	70	26	17	12	7	4.7	0.6	0.6	-0.1
Gas	1	83	146	221	270	2	24	26	26	22	18.7	5.8	5.0	4.0
Transformation, Own Use & Losses	**7**	**75**	**100**	**126**	**147**	**100**	**100**	**100**	**100**	**100**	**8.6**	**2.9**	**2.6**	**2.3**
Total Final Consumption	**190**	**702**	**926**	**1182**	**1443**	**100**	**100**	**100**	**100**	**100**	**4.6**	**2.8**	**2.6**	**2.4**
Coal	98	254	280	305	325	52	36	30	26	23	3.3	1.0	0.9	0.8
Oil	90	420	604	819	1041	47	60	65	69	72	5.5	3.7	3.4	3.1
Gas	2	28	42	58	76	1	4	4	5	5	9.3	4.2	3.8	3.5
Industry	**125**	**370**	**441**	**508**	**570**	**100**	**100**	**100**	**100**	**100**	**3.8**	**1.8**	**1.6**	**1.5**
Coal	94	240	263	285	303	76	65	60	56	53	3.3	0.9	0.9	0.8
Oil	29	106	141	171	200	23	29	32	34	35	4.6	2.9	2.4	2.1
Gas	2	24	37	51	67	1	6	8	10	12	9.9	4.4	3.9	3.5
Transportation	**35**	**232**	**354**	**512**	**680**	**100**	**100**	**100**	**100**	**100**	**6.7**	**4.3**	**4.0**	**3.6**
Oil	35	232	354	512	680	99	100	100	100	100	6.7	4.3	4.0	3.6
Other fuels	0.5	0.5	0.02	0.02	0.02	1	0	0	0	0	0.3	-29.3	-15.9	-10.9
Other Sectors	**28**	**97**	**127**	**158**	**188**	**100**	**100**	**100**	**100**	**100**	**4.4**	**2.8**	**2.5**	**2.2**
Coal	3	14	16	19	21	11	14	13	12	11	5.2	1.7	1.7	1.5
Oil	24	80	106	133	158	87	83	84	84	84	4.2	2.9	2.6	2.3
Gas	1	3	5	7	9	2	3	4	4	5	6.3	4.3	3.8	3.5
Non-Energy Use	**2**	**3**	**3**	**4**	**5**	**100**	**100**	**100**	**100**	**100**	**0.9**	**2.8**	**2.7**	**2.4**

Reference Scenario: Indonesia

	Energy Demand (Mtoe)					Shares (%)					Growth Rates (% per annum)			
	1971	2000	2010	2020	2030	1971	2000	2010	2020	2030	1971-2000	2000-2010	2000-2020	2000-2030
Total Primary Energy Supply	**9**	**98**	**152**	**213**	**276**	**100**	**100**	**100**	**100**	**100**	**8.7**	**4.5**	**4.0**	**3.5**
Coal	0	14	24	40	63	1	14	16	19	23	17.4	6.0	5.5	5.2
Oil	8	53	73	96	118	94	54	48	45	43	6.6	3.2	3.0	2.7
Gas	0	28	45	64	78	3	29	30	30	28	17.8	4.7	4.2	3.4
Nuclear	0	0	0	0	0	0	0	0	0	0	-	-	-	-
Hydro	0	1	2	2	2	1	1	1	1	1	6.6	7.2	4.7	3.8
Other Renewables	0	2	8	12	16	0	2	5	6	6	-	13.5	8.6	6.7
Power Generation	**1**	**22**	**45**	**73**	**104**	**100**	**100**	**100**	**100**	**100**	**10.4**	**7.3**	**6.1**	**5.2**
Coal	0	9	18	31	53	0	38	39	43	51	-	7.5	6.7	6.3
Oil	1	4	4	5	5	90	19	10	7	5	4.7	0.4	0.8	0.5
Gas	0	6	14	23	27	0	29	30	31	26	-	7.7	6.5	4.9
Nuclear	0	0	0	0	0	0	0	0	0	0	-	-	-	-
Hydro	0	1	2	2	2	10	4	3	3	2	6.6	7.2	4.7	3.8
Other Renewables	0	2	8	12	16	0	10	18	16	15	-	13.5	8.6	6.7
Other Transformation, Own Use & Losses	**1**	**17**	**26**	**34**	**42**						**10.1**	**4.3**	**3.6**	**3.0**
of which electricity	*0*	*1*	*2*	*4*	*6*						*11.2*	*7.3*	*6.2*	*5.4*
Total Final Consumption	**7**	**67**	**96**	**132**	**169**	**100**	**100**	**100**	**100**	**100**	**8.2**	**3.7**	**3.5**	**3.2**
Coal	0	6	7	8	9	1	9	7	6	5	15.3	1.8	1.8	1.6
Oil	6	45	64	85	106	95	68	66	64	63	7.0	3.5	3.2	2.9
Gas	0	9	12	17	21	1	13	13	13	13	16.7	3.4	3.2	2.9
Electricity	0	7	13	22	33	2	10	14	17	19	13.7	6.8	6.0	5.4
Heat	0	0	0	0	0	0	0	0	0	0	-	-	-	-
Renewables	0	0	0	0	0	0	0	0	0	0	-	-	-	-

Industry	**2**	**23**	**32**	**41**	**50**	**100**	**100**	**100**	**100**	**100**	**9.5**	**3.2**	**2.9**	**2.6**
Coal	0	3	4	5	6	4	15	13	12	11	14.9	1.8	1.8	1.6
Oil	1	9	12	15	17	88	40	38	36	33	6.6	2.9	2.3	2.0
Gas	0	8	10	13	16	6	33	32	31	31	16.1	2.8	2.7	2.4
Electricity	0	3	5	8	12	3	13	17	21	24	15.6	6.0	5.4	4.8
Heat	0	0	0	0	0	0	0	0	0	0	-	-	-	-
Renewables	0	0	0	0	0	0	0	0	0	0	-	-	-	-
Transportation	**3**	**21**	**32**	**45**	**59**	**100**	**100**	**100**	**100**	**100**	**7.3**	**4.0**	**3.8**	**3.4**
Oil	3	21	32	45	59	99	100	100	100	100	7.4	4.0	3.8	3.4
Other fuels	0	0	0	0	0	1	0	0	0	0	-	-	-	-
Other Sectors	**2**	**21**	**31**	**44**	**58**	**100**	**100**	**100**	**100**	**100**	**8.0**	**4.0**	**3.7**	**3.4**
Coal	0	2	3	3	4	0	11	9	7	6	-	1.8	1.8	1.6
Oil	2	14	19	24	29	96	66	60	55	51	6.6	3.1	2.8	2.5
Gas	0	1	2	4	6	0	6	8	9	10	-	5.9	5.6	4.9
Electricity	0	4	7	12	19	4	17	23	28	33	13.0	7.4	6.5	5.8
Heat	0	0	0	0	0	0	0	0	0	0	-	-	-	-
Renewables	0	0	0	0	0	0	0	0	0	0	-	-	-	-
Non-Energy Use	**0**	**1**	**2**	**2**	**3**						**8.4**	**4.5**	**4.1**	**3.7**
Electricity Generation (TWh)	**3**	**93**	**180**	**300**	**445**	**100**	**100**	**100**	**100**	**100**	**12.5**	**6.8**	**6.1**	**5.4**
Coal	0	29	68	128	238	0	31	38	43	54	-	9.0	7.8	7.3
Oil	2	20	20	23	23	53	22	11	8	5	9.1	0.1	0.7	0.4
Gas	0	32	63	111	136	0	34	35	37	30	-	7.1	6.4	5.0
Hydrogen - Fuel Cell	0	0	0	0	0	0	0	0	0	0	-	-	-	-
Nuclear	0	0	0	0	0	0	0	0	0	0	-	-	-	-
Hydro	1	9	18	23	28	47	10	10	8	6	6.6	7.2	4.7	3.8
Other Renewables	0	3	10	15	21	0	3	6	5	5	-	14.4	9.0	7.1
Biomass (NOT included above)	27	47	49	44	37	76	32	24	17	12	1.8	0.5	-0.2	-0.7
Total Primary Energy Supply (including Biomass)	**36**	**145**	**201**	**257**	**313**	**100**	**100**	**100**	**100**	**100**	**4.9**	**3.3**	**2.9**	**2.6**

Reference Scenario: Indonesia

	Capacity (GW)				Shares (%)				Growth Rates (% per annum)		
	1999	2010	2020	2030	1999	2010	2020	2030	1999-2010	1999-2020	1999-2030
Total Capacity	**36**	**56**	**83**	**124**	**100**	**100**	**100**	**100**	**4.0**	**4.1**	**4.1**
Coal	5	11	21	44	13	20	25	36	7.8	7.2	7.4
Oil	16	18	20	21	45	32	24	17	0.9	0.9	0.9
Gas	10	19	33	46	28	35	40	37	6.1	5.8	5.0
Hydrogen - Fuel Cell	0	0	0	0	0	0	0	0	-	-	-
Nuclear	0	0	0	0	0	0	0	0	-	-	-
Hydro	4	6	7	9	12	10	9	7	3.1	2.7	2.4
Other renewables	1	2	2	4	2	3	3	3	6.0	5.1	4.9

	Capacity (GW)				Shares (%)				Growth Rates (% per annum)		
	1999	2010	2020	2030	1999	2010	2020	2030	1999-2010	1999-2020	1999-2030
Other Renewables	**1**	**2**	**2**	**4**	**100**	**100**	**100**	**100**	**6.0**	**5.1**	**4.9**
Biomass	0	1	1	1	37	32	29	27	4.6	4.1	3.9
Wind	0	0	0	0	0	0	4	8	-	-	-
Geothermal	1	1	2	2	63	68	66	56	6.7	5.4	4.5
Solar	0	0	0	0	0	0	0	8	-	-	-
Tide/Wave	0	0	0	0	0	0	0	0	-	-	-

	Electricity Generation (TWh)				Shares (%)				Growth Rates (% per annum)		
	2000	2010	2020	2030	2000	2010	2020	2030	2000-2010	2000-2020	2000-2030
Other Renewables	**3**	**10**	**15**	**21**	**100**	**100**	**100**	**100**	**14.4**	**9.0**	**7.1**
Biomass	0	2	3	4	0	22	21	21	-	-	-
Wind	0	0	0	1	0	0	2	4	-	-	-
Geothermal	3	8	12	15	100	78	78	73	11.6	7.7	6.0
Solar	0	0	0	1	0	0	0	3	-	-	-
Tide/Wave	0	0	0	0	0	0	0	0	-	-	-

Reference Scenario: Indonesia

	CO_2 Emissions (Mt)					Shares (%)					Growth Rates (% per annum)			
	1971	2000	2010	2020	2030	1971	2000	2010	2020	2030	1971-2000	2000-2010	2000-2020	2000-2030
Total CO_2 Emissions	**25**	**269**	**396**	**560**	**739**	**100**	**100**	**100**	**100**	**100**	**8.6**	**3.9**	**3.7**	**3.4**
change since 1990 (%)		*100.0*	*194.3*	*315.9*	*448.8*									
Coal	1	58	98	157	246	2	22	25	28	33	17.8	5.4	5.1	4.9
Oil	24	153	206	271	334	97	57	52	48	45	6.6	3.1	2.9	2.6
Gas	0	59	92	132	159	1	22	23	24	21	20.0	4.6	4.1	3.4
Power Generation	**4**	**62**	**114**	**191**	**286**	**100**	**100**	**100**	**100**	**100**	**10.3**	**6.3**	**5.8**	**5.2**
Coal	0	33	68	122	207	0	54	60	64	72	-	7.5	6.7	6.3
Oil	4	13	14	16	15	100	22	12	8	5	4.6	0.4	0.8	0.5
Gas	0	15	32	53	64	0	24	28	28	22	-	7.7	6.5	4.9
Transformation, Own Use & Losses	**2**	**37**	**48**	**62**	**74**	**100**	**100**	**100**	**100**	**100**	**10.8**	**2.8**	**2.7**	**2.4**
Total Final Consumption	**19**	**171**	**234**	**307**	**379**	**100**	**100**	**100**	**100**	**100**	**7.8**	**3.2**	**3.0**	**2.7**
Coal	0	25	30	35	40	2	14	13	11	10	15.5	1.8	1.8	1.6
Oil	19	132	186	248	310	97	77	79	81	82	7.0	3.5	3.2	2.9
Gas	0	14	19	24	30	1	8	8	8	8	15.3	2.7	2.7	2.5
Industry	**5**	**55**	**71**	**86**	**100**	**100**	**100**	**100**	**100**	**100**	**8.7**	**2.6**	**2.3**	**2.0**
Coal	0	13	16	19	22	5	24	23	22	22	14.6	1.8	1.8	1.6
Oil	4	28	38	45	52	90	52	53	52	51	6.6	2.9	2.3	2.0
Gas	0	13	17	22	27	5	24	24	26	27	14.9	2.8	2.7	2.4
Transportation	**8**	**62**	**91**	**129**	**169**	**100**	**100**	**100**	**100**	**100**	**7.4**	**4.0**	**3.7**	**3.4**
Oil	8	61	91	129	169	98	99	100	100	100	7.5	4.0	3.8	3.4
Other fuels	0.1	0.5	0	0	0	2	1	0	0	0	4.8	-	-	-
Other Sectors	**6**	**53**	**70**	**90**	**108**	**100**	**100**	**100**	**100**	**100**	**7.5**	**2.8**	**2.6**	**2.4**
Coal	0	11	13	16	18	0	21	19	18	17	-	1.8	1.8	1.6
Oil	6	41	56	72	87	100	78	79	80	81	6.6	3.1	2.8	2.5
Gas	0	1	1	2	3	0	1	2	2	3	-	5.9	5.6	4.9
Non-Energy Use	**0**	**1**	**1**	**2**	**2**	**100**	**100**	**100**	**100**	**100**	**4.5**	**3.2**	**2.9**	**2.5**

Reference Scenario: South Asia

	Energy Demand (Mtoe)					Shares (%)					Growth Rates (% per annum)			
	1971	2000	2010	2020	2030	1971	2000	2010	2020	2030	1971-2000	2000-2010	2000-2020	2000-2030
Total Primary Energy Supply	**72**	**354**	**499**	**695**	**932**	**100**	**100**	**100**	**100**	**100**	**5.7**	**3.5**	**3.4**	**3.3**
Coal	38	168	206	265	347	53	47	41	38	37	5.2	2.1	2.3	2.5
Oil	27	129	184	260	357	38	36	37	37	38	5.5	3.7	3.6	3.5
Gas	3	44	84	134	179	5	12	17	19	19	9.4	6.7	5.7	4.8
Nuclear	0	5	7	12	17	0	1	1	2	2	9.8	3.8	4.5	4.1
Hydro	3	9	14	19	23	4	3	3	3	2	4.1	4.8	3.8	3.2
Other Renewables	0	0	3	5	9	0	0	1	1	1	-	31.5	18.1	14.0
Power Generation	**15**	**163**	**234**	**334**	**455**	**100**	**100**	**100**	**100**	**100**	**8.7**	**3.7**	**3.7**	**3.5**
Coal	8	124	156	208	283	57	76	67	62	62	9.7	2.4	2.6	2.8
Oil	2	9	13	19	25	12	5	6	6	6	5.5	4.6	4.1	3.7
Gas	1	16	40	71	98	9	10	17	21	22	9.1	9.4	7.7	6.2
Nuclear	0	5	7	12	17	2	3	3	4	4	9.8	3.8	4.5	4.1
Hydro	3	9	14	19	23	19	5	6	6	5	4.1	4.8	3.8	3.2
Other Renewables	0	0	3	5	9	0	0	1	1	2	-	31.5	18.1	14.0
Other Transformation, Own Use & Losses	**8**	**41**	**45**	**58**	**71**						**5.9**	**1.0**	**1.7**	**1.8**
of which electricity	*2*	*18*	*26*	*36*	*47*						*8.9*	*3.7*	*3.5*	*3.3*
Total Final Consumption	**56**	**205**	**306**	**432**	**590**	**100**	**100**	**100**	**100**	**100**	**4.6**	**4.1**	**3.8**	**3.6**
Coal	26	34	42	49	56	47	17	14	11	9	1.0	2.0	1.8	1.6
Oil	23	111	164	234	325	41	54	54	54	55	5.6	4.0	3.8	3.7
Gas	2	23	39	56	72	4	11	13	13	12	8.8	5.4	4.5	3.8
Electricity	5	37	60	94	137	9	18	20	22	23	7.0	5.1	4.8	4.5
Heat	0	0	0	0	0	0	0	0	0	0	-	-	-	-
Renewables	0	0	0	0	0	0	0	0	0	0	-	-	-	-

Industry	**22**	**86**	**122**	**161**	**205**	**100**	**100**	**100**	**100**	**100**	**4.9**	**3.6**	**3.2**	**2.9**
Coal	11	27	35	42	50	53	31	29	26	24	3.0	2.6	2.3	2.0
Oil	5	25	34	45	58	23	29	28	28	28	5.8	3.0	2.9	2.8
Gas	2	18	30	42	53	9	21	25	26	26	8.1	5.2	4.2	3.6
Electricity	3	15	23	32	44	15	18	19	20	22	5.5	4.0	3.7	3.6
Heat	0	0	0	0	0	0	0	0	0	0	-	-	-	-
Renewables	0	0	0	0	0	0	0	0	0	0	-	-	-	-
Transportation	**17**	**57**	**89**	**139**	**208**	**100**	**100**	**100**	**100**	**100**	**4.3**	**4.6**	**4.6**	**4.4**
Oil	9	56	88	138	207	52	99	99	99	99	6.6	4.6	4.6	4.5
Other fuels	8	1	1	2	2	48	1	1	1	1	-8.0	5.5	3.8	2.7
Other Sectors	**15**	**56**	**88**	**124**	**169**	**100**	**100**	**100**	**100**	**100**	**4.7**	**4.6**	**4.1**	**3.7**
Coal	7	7	7	7	6	45	13	8	5	4	0.3	-0.5	-0.5	-0.5
Oil	7	23	36	43	52	44	42	41	35	31	4.5	4.3	3.2	2.7
Gas	0	5	9	14	19	1	9	10	11	11	14.0	6.2	5.5	4.7
Electricity	2	20	36	60	91	10	37	41	48	54	9.3	5.9	5.5	5.1
Heat	0	0	0	0	0	0	0	0	0	0	-	-	-	-
Renewables	0	0	0	0	0	0	0	0	0	0	-	-	-	-
Non-Energy Use	**3**	**6**	**7**	**8**	**8**						**3.0**	**0.9**	**0.9**	**1.0**
Electricity Generation (TWh)	**76**	**635**	**1001**	**1505**	**2142**	**100**	**100**	**100**	**100**	**100**	**7.6**	**4.7**	**4.4**	**4.1**
Coal	32	420	556	792	1182	42	66	56	53	55	9.3	2.8	3.2	3.5
Oil	5	32	53	78	104	7	5	5	5	5	6.6	5.1	4.6	4.0
Gas	5	58	187	350	490	6	9	19	23	23	9.1	12.5	9.4	7.4
Hydrogen - Fuel Cell	0	0	0	0	0	0	0	0	0	0	-	-	-	-
Nuclear	1	19	28	47	65	2	3	3	3	3	9.8	3.8	4.5	4.1
Hydro	33	104	166	218	266	43	16	17	14	12	4.1	4.8	3.8	3.2
Other Renewables	0	2	11	20	35	0	0	1	1	2	-	20.9	13.4	10.8

Biomass (NOT included above)	141	241	266	281	285	66	40	35	29	23	1.9	1.0	0.8	0.6
Total Primary Energy Supply (including Biomass)	**213**	**595**	**765**	**975**	**1216**	**100**	**100**	**100**	**100**	**100**	**3.6**	**2.5**	**2.5**	**2.4**

Reference Scenario: South Asia

	Capacity (GW)				Shares (%)				Growth Rates (% per annum)		
	1999	2010	2020	2030	1999	2010	2020	2030	1999-2010	1999-2020	1999-2030
Total Capacity	**132**	**215**	**323**	**459**	**100**	**100**	**100**	**100**	**4.5**	**4.3**	**4.1**
Coal	65	94	133	199	49	44	41	43	3.5	3.5	3.7
Oil	15	22	30	39	11	10	9	9	3.3	3.3	3.1
Gas	19	46	87	127	14	21	27	28	8.4	7.5	6.3
Hydrogen - Fuel Cell	0	0	0	0	0	0	0	0	-	-	-
Nuclear	3	4	7	9	2	2	2	2	4.3	4.6	4.2
Hydro	30	45	59	72	22	21	18	16	3.9	3.4	2.9
Other renewables	1	4	7	12	1	2	2	3	11.1	8.6	7.6

	Capacity (GW)				Shares (%)				Growth Rates (% per annum)		
	1999	2010	2020	2030	1999	2010	2020	2030	1999-2010	1999-2020	1999-2030
Other Renewables	**1**	**4**	**7**	**12**	**100**	**100**	**100**	**100**	**11.1**	**8.6**	**7.6**
Biomass	0	1	2	5	15	27	33	38	17.6	13.0	11.0
Wind	1	3	4	6	85	68	62	54	8.8	6.9	6.0
Geothermal	0	0	0	0	0	0	0	0	-	-	-
Solar	0	0	0	1	0	5	5	8	94.2	46.1	33.3
Tide/Wave	0	0	0	0	0	0	0	0	-	-	-

	Electricity Generation (TWh)				Shares (%)				Growth Rates (% per annum)		
	2000	2010	2020	2030	2000	2010	2020	2030	2000-2010	2000-2020	2000-2030
Other Renewables	**2**	**11**	**20**	**35**	**100**	**100**	**100**	**100**	**20.9**	**13.4**	**10.8**
Biomass	0	4	9	18	4	40	47	52	51.0	27.7	20.4
Wind	2	6	10	15	95	57	50	42	14.8	9.8	7.8
Geothermal	0	0	0	0	0	0	0	0	-	-	-
Solar	0	0	1	2	0	3	3	5	46.3	25.0	19.8
Tide/Wave	0	0	0	0	0	0	0	0	-	-	-

Reference Scenario: South Asia

	CO_2 Emissions (Mt)					Shares (%)					Growth Rates (% per annum)			
	1971	2000	2010	2020	2030	1971	2000	2010	2020	2030	1971-2000	2000-2010	2000-2020	2000-2030
Total CO_2 Emissions	**226**	**1071**	**1490**	**2045**	**2738**	**100**	**100**	**100**	**100**	**100**	**5.5**	**3.4**	**3.3**	**3.2**
change since 1990 (%)		*61.9*	*125.3*	*209.2*	*313.9*									
Coal	150	638	798	1032	1355	66	60	54	50	49	5.1	2.3	2.4	2.5
Oil	69	342	510	720	991	30	32	34	35	36	5.7	4.1	3.8	3.6
Gas	7	90	182	293	392	3	8	12	14	14	9.1	7.2	6.1	5.0
Power Generation	**41**	**547**	**744**	**1037**	**1410**	**100**	**100**	**100**	**100**	**100**	**9.3**	**3.1**	**3.2**	**3.2**
Coal	33	483	609	811	1101	79	88	82	78	78	9.7	2.4	2.6	2.8
Oil	6	27	43	61	80	14	5	6	6	6	5.5	4.6	4.1	3.7
Gas	3	38	92	166	229	7	7	12	16	16	9.1	9.4	7.7	6.2
Transformation, Own Use & Losses	**7**	**29**	**34**	**35**	**35**	**100**	**100**	**100**	**100**	**100**	**4.9**	**1.8**	**1.0**	**0.6**
Total Final Consumption	**177**	**494**	**712**	**973**	**1292**	**100**	**100**	**100**	**100**	**100**	**3.6**	**3.7**	**3.4**	**3.3**
Coal	114	151	184	216	248	64	30	26	22	19	1.0	2.0	1.8	1.7
Oil	59	292	440	631	884	33	59	62	65	68	5.7	4.2	3.9	3.8
Gas	4	52	89	126	161	2	11	12	13	12	8.9	5.5	4.5	3.8
Industry	**73**	**220**	**302**	**386**	**474**	**100**	**100**	**100**	**100**	**100**	**3.9**	**3.2**	**2.9**	**2.6**
Coal	56	122	158	191	225	77	56	52	49	47	2.7	2.6	2.3	2.0
Oil	13	57	77	103	133	18	26	26	27	28	5.2	3.1	3.0	2.9
Gas	4	40	67	93	116	6	18	22	24	24	8.2	5.2	4.2	3.6
Transportation	**54**	**160**	**251**	**393**	**590**	**100**	**100**	**100**	**100**	**100**	**3.8**	**4.6**	**4.6**	**4.4**
Oil	23	160	251	393	590	42	100	100	100	100	6.9	4.6	4.6	4.4
Other fuels	31	0	0	0	0	58	0	0	0	0	-17.0	0.0	0.0	0.0
Other Sectors	**48**	**109**	**152**	**186**	**220**	**100**	**100**	**100**	**100**	**100**	**2.9**	**3.4**	**2.7**	**2.4**
Coal	27	28	26	25	23	57	26	17	13	10	0.1	-0.7	-0.7	-0.7
Oil	20	69	105	128	152	43	63	69	69	69	4.3	4.3	3.1	2.7
Gas	0	12	21	33	45	1	11	14	18	21	14.1	6.2	5.5	4.7
Non-Energy Use	**3**	**6**	**7**	**8**	**9**	**100**	**100**	**100**	**100**	**100**	**2.9**	**1.1**	**1.1**	**1.1**

Reference Scenario: India

	Energy Demand (Mtoe)					Shares (%)					Growth Rates (% per annum)			
	1971	2000	2010	2020	2030	1971	2000	2010	2020	2030	1971-2000	2000-2010	2000-2020	2000-2030
Total Primary Energy Supply	**62**	**300**	**413**	**567**	**750**	**100**	**100**	**100**	**100**	**100**	**5.6**	**3.2**	**3.2**	**3.1**
Coal	37	165	202	260	341	60	55	49	46	45	5.3	2.1	2.3	2.4
Oil	22	102	145	201	271	35	34	35	35	36	5.5	3.5	3.4	3.3
Gas	1	22	46	75	97	1	7	11	13	13	13.2	7.8	6.4	5.1
Nuclear	0	4	6	11	16	0	1	2	2	2	9.6	3.8	4.8	4.3
Hydro	2	6	11	15	18	4	2	3	3	2	3.4	5.7	4.2	3.5
Other Renewables	0	0	2	4	8	0	0	1	1	1	-	32.8	19.1	14.5
Power Generation	**13**	**145**	**203**	**286**	**385**	**100**	**100**	**100**	**100**	**100**	**8.7**	**3.4**	**3.4**	**3.3**
Coal	8	124	156	206	280	65	85	77	72	73	9.7	2.3	2.6	2.8
Oil	2	2	4	4	4	12	2	2	1	1	1.3	4.8	3.0	2.1
Gas	0	8	24	45	60	2	6	12	16	16	12.2	11.5	8.9	6.9
Nuclear	0	4	6	11	16	2	3	3	4	4	9.6	3.8	4.8	4.3
Hydro	2	6	11	15	18	19	4	5	5	5	3.4	5.7	4.2	3.5
Other Renewables	0	0	2	4	8	0	0	1	1	2	-	32.8	19.1	14.5
Other Transformation, Own Use & Losses	**7**	**36**	**39**	**49**	**60**						**5.9**	**0.6**	**1.5**	**1.7**
of which electricity	*1*	*16*	*22*	*30*	*40*						*9.0*	*3.5*	*3.4*	*3.2*
Total Final Consumption	**49**	**165**	**244**	**341**	**459**	**100**	**100**	**100**	**100**	**100**	**4.3**	**4.0**	**3.7**	**3.5**
Coal	25	32	39	46	52	52	20	16	13	11	0.8	2.0	1.8	1.6
Oil	18	92	135	191	262	38	55	55	56	57	5.7	4.0	3.7	3.6
Gas	0	10	19	26	30	1	6	8	8	7	12.7	6.1	4.7	3.6
Electricity	5	31	51	79	115	9	19	21	23	25	6.8	5.1	4.8	4.5
Heat	0	0	0	0	0	0	0	0	0	0	-	-	-	-
Renewables	0	0	0	0	0	0	0	0	0	0	-	-	-	-

Industry	**19**	**70**	**98**	**128**	**159**	**100**	**100**	**100**	**100**	**100**	**4.7**	**3.4**	**3.1**	**2.8**
Coal	11	25	33	39	46	58	36	33	31	29	3.0	2.6	2.3	2.1
Oil	5	22	29	38	49	25	32	30	30	31	5.6	2.7	2.8	2.7
Gas	0	10	17	23	27	2	14	17	18	17	12.7	6.1	4.6	3.5
Electricity	3	13	20	27	37	16	19	20	21	23	5.3	3.9	3.6	3.5
Heat	0	0	0	0	0	0	0	0	0	0	-	-	-	-
Renewables	0	0	0	0	0	0	0	0	0	0	-	-	-	-
Transportation	**15**	**44**	**70**	**108**	**160**	**100**	**100**	**100**	**100**	**100**	**3.8**	**4.6**	**4.6**	**4.4**
Oil	7	44	68	107	158	46	98	98	99	99	6.6	4.6	4.6	4.4
Other fuels	8	1	1	2	2	54	2	2	1	1	-8.0	5.6	3.8	2.7
Other Sectors	**13**	**46**	**71**	**99**	**133**	**100**	**100**	**100**	**100**	**100**	**4.5**	**4.5**	**4.0**	**3.7**
Coal	7	7	7	7	6	52	16	10	7	5	0.3	-0.5	-0.5	-0.5
Oil	5	21	32	40	48	38	45	46	40	36	5.1	4.7	3.4	2.9
Gas	0	1	1	2	3	0	2	2	2	2	13.5	6.5	5.9	4.9
Electricity	1	17	30	50	76	10	37	43	51	57	9.3	5.9	5.5	5.1
Heat	0	0	0	0	0	0	0	0	0	0	-	-	-	-
Renewables	0	0	0	0	0	0	0	0	0	0	-	-	-	-
Non-Energy Use	**2**	**5**	**6**	**6**	**7**						**3.1**	**0.7**	**0.7**	**0.8**
Electricity Generation (TWh)	**66**	**542**	**848**	**1267**	**1804**	**100**	**100**	**100**	**100**	**100**	**7.5**	**4.6**	**4.3**	**4.1**
Coal	32	420	552	784	1169	48	77	65	62	65	9.3	2.8	3.2	3.5
Oil	4	5	11	14	15	6	1	1	1	1	0.7	7.8	5.0	3.5
Gas	1	25	122	237	321	1	5	14	19	18	12.3	17.4	12.0	9.0
Hydrogen - Fuel Cell	0	0	0	0	0	0	0	0	0	0	-	-	-	-
Nuclear	1	17	24	43	60	2	3	3	3	3	9.6	3.8	4.8	4.3
Hydro	28	74	129	171	208	42	14	15	13	12	3.4	5.7	4.2	3.5
Other Renewables	0	1	9	18	31	0	0	1	1	2	-	20.1	13.1	10.6

Biomass (NOT included above)	121	198	215	221	221	66	40	34	28	23	1.7	0.8	0.6	0.4
Total Primary Energy Supply (including Biomass)	**183**	**498**	**628**	**788**	**971**	**100**	**100**	**100**	**100**	**100**	**3.5**	**2.3**	**2.3**	**2.2**

Reference Scenario: India

	Capacity (GW)				Shares (%)				Growth Rates (% per annum)		
	1999	2010	2020	2030	1999	2010	2020	2030	1999-2010	1999-2020	1999-2030
Total Capacity	**108**	**174**	**259**	**366**	**100**	**100**	**100**	**100**	**4.4**	**4.2**	**4.0**
Coal	64	93	132	197	59	54	51	54	3.4	3.5	3.7
Oil	5	7	7	7	5	4	3	2	2.0	1.5	1.1
Gas	12	30	59	83	11	17	23	23	8.6	7.8	6.4
Hydrogen - Fuel Cell	0	0	0	0	0	0	0	0	-	-	-
Nuclear	2	4	6	8	2	2	2	2	4.5	5.1	4.5
Hydro	23	37	49	59	21	21	19	16	4.3	3.6	3.1
Other renewables	1	3	6	11	1	2	2	3	9.6	7.9	7.1

	Capacity (GW)				Shares (%)				Growth Rates (% per annum)		
	1999	2010	2020	2030	1999	2010	2020	2030	1999-2010	1999-2020	1999-2030
Other Renewables	**1**	**3**	**6**	**11**	**100**	**100**	**100**	**100**	**9.6**	**7.9**	**7.1**
Biomass	0	1	2	4	15	26	34	39	15.5	12.3	10.6
Wind	1	2	4	6	85	70	62	54	7.7	6.3	5.5
Geothermal	0	0	0	0	0	0	0	0	-	-	-
Solar	0	0	0	1	0	4	5	7	330.7	122.1	77.1
Tide/Wave	0	0	0	0	0	0	0	0	-	-	-

	Electricity Generation (TWh)				Shares (%)				Growth Rates (% per annum)		
	2000	2010	2020	2030	2000	2010	2020	2030	2000-2010	2000-2020	2000-2030
Other Renewables	**1**	**9**	**18**	**31**	**100**	**100**	**100**	**100**	**20.1**	**13.1**	**10.6**
Biomass	0	4	8	16	0	38	47	53	-	-	-
Wind	1	5	9	13	100	59	50	42	13.9	9.2	7.5
Geothermal	0	0	0	0	0	0	0	0	-	-	-
Solar	0	0	1	1	0	3	3	5	-	-	-
Tide/Wave	0	0	0	0	0	0	0	0	-	-	-

Reference Scenario: India

	CO_2 Emissions (Mt)					Shares (%)					Growth Rates (% per annum)			
	1971	2000	2010	2020	2030	1971	2000	2010	2020	2030	1971-2000	2000-2010	2000-2020	2000-2030
Total CO_2 Emissions	**203**	**937**	**1279**	**1726**	**2280**	**100**	**100**	**100**	**100**	**100**	**5.4**	**3.2**	**3.1**	**3.0**
change since 1990 (%)		*60.7*	*119.2*	*195.8*	*290.8*									
Coal	147	629	785	1013	1328	72	67	61	59	58	5.1	2.2	2.4	2.5
Oil	55	266	395	549	743	27	28	31	32	33	5.6	4.0	3.7	3.5
Gas	1	42	99	164	209	1	5	8	9	9	12.4	8.8	7.0	5.5
Power Generation	**38**	**508**	**673**	**922**	**1243**	**100**	**100**	**100**	**100**	**100**	**9.3**	**2.9**	**3.0**	**3.0**
Coal	32	482	605	803	1089	85	95	90	87	88	9.7	2.3	2.6	2.8
Oil	5	7	11	13	13	13	1	2	1	1	1.3	4.8	3.0	2.1
Gas	1	19	57	105	141	2	4	8	11	11	12.2	11.5	8.9	6.9
Transformation, Own Use & Losses	**7**	**27**	**32**	**33**	**32**	**100**	**100**	**100**	**100**	**100**	**5.0**	**1.8**	**1.0**	**0.6**
Total Final Consumption	**159**	**403**	**574**	**772**	**1005**	**100**	**100**	**100**	**100**	**100**	**3.3**	**3.6**	**3.3**	**3.1**
Coal	111	142	174	203	233	70	35	30	26	23	0.8	2.0	1.8	1.7
Oil	47	237	357	510	704	29	59	62	66	70	5.8	4.2	3.9	3.7
Gas	1	23	42	58	68	0	6	7	8	7	12.6	6.1	4.7	3.7
Industry	**66**	**183**	**249**	**314**	**375**	**100**	**100**	**100**	**100**	**100**	**3.6**	**3.1**	**2.7**	**2.4**
Coal	54	114	148	179	210	81	62	59	57	56	2.6	2.6	2.3	2.1
Oil	12	47	62	81	104	18	26	25	26	28	5.0	2.7	2.8	2.7
Gas	1	22	39	53	62	1	12	16	17	16	12.6	6.1	4.6	3.5
Transportation	**49**	**125**	**196**	**306**	**454**	**100**	**100**	**100**	**100**	**100**	**3.3**	**4.6**	**4.6**	**4.4**
Oil	18	125	196	306	454	37	100	100	100	100	6.9	4.6	4.6	4.4
Other fuels	31	0	0	0	0	63	0	0	0	0	-100.0	0.0	0.0	0.0
Other Sectors	**41**	**90**	**124**	**146**	**169**	**100**	**100**	**100**	**100**	**100**	**2.7**	**3.3**	**2.5**	**2.1**
Coal	27	28	26	24	23	65	31	21	17	13	0.2	-0.7	-0.7	-0.7
Oil	15	60	94	117	140	35	67	76	80	82	5.0	4.7	3.4	2.9
Gas	0	2	3	5	7	0	2	2	4	4	13.6	6.5	5.9	4.9
Non-Energy Use	**2**	**5**	**5**	**5**	**6**	**100**	**100**	**100**	**100**	**100**	**2.5**	**0.6**	**0.7**	**0.8**

Reference Scenario: Latin America

	Energy Demand (Mtoe)					Shares (%)					Growth Rates (% per annum)			
	1971	2000	2010	2020	2030	1971	2000	2010	2020	2030	1971-2000	2000-2010	2000-2020	2000-2030
Total Primary Energy Supply	**140**	**388**	**525**	**705**	**937**	**100**	**100**	**100**	**100**	**100**	**3.6**	**3.1**	**3.0**	**3.0**
Coal	7	23	27	33	44	5	6	5	5	5	4.2	1.9	2.0	2.3
Oil	107	219	279	355	450	76	56	53	50	48	2.5	2.5	2.4	2.4
Gas	19	90	144	215	320	13	23	27	31	34	5.6	4.7	4.4	4.3
Nuclear	0	3	5	7	7	0	1	1	1	1	-	4.1	3.3	2.2
Hydro	6	47	61	76	89	4	12	12	11	10	7.2	2.5	2.4	2.1
Other Renewables	1	5	9	18	27	1	1	2	3	3	4.7	6.0	6.4	5.7
Power Generation	**27**	**111**	**159**	**224**	**307**	**100**	**100**	**100**	**100**	**100**	**4.9**	**3.6**	**3.6**	**3.4**
Coal	2	8	11	14	21	6	7	7	6	7	5.6	2.9	2.9	3.4
Oil	14	24	23	23	18	50	22	15	10	6	2.0	-0.4	-0.3	-0.9
Gas	5	23	50	86	143	17	21	31	38	47	5.8	8.0	6.7	6.3
Nuclear	0	3	5	7	7	0	3	3	3	2	-	4.1	3.3	2.2
Hydro	6	47	61	76	89	23	42	38	34	29	7.2	2.5	2.4	2.1
Other Renewables	1	5	9	18	27	5	5	6	8	9	4.7	6.0	6.4	5.7
Other Transformation, Own Use & Losses	**28**	**56**	**72**	**92**	**119**						**2.4**	**2.6**	**2.5**	**2.6**
of which electricity	*2*	*13*	*18*	*25*	*34*						*6.8*	*3.4*	*3.3*	*3.2*
Total Final Consumption	**97**	**290**	**391**	**523**	**695**	**100**	**100**	**100**	**100**	**100**	**3.9**	**3.1**	**3.0**	**3.0**
Coal	3	10	11	14	16	3	3	3	3	2	4.0	1.7	1.8	1.8
Oil	78	183	240	311	404	81	63	61	59	58	3.0	2.7	2.7	2.7
Gas	6	41	61	89	125	6	14	16	17	18	6.9	4.0	3.9	3.8
Electricity	10	56	79	110	150	10	19	20	21	22	6.3	3.5	3.4	3.3
Heat	0	0	0	0	0	0	0	0	0	0	-	-	-	-
Renewables	0	0	0	0	0	0	0	0	0	0	-	-	-	-

Industry	**32**	**105**	**139**	**184**	**241**	**100**	**100**	**100**	**100**	**100**	**4.2**	**2.8**	**2.9**	**2.8**
Coal	3	9	11	13	16	8	9	8	7	7	4.5	1.7	1.8	1.8
Oil	20	42	52	63	76	64	40	37	34	32	2.6	2.0	2.0	2.0
Gas	4	28	41	59	83	13	27	30	32	34	6.9	3.8	3.8	3.7
Electricity	5	25	35	48	66	15	24	25	26	27	5.8	3.5	3.4	3.3
Heat	0	0	0	0	0	0	0	0	0	0	-	-	-	-
Renewables	0	0	0	0	0	0	0	0	0	0	-	-	-	-
Transportation	**40**	**103**	**144**	**198**	**271**	**100**	**100**	**100**	**100**	**100**	**3.3**	**3.4**	**3.3**	**3.3**
Oil	40	101	140	191	261	99	98	97	96	96	3.2	3.3	3.2	3.2
Other fuels	0	2	4	7	10	1	2	3	4	4	7.1	7.1	6.6	5.6
Other Sectors	**21**	**72**	**97**	**127**	**167**	**100**	**100**	**100**	**100**	**100**	**4.4**	**3.0**	**2.9**	**2.8**
Coal	0	0	0	0	0	1	0	0	0	0	-3.7	-1.7	-1.7	-1.7
Oil	14	30	37	43	51	68	42	38	34	30	2.6	2.1	1.9	1.8
Gas	2	11	16	23	32	9	15	17	18	19	6.4	3.9	3.7	3.6
Electricity	5	31	44	61	84	22	43	45	48	50	6.8	3.6	3.5	3.4
Heat	0	0	0	0	0	0	0	0	0	0	-	-	-	-
Renewables	0	0	0	0	0	0	0	0	0	0	-	-	-	-
Non-Energy Use	**4**	**10**	**12**	**14**	**16**						**3.4**	**1.7**	**1.8**	**1.7**
Electricity Generation (TWh)	**135**	**804**	**1135**	**1566**	**2134**	**100**	**100**	**100**	**100**	**100**	**6.4**	**3.5**	**3.4**	**3.3**
Coal	4	33	45	62	99	3	4	4	4	5	7.3	3.1	3.1	3.7
Oil	41	94	92	90	74	30	12	8	6	3	2.9	-0.2	-0.2	-0.8
Gas	14	98	247	454	822	11	12	22	29	39	6.9	9.6	7.9	7.3
Hydrogen - Fuel Cell	0	0	0	0	0	0	0	0	0	0	-	-	-	-
Nuclear	0	13	20	25	25	0	2	2	2	1	-	4.1	3.3	2.2
Hydro	73	550	705	889	1040	54	68	62	57	49	7.2	2.5	2.4	2.1
Other Renewables	2	15	26	45	74	2	2	2	3	3	6.8	5.4	5.5	5.3

Biomass (NOT included above)	61	76	79	85	92	30	16	13	11	9	0.8	0.4	0.5	0.6
Total Primary Energy Supply (including Biomass)	**202**	**464**	**605**	**789**	**1029**	**100**	**100**	**100**	**100**	**100**	**2.9**	**2.7**	**2.7**	**2.7**

Reference Scenario: Latin America

	Capacity (GW)				Shares (%)				Growth Rates (% per annum)		
	1999	2010	2020	2030	1999	2010	2020	2030	1999-2010	1999-2020	1999-2030
Total Capacity	**168**	**256**	**357**	**492**	**100**	**100**	**100**	**100**	**3.9**	**3.6**	**3.5**
Coal	6	9	12	19	3	3	3	4	3.7	3.5	3.9
Oil	30	32	32	26	18	12	9	5	0.5	0.2	-0.5
Gas	22	68	124	219	13	26	35	45	10.9	8.7	7.7
Hydrogen - Fuel Cell	0	0	0	0	0	0	0	0	-	-	-
Nuclear	2	3	4	4	1	1	1	1	5.6	4.2	2.8
Hydro	106	139	175	205	63	54	49	42	2.5	2.4	2.1
Other renewables	3	6	11	19	2	2	3	4	5.8	5.9	5.8

	Capacity (GW)				Shares (%)				Growth Rates (% per annum)		
	1999	2010	2020	2030	1999	2010	2020	2030	1999-2010	1999-2020	1999-2030
Other Renewables	**3**	**6**	**11**	**19**	**100**	**100**	**100**	**100**	**5.8**	**5.9**	**5.8**
Biomass	3	4	6	8	86	66	50	42	3.3	3.2	3.4
Wind	0	1	4	8	2	21	32	41	30.4	20.5	16.5
Geothermal	0	1	2	3	12	13	18	16	6.9	8.2	6.8
Solar	0	0	0	0	0	0	0	1	-	-	-
Tide/Wave	0	0	0	0	0	0	0	0	-	-	-

	Electricity Generation (TWh)				Shares (%)				Growth Rates (% per annum)		
	2000	2010	2020	2030	2000	2010	2020	2030	2000-2010	2000-2020	2000-2030
Other Renewables	**15**	**26**	**45**	**74**	**100**	**100**	**100**	**100**	**5.4**	**5.5**	**5.3**
Biomass	13	18	24	35	86	69	54	48	3.1	3.1	3.3
Wind	0	3	9	20	2	12	19	27	23.7	16.9	14.0
Geothermal	2	5	12	18	12	19	27	25	10.3	9.9	8.0
Solar	0	0	0	0	0	0	0	0	-	-	-
Tide/Wave	0	0	0	0	0	0	0	0	-	-	-

Reference Scenario: Latin America

	CO₂ Emissions (Mt)					Shares (%)					Growth Rates (% per annum)			
	1971	2000	2010	2020	2030	1971	2000	2010	2020	2030	1971-2000	2000-2010	2000-2020	2000-2030
Total CO_2 Emissions	**360**	**877**	**1185**	**1571**	**2104**	**100**	**100**	**100**	**100**	**100**	**3.1**	**3.1**	**3.0**	**3.0**
change since 1990 (%)		*47.0*	*98.7*	*163.5*	*252.8*									
Coal	24	82	106	132	180	7	9	9	8	9	4.3	2.6	2.4	2.7
Oil	294	612	778	979	1234	82	70	66	62	59	2.6	2.4	2.4	2.4
Gas	41	183	301	460	691	11	21	25	29	33	5.3	5.1	4.7	4.5
Power Generation	**61**	**160**	**243**	**338**	**490**	**100**	**100**	**100**	**100**	**100**	**3.4**	**4.3**	**3.8**	**3.8**
Coal	7	35	51	67	102	12	22	21	20	21	5.6	3.9	3.3	3.7
Oil	43	75	80	77	62	71	47	33	23	13	1.9	0.5	0.1	-0.7
Gas	11	50	112	194	327	17	31	46	57	67	5.5	8.5	7.0	6.5
Transformation, Own Use & Losses	**45**	**78**	**92**	**112**	**141**	**100**	**100**	**100**	**100**	**100**	**1.9**	**1.6**	**1.8**	**2.0**
Total Final Consumption	**254**	**638**	**850**	**1121**	**1473**	**100**	**100**	**100**	**100**	**100**	**3.2**	**2.9**	**2.9**	**2.8**
Coal	15	43	50	61	73	6	7	6	5	5	3.6	1.6	1.8	1.8
Oil	225	503	662	859	1176	89	79	78	77	76	2.8	2.8	2.7	2.7
Gas	13	92	137	202	284	5	14	16	18	19	7.0	4.1	4.0	3.8
Industry	**87**	**214**	**274**	**353**	**450**	**100**	**100**	**100**	**100**	**100**	**3.2**	**2.5**	**2.5**	**2.5**
Coal	13	42	49	60	72	15	19	18	17	16	4.1	1.7	1.8	1.8
Oil	65	111	134	162	195	75	52	49	46	43	1.9	1.9	1.9	1.9
Gas	9	62	90	131	183	11	29	33	37	41	6.8	3.8	3.8	3.7
Transportation	**114**	**299**	**417**	**573**	**782**	**100**	**100**	**100**	**100**	**100**	**3.4**	**3.4**	**3.3**	**3.3**
Oil	113	293	405	551	751	99	98	97	96	96	3.3	3.3	3.2	3.2
Other fuels	1	6	12	22	31	1	2	3	4	4	7.4	6.6	6.4	5.5
Other Sectors	**46**	**109**	**140**	**173**	**214**	**100**	**100**	**100**	**100**	**100**	**3.0**	**2.5**	**2.3**	**2.3**
Coal	2	1	1	0	0	3	1	0	0	0	-2.2	-3.6	-2.6	-2.3
Oil	41	84	104	123	144	89	77	74	71	67	2.6	2.2	1.9	1.8
Gas	4	24	35	49	70	8	22	25	29	33	6.6	3.9	3.7	3.6
Non-Energy Use	**7**	**16**	**19**	**23**	**27**	**100**	**100**	**100**	**100**	**100**	**3.0**	**1.7**	**1.7**	**1.7**

Reference Scenario: Brazil

	Energy Demand (Mtoe)					Shares (%)					Growth Rates (% per annum)			
	1971	2000	2010	2020	2030	1971	2000	2010	2020	2030	1971-2000	2000-2010	2000-2020	2000-2030
Total Primary Energy Supply	**34**	**138**	**194**	**255**	**332**	**100**	**100**	**100**	**100**	**100**	**4.9**	**3.4**	**3.1**	**3.0**
Coal	2	14	16	19	24	7	10	8	7	7	6.0	1.7	1.7	1.9
Oil	28	88	117	148	186	81	63	60	58	56	4.0	2.9	2.6	2.5
Gas	0	7	21	36	62	0	5	11	14	19	15.7	11.1	8.1	7.3
Nuclear	0	2	3	5	5	0	1	2	2	2	-	7.5	6.3	4.2
Hydro	4	26	34	44	51	11	19	17	17	15	7.0	2.5	2.6	2.2
Other Renewables	0	2	3	4	5	0	1	1	1	2	9.2	3.4	3.2	3.6
Power Generation	**6**	**37**	**58**	**80**	**110**	**100**	**100**	**100**	**100**	**100**	**6.5**	**4.5**	**3.9**	**3.7**
Coal	1	3	4	5	7	10	8	7	6	6	5.7	3.1	2.3	2.9
Oil	2	4	4	4	3	25	11	7	5	2	3.7	-0.7	-0.8	-1.7
Gas	0	1	10	19	39	0	1	18	24	35	-	34.6	19.7	15.4
Nuclear	0	2	3	5	5	0	4	6	7	5	-	7.5	6.3	4.2
Hydro	4	26	34	44	51	62	70	58	54	46	7.0	2.5	2.6	2.2
Other Renewables	0	2	3	4	5	2	5	5	4	5	9.2	3.4	3.2	3.6
Other Transformation, Own Use & Losses	**4**	**13**	**18**	**23**	**29**						**4.3**	**3.7**	**3.0**	**2.8**
of which electricity	*1*	*3*	*5*	*7*	*11*						*4.1*	*6.7*	*5.5*	*4.9*
Total Final Consumption	**29**	**119**	**161**	**211**	**272**	**100**	**100**	**100**	**100**	**100**	**4.9**	**3.1**	**2.9**	**2.8**
Coal	1	6	7	9	11	2	5	4	4	4	7.6	1.9	2.0	2.0
Oil	25	80	107	137	175	85	68	67	65	64	4.2	3.0	2.7	2.6
Gas	0	5	8	13	19	1	4	5	6	7	12.6	5.8	5.2	4.7
Electricity	4	27	38	51	68	12	23	24	24	25	7.2	3.4	3.2	3.0
Heat	0	0	0	0	0	0	0	0	0	0	-	-	-	-
Renewables	0	0	0	0	0	0	0	0	0	0	-	-	-	-

Industry	**9**	**45**	**59**	**77**	**98**	**100**	**100**	**100**	**100**	**100**	**5.6**	**2.8**	**2.8**	**2.7**
Coal	1	6	7	9	11	7	13	12	11	11	7.8	1.9	2.0	2.0
Oil	7	22	28	36	45	72	49	48	46	45	4.2	2.5	2.5	2.4
Gas	0	4	8	12	17	0	10	13	15	17	17.9	5.8	5.1	4.6
Electricity	2	13	16	21	27	21	28	28	27	27	6.7	2.6	2.6	2.5
Heat	0	0	0	0	0	0	0	0	0	0	-	-	-	-
Renewables	0	0	0	0	0	0	0	0	0	0	-	-	-	-
Transportation	**14**	**42**	**59**	**78**	**104**	**100**	**100**	**100**	**100**	**100**	**3.8**	**3.4**	**3.2**	**3.1**
Oil	14	42	58	77	102	100	99	99	99	98	3.8	3.4	3.1	3.0
Other fuels	0	0	1	1	2	0	1	1	1	2	6.9	5.1	5.4	5.3
Other Sectors	**4**	**27**	**38**	**49**	**63**	**100**	**100**	**100**	**100**	**100**	**6.5**	**3.5**	**3.0**	**2.8**
Coal	0	0	0	0	0	0	0	0	0	0	-	-	-	-
Oil	3	12	16	19	21	60	45	42	38	34	5.4	2.6	2.0	1.8
Gas	0	0	0	1	1	3	1	1	1	1	2.3	4.8	4.7	4.1
Electricity	2	15	22	30	41	38	54	57	61	65	7.8	4.1	3.7	3.4
Heat	0	0	0	0	0	0	0	0	0	0	-	-	-	-
Renewables	0	0	0	0	0	0	0	0	0	0	-	-	-	-
Non-Energy Use	**1**	**4**	**5**	**6**	**7**						**4.6**	**1.4**	**1.4**	**1.4**
Electricity Generation (TWh)	**52**	**349**	**505**	**685**	**909**	**100**	**100**	**100**	**100**	**100**	**6.8**	**3.8**	**3.4**	**3.2**
Coal	2	10	15	18	30	3	3	3	3	3	6.4	4.1	3.0	3.7
Oil	6	17	16	16	12	12	5	3	2	1	3.6	-0.4	-0.4	-1.2
Gas	0	2	56	104	226	0	1	11	15	25	-	37.2	20.8	16.4
Hydrogen - Fuel Cell	0	0	0	0	0	0	0	0	0	0	-	-	-	-
Nuclear	0	6	12	21	21	0	2	2	3	2	-	7.5	6.3	4.2
Hydro	43	305	391	507	589	84	87	78	74	65	7.0	2.5	2.6	2.2
Other Renewables	1	9	13	19	32	1	3	3	3	4	9.5	3.8	3.8	4.3
Biomass (NOT included above)	35	41	42	44	48	51	23	18	15	13	0.5	0.2	0.4	0.5
Total Primary Energy Supply (including Biomass)	**70**	**179**	**236**	**299**	**380**	**100**	**100**	**100**	**100**	**100**	**3.3**	**2.8**	**2.6**	**2.5**

Reference Scenario: Brazil

	Capacity (GW)				Shares (%)				Growth Rates (% per annum)		
	1999	2010	2020	2030	1999	2010	2020	2030	1999-2010	1999-2020	1999-2030
Total Capacity	**68**	**102**	**141**	**188**	**100**	**100**	**100**	**100**	**3.8**	**3.5**	**3.3**
Coal	2	3	3	5	3	3	2	3	3.3	2.7	3.5
Oil	4	5	5	4	6	5	4	2	1.6	0.8	-0.2
Gas	0	12	24	51	1	12	17	27	38.0	22.2	17.3
Hydrogen - Fuel Cell	0	0	0	0	0	0	0	0	-	-	-
Nuclear	1	2	3	3	1	2	2	2	10.6	8.0	5.3
Hydro	59	77	100	116	87	75	71	62	2.5	2.5	2.2
Other renewables	2	3	5	9	3	3	4	5	4.4	4.6	5.1

	Capacity (GW)				Shares (%)				Growth Rates (% per annum)		
	1999	2010	2020	2030	1999	2010	2020	2030	1999-2010	1999-2020	1999-2030
Other Renewables	**2**	**3**	**5**	**9**	**100**	**100**	**100**	**100**	**4.4**	**4.6**	**5.1**
Biomass	2	3	4	5	100	90	70	55	3.5	2.9	3.1
Wind	0	0	2	4	0	10	30	44	-	-	-
Geothermal	0	0	0	0	0	0	0	0	-	-	-
Solar	0	0	0	0	0	0	0	1	-	-	-
Tide/Wave	0	0	0	0	0	0	0	0	-	-	-

	Electricity Generation (TWh)				Shares (%)				Growth Rates (% per annum)		
	2000	2010	2020	2030	2000	2010	2020	2030	2000-2010	2000-2020	2000-2030
Other Renewables	**9**	**13**	**19**	**32**	**13**	**13**	**14**	**17**	**3.8**	**3.8**	**4.3**
Biomass	9	12	16	22	13	12	9	12	3.2	2.7	3.0
Wind	0	1	4	10	0	1	3	5	93.5	50.8	35.8
Geothermal	0	0	0	0	0	0	0	0	-	-	-
Solar	0	0	0	0	0	0	0	0	-	-	-
Tide/Wave	0	0	0	0	0	0	0	0	-	-	-

Reference Scenario: Brazil

	CO_2 Emissions (Mt)					Shares (%)					Growth Rates (% per annum)			
	1971	2000	2010	2020	2030	1971	2000	2010	2020	2030	1971-2000	2000-2010	2000-2020	2000-2030
Total CO_2 Emissions	**91**	**303**	**438**	**569**	**760**	**100**	**100**	**100**	**100**	**100**	**4.2**	**3.7**	**3.2**	**3.1**
change since 1990 (%)		*57.0*	*126.6*	*194.4*	*293.4*									
Coal	7	44	60	71	94	8	15	14	12	12	6.4	3.0	2.4	2.5
Oil	83	243	328	412	516	92	80	75	73	68	3.8	3.0	2.7	2.5
Gas	0	16	51	85	150	0	5	12	15	20	15.8	12.0	8.6	7.7
Power Generation	**7**	**28**	**71**	**97**	**158**	**100**	**100**	**100**	**100**	**100**	**4.8**	**9.5**	**6.3**	**5.9**
Coal	3	14	24	28	42	35	49	34	29	26	6.0	5.5	3.5	3.7
Oil	5	13	19	18	12	65	46	27	18	8	3.5	3.8	1.5	-0.2
Gas	0	1	28	51	104	0	4	39	53	66	-	36.4	20.5	15.9
Transformation, Own Use & Losses	**5**	**20**	**23**	**28**	**34**	**100**	**100**	**100**	**100**	**100**	**4.6**	**1.8**	**1.8**	**1.8**
Total Final Consumption	**78**	**255**	**344**	**444**	**568**	**100**	**100**	**100**	**100**	**100**	**4.2**	**3.0**	**2.8**	**2.7**
Coal	4	27	32	39	48	5	10	9	9	8	6.7	1.8	2.0	2.0
Oil	74	218	294	378	483	95	85	86	85	85	3.8	3.0	2.8	2.7
Gas	0	11	17	27	38	0	4	5	6	7	20.2	5.2	4.8	4.4
Industry	**25**	**87**	**114**	**148**	**188**	**100**	**100**	**100**	**100**	**100**	**4.4**	**2.8**	**2.7**	**2.6**
Coal	4	26	32	39	47	14	30	28	26	25	7.1	1.9	2.0	2.0
Oil	21	51	66	83	104	85	59	58	56	55	3.1	2.5	2.5	2.4
Gas	0	10	17	26	37	0	11	15	17	19	19.8	5.8	5.1	4.6
Transportation	**42**	**126**	**175**	**232**	**307**	**100**	**100**	**100**	**100**	**100**	**3.9**	**3.3**	**3.1**	**3.0**
Oil	42	125	175	232	307	100	100	100	100	100	3.8	3.4	3.1	3.0
Other fuels	0	0	0	0	0	0	0	0	0	0	-	-	-	-
Other Sectors	**8**	**35**	**47**	**54**	**62**	**100**	**100**	**100**	**100**	**100**	**5.4**	**2.8**	**2.1**	**1.9**
Coal	0.3	0	0	0	0	3	0	0	0	0	-2.4	-	-	-
Oil	8	35	46	53	61	97	99	99	98	98	5.4	2.8	2.1	1.9
Gas	0	0	1	1	1	0	1	1	2	2	-	4.8	4.7	4.1
Non-Energy Use	**4**	**7**	**8**	**10**	**11**	**100**	**100**	**100**	**100**	**100**	**2.4**	**1.4**	**1.4**	**1.4**

Reference Scenario: Middle East

	Energy Demand (Mtoe)					Shares (%)					Growth Rates (% per annum)			
	1971	2000	2010	2020	2030	1971	2000	2010	2020	2030	1971-2000	2000-2010	2000-2020	2000-2030
Total Primary Energy Supply	**50**	**369**	**487**	**610**	**741**	**100**	**100**	**100**	**100**	**100**	**7.1**	**2.8**	**2.5**	**2.3**
Coal	0	7	9	12	14	0	2	2	2	2	12.2	3.3	3.0	2.6
Oil	38	196	250	306	371	76	53	51	50	50	5.8	2.5	2.3	2.2
Gas	11	166	224	287	351	23	45	46	47	47	9.7	3.0	2.8	2.5
Nuclear	0	0	2	2	2	0	0	0	0	0	-	-	-	-
Hydro	0	1	2	3	3	1	0	0	0	0	5.0	5.2	3.5	2.7
Other Renewables	0	0	0	0	1	0	0	0	0	0	-	22.0	14.3	12.3
Power Generation	**8**	**103**	**147**	**187**	**229**	**100**	**100**	**100**	**100**	**100**	**9.4**	**3.6**	**3.0**	**2.7**
Coal	0	5	7	10	12	0	5	5	5	5	-	3.8	3.4	2.9
Oil	6	43	55	65	76	78	42	38	35	33	7.1	2.5	2.0	1.9
Gas	1	53	80	108	137	17	52	55	58	60	13.6	4.1	3.6	3.2
Nuclear	0	0	2	2	2	0	0	1	1	1	-	-	-	-
Hydro	0	1	2	3	3	4	1	2	1	1	5.0	5.2	3.5	2.7
Other Renewables	0	0	0	0	1	0	0	0	0	0	-	22.0	14.3	12.3
Other Transformation, Own Use & Losses	**14**	**50**	**70**	**90**	**113**						**4.5**	**3.4**	**3.0**	**2.7**
of which electricity	*0*	*7*	*10*	*14*	*17*						*12.0*	*3.7*	*3.2*	*2.9*
Total Final Consumption	**31**	**256**	**327**	**410**	**499**	**100**	**100**	**100**	**100**	**100**	**7.6**	**2.5**	**2.4**	**2.3**
Coal	0	1	1	1	1	0	0	0	0	0	6.3	1.1	1.2	1.2
Oil	21	144	179	218	265	69	56	55	53	53	6.8	2.2	2.1	2.0
Gas	7	78	100	127	150	24	31	31	31	30	8.5	2.5	2.4	2.2
Electricity	2	33	48	64	83	7	13	15	16	17	10.0	3.9	3.4	3.2
Heat	0	0	0	0	0	0	0	0	0	0	-	-	-	-
Renewables	0	0	0	0	0	0	0	0	0	0	-	-	-	-

Industry	**12**	**84**	**105**	**125**	**144**	**100**	**100**	**100**	**100**	**100**	**6.9**	**2.3**	**2.0**	**1.8**
Coal	0	1	1	1	1	1	1	1	1	1	6.3	1.1	1.2	1.2
Oil	5	33	38	44	49	42	39	36	35	34	6.7	1.6	1.5	1.4
Gas	6	44	58	70	81	50	52	55	56	56	7.1	2.8	2.4	2.1
Electricity	1	6	9	11	13	7	8	8	8	9	7.1	3.1	2.6	2.5
Heat	0	0	0	0	0	0	0	0	0	0	-	-	-	-
Renewables	0	0	0	0	0	0	0	0	0	0	-	-	-	-
Transportation	**10**	**64**	**81**	**104**	**133**	**100**	**100**	**100**	**100**	**100**	**6.7**	**2.4**	**2.5**	**2.5**
Oil	10	64	81	104	133	100	100	100	100	100	6.7	2.4	2.5	2.5
Other fuels	0	0	0	0	0	0	0	0	0	0	-	1.5	1.5	1.5
Other Sectors	**8**	**101**	**131**	**169**	**208**	**100**	**100**	**100**	**100**	**100**	**9.2**	**2.7**	**2.6**	**2.4**
Coal	0	0	0	0	0	0	0	0	0	0	-	-	-	-
Oil	5	40	50	59	69	67	40	38	35	33	7.3	2.1	1.9	1.8
Gas	1	34	42	57	69	18	34	32	34	33	-	2.1	2.6	2.4
Electricity	1	26	39	53	70	15	26	30	32	33	11.2	4.1	3.6	3.3
Heat	0	0	0	0	0	0	0	0	0	0	-	-	-	-
Renewables	0	0	0	0	0	0	0	0	0	0	-	-	-	-
Non-Energy Use	**1**	**8**	**10**	**11**	**14**						**6.5**	**2.4**	**2.1**	**2.0**
Electricity Generation (TWh)	**27**	**462**	**675**	**899**	**1161**	**100**	**100**	**100**	**100**	**100**	**10.2**	**3.9**	**3.4**	**3.1**
Coal	0	24	35	47	58	0	5	5	5	5	-	3.9	3.6	3.0
Oil	20	195	249	290	340	72	42	37	32	29	8.2	2.5	2.0	1.9
Gas	4	228	358	521	716	15	49	53	58	62	14.9	4.6	4.2	3.9
Hydrogen - Fuel Cell	0	0	0	0	0	0	0	0	0	0	-	-	-	-
Nuclear	0	0	6	6	6	0	0	1	1	1	-	-	-	-
Hydro	4	16	26	31	35	14	3	4	3	3	5.0	5.2	3.5	2.7
Other Renewables	0	0	1	3	6	0	0	0	0	1	-	22.0	14.3	12.3

Biomass (NOT included above)	1	1	1	1	1	1	0	0	0	0	2.5	1.1	1.1	1.1
Total Primary Energy Supply (including Biomass)	**50**	**370**	**488**	**611**	**743**	**100**	**100**	**100**	**100**	**100**	**7.1**	**2.8**	**2.5**	**2.3**

Reference Scenario: Middle East

	Capacity (GW)				Shares (%)				Growth Rates (% per annum)		
	1999	2010	2020	2030	1999	2010	2020	2030	1999-2010	1999-2020	1999-2030
Total Capacity	**120**	**160**	**214**	**277**	**100**	**100**	**100**	**100**	**2.7**	**2.8**	**2.7**
Coal	4	6	8	11	4	4	4	4	2.8	3.2	2.9
Oil	53	65	79	96	44	41	37	34	1.8	1.9	1.9
Gas	56	77	113	154	47	48	53	56	2.9	3.3	3.3
Hydrogen - Fuel Cell	0	0	0	0	0	0	0	0	-	-	-
Nuclear	0	1	1	1	0	1	0	0	-	-	-
Hydro	6	10	12	14	5	6	6	5	5.3	3.6	2.8
Other renewables	0	1	1	3	0	0	1	1	31.3	19.1	15.6

	Capacity (GW)				Shares (%)				Growth Rates (% per annum)		
	1999	2010	2020	2030	1999	2010	2020	2030	1999-2010	1999-2020	1999-2030
Other Renewables	**0**	**1**	**1**	**3**	**100**	**100**	**100**	**100**	**31.3**	**19.1**	**15.6**
Biomass	0	0	0	0	0	0	0	0	-	-	-
Wind	0	1	1	2	99	89	91	80	30.0	18.6	14.8
Geothermal	0	0	0	0	0	0	0	0	-	-	-
Solar	0	0	0	1	1	11	9	20	64.6	33.0	27.8
Tide/Wave	0	0	0	0	0	0	0	0	-	-	-

	Electricity Generation (TWh)				Shares (%)				Growth Rates (% per annum)		
	2000	2010	2020	2030	2000	2010	2020	2030	2000-2010	2000-2020	2000-2030
Other Renewables	**0**	**1**	**3**	**6**	**100**	**100**	**100**	**100**	**22.0**	**14.3**	**12.3**
Biomass	0	0	0	0	0	0	0	0	-	-	-
Wind	0	1	2	5	93	90	92	82	21.6	14.2	11.8
Geothermal	0	0	0	0	0	0	0	0	-	-	-
Solar	0	0	0	1	7	10	8	18	26.6	15.4	16.1
Tide/Wave	0	0	0	0	0	0	0	0	-	-	-

Reference Scenario: Middle East

	CO_2 Emissions (Mt)					Shares (%)					Growth Rates (% per annum)			
	1971	2000	2010	2020	2030	1971	2000	2010	2020	2030	1971-2000	2000-2010	2000-2020	2000-2030
Total CO_2 Emissions	**122**	**978**	**1256**	**1557**	**1879**	**100**	**100**	**100**	**100**	**100**	**7.4**	**2.5**	**2.4**	**2.2**
change since 1990 (%)		*69.7*	*118.0*	*170.2*	*226.2*									
Coal	1	25	34	45	54	1	3	3	3	3	12.5	3.3	3.0	2.6
Oil	95	579	720	866	1035	78	59	57	56	55	6.4	2.2	2.0	2.0
Gas	26	374	502	646	791	21	38	40	42	42	9.6	3.0	2.8	2.5
Power Generation	**22**	**280**	**389**	**493**	**602**	**100**	**100**	**100**	**100**	**100**	**9.2**	**3.4**	**2.9**	**2.6**
Coal	0	20	29	39	47	0	7	7	8	8	-	3.8	3.4	2.9
Oil	19	135	173	202	236	86	48	44	41	39	7.0	2.5	2.0	1.9
Gas	3	125	188	252	319	14	45	48	51	53	13.6	4.1	3.6	3.2
Transformation, Own Use & Losses	**26**	**118**	**142**	**170**	**203**	**100**	**100**	**100**	**100**	**100**	**5.3**	**1.9**	**1.8**	**1.8**
Total Final Consumption	**74**	**580**	**724**	**893**	**1074**	**100**	**100**	**100**	**100**	**100**	**7.3**	**2.2**	**2.2**	**2.1**
Coal	1	5	5	6	7	1	1	1	1	1	6.3	1.1	1.2	1.2
Oil	57	396	489	596	721	76	68	67	67	67	6.9	2.1	2.1	2.0
Gas	17	180	230	291	346	23	31	32	33	32	8.5	2.5	2.5	2.2
Industry	**30**	**198**	**246**	**290**	**332**	**100**	**100**	**100**	**100**	**100**	**6.7**	**2.2**	**1.9**	**1.7**
Coal	1	5	5	6	7	3	2	2	2	2	6.3	1.1	1.2	1.2
Oil	16	94	109	126	141	52	47	44	43	42	6.3	1.6	1.5	1.4
Gas	14	99	131	159	184	45	50	53	55	55	7.1	2.8	2.4	2.1
Transportation	**23**	**169**	**215**	**276**	**352**	**100**	**100**	**100**	**100**	**100**	**7.1**	**2.4**	**2.5**	**2.5**
Oil	23	169	215	276	352	100	100	100	100	100	7.1	2.4	2.5	2.5
Other fuels	0	0	0	0	0	0	0	0	0	0	-	-	-	-
Other Sectors	**19**	**201**	**247**	**308**	**368**	**100**	**100**	**100**	**100**	**100**	**8.5**	**2.1**	**2.2**	**2.0**
Coal	0	0	0	0	0	0	0	0	0	0	-	-	-	-
Oil	16	121	149	176	206	83	60	60	57	56	7.2	2.1	1.9	1.8
Gas	3	80	99	133	162	17	40	40	43	44	11.7	2.1	2.6	2.4
Non-Energy Use	**2**	**12**	**16**	**19**	**22**	**100**	**100**	**100**	**100**	**100**	**6.6**	**2.4**	**2.1**	**2.0**

Reference Scenario: Africa

	Energy Demand (Mtoe)					Shares (%)					Growth Rates (% per annum)			
	1971	2000	2010	2020	2030	1971	2000	2010	2020	2030	1971-2000	2000-2010	2000-2020	2000-2030
Total Primary Energy Supply	**75**	**247**	**341**	**479**	**681**	**100**	**100**	**100**	**100**	**100**	**4.2**	**3.3**	**3.4**	**3.4**
Coal	36	91	105	131	174	48	37	31	27	25	3.2	1.5	1.9	2.2
Oil	35	99	138	191	262	46	40	41	40	38	3.7	3.4	3.3	3.3
Gas	2	47	84	137	211	3	19	25	29	31	10.7	6.0	5.5	5.2
Nuclear	0	3	3	3	3	0	1	1	1	1	-	0.0	0.3	0.2
Hydro	2	6	8	9	11	3	2	2	2	2	4.1	2.2	2.1	2.0
Other Renewables	0	1	3	8	20	0	1	1	2	3	-	8.7	8.9	9.2
Power Generation	**26**	**95**	**144**	**221**	**341**	**100**	**100**	**100**	**100**	**100**	**4.5**	**4.2**	**4.3**	**4.3**
Coal	21	51	64	86	124	81	54	44	39	36	3.1	2.2	2.6	3.0
Oil	3	14	19	26	36	10	15	13	12	11	6.0	2.8	3.0	3.1
Gas	0	19	47	88	146	1	20	32	40	43	14.5	9.6	8.0	7.1
Nuclear	0	3	3	3	3	0	3	2	2	1	-	0.0	0.3	0.2
Hydro	2	6	8	9	11	7	6	5	4	3	4.1	2.2	2.1	2.0
Other Renewables	0	1	3	8	20	0	2	2	4	6	-	8.7	8.9	9.2
Other Transformation, Own Use & Losses	**1**	**40**	**50**	**66**	**86**						**13.6**	**2.1**	**2.5**	**2.6**
of which electricity	1	*8*	*12*	*19*	*30*						*6.9*	*4.7*	*4.7*	*4.7*
Total Final Consumption	**57**	**149**	**206**	**287**	**403**	**100**	**100**	**100**	**100**	**100**	**3.4**	**3.3**	**3.3**	**3.4**
Coal	19	18	21	24	27	33	12	10	8	7	-0.2	1.5	1.5	1.4
Oil	31	87	117	156	211	54	59	57	55	52	3.7	3.0	2.9	3.0
Gas	1	14	22	32	46	1	9	10	11	11	12.0	4.4	4.2	4.0
Electricity	7	30	47	75	119	12	20	23	26	29	5.1	4.6	4.7	4.7
Heat	0	0	0	0	0	0	0	0	0	0	-	-	-	-
Renewables	0	0	0	0	0	0	0	0	0	0	-	-	-	-

Industry	**22**	**53**	**69**	**89**	**116**	**100**	**100**	**100**	**100**	**100**	**3.1**	**2.7**	**2.6**	**2.6**
Coal	10	15	18	21	23	48	29	26	23	20	1.4	1.5	1.5	1.4
Oil	7	13	15	16	18	30	25	22	18	16	2.5	1.4	1.0	1.0
Gas	0	10	14	20	28	2	18	21	23	24	11.4	4.0	3.8	3.6
Electricity	4	14	21	32	47	20	27	31	36	40	4.2	4.0	4.0	4.0
Heat	0	0	0	0	0	0	0	0	0	0	-	-	-	-
Renewables	0	0	0	0	0	0	0	0	0	0	-	-	-	-
Transportation	**20**	**47**	**64**	**88**	**121**	**100**	**100**	**100**	**100**	**100**	**3.0**	**3.1**	**3.1**	**3.2**
Oil	16	46	63	86	119	80	98	98	98	98	3.8	3.1	3.1	3.2
Other fuels	4	1	2	2	3	20	2	2	2	2	-4.3	3.6	3.2	3.0
Other Sectors	**13**	**43**	**65**	**100**	**155**	**100**	**100**	**100**	**100**	**100**	**4.1**	**4.3**	**4.4**	**4.4**
Coal	4	2	3	3	4	27	5	4	3	2	-1.7	1.7	1.8	1.8
Oil	7	22	31	45	64	54	51	48	45	41	3.9	3.6	3.7	3.7
Gas	0	4	6	10	16	1	9	9	10	10	13.3	5.4	5.2	5.1
Electricity	2	15	25	42	71	18	35	38	42	46	6.5	5.3	5.4	5.4
Heat	0	0	0	0	0	0	0	0	0	0	-	-	-	-
Renewables	0	0	0	0	0	0	0	0	0	0	-	-	-	-
Non-Energy Use	**2**	**6**	**8**	**9**	**10**						**3.7**	**2.7**	**2.1**	**1.9**
Electricity Generation (TWh)	**90**	**435**	**684**	**1091**	**1733**	**100**	**100**	**100**	**100**	**100**	**5.6**	**4.6**	**4.7**	**4.7**
Coal	56	207	268	382	591	62	48	39	35	34	4.6	2.6	3.1	3.6
Oil	11	63	83	114	160	12	15	12	10	9	6.4	2.8	3.0	3.1
Gas	1	78	225	456	790	1	18	33	42	46	16.3	11.1	9.2	8.0
Hydrogen - Fuel Cell	0	0	0	0	0	0	0	0	0	0	-	-	-	-
Nuclear	0	13	13	13	13	0	3	2	1	1	-	0.0	0.3	0.2
Hydro	22	71	88	107	128	25	16	13	10	7	4.1	2.2	2.1	2.0
Other Renewables	0	2	7	19	50	0	1	1	2	3	-	10.6	10.6	10.5
Biomass (NOT included above)	124	253	293	316	333	62	51	46	40	33	2.5	1.5	1.1	0.9
Total Primary Energy Supply (including Biomass)	**200**	**500**	**634**	**795**	**1014**	**100**	**100**	**100**	**100**	**100**	**3.2**	**2.4**	**2.3**	**2.4**

Reference Scenario: Africa

	Capacity (GW)				Shares (%)				Growth Rates (% per annum)		
	1999	2010	2020	2030	1999	2010	2020	2030	1999-2010	1999-2020	1999-2030
Total Capacity	**100**	**160**	**254**	**400**	**100**	**100**	**100**	**100**	**4.4**	**4.5**	**4.6**
Coal	36	46	64	99	36	28	25	25	2.2	2.8	3.3
Oil	24	36	54	81	24	22	21	20	3.9	4.0	4.1
Gas	17	49	96	165	17	30	38	41	9.8	8.5	7.6
Hydrogen - Fuel Cell	0	0	0	0	0	0	0	0	-	-	-
Nuclear	2	2	2	2	2	1	1	0	0.0	0.3	0.2
Hydro	21	27	32	39	21	17	13	10	2.3	2.1	2.0
Other renewables	0	2	5	15	0	1	2	4	13.6	12.7	12.0

	Capacity (GW)				Shares (%)				Growth Rates (% per annum)		
	1999	2010	2020	2030	1999	2010	2020	2030	1999-2010	1999-2020	1999-2030
Other Renewables	**0**	**2**	**5**	**15**	**100**	**100**	**100**	**100**	**13.6**	**12.7**	**12.0**
Biomass	0	1	2	5	86	44	37	33	7.0	8.2	8.7
Wind	0	1	2	5	2	39	37	33	47.5	28.9	22.3
Geothermal	0	0	0	1	12	11	7	7	12.7	10.1	9.9
Solar	0	0	1	4	0	6	19	27	98.9	60.0	43.8
Tide/Wave	0	0	0	0	0	0	0	0	-	-	-

	Electricity Generation (TWh)				Shares (%)				Growth Rates (% per annum)		
	2000	2010	2020	2030	2000	2010	2020	2030	2000-2010	2000-2020	2000-2030
Other Renewables	**2**	**7**	**19**	**50**	**100**	**100**	**100**	**100**	**10.6**	**10.6**	**10.5**
Biomass	2	4	9	22	73	51	47	44	6.7	8.2	8.6
Wind	0	2	5	12	7	25	26	25	25.5	18.0	15.2
Geothermal	0	1	3	7	19	20	15	14	11.6	9.4	9.4
Solar	0	0	2	9	1	3	12	18	27.0	26.5	22.5
Tide/Wave	0	0	0	0	0	0	0	0	-	-	-

Reference Scenario: Africa

	CO_2 Emissions (Mt)					Shares (%)					Growth Rates (% per annum)			
	1971	2000	2010	2020	2030	1971	2000	2010	2020	2030	1971-2000	2000-2010	2000-2020	2000-2030
Total CO_2 Emissions	**266**	**676**	**931**	**1309**	**1874**	**100**	**100**	**100**	**100**	**100**	**3.3**	**3.3**	**3.4**	**3.5**
change since 1990 (%)		*24.8*	*72.0*	*141.8*	*246.2*									
Coal	161	275	336	436	599	60	41	36	33	32	1.9	2.0	2.3	2.6
Oil	100	304	415	573	807	38	45	45	44	43	3.9	3.2	3.2	3.3
Gas	5	96	180	300	469	2	14	19	23	25	10.4	6.5	5.9	5.4
Power Generation	**93**	**289**	**418**	**622**	**937**	**100**	**100**	**100**	**100**	**100**	**4.0**	**3.8**	**3.9**	**4.0**
Coal	83	200	248	333	482	90	69	59	54	51	3.1	2.2	2.6	3.0
Oil	9	46	61	83	114	9	16	15	13	12	6.0	2.8	3.0	3.1
Gas	1	44	109	206	341	1	15	26	33	36	14.5	9.6	8.0	7.1
Transformation, Own Use & Losses	**8**	**49**	**68**	**102**	**160**	**100**	**100**	**100**	**100**	**100**	**6.4**	**3.4**	**3.7**	**4.0**
Total Final Consumption	**166**	**338**	**445**	**586**	**778**	**100**	**100**	**100**	**100**	**100**	**2.5**	**2.8**	**2.8**	**2.8**
Coal	78	76	88	102	116	47	22	20	17	15	-0.1	1.5	1.5	1.4
Oil	87	234	314	421	571	53	69	71	72	73	3.5	3.0	3.0	3.0
Gas	1	27	42	63	91	1	8	10	11	12	12.8	4.5	4.2	4.1
Industry	**68**	**123**	**148**	**174**	**204**	**100**	**100**	**100**	**100**	**100**	**2.1**	**1.9**	**1.7**	**1.7**
Coal	47	67	78	90	101	69	55	53	52	50	1.3	1.5	1.5	1.4
Oil	21	39	44	47	52	30	31	30	27	26	2.2	1.4	1.0	1.0
Gas	1	17	26	36	50	1	14	17	21	25	11.8	4.0	3.8	3.6
Transportation	**57**	**124**	**169**	**231**	**319**	**100**	**100**	**100**	**100**	**100**	**2.7**	**3.1**	**3.1**	**3.2**
Oil	42	123	167	228	315	75	99	99	99	99	3.7	3.1	3.1	3.2
Other fuels	14	2	2	3	4	25	1	1	1	1	-7.2	4.1	3.6	3.4
Other Sectors	**36**	**81**	**115**	**166**	**239**	**100**	**100**	**100**	**100**	**100**	**2.9**	**3.6**	**3.7**	**3.7**
Coal	14	9	10	12	14	39	11	9	7	6	-1.7	1.7	1.8	1.8
Oil	22	64	91	131	187	60	79	79	79	78	3.8	3.6	3.7	3.7
Gas	0	8	14	23	37	0	10	12	14	16	14.7	5.4	5.2	5.1
Non-Energy Use	**6**	**9**	**12**	**14**	**17**	**100**	**100**	**100**	**100**	**100**	**1.9**	**2.7**	**2.1**	**1.9**

APPENDIX 1
DESCRIPTION OF THE WORLD ENERGY MODEL 2002

Objectives

Since 1993, the IEA has provided long-term energy projections using a World Energy Model (WEM). For the *WEO 2002* the WEM has undergone a significant transformation and has been enhanced in a number of different ways. Specifically, the time horizon has been extended to 2030 and the model now includes:

- 18 separately modelled countries and regions, including new, separate models for Mexico, Korea, Indonesia, the European Union and other OECD Europe (Figure A1.2);
- for the OECD regions, a substantially more detailed sectoral representation of the industry sector, and projections of demand by end-use or mode in the transport, residential and services sectors;
- a world refinery model that analyses the regional implications of growing oil product demand on new capacity requirements and product trade;
- improvements in the modelling of technology and renewables in the power generation sector;
- projections of the electrification rate and biomass use in the developing world.

The WEM used to produce this *Outlook* is the seventh version of the model. The WEM is a tool to analyse:

- *Global energy prospects:* Trends in demand, supply availability and constraints, international trade and energy balances by sector and fuel to 2030.
- *Environmental impact of energy use:* CO_2 emissions from fuel combustion are derived from the detailed projections of energy consumption, while emissions trading can be simulated to arrive at a price for tradable permits.
- *Effects of policy actions or technological changes:* Alternative policy scenarios can be devised and run to analyse the impact of policy actions and developments in technologies on energy demand and emissions.

Model Structure

The WEM is a mathematical model made up of five main modules: *final energy demand; power generation; refinery and other transformation; fossil fuel supply and emissions trading*. Figure A1.1 provides a simplified overview of the structure of the model.

Figure A1.1: **World Energy Model Overview**

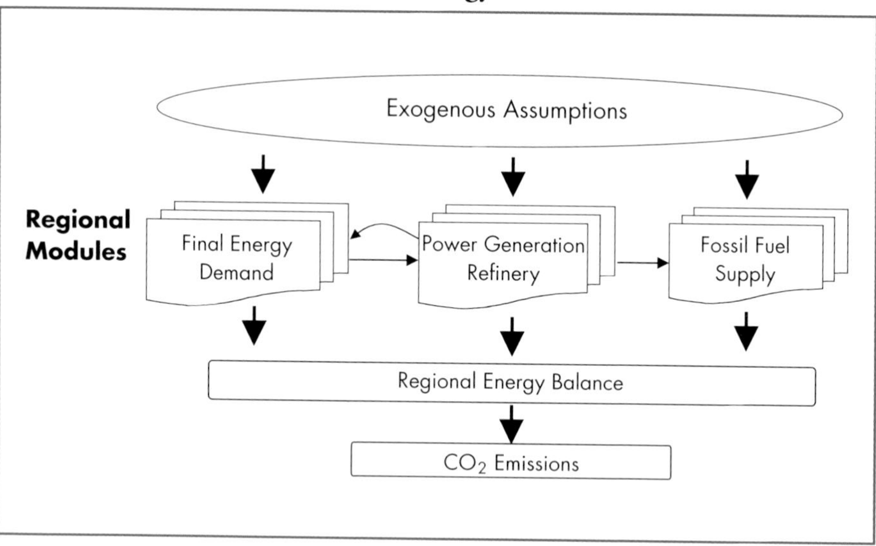

The main exogenous variables are GDP, demographics, household size, international fossil fuel prices and technological developments. The level of electricity consumption and electricity prices dynamically link the final energy demand and power generation modules. Primary demand for fossil fuels serves as input for the supply modules. Complete energy balances are compiled at a regional level and the CO_2 emissions of each region are then calculated using derived carbon factors.[1]

Technical Aspects

The development and running of the WEM requires access to large quantities of historical data on economic and energy variables. Most of the data are obtained from the IEA's own databases of energy and economic

1. The emissions trading module, although not run in any scenario in the *WEO 2002*, uses marginal abatement cost curves, obtained by an iterative process of running the WEM with different carbon values.

Note: Annex B Other Transition Economies are modelled separately.

statistics. A significant amount of additional data from a wide range of external sources is also utilised.

The parameters of each module's equations are estimated econometrically, usually with data for the period 1971-2000. Shorter periods are sometimes used where data are unavailable or significant structural breaks are identified. To take into account expected structural, policy or technological changes, adjustments to these parameters are sometimes made over the projection period, using econometric and other modelling techniques. In regions such as the transition economies, where most data are only available from 1992, it is not possible to use econometric estimation. The results are prepared by using assumptions based on cross-country analyses or expert judgement.

Simulations are carried out on an annual basis. Demand modules can be isolated and simulations run separately. This is particularly useful in the adjustment process and in the sensitivity analyses of specific factors.

The WEM makes use of a wide range of software, including specific database management tools, econometric software and simulation programmes.

Description of the Modules

Final Energy Demand

For the *WEO 2002* the OECD regions have been modelled in greater sectoral and end-use detail than in previous editions, specifically:

- *Industry:* energy demand is separated into six sub-sectors allowing a more detailed analysis of the trends and drivers of the industrial sector.
- *Residential:* energy demand is separated into five end-uses by fuel.
- *Services:* energy demand is modelled as three end-uses by fuel.
- *Transport:* energy demand is modelled in detail by mode and fuel, again enhancing the analysis of this sector.

This level of detail in the data is not available for non-OECD regions. As a result, the non-OECD country/region models do not match the level of detail of the OECD. However, in most cases the level of sectoral disaggregation in the non-OECD countries/regions has been increased for the *WEO 2002*.

Total final energy demand is the sum of energy consumption in each final demand sector. In each sub-sector or end-use, at least six types of energy are shown: coal, oil, gas, electricity, heat and renewables. This aggregation conceals more detail; for example, the different oil products are modelled separately as an input to the refinery model, and renewables are split into biomass and "other".

Within each sub-sector or end-use, energy demand is estimated as the product of an energy intensity and activity variable. For example, the projection of the unitary consumption of gas by a single household for water heating is multiplied by the projection of the number of households with gas water heating to arrive at the total residential sector consumption of gas for water heating.

In most of the equations, energy demand is a function of the following explanatory variables:

- *Activity variables:* This is often a GDP or GDP per capita variable. However, in many cases, a specific activity variable, which is usually driven by GDP, is used. For example, in the OECD regions demand in each industrial sub-sector is a function of the economic output of that sector. Energy demand in the services sector is a function of floor area and the number of employees in the services sector. In the transport sector, vehicle stock, passenger-kilometres and tonne-kilometres are used. In the non-OECD regions demand specific activity variables are used less often; although examples where they are used include agricultural and iron and steel output.
- *Price:* End-user prices are calculated from the exogenous international energy prices. They take into account both variable and fixed taxes, and also transformation and distribution costs. For each sector, a representative price (usually a weighted average) is derived. This takes account of the product mix in final consumption and differences between countries. This representative price is then used as an explanatory variable directly, lagged or as a moving average.
- *Other variables:* Other variables are used to take into account structural and technological changes, saturation effects, or other important drivers (such as the gap between test and on-road fuel efficiency).

Detailed capital stock models are integrated into the WEM model in the OECD regions in order to model the impact that low capital stock turnover has on the penetration of more efficient equipment.

Industry in the OECD Regions

The industrial sector in the OECD regions is split into six sub-sectors: *iron and steel, chemicals, paper and pulp, food and beverages, non-metallic minerals and other industry.*

The intensity of fuel consumption per unit of each sub-sector's output is projected on an econometric basis. The output level of each sector is modelled separately and is combined with the projections of fuel intensity to derive the consumption of each fuel by sub-sector. This allows a more detailed analysis of the drivers of demand in the industrial sector and the impact of structural change on fuel consumption trends.

The increased disaggregation also facilitates the modelling of alternative scenarios, where end-use shares and technology descriptions are applied in conjunction with capital stock turnover models in order to analyse in detail the impact of alternative policies or technology choices on the sector.

Transport in the OECD Regions

In *WEO 2000,* separate detailed bottom-up models for the OECD transport sector were developed for the alternative scenario. For *WEO 2002* the WEM now fully incorporates a detailed bottom-up approach for the transport sector in the OECD regions (see Figure A1.3).

Transport energy demand is split between *passenger* and *freight* travel for *light duty vehicles, buses, trucks, rail, aviation* and *navigation.* Vehicle stock models track the levels of passenger cars and light trucks for diesel, gasoline, hybrids, and alternative fuel vehicles, as well as freight trucks. The gap between test and on-road fuel efficiency is also projected.

For each region, activity levels for each mode of transport are estimated econometrically as a function of population, GDP and fuel prices. Additional assumptions to reflect passenger vehicle ownership saturation are also made. Transport activity is linked to price through an elasticity of fuel cost per km. This is estimated in all modes except for passenger bus and rail, and inland navigation. In the case of passenger vehicles, this elasticity variable accounts for the rebound effect of increased transport demand resulting from improved fuel intensity.

Modal energy intensity is projected taking into account changes in energy efficiency and fuel prices. A more detailed approach is used for cars and light trucks. Stock turnover is explicitly modelled in order to allow for the effects of fuel efficiency regulation of new cars on fleet energy intensity.

Figure A1.3 : **Structure of the Transport Sector Demand Module**

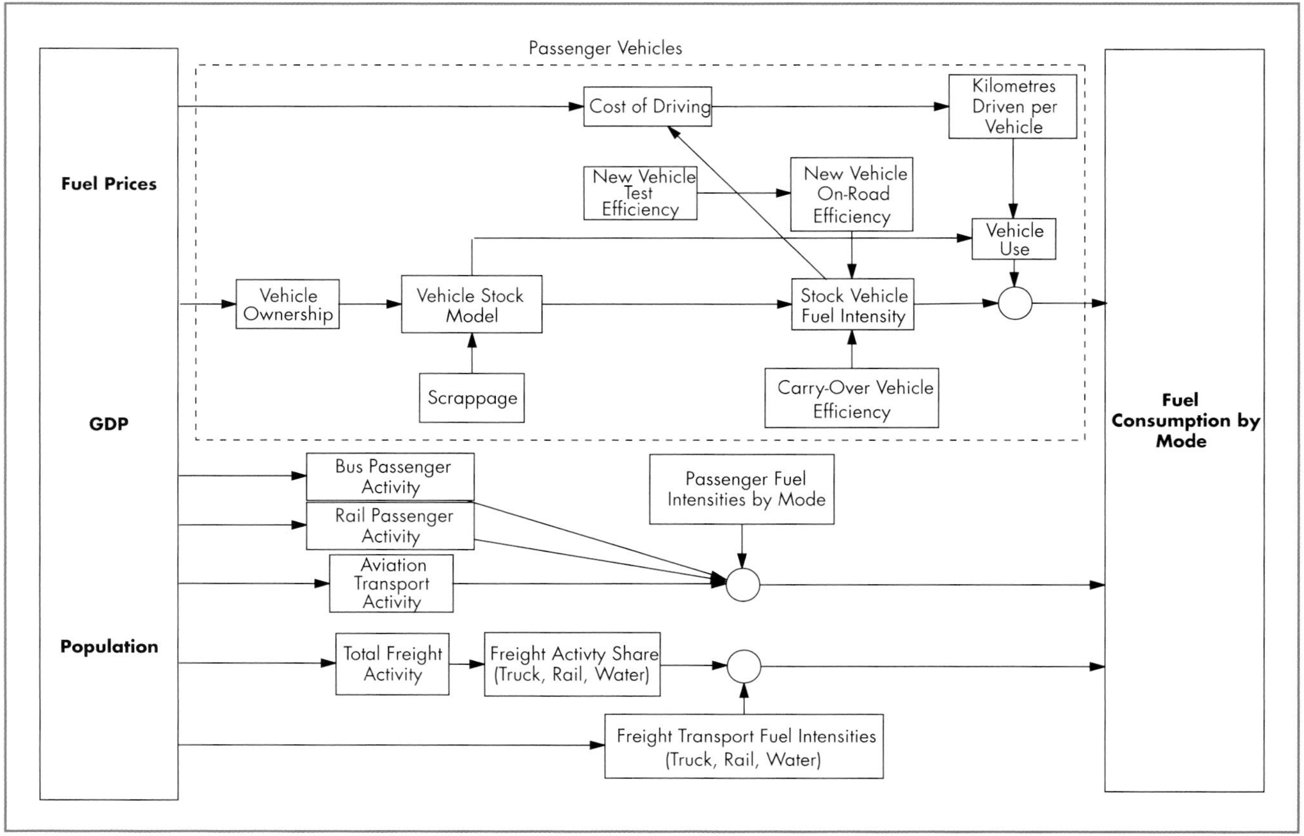

Fuel efficiency regulation for new cars and light trucks can thus be modelled explicitly.

Residential and Services in the OECD Regions

The *WEO 2002* models energy demand in the residential and services sectors by end-use, significantly increasing the detail of the projections for these sectors over previous editions (Figure A1.4).

In the residential sector the number of households using each fuel for water heating and space heating is projected econometrically, with some saturation limits on shares. The fuel intensity per household (that actually uses a fuel for each end-use) for space and water heating is then estimated econometrically.

Lighting intensity and appliance intensity per household are then projected separately, and combined with total household numbers to yield electricity demand for these end-uses. Detailed capital stock models analyse the impact compared to the Reference Scenario of alternative equipment standards and energy efficiency measures on individual appliances and heating and cooling plant.

The service sector model splits consumption by fuel into three end-uses. Heating, hot water and cooking (HHC) are considered together to analyse the inter-fuel competition in these end-uses. The remaining electricity consumption is in personal computers (PC) and related equipment and other electricity end-use, including ventilation, space cooling and lighting. The total fuel demand for HHC is projected per square metre of floor area. Floor area in services is estimated as a function of value-added in the service sector, which in turn is projected from the GDP assumptions. Two components then allocate the total demand for HHC to fuels: an "existing stock" model determines energy consumption by fuel based on historical shares, while a portion of demand is allocated to "new stock" where fuel shares are a function of relative prices and existing shares of each fuel.

Projections of PC-related electricity use and per square metre electricity use for the other electricity end-uses not already covered are combined with projections for PCs and floor area in services to calculate their total electricity demand. The estimation of PC numbers is based on the growth in services sector employment. The number of employees in the services sector is a function of the active population.

Figure A1.4: **Structure of the Residential and Service Sector Demand Modules**

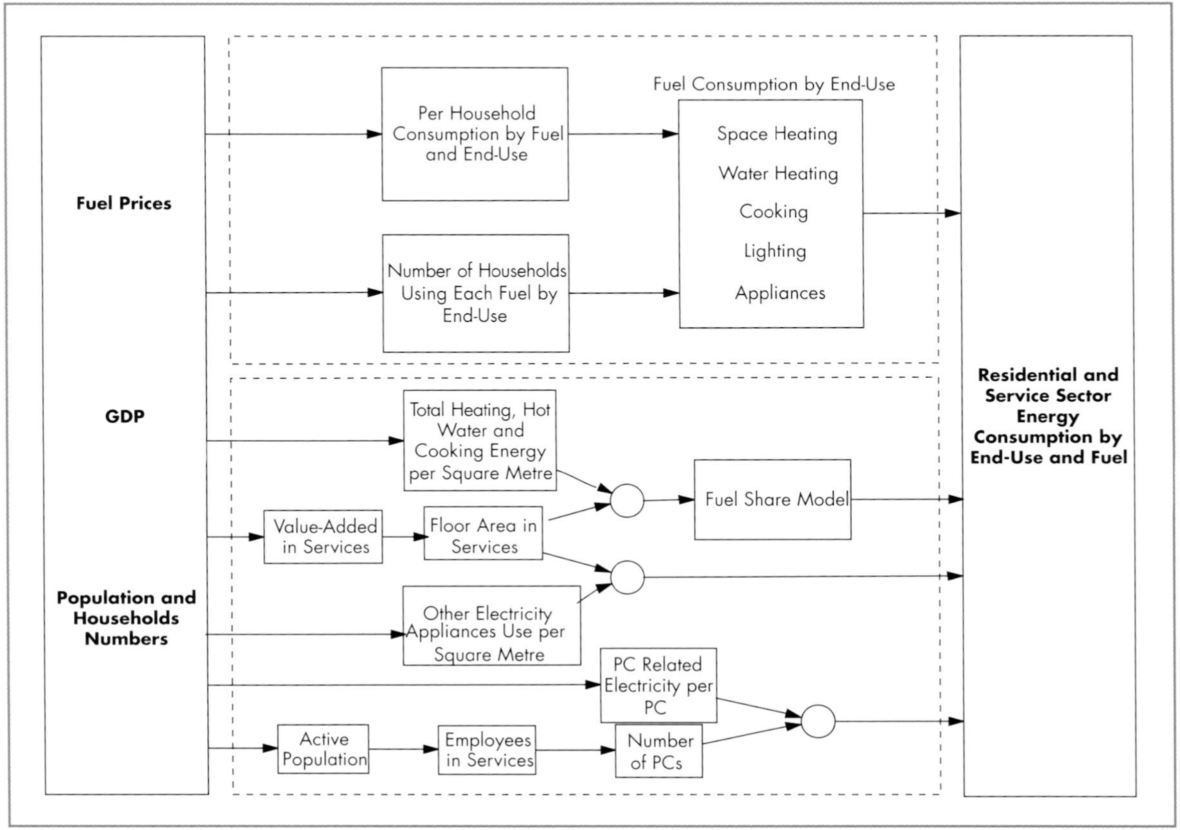

Power Generation and Heat Plants

The purpose of the power generation module is to calculate the following:

- electricity generation by type of plant to meet electricity demand;
- fuel consumption of the power generation sector;
- new generating capacity needed;
- type of any new plant to be built;
- system marginal cost of generation.

The structure of the power generation module is depicted in Figure A1.5. Peak load is calculated by using the demand for electricity together with an assumed load curve. The need for new generating capacity is calculated by adding a minimum reserve plant margin to peak load and comparing that with the capacity of existing plants minus plant retirements using assumed plant lives. An allowance is made for assumed plant availability. If new plant is needed, the model makes its choice on the basis of levelised cost. The levelised generating cost (expressed as monetary value per kWh) combines capital, operating and fuel costs over the whole operating life of a plant using a given discount rate, plant efficiency and plant utilisation rate. The model uses 15 different types of plant:

- coal, oil and gas steam boilers;
- combined-cycle gas turbine (CCGT);
- open-cycle gas turbine (GT);
- integrated gasification combined cycle (IGCC);
- oil and gas internal combustion;
- fuel cell;
- nuclear;
- biomass;
- geothermal;
- wind (onshore);
- wind (offshore);
- hydro (conventional);
- hydro (pumped storage);
- solar (photovoltaics);
- solar (thermal).

The combined heat and power (CHP) option is considered for fossil fuels and biomass plants. Distributed generation is treated separately within the model.

Figure A1.5: **Structure of the Power Generation Module**

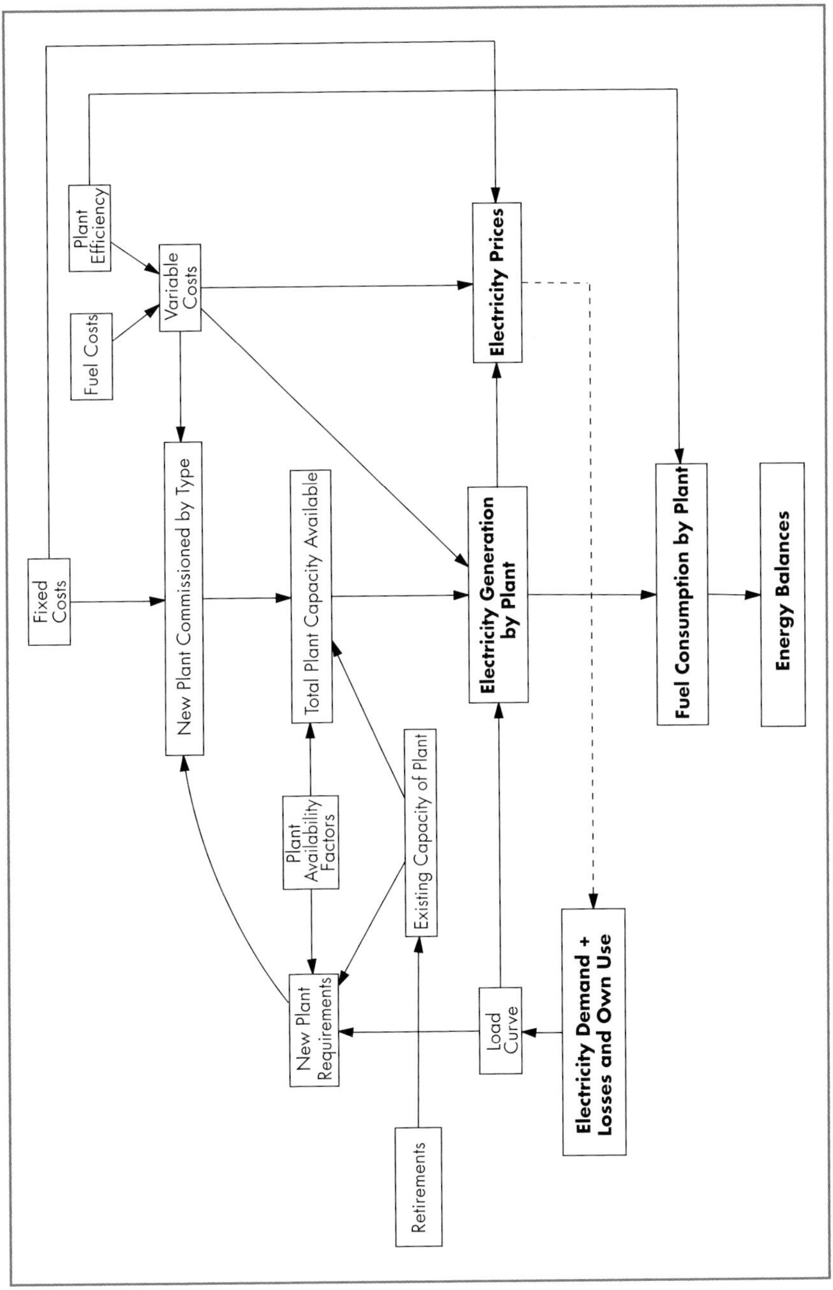

Fixed Costs
Fuel Costs
Plant Efficiency
Variable Costs
New Plant Commissioned by Type
New Plant Requirements
Plant Availability Factors
Total Plant Capacity Available
Retirements
Existing Capacity of Plant
Load Curve
Electricity Generation by Plant
Electricity Prices
Electricity Demand + Losses and Own Use
Fuel Consumption by Plant
Energy Balances

Capacities for nuclear are calculated mainly from exogenous assumptions, but are influenced by international fossil fuel prices to take account of price incentives to develop such plants.

Unlike in the *WEO 2000*, the new capacity for renewables is modelled as a function of its economic factors with detailed supply curves for the OECD regions.

Fossil fuel prices and efficiencies are used to load plants in ascending order of their short-run marginal operating costs, allowing for assumed plant availability. Once the mix of generation plants has been determined, the fuel requirements are then deduced by plant type using an assumed efficiency.

The marginal generation cost of the system is calculated, and this cost is then fed back to the demand model to determine the final electricity price.

CO_2 Emissions

For each region, sector and fuel, CO_2 emissions are calculated by multiplying energy demand by an implied carbon emission factor from IEA data. Implied emission factors for coal, oil and gas differ between sectors and regions, reflecting the product mix. They have been calculated from year 2000 emissions data in the OECD regions, Brazil, China, India, Indonesia and Russia.

Fossil Fuel Supply

Oil Module

The purpose of this module is to determine the level of oil production in each region. Production is split into three categories:

- non-OPEC;
- OPEC;
- non-conventional oil production.

Total oil demand is the sum of regional oil demand, world bunkers and stock variation. OPEC conventional oil production is assumed to fill the gap between non-OPEC production and non-conventional and total world oil demand (Figure A1.6).

The derivation of conventional (crude oil and natural gas liquids) non-OPEC production uses a combination of two different approaches. A short-term approach estimates production profiles based on a field-by-field analysis. A long-term approach involves the determination of production

Figure A1.6: **Structure of Oil Supply Module**

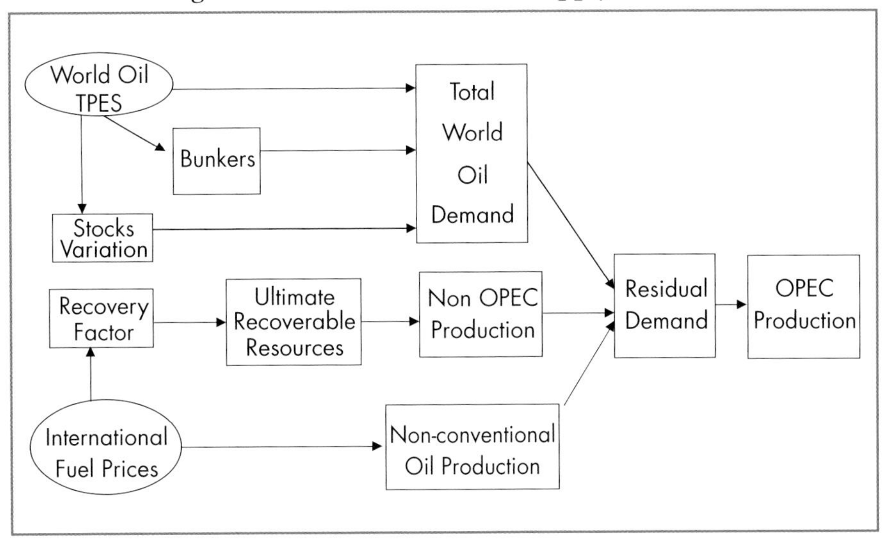

according to the level of ultimate recoverable resources and a depletion rate estimated by using historical data. Ultimate recoverable resources depend on a recovery factor. This recovery factor reflects reserve growth, which results from improvements in drilling, exploration and production technologies. The trend in the recovery rate is, in turn, a function of the oil price and a technological improvement factor. The level of non-conventional oil supply is directly linked to the evolution of the oil price. Higher oil prices bring forth greater non-conventional oil supply over time.

Refinery Module

The purpose of this module is to calculate, for each WEO region, the following:

- new crude oil distillation unit capacity (CDU);
- refinery output by product category (light, middle, heavy and other);
- fuel consumption of the refining sector.

The regional refinery output projections are based on demand for refined products, derived from the demand module. CDU projections are based on refinery output projections, past trends in refinery construction, currently announced plans for additional CDU capacity and existing surplus CDU capacity. The module accounts for the evolution of demand product mix, of technological improvements in the refining process and for the contribution of technology such as gas-to-liquids. The balance between world refined product demand and supply is calculated according to regional product spare capacity.

Gas Module

The gas module is based on a resources approach, but with some important differences from the oil module. In particular, three regional gas markets — America, Europe and Asia — are considered, whereas oil is modelled as a single international market. Two country types are modelled: net importers and net exporters. Once gas production from each net importer region is estimated, taking into account ultimate recoverable resources and a depletion rate, the remaining regional demand is allocated to the net exporter regions according to their export potential.

Coal Module

Sufficient reserves of coal exist to meet world demand and coal reserves are much more evenly distributed throughout the world than oil and gas reserves. Because of the wide diversity of existing and potential coal suppliers, availability of coal supply is not an issue. The current WEM does not, therefore, model coal supply explicitly but information on coal production prospects is provided in the different chapters.

APPENDIX 2
DEFINITIONS

This appendix provides general information on definitions used throughout *WEO 2002*. Readers interested in obtaining more detailed information should consult the annual IEA publications *Energy Balances of OECD Countries, Energy Balances of Non-OECD Countries, Coal Information, Oil Information* and *Gas Information.*

Coal

Coal includes all coal: both coal (including hard coal and lignite) and derived fuels (including patent fuel, coke-oven coke, gas coke, coke-oven gas and blast-furnace gas). Peat is also included in this category.

Oil

Oil includes crude oil, natural gas liquids, refinery feedstocks and additives, other hydrocarbons and petroleum products (refinery gas, ethane, liquefied petroleum gas, aviation gasoline, motor gasoline, jet fuels, kerosene, gas/diesel oil, heavy fuel oil, naphtha, white spirit, lubricants, bitumen, paraffin waxes, petroleum coke and other petroleum products).

Light Petroleum Products

Light petroleum products include liquefied petroleum gas, naphtha and gasoline.

Middle Distillates

Middle distillates include jet fuel, kerosene, diesel and heating oil.

Heavy Petroleum Products

Heavy petroleum products include heavy fuel oil.

Other Petroleum Products

Other petroleum products include refinery gas, ethane, lubricants, bitumen, petroleum coke and waxes.

Gas

Gas includes natural gas (both associated and non-associated with petroleum deposits but excluding natural gas liquids) and gas works gas.

Biomass

Biomass includes solid biomass and animal products, gas and liquids derived from biomass, industrial waste and municipal waste.

Other Renewables

Other renewables include geothermal, solar, wind, tide, and wave energy for electricity generation. Direct use of geothermal and solar heat is also included in this category. For OECD countries, other renewables include biomass. Biomass is indicated separately for non-OECD regions, except for electricity output, which includes biomass for all regions.

Heat

Heat is heat produced for sale. The large majority of the heat included in this category comes from the combustion of fuels, although some small amounts are produced from electrically-powered heat pumps and boilers.

Nuclear

Nuclear refers to the primary heat equivalent of the electricity produced by a nuclear plant with an average thermal efficiency of 33%.

Hydro

Hydro refers to the energy content of the electricity produced in hydropower plants assuming 100% efficiency.

Hydrogen-Fuel Cell

A hydrogen fuel cell is a high efficiency electrochemical energy conversion device that generates electricity and produces heat, with the help of catalysts.

Total Primary Energy Supply

Total primary energy supply (TPES) is equivalent to primary energy demand. This represents inland demand only and excludes international marine bunkers, except for world energy demand.

International Marine Bunkers

International marine bunkers cover those quantities delivered to sea-going ships of all flags, including warships. Consumption by ships plying in inland and coastal waters is not included.

Net Imports

The difference between domestic demand and production.

Net Inter-regional Trade

The net exchange between regions. It excludes the trade between countries within each region.

Power Generation

Power generation refers to fuel use in electricity plants, heat plants and combined heat and power (CHP) plants. Both public plants and small plants that produce fuel for their own use (autoproducers) are included.

Total Final Consumption

Total final consumption (TFC) is the sum of consumption by the different end-use sectors. TFC is broken down into energy demand in the following sectors: industry, transport, other (includes agriculture, residential, commercial and public services) and non-energy use. Industry includes manufacturing, construction and mining industries. In final consumption, petrochemical feedstocks appear under *industry use.* Other non-energy uses are shown under *non-energy use.*

Other Transformation, Own Use and Losses

Other transformation, own use and losses covers the use of energy by transformation industries and the energy losses in converting primary energy into a form that can be used in the final consuming sectors. It includes energy use and loss by gas works, petroleum refineries, coal and gas transformation and liquefaction. It also includes energy used in coal mines, in oil and gas extraction, and in electricity and heat production. Transfers and statistical differences are also included in this category

Electricity Generation

Electricity generation shows the total amount of electricity generated by power plants. It includes own-use and transmission and distribution losses.

REGIONAL DEFINITIONS

OECD Europe

OECD Europe consists of Austria, Belgium, the Czech Republic, Denmark, Finland, France, Germany, Greece, Hungary, Iceland, Ireland, Italy, Luxembourg, the Netherlands, Norway, Poland, Portugal, Spain, Sweden, Switzerland, Turkey and the United Kingdom.

OECD North America

OECD North America consists of the United States, Canada and Mexico.

OECD Pacific

OECD Pacific consists of Japan, Korea, Australia and New Zealand.

Transition Economies

The transition economies include Albania, Armenia, Azerbaijan, Belarus, Bosnia-Herzegovina, Bulgaria, Croatia, Estonia, the Federal Republic of Yugoslavia, the former Yugoslav Republic of Macedonia, Georgia, Kazakhstan, Kyrgyzstan, Latvia, Lithuania, Moldova, Romania, Russia, the Slovak Republic, Slovenia, Tajikistan, Turkmenistan, Ukraine and Uzbekistan. For statistical reasons, this region also includes Cyprus, Gibraltar and Malta.

China

China refers to the People's Republic of China and Hong Kong.

East Asia

East Asia includes Bhutan, Brunei, Chinese Taipei, Fiji, French Polynesia, Indonesia, Kiribati, Democratic People's Republic of Korea, Malaysia, Maldives, Myanmar, New Caledonia, Papua New Guinea, the Philippines, Samoa, Singapore, Solomon Islands, Thailand, Vietnam and Vanuatu.

South Asia

South Asia consists of Afghanistan, Bangladesh, India, Nepal, Pakistan and Sri Lanka.

Latin America

Latin America includes Antigua and Barbuda, Argentina, Bahamas, Barbados, Belize, Bermuda, Bolivia, Brazil, Chile, Colombia, Costa Rica, Cuba, Dominica, the Dominican Republic, Ecuador, El Salvador, French Guiana, Grenada, Guadeloupe, Guatemala, Guyana, Haiti, Honduras, Jamaica, Martinique, Netherlands Antilles, Nicaragua, Panama, Paraguay, Peru, St. Kitts-Nevis-Anguilla, Saint Lucia, St. Vincent-Grenadines and Suriname, Trinidad and Tobago, Uruguay and Venezuela.

Africa

Africa comprises Algeria, Angola, Benin, Botswana, Burkina Faso, Burundi, Cameroon, Cape Verde, the Central African Republic, Chad, Congo, the Democratic Republic of Congo, Cote d'Ivoire, Djibouti, Egypt, Equatorial Guinea, Eritrea, Ethiopia, Gabon, Gambia, Ghana, Guinea, Guinea-Bissau, Kenya, Lesotho, Liberia, Libya, Madagascar, Malawi, Mali, Mauritania, Mauritius, Morocco, Mozambique, Namibia, Niger, Nigeria, Rwanda, Sao Tome and Principe, Senegal, Seychelles, Sierra Leone, Somalia, South Africa, Sudan, Swaziland, the United Republic of Tanzania, Togo, Tunisia, Uganda, Zambia and Zimbabwe.

Middle East

The Middle East is defined as Bahrain, Iran, Iraq, Israel, Jordan, Kuwait, Lebanon, Oman, Qatar, Saudi Arabia, Syria, the United Arab Emirates and Yemen. It includes the neutral zone between Saudi Arabia and Iraq.

*

* *

In addition to the WEO regions, the following groupings are also referred to in the text.

European Union (EU15)

Austria, Belgium, Denmark, Finland, France, Germany, Greece, Ireland, Italy, Luxembourg, the Netherlands, Portugal, Spain, Sweden and the United Kingdom.

Annex B Countries

Australia, Austria, Belgium, Bulgaria, Canada, Croatia, the Czech Republic, Denmark, Estonia, Finland, France, Germany, Greece, Hungary, Iceland, Ireland, Italy, Japan, Latvia, Lithuania, Luxembourg, the Netherlands, New Zealand, Norway, Poland, Portugal, Romania, Russia, Slovakia, Slovenia, Spain, Sweden, Switzerland, Ukraine, the United Kingdom and the United States of America.

Asia

OECD Pacific, China, East Asia and South Asia.

Association of South-East Asian Nations (ASEAN)

Brunei Darussalam, Cambodia, Indonesia, Laos, Malaysia, Myanmar, Philippines, Singapore, Thailand and Vietnam.

Developing Asia

China, East Asia and South Asia.

Asia-Pacific Economic Co-operation (APEC)

Australia, Brunei Darussalam, Canada, Chile, China, Indonesia, Japan, Korea, Malaysia, Mexico, New Zealand, Papua New Guinea, Peru, Philippines, Russia, Singapore, Chinese Taipei, Thailand, the United States of America and Vietnam.

Organization of Petroleum Exporting Countries (OPEC)

Algeria, Indonesia, Iran, Iraq, Kuwait, Libya, Nigeria, Qatar, Saudi Arabia, United Arab Emirates and Venezuela.

Sub-Saharan Africa

Includes all African countries except North Africa (Algeria, Egypt, Libya, Morocco and Tunisia).

ABBREVIATIONS AND ACRONYMS

In this book, acronyms are frequently substituted for a number of terms used within the International Energy Agency. This glossary provides a quick and central reference for many of the abbreviations used.

ADB	Asian Development Bank
APERC	Asia-Pacific Energy Research Center
ASEAN	Association of South-East Asian Nations
bcm	billion cubic metres
b/d	barrels per day
boe	barrels of oil equivalent
CAFE	Corporate Average Fuel Economy
CBM	coal-bed methane
CCGT	combined-cycle gas turbine
CDU	crude distillation unit
CHP	combined production of heat and power; sometimes, when referring to industrial CHP, the term co-generation is used
CO_2	carbon dioxide
CRW	combustible renewables and waste
DG	distributed generation
DoE	Department of Energy
EC	European Commission
EU	European Union
FDI	foreign direct investment
FSU	former Soviet Union
FYROM	former Yugoslav Republic of Macedonia
GDP	gross domestic product
GHG	greenhouse gas
GTL	gas-to-liquids
GW	gigawatt (1 watt $\times 10^9$)
ICT	information and communication technology
IEA	International Energy Agency
IMF	International Monetary Fund

IPP	independent power producer
kb/d	thousand barrels per day
kW	kilowatt (1 watt × 1,000)
kWh	kilowatt-hour
LNG	liquefied natural gas
LPG	liquefied petroleum gas
mb/d	million barrels per day
MBtu	million British thermal units
mcm/d	million cubic metres per day
MOCIE	Ministry of Commerce, Industry and Energy (Korea)
MSC	multiple service contract
Mt	million tonnes
Mtoe	million tonnes of oil equivalent
MW	megawatt (1 watt × 10^6)
MWh	megawatt-hour
NAFTA	North American Free Trade Agreement
NGL	natural gas liquid
NO_x	nitrogen oxides
OECD	Organisation for Economic Co-operation and Development
OPEC	Organization of Petroleum Exporting Countries
PPP	purchasing power parity; the rate of currency conversion that equalises the purchasing power of different currencies. PPPs compare costs in different currencies of a fixed basket of traded and non-traded goods and services and yield a widely based measure of standard of living
PPA	power purchasing agreement
SO_2	sulphur dioxide
SOE	state-owned enterprise
tcf	thousand cubic feet
tce	tonne of coal equivalent
tcm	trillion cubic metres
TFC	total final consumption
toe	tonne of oil equivalent
tonne	metric ton
TPES	total primary energy supply
TW	terawatt (1 watt × 10^{12})

TWh	terawatt-hour
WEM	World Energy Model
WEO	World Energy Outlook
WHO	World Health Organization
WTO	World Trade Organization

REFERENCES

CHAPTER 1

Asian Development Bank (ADB) (2001), *Asian Development Outlook 2001*, New York: Oxford University Press.

International Energy Agency (IEA) (forthcoming), *Dealing with Climate Change: Policies and Measures in IEA Member Countries*, Paris: OECD.

IEA (2001a), *Dealing with Climate Change: Policies and Measures in IEA Member Countries*, Paris: OECD.

IEA (2001b), *Oil Price Volatility: Trends and Consequences*, Economic Analysis Division *Working Paper* (not published).

IEA (2001c), *World Energy Outlook 2001 Insights: Assessing Today's Supplies to Fuel Tomorrow's Growth*, Paris: OECD.

IEA (2000), *World Energy Outlook 2000*, Paris: OECD.

International Monetary Fund (IMF) (2002), *World Economic Outlook: A Survey by the Staff of the International Monetary Fund*, Washington: IMF.

Organisation for Economic Co-operation and Development (OECD) (2002), *OECD Economic Outlook 71*, April, Paris: OECD.

United Nations Population Division (2001), *World Population Prospects: The 2000 Revision*, New York: United Nations.

World Bank (2001), *Global Economic Prospects and the Developing Countries*, Washington: World Bank.

CHAPTER 2

Barreto, L, Makihira, A. and Riahi, K. (2002), The Hydrogen Economy in the 21st Century: A Sustainable Development Scenario, *International Journal of Hydrogen Energy* (forthcoming), Florida: International Association for Hydrogen Energy.

IEA (2000), *World Energy Outlook 2000*, Paris: OECD.

Saghir, J. (2002), *The World Bank's Vision of the Future Energy Environment in the Developing World*, Presentation at CERA Week 2002 Electric Power Plenary, Houston, February.

CHAPTER 3

American Petroleum Institute (API) (2001), *US Refining Industry: A System Stretched to the Limit*, Washington: API.

Cedigaz (2001), *Natural Gas in the World: 2001 Survey*, Paris: Institut Français du Pétrole.

DRI-WEFA (2001), *World Energy Service: World Outlook 2000*, Massachusetts: DRI-WEFA.

Huber, C., Haas, R., Resch, G., Faber, T., Green, J., Erge, T., Ruijgrok, W., and Twidell, J. (2001), *Final Report of the Project ElGreen*, Austria: Energy Economics Group, Vienna University of Technology.

Institut Français du Pétrole (2002), *Panorama 2002*, Paris: Institut Français du Pétrole.

IEA (2002), *Distributed Generation in Liberalised Electricity Markets*, Paris: OECD.

IEA (2001), *World Energy Outlook 2001 Insights: Assessing Today's Supplies to Fuel Tomorrow's Growth*, Paris: OECD.

IEA (1999), *World Energy Outlook Insights 1999: Looking at Energy Subsidies: Getting the Prices Right*, Paris: OECD.

Petroleum Economics Ltd. (2002), *World Long Term Oil and Energy Outlook*, June, Surrey: PEL Market Services.

Platts–UDI (2002), *World Electric Power Plants Database* (June), New York: McGraw-Hill.

Shell International (2001), *Energy Needs, Choices and Possibilities: Scenarios to 2050*, London: Shell International Limited.

Shihab-Eldin, Adnan, Lounnas, Rezki and Brennand, Garry (2001), Oil Outlook to 2020, *OPEC Review*, December, Oxford: Blackwell Publishers.

Simmons, Matthew R. (2002), The World's Giant Oil Fields, *Hubert Centre Newsletter* (January), Colorado: Colorado School of Mines.

United States Department of Energy/Energy Information Agency (DOE/EIA) (2002), *Annual Energy Outlook 2002,* Washington.

DOE/EIA (2001), *Annual Energy Outlook 2001,* Washington.

United States Geological Survey (USGS) (2000), *World Petroleum Assessment 2000*, Washington.

CHAPTER 4

Baker, George (2002), Mexican energy sector reforms include foreign operators' participation, in E & D, *Oil and Gas Journal*, vol. 100, No. 6, February 11, Houston: Pennwell.

Cedigaz (2001), *Natural Gas in the World: 2001 Survey*, Paris: Institut Français du Pétrole.

DOE/EIA (2001a), *Annual Energy Outlook 2001,* Washington.

DOE/EIA (2001b), *U.S. Natural Gas Markets: Mid-Term Prospects for Natural Gas Supply*, Washington.

IEA (2002), *Energy Prices and Taxes – Quarterly Statistics, 3rd Quarter*, Paris: OECD.

IEA (2001), *World Energy Outlook 2001 Insights: Assessing Today's Supplies to Fuel Tomorrow's Growth*, Paris: OECD.

IEA (2000), *World Energy Outlook 2000*, Paris: OECD.

Mexican Ministry of Energy (MME) (2001a), *Programa Sectorial de Energia 2001-2006*, Mexico City: Secretaria de Energia.

MME (2001b), *Prospectiva del Mercado de Gas Natural 2001-2010,* Mexico City: Secretaria de Energia.
MME (2001c), *Prospectiva del Sector Electrico 2001-2010,* Mexico City: Secretaria de Energia.
MME (2000), *Prospectiva del Sector Energia 2000-2009,* Mexico City: Secretaria de Energia.
Mexican Ministry of Finance and Public Credit (2002), *Criterios Generales de Política Económica 2002,* Mexico City: Secretaria de Hacienda Credito Publico.
National Climate Change Process (NCCP) (2000), *Canada's First National Climate Change Business Plan,* Ottawa: NCCP.
OECD (2002), *Economic Survey – Mexico,* Paris: OECD.
Simmons, Matthew R. (2002), *Unlocking the Natural Gas Riddle,* Houston: Simmons and Company International.
USGS (2000), *World Petroleum Assessment 2000,* Washington.
United States National Academy of Sciences (2001), *Effectiveness and Impact of Corporate Average Fuel Economy Standards,* Washington: National Academy Press.
United States National Energy Policy Development Group (2001), *Reliable, Affordable, and Environmentally Sound Energy for America's Future: National Energy Policy Report,* Washington: US Government Printing Office.
World Bank (2001), *World Development Indicators 2001,* Washington: World Bank.
World Energy Council (WEC) (2001), *Survey of Energy Resources 2001,* London: WEC.

CHAPTER 5

European Commission (EC) (2001a), *First Report on the Implementation of the Internal Electricity and Gas Market,* Brussels: EC.
EC (2001b), *Electricity Liberalisation Indicators in Europe,* Brussels: EC.
EC (2000), *A Green Paper: Towards a European Strategy for the Security of Energy Supply,* Brussels: EC.
IEA (2001), *Coal Information 2001,* Paris: OECD.

CHAPTER 6

Asia-Pacific Energy Research Center (2000), *Natural Gas Pipeline Development in Northeast Asia:* Tokyo: Institute of Energy Economics.
Cedigaz (2001), *Natural Gas in the World: 2001 Survey,* Paris: Institut Français du Pétrole.
IEA (2002), *Energy Policies of IEA Countries: Korea,* Paris: OECD.
IEA (2001a), *Coal Information 2001,* Paris: OECD.
IEA (2001b), *Energy Policies of IEA Countries: Australia,* Paris: OECD.
IEA (2001c), *Energy Policies of IEA Countries: New Zealand,* Paris: OECD.

IEA (2001d), *Natural Gas Information 2001*, Paris: OECD.

Ministry of Economy, Trade and Industry (METI) (2001a), *Comprehensive Review of Japanese Energy Policy*, Tokyo: METI.

METI (2001b), *Electricity Supply Plan*, Tokyo: METI.

OECD (2002), *Economic Outlook 71*, Paris: OECD.

Radler, Marilyn (2000), World Crude and Natural Gas Reserves Rebound in 2000, *Oil and Gas Journal*, Vol. 98, No. 51, December 18, Houston: Pennwell.

Republic of Korea Ministry of Commerce, Industry and Energy (MOCIE) (1999), *Basic Plan for Restructuring of the Electricity Supply Industry*, Seoul: MOCIE.

USGS (2000), *World Petroleum Assessment 2000*, Washington: USGS.

CHAPTER 7

ADB (2000), *Country Economic Review: People's Republic of China*, New York: Oxford University Press.

Brennand, Timothy (2001), Natural Gas, a Fuel of Choice in China, *Energy for Sustainable Development*, No. 4, December.

China State Statistical Bureau (Various Issues), *China Statistical Yearbook*, Beijing: China Statistical Publishing House.

Farinelli, Ugo, Yokobori, Keiichi, and Zhou, Fengqi (2001), Energy Efficiency in China, *Energy for Sustainable Development*, No. 4, December.

IEA (forthcoming), *Developing China's Gas Market: The Energy Policy Challenges*, Paris: OECD.

IEA (2000), *World Energy Outlook 2000*, Paris: OECD.

Koyama, Ken (IEEJ) (2000), *Energy Strategies in China and India and Their Implications*, Tokyo: IEEJ.

Li, Qunren and Mao, Yushi (2001), China's Transportation and Its Energy Use, *Energy for Sustainable Development*, No. 4, December.

Logan, Jeffrey (2001), Diverging Energy and Economic Growth in China: Where Has All the Coal Gone ? *Pacific and Asian Journal of Energy*, Vol. 11, June, New Delhi: Tata Energy Research Institute.

OECD (2001), *China: Regional Disparities and Trade and Investment Liberalisation*, Paris: OECD.

OECD (2002), *China in the World Economy*, Paris: OECD.

PIRA Energy Group (PIRA) (2002), *World Trade Organisation and Significant Potential Impacts on the Oil Industry: Cases of China, Russia, Saudi Arabia, and OPEC*, New York: PIRA Energy Group.

Sinton, Jonathan and Fridley, David (2000), What Goes Up: Recent Trends in China's Energy Consumption, *Energy Policy*, Vol. 28, No. 10, January.

World Bank (2002), *World Development Indicators 2001*, Washington: World Bank.

World Bank (2001), *Fostering Competition in China's Power Markets*, Washington: World Bank.

World Bank and the Institute of Economic System and Management (2001), *Modernising China's Oil and Gas Sector: Structure Reform and Regulation*, Beijing: World Bank.

Wu, Kang (2001a), China's Crude and Product Trade Patterns in 2002: The Impact of WTO, *FACTS Energy Alert*, No. 262, November: FACTS.

Wu, Kang (2001b), China's National Energy Policies for Oil and Gas under the Tenth Five Year Plan, *FACTS Energy Alert*, No. 33, September: FACTS.

Zhou, Fengqi (2001*), Policies and Measures on Environmental Protection in China,* Presentation at APERC's Advisory Board Meeting, January 31, Tokyo.

CHAPTER 8

Cedigaz (2001), *Natural Gas in the World: 2001 Survey*, Paris: Institut Français du Pétrole.

IEA (2002), *Russia Energy Survey 2002*, Paris: OECD.

IEA (2001a), *Coal Information*, Paris: OECD.

IEA (2001b), *World Energy Outlook 2001 Insights: Assessing Today's Supplies to Fuel Tomorrow's Growth*, Paris: OECD.

OECD (2002), *Russia Economic Survey*, Paris: OECD.

USGS (2000), *World Petroleum Assessment 2000*, Washington: USGS.

World Energy Council (WEC) (2001), *Survey of Energy Resources 2001*, London: WEC.

CHAPTER 9

IEA (2002), *Electricity in India*, Paris: OECD.

IEA/Coal Industry Advisory Board (2002), *Coal in the Energy Supply of India*, Paris: OECD.

Government of India Planning Commission (2001), *Annual Report on the Workings of State Electricity Boards and Electricity Departments*, New Delhi.

Tata Energy Research Institute (TERI) (2002), *TEDDY 2001/2002: TERI Energy Data Directory and Yearbook*, New Delhi: TERI.

TERI (1999), *Green India 2047*, New Delhi: TERI.

United Nations Development Program (UNDP), United Nations Department of Economic and Social Affairs and World Energy Council (2002), *World Energy Assessment*. New York: United Nations.

CHAPTER 10

Cedigaz (2001), *Natural Gas in the World: 2001 Survey*, Paris: Institut Français du Pétrole.

DOE/EIA (2001), *Brazil Country Study*, Washington.

Enever, Andrew (2001), Bolivia – Has Gas, Wants It to Travel, *International Gas Report,* Issue 433, 17 September, London: Financial Times Energy.

IEA (forthcoming), South American Gas: Daring to Tap the Bounty, Paris: OECD.

IEA (2000), *World Energy Outlook 2000,* Paris: OECD.

International Road Federation (IRF) (2002), *World Road Statistics 2002 edition,* Geneva.

OECD (2002), *OECD Economic Outlook 71,* June, Paris: OECD.

OECD (2001), *Economic Survey of Brazil,* Paris: OECD.

Platts (2002a), *International Private Power,* First Quarter, New York: McGraw-Hill.

Platts (2002b), Power in Latin America, How much new generation does Brazil really need ?, *Platts,* Issue 83, 22 March, New York: McGraw-Hill.

Thouin, Pierre (2001), Prospects for Natural Gas in South America, *Latin American Energy Organisation (OLADE) Energy Magazine,* October-November-December, Quito: OLADE.

USGS (2000), *World Petroleum Assessment 2000,* Washington: USGS.

WEC (2001), *Survey of Energy Resources 2001,* London: WEC.

CHAPTER 11

Australian Bureau of Agricultural and Resource Economics (ABARE) (2002), *Global Coal Markets: Prospects to 2010,* Canberra: ABARE.

Cedigaz (2001), *Natural Gas in the world: 2001 Survey,* Paris: Institut Français du pétrole.

DOE/EIA (2002), *Country Analysis Brief: Indonesia,* Washington: US DOE.

IEA (2001a), *Coal Information 2001,* Paris: OECD.

IEA (2001b), *Natural Gas Information 2001,* Paris: OECD.

IEA (1999), *World Energy Outlook Insights 1999: Looking at Energy Subsidies-Getting the Prices Right,* Paris: OECD.

Organisation of Petroleum Exporting Countries (OPEC) (2001), New Indonesian Oil and Gas Bill Due to Be Passed Shortly, *OPEC Bulletin,* October, Vienna: OPEC.

USGS (2000), *World Petroleum Assessment 2000,* Washington.

World Bank (2000), *Indonesia: Oil and Gas Sector Study,* June, Washington.

WEC (2001), *Survey of Energy Resources 2001,* London: WEC.

CHAPTER 12

Australian Greenhouse Office (2000), *National Greenhouse Strategy: 2000 Progress Report,* Commonwealth of Australia.

Barreto, L, A. Makihira and K. Riahi (2002), The Hydrogen Economy in the 21st Century: A Sustainable Development Scenario, *International Journal of Hydrogen Energy,* (forthcoming), Florida: International Association for Hydrogen Energy.

European Commission (2000), *A Green Paper: Towards a European Strategy for the Security of Energy Supply*, Brussels: EC.
Global Warming Prevention Headquarters (2002), *The New Climate Change Policy Program*, Government of Japan.
Ministry of Economy, Trade, and Industry (2001), *Energy Conservation Subcommittee Report*, Government of Japan.
United States National Academy of Sciences (2001), *Effectiveness and Impact of Corporate Average Fuel Economy Standards*, Washington: National Academy Press.
United States National Energy Policy Development Group (2001), *Reliable, Affordable, and Environmentally Sound Energy for America's Future: National Energy Policy Report*, Washington: U.S Government Printing Office.

CHAPTER 13

Barnett, Andrew (2000), *Energy and the Fight against Poverty*, June.
Bhasin, Reena (2001), *Urban Poverty and Urbanisation*, New Delhi: Deep & Deep Publications.
Davis, Mark (1998), Rural Household Energy Consumption: The Effects of Access to Electricity – Evidence from South Africa, *Energy Policy*, Vol. 26, No. 3, pp. 207-217, United Kingdom: Elsevier.
Department of Trade and Industry (DTI) (2002), *Fuel Poverty*, United Kingdom.
Hulscher, Dr. W. S. (1997), Fuel Complementation Rather than Substitution, *Wood Energy News*, Vol. 12, No. 2, October, Bangkok: FAO-RWEDP.
International Energy Agency (IEA) (2002), *Electricity in India*, Paris: OECD.
International Energy Agency (IEA) (2001), *World Energy Outlook Insight: Assessing Today's Supplies to Fuel Tomorrow's Growth*, Paris: OECD.
International Energy Agency (IEA) (1999), *World Energy Outlook Insight: Looking at Energy Subsidies-Getting the Prices Right*, Paris: OECD.
Katyega, Maneno J.J. (2001), Improving Modern Energy Access to the Urban Poor in Tanzania, *Journal of Energy Southern Africa*, August.
Masera, Omar, Saatkamp, Barbara and Kammen, Daniel (2000), From Linear Fuel Switching to Multiple Cooking Strategies: A Critique and Alternative to the Energy Ladder Model, *World Development*, Vol. 28, No. 12, pp. 2083-2103, United Kingdom: Elsevier.
OECD (2002), *African Economic Outlook*, Paris: OECD.
Ping, Zhou Jia (2001), China's New and Renewable Energy Situation, Energie Verwertungsagentur, November.
Tata Energy Research Institute (TERI) (2002), *Empowering the Indian City: Scenarios and Solutions*, New Delhi: TERI.
Tata Energy Research Institute (TERI) (2001), *National Sample Survey: 55th Round*, New Delhi: TERI.
Tata Energy Research Institute (TERI), India, Energy Research Institute (ERI), China, Wageningen Agricultural University, the Netherlands, International

Institute for Applied Systems Analysis (IIASA), Austria (1999), Potential for Use of Renewable Sources of Energy in Asia and Their Cost Effectiveness in Air Pollution Abatement, Final Report on Work Package 1, December.

United Nations Development Program (UNDP) (2000), *Bioenergy Primer: Modernised Biomass Energy for Sustainable Development*, New York: UNDP.

United Nations Development Program, United Nations Department of Economic and Social Affairs and World Energy Council (2002), *World Energy Assessment*, New York: UNDP.

United Nations Population Division (2001), *World Population Prospects,* New York: United Nations.

World Bank (2001), *Global Economic Prospects*, Washington, DC: World Bank.

World Bank (1996), *Rural Energy and Development: Improving Energy Supplies to 2 Billion People*, Washington, DC: World Bank.

World Bank (1995), *Republic of Chad: Household Energy Project*, Washington, DC: World Bank.

The World Bank Group (2002), Poverty Reduction Strategy Source Book, Washington, DC: World Bank.

World Energy Council and United Nations Food and Agriculture Organisation (WEC/FAO) (1999), *The Challenge of Rural Energy Poverty in Developing Countries*, October, United Kingdom: WEC.

APPENDIX 1

IEA (1999), *CO_2 Emissions from Fuel Combustion*, Paris: OECD.